# THE DRAMA OF THE
# COMMONS

Committee on the Human Dimensions of Global Change

Elinor Ostrom, Thomas Dietz, Nives Dolšak,
Paul C. Stern, Susan Stonich, and Elke U. Weber, Editors

Division of Behavioral and Social Sciences and Education
National Research Council

NATIONAL ACADEMY PRESS
Washington, DC

**NATIONAL ACADEMY PRESS • 2101 Constitution Avenue, N.W. • Washington, DC 20418**

NOTICE: The project that is the subject of this report was approved by the Governing Board of the National Research Council, whose members are drawn from the councils of the National Academy of Sciences, the National Academy of Engineering, and the Institute of Medicine. The members of the committee responsible for the report were chosen for their special competences and with regard for appropriate balance.

The study was supported by Grant No. BCS-9906253 between the National Academy of Sciences and the National Science Foundation and a grant from the Rockefeller Brothers Fund to Indiana University. Any opinions, findings, conclusions, or recommendations expressed in this publication are those of the author(s) and do not necessarily reflect the view of the organizations or agencies that provided support for this project.

Additional copies of this report are available from National Academy Press, 2101 Constitution Avenue, N.W., Washington, DC 20418
Call (800) 624-6242 or (202) 334-3313 (in the Washington metropolitan area)

This report is also available online at **http://www.nap.edu**

Printed in the United States of America

# THE NATIONAL ACADEMIES

National Academy of Sciences
National Academy of Engineering
Institute of Medicine
National Research Council

The **National Academy of Sciences** is a private, nonprofit, self-perpetuating society of distinguished scholars engaged in scientific and engineering research, dedicated to the furtherance of science and technology and to their use for the general welfare. Upon the authority of the charter granted to it by the Congress in 1863, the Academy has a mandate that requires it to advise the federal government on scientific and technical matters. Dr. Bruce M. Alberts is president of the National Academy of Sciences.

The **National Academy of Engineering** was established in 1964, under the charter of the National Academy of Sciences, as a parallel organization of outstanding engineers. It is autonomous in its administration and in the selection of its members, sharing with the National Academy of Sciences the responsibility for advising the federal government. The National Academy of Engineering also sponsors engineering programs aimed at meeting national needs, encourages education and research, and recognizes the superior achievements of engineers. Dr. Wm. A. Wulf is president of the National Academy of Engineering.

The **Institute of Medicine** was established in 1970 by the National Academy of Sciences to secure the services of eminent members of appropriate professions in the examination of policy matters pertaining to the health of the public. The Institute acts under the responsibility given to the National Academy of Sciences by its congressional charter to be an adviser to the federal government and, upon its own initiative, to identify issues of medical care, research, and education. Dr. Kenneth I. Shine is president of the Institute of Medicine.

The **National Research Council** was organized by the National Academy of Sciences in 1916 to associate the broad community of science and technology with the Academy's purposes of furthering knowledge and advising the federal government. Functioning in accordance with general policies determined by the Academy, the Council has become the principal operating agency of both the National Academy of Sciences and the National Academy of Engineering in providing services to the government, the public, and the scientific and engineering communities. The Council is administered jointly by both Academies and the Institute of Medicine. Dr. Bruce M. Alberts and Dr. Wm. A. Wulf are chairman and vice chairman, respectively, of the National Research Council.

# Preface

"The commons" has long been a pivotal idea in environmental studies, and the resources and institutions described by that term have long been recognized as central to many environmental problems, especially problems of global environmental change. Since its birth in 1989, the Committee on the Human Dimensions of Global Change of the National Research Council has recognized the importance of commons and commons research (Global Environmental Change: Understanding the Human Dimensions, National Academy Press, 1992). Not only is the topic important in its own right, the commons is also a central theme in studies of international cooperation, environmental decision making, and the design of resource management institutions. Its importance is highlighted in the International Human Dimensions Programme's science plans on Land Use and Land Cover Change (www.uni-bonn.de/ihdp/lucc) and Institutional Dimensions of Global Environmental Change (www.dartmouth.edu/~idgec). So the commons is at the center of the international research agenda on the human dimensions of global change.

The importance of the topic is one reason the National Research Council has undertaken a review of knowledge about the commons at this time. Another reason is that it has been 15 years since the Council completed the work of its Panel on the Study of Common Property Resource Management. That work, as discussed in Chapter 1, marked a turning point in the history of research on commons—it marked the emergence of a self-conscious interdisciplinary and international research community focused on understanding commons. After 15 productive years of research since that early synthetic effort, we felt it appropriate to reexamine and reintegrate what had been learned.

The committee is very pleased to have received support from the U.S. National Science Foundation to conduct this study. We began by commissioning a series of papers that were presented at the 8th Meeting of the International Association for the Study of Common Property in June 2000 at Indiana University. That meeting provided an excellent venue for discussing the work in progress with an international, interdisciplinary group of experts on the commons. Support from the Rockefeller Brothers Fund allowed us to hold a follow-up meeting of the authors and editors at the Pocantico Center in Tarrytown, New York in September 2000.

We believe the result of our project is a rich series of papers that review what we know about the commons, integrate what in the past have been somewhat disparate literatures, and point directions for the future. We hope this volume achieves several goals. First, for those not familiar with the rich literature since Hardin's seminal 1968 paper, we hope it provides a sound grounding in what we have learned and shows how and where knowledge has advanced since Hardin proposed his model. Second, for researchers already working in the field, we hope it provides a broad state-of-the-art review and shows connections and gaps in knowledge that may not have been obvious in the past. Third, for researchers and those funding research, we believe it conveys a sense of pride in what has been accomplished with relatively modest funding and indicates priorities for future work. Finally, although not a management handbook, we hope it provides some guidance to those who design and manage institutions dealing with the commons and makes it easier for them to base their decisions on the best available science.

On behalf of the committee, I wish to thank the National Science Foundation and the Rockefeller Brothers Fund for their support of this project and the staff and students of the Workshop in Political Theory and Policy Analysis at Indiana University, who hosted the project participants in Indiana and have provided assistance at various stages in the project. The committee's gratitude goes to Brian Tobachnick, who managed the logistics of the project during its early stages and to Deborah Johnson, who carried it the rest of the way. We also owe a debt to Laura Penny, who did the copy editing, and to Yvonne Wise, who managed the review and editorial processes.

I wish to thank the following individuals for their participation in the review of the papers in this volume: James Acheson, Indiana University; Robert Axelrod, University of Michigan; Susan Buck, University of North Carolina, Greensboro; Susan Hanna, Oregon State University; Peter Haas, University of Massachusetts; Kai Lee, Williams College, Williamstown, Massachusetts; Gary Libecap, University of Arizona; Margaret McKean, Duke University; Ruth Meinzen-Dick, International Food Policy Research Institute, Washington, DC; Ronald Mitchell, Stanford University; Emilio Moran, Indiana University; Granger Morgan, Carnegie Mellon University; Edward Parson, John F. Kennedy School of Government, Harvard University; Pauline Peters, Harvard University; Charles Plott, California Technical Institute; Lore Ruttan, Indiana University; Edella Schlager, University of Arizona; Robert Stavins, Harvard University; Mark Van Vugt, University of Southhampton, England; James Walker, Indiana University; and Rick Wilson, Rice University.

Although the individuals listed provided constructive comments and suggestions, it must be emphasized that the final responsibility for the content of this book rests with the authors and editors.

Thomas Dietz, *Chair*
Committee on the Human Dimensions
of Global Change

# Contents

# INTRODUCTION

# 1

# The Drama of the Commons

*Thomas Dietz, Nives Dolšak, Elinor Ostrom, and Paul C. Stern*

The "tragedy of the commons" is a central concept in human ecology and the study of the environment. The prototypical scenario is simple. There is a resource—usually referred to as a common-pool resource—to which a large number of people have access. The resource might be an oceanic ecosystem from which fish are harvested, the global atmosphere into which greenhouse gases are released, or a forest from which timber is harvested. Overuse of the resource creates problems, often destroying its sustainability. The fish population may collapse, climate change may ensue, or the forest might cease regrowing enough trees to replace those cut. Each user faces a decision about how much of the resource to use—how many fish to catch, how much greenhouse gases to emit, or how many trees to cut. If all users restrain themselves, then the resource can be sustained. But there is a dilemma. If you limit your use of the resources and your neighbors do not, then the resource still collapses and you have lost the short-term benefits of taking your share (Hardin, 1968).

The logic of the tragedy of the commons seems inexorable. As we discuss, however, that logic depends on a set of assumptions about human motivation, about the rules governing the use of the commons, and about the character of the common resource. One of the important contributions of the past 30 years of research has been to clarify the concepts involved in the tragedy of the commons. Things are not as simple as they seem in the prototypical model. Human motivation is complex, the rules governing real commons do not always permit free access to everyone, and the resource systems themselves have dynamics that influence their response to human use. The result is often not the tragedy described by Hardin but what McCay (1995, 1996; McCay and Acheson, 1987b; see also

Rose, 1994) has described as a "comedy"—a drama for certain, but one with a happy ending.

Three decades of empirical research have revealed many rich and complicated histories of commons management. Sometimes these histories tell of Hardin's tragedy. Sometimes the outcome is more like McCay's comedy. Often the results are somewhere in between, filled with ambiguity. But drama is always there. That is why we have chosen to call this book *The Drama of the Commons*—because the commons entails history, comedy, and tragedy.

Research on the commons would be warranted entirely because of its practical importance. Nearly all environmental issues have aspects of the commons in them. Important theoretical reasons exist for studying the commons as well. At the heart of all social theory is the contrast between humans as motivated almost exclusively by narrow self-interest and humans as motivated by concern for others or for society as a whole.[1] The rational actor model that dominates economic theory, but is also influential in sociology, political science, anthropology, and psychology, posits strict self-interest. As Adam Smith put it, "We are not ready to suspect any person of being defective in selfishness" (Smith, 1977[1804]:446). This assumption is what underpins Hardin's analysis.

Opposing views, however, have always assumed that humans take account of the interests of the group. For example, functionalist theory in sociology and anthropology, especially the human ecological arguments of Rappaport and Vayda (Rappaport, 1984; Vayda and Rappaport, 1968), argued that the "tragedy of the commons" could be averted by mechanisms that cause individuals to act in the interests of the collective good rather than with narrow self-interest. Nor has this debate been restricted to the social sciences. In evolutionary theory, arguments for adaptations that give advantage to the population or the species at cost to the individual have been under criticism at least since the 1960s (Williams, 1966). But strong arguments remain for the presence of altruism (Sober and Wilson, 1998).

If we assume narrow self-interest and one-time interactions, then the tragedy of the commons is one of a set of paradoxes that follow. Another is the classical prisoners' dilemma. In the canonical formulation, two co-conspirators are captured by the police. If neither informs on the other, they both face light sentences. If both inform, they both face long jail terms. If one informs and the other doesn't, the informer receives a very light sentence or is set free while the noninformer receives a very heavy sentence. Faced with this set of payoffs, the narrow self-interest of each will cause both to inform, producing a result less desirable to each than if they both had remained silent.

Olson (1965) made us aware that the organization of groups to pursue collective ends, such as political and policy outcomes, was vulnerable to a paradox, often called the "free-rider problem," that had previously been identified in regard to other "public goods" (Samuelson, 1954). A public good is something to which everyone has access but, unlike a common-pool resource, one person's use of the resource does not necessarily diminish the potential for use by another. Public radio stations, scientific knowledge, and world peace are public goods in that we all enjoy the benefits without reducing the quantity or quality of the good. The problem is that, in a large group, an individual will enjoy the benefits of the public good whether or not he or she contributes to producing it. You can listen to public radio whether or not you pledge and make a contribution. And in a large population, whether or not you contribute has no real impact on the quantity of the public good. So a person following the dictates of narrow self-interest will avoid the costs of contributing. Such a person can continue to enjoy the benefits from the contributions provided by others. But if everyone follows this logic, the public good will not be supplied, or will be supplied in less quantity or quality than is ideal.

Here we see the importance of the tragedy of the commons and its kin. All of the analyses just sketched presume that self-interest is the only motivator and that social mechanisms to control self-interest, such as communication, trust, and the ability to make binding agreements, are lacking or ineffective. These conditions certainly describe some interactions. People sometimes do, however, move beyond individual self-interest. Communication, trust, the anticipation of future interactions, and the ability to build agreements and rules sometimes control behavior well enough to prevent tragedy. So the drama of the commons does not always play out as tragedy.

This volume examines what has been learned over decades of research into how the drama of the commons plays out. It should be of interest to people concerned with important commons such as ecosystems, water supplies, and the atmosphere. In addition, commons situations provide critically important test beds for addressing many of the central questions of the social sciences. How does our identity relate to the resources in our environment? How do we manage to live together? How do societies control individuals' egoistic and antisocial impulses? Which social arrangements persist and which do not? In looking at the long sweep of human history and the thousands of social forms spread across it, these questions may become unmanageable to study in a systematic manner. The commons, however, provides a tractable and yet important context in which to address these questions. Just as evolutionary and developmental biology progressed by studying the fruitfly, *Drosophila melanogaster*, an organism well suited to the tools available, we suggest that studies of the commons and related problems are an ideal test bed for many key questions in the social sciences.[2]

As is evident in the chapters of this volume, commons research already draws on most of the methodological traditions of the social sciences. There are elegant mathematical models, carefully designed laboratory experiments, and meticulous historical and comparative case studies. The statistical tools applicable to large or moderate-sized data sets also are being brought to bear. As we will detail, research on the commons attracts scientists from a great diversity of disciplines and from all regions of the world. Advances in the social sciences are likely to come from just such an admixture of methods and perspectives focused on a problem that touches on core theoretical issues of great practical importance.

This volume presents a series of papers that review and synthesize what we know about the commons, integrating what in the past have been somewhat disparate literatures and pointing directions for the future. It has several goals. First, for those not familiar with the rich literature since Hardin's 1968 article, it is intended to provide a sound grounding in what has been learned. Second, for researchers in the field, it offers a state-of-the-art review that spans the field and shows connections that may not have been obvious in the past. Third, for researchers and those funding research, it conveys a sense of what has been accomplished with relatively modest funding and indicates the priorities for future work. Finally, although it is not a management handbook, it provides some guidance to those who design and manage institutions dealing with the commons by compiling the best available science for informing their choices.

This chapter offers a brief history of research on the commons, starting with Hardin's influence but also acknowledging his predecessors. It describes the synthetic work that occurred in the mid-1980s. Building on that work, it clarifies the key concepts involved in understanding the commons. One of the major contributions of commons scholarship has been to make much clearer which concepts must be brought to bear and which distinctions made in understanding the commons. These include the crucial distinction between the resource itself, the arrangements humans use to govern access to the resources, and the key properties of the resource and the arrangements that drive the drama. The chapter concludes by sketching the plan of the book.

## A SHORT INTELLECTUAL HISTORY OF THE FIELD

### A Point of Departure

Hardin's influential 1968 article in *Science* on "The Tragedy of the Commons" is one of the most often-cited scientific papers written in the second half of the twentieth century. The article stimulated immense intellectual interest across both the natural and social sciences,[3] extensive debate, and a new interdisciplinary field of study. Scientific interest in the commons grew throughout the 1970s and early 1980s largely in reaction to Hardin's article and the frightening news stories about sharp population declines of many species, particularly those from the ocean. Interest was fanned by the debate about limits to growth, and the increasing awareness of deforestation in tropical regions of the world.

Prior to the publication of Hardin's article, titles such as "commons," "common-pool resources," or "common property" appeared only 17 times in the academic literature published in English and cataloged in the "Common-Pool Resource Bibliography" maintained by Hess at Indiana University.[4] Between that time and 1984, before the Annapolis, Maryland conference organized by the National Research Council (NRC) Panel on Common Property Resource Management, the number of such titles had grown to 115. The Annapolis conference in 1985 brought together a large number of scientists from different fields and different nations to examine common-pool resources and their management.[5] The conference provided an opportunity for scholars to synthesize what was known in disparate disciplines as of 1985—which we summarize briefly in this chapter. This conference and several others held at about the same time stimulated even greater interest in the commons. From 1985 to 1990, the number of scholarly works on the commons more than doubled to 275. In the next 5 years (1991-1995), they nearly doubled again to 444 articles. Between 1996 and 2000, 573 new articles appeared on the commons. In 1990, the International Association for the Study of Common Property (IASCP) was officially established. Its first meeting at Duke University was attended by 150 scholars from multiple disciplines. As can be seen from Figure 1-1, a substantial increase of interest in this field has brought an ever greater number of scholars to the IASCP meetings. By 2000, more than 600 scholars attended these meetings.

A key characteristic in the field, in addition to its rapid growth, is the extraordinary extent of interdisciplinary and international participation. For example, scholars from a dozen disciplines and 52 countries attended the 2000 meeting of the IASCP. Although such broad participation challenges all involved to find

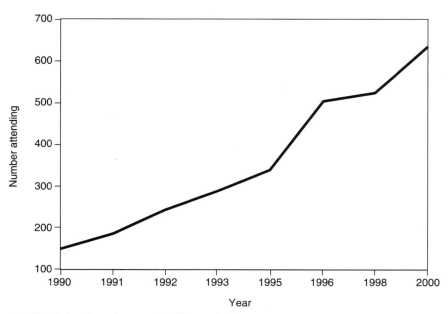

FIGURE 1-1    Attendance at IASCP meetings.

shared concepts and common technical language, the results have been well worth the effort.

## Early Work on the Commons

Although Hardin's article was the fulcrum for recent work on common-pool resources, scholars long before Hardin had expressed pessimism about the sustainable management of these resources. Aristotle observed that "what is common to the greatest number has the least care bestowed upon it. Everyone thinks chiefly of his own, hardly at all of the common interest" (*Politics*, Book II, ch. 3). The French naturalist, Marcet (1819) wrote in *Conversations on Political Economy* (1819, cited in Baumol and Oates, 1988) that open access to natural resources results in overexploitation of those resources and harvesting of the resources prior to their harvest time. Lloyd (1977 [1833]), whose work strongly influenced Hardin, similarly argued that a common-pool resource will be overused because of the higher value of present benefits of use compared to potential future costs of unrestricted use, especially when each individual user bears only a fraction of those costs but gains the entirety of present benefits. Further, Lloyd argued that an individual's decisions regarding whether to withdraw another unit from a common-pool resource (in Lloyd's analysis, whether to have another child) depends on the institutions that define the benefits and costs of such action.

Less pessimistic voices were raised earlier as well. In his classic study of Indian villages, the township in England and Scotland, and the complex, early village structures of Germany (the *Mark*) and Russia (the *mir*), Maine (1871) argued that village communities occur everywhere and facilitate their subsistence by allocating agricultural lands as private property and forest and pastures surrounding arable lands as common property. In describing the German version, Maine (1871:10) asserted: "The Township (I state the matter in my own way) was an organized, self-acting group of Teutonic families, exercising a common proprietorship over a definite tract of land, its Mark, cultivating its domain on a common system, and sustaining itself by the product." In an in-depth analysis of Maine's work, Grossi (1981) argues that Maine had identified how village communities in many settings had developed a keen sense of private property for agricultural plots combined with a common-property system for forested and pasture lands. Malinowski (1926) cautioned readers not to believe that any kind of property regime—including common property with joint owners—was a "simple" system that could be characterized as having only one set of consequences. He pointed out that:

> Ownership, therefore, can be defined neither by such words as "communism" nor "individualism," nor by reference to "joint-stock company" system or "personal enterprise," but by the concrete facts and conditions of use. It is the sum of duties, privileges, and mutualities which bind the joint owners to the object and to each other. (1926:21)

## Early Formal Analyses of the Commons by Resource Economists

The influential work of Gordon (1954) and Schaefer (1957) drew attention to the economic factors in the management of one type of common-pool resource—fisheries. Gordon and Schaefer modeled the effect of fishing effort (the quantity of fish harvested from a fishery) on ecologically sustainable yields as well as calculating the economic results of varying levels of effort. The so-called Gordon-Schaefer model has dominated the study and execution of fisheries management since the 1950s. Both scholars assumed that at low levels of fishing effort in a newly opened fishery, yield increases rapidly as a function of effort but with diminishing returns as more effort is needed to harvest additional units of fish. Beyond the "maximum sustainable yield," however, further increases in harvesting would result in a decrease of total harvest and revenue because replenishment of the fish stock was presumed to depend on the size of the current fish stock, which falls below the level necessary for full replacement once fishing extracts more than this yield. By including the revenue occurring from fishing (yield times the fish price) and the costs of fishing effort, they defined the "maximum economic yield," that is, the fishing effort at which the difference between fishing revenue and costs is maximum, and the level of the fishing effort under open access. The relationships they described are illustrated in Figure 1-2.

As shown in Figure 1-2, the underlying relationship between fishing effort measured on the horizontal axis and cost measured on the vertical axis is linear, while the relationship to revenue, also measured on the vertical axis, is curvilinear. This is due to the presumed basic biological relationships involved in determining maximum sustainable yield. Yield increases with effort until the maximum sustainable yield is reached; beyond that, the fish stock can replenish only at a lower rate—the population is simply drawn down. Whether the population is sustainable depends on the behavior of the harvesters.

If no rules exist related to access or amount of harvest (an open access situation), the equilibrium is a harvest rate that is larger than either the maximum sustainable yield (in biological terms) or the maximum economic yield (the harvest that yields the maximum difference between prices obtained and costs of fishing effort) (see Figure 1-2). This is because each fisher takes into account only the costs of his or her own effort and not the increased costs that individual effort imposes on others. The maximum economic yield (achievable if the rules regulating access and harvesting practices limit effort to the economically optimal strategy) turns out to be less than the biologically maximum sustainable yield. Based on this analysis, resource economists argued strongly that fisheries and other common-pool resources would be better managed by a single owner—preferably a private owner. Government ownership was, however, consistent with their argument. The single owner could then determine the maximum economic yield and manage the resource so as to obtain that yield (see, e.g., Crutchfield, 1964; Demsetz, 1967; Johnson, 1972).

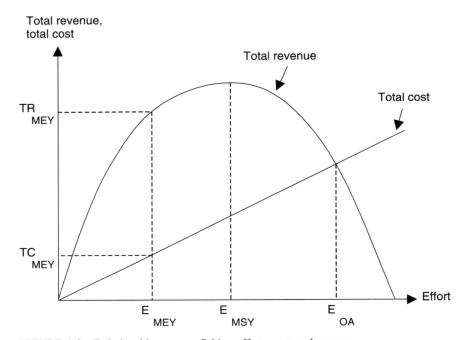

FIGURE 1-2   Relationships among fishing effort, cost, and revenue.
SOURCE:  Townsend and Wilson (1987:317). Reprinted with permission.
NOTE:  Total revenue, TR; total cost, TC; level of fishing effort; E; maximum economic
yield, MEY; maximum sustainable yield; MSY; open access, OA. Profit is revenue minus
cost and is represented by the vertical distance between the total revenue and total cost
curves at any particular level of effort.


Gordon's and Schaefer's work emphasized the use of biological science and
microeconomics in policy design. However, the science of fish population dynamics was not as well established as the Gordon-Schaefer model presumed. In
particular, not all scientists accepted the underlying presumption of the "maximum sustainable yield" concept, that the stocks of adult fish and the regeneration
rate in one time period depended only on the catching effort of the prior period.
Gordon himself noted this. "Large broods, however, do not appear to depend on
large numbers of adult spawners, and this lends support to the belief that the fish
population is entirely unaffected by the activity of man" (Gordon, 1954:126).
Wilson (Chapter 10) discusses why alternative views were ignored for so many
years and argues that the quality of knowledge, scientific uncertainty, and the
knowledge of nonscientists are important variables in common-pool resource
management.

Many policy innovations of the 1960s and 1970s were based on the early work of resource economists and consistent with Hardin's thesis that "freedom in a commons brings ruin to all" (Hardin, 1968:1244). This literature stressed the importance of unitary ownership—including privatization as well as government ownership. However, the major policy innovation of this era was legislation in many countries—particularly developing countries—that transferred forests, pasture land, in-shore fisheries, and other natural resources from their previous property rights regimes to government ownership (see Arnold and Campbell, 1986).

Extensive research and experience since 1968 shows that these transfers of property rights were sometimes disastrous for the resources they were intended to protect. Instead of creating a single owner with a long-term interest in the resource, nationalizing common-pool resources typically led to (1) a rejection of any existing indigenous institutions—making the actions of local stewards to sustain a resource illegal; (2) poor monitoring of resource boundaries and harvesting practices because many governments did not have the resources to monitor the resources to which they asserted ownership; and (3) de facto open access conditions and a race to use of the resources. Thus, the presumption that government ownership was one of two universally applicable "solutions" to the "tragedy" was seriously challenged by these historical experiences.

## Hardin's Model and Its Limitations

Hardin argued that a "man is locked into a system that compels him to increase his herd without limit—in a world that is limited" (Hardin, 1968:1244). He further asserted that having a conscience was self-eliminating.[6] Those who restrain their use of a common-pool resource lose out economically in comparison to those who continue unrestrained use. Thus, evolutionary processes will select for those who exercise unrestrained use and against those who restrain their own harvesting. Hardin's solution was "mutually agreed upon" coercion. Two inferences were usually drawn from this formulation. One is that only what psychologists call aversive (coercive) controls can be effective, suggesting that effective rules cannot be based on creating internalized norms or obligations in resource users. The other is that agreements on rules must be reached only through the state (usually, the national government), suggesting that local governments and informal and nongovernmental institutions cannot develop effective ways to prevent or remedy situations that lead to tragedy (Gibson, 2001).

Challenges to the conceptual underpinnings, to the empirical validity, to the theoretical adequacy, and to the generalizability of Hardin's model and the related work in resource economics were articulated throughout the 1970s and early 1980s. A key challenge to the Hardin model came from researchers familiar with diverse common property institutions in the field. They argued that Hardin had seriously confused the concept of common *property* with open access conditions where no rules existed to limit entry and use. As Ciriacy-Wantrup and Bishop (1975:715) expressed it, "common property is not everyone's property." They and other researchers (see, e.g., Thompson, 1975) stressed that where common property existed, users had developed rich webs of use rights that identified who had a long-term interest in the resource and thus an incentive to try to avoid overuse. Few asserted that all common property regimes were optimally efficient or fair. Rather, the specifics of a particular regime had to be examined before presuming that an external authority should step in, violate local customs, and impose a new set of rules that were unlikely to be viewed locally as legitimate.

Another type of challenge came from game theorists. Early attempts to formalize commons situations using game theory typically posed the problem as a prisoners' dilemma (PD) of the form described earlier, but extended the analysis from the classical two-player case to the N-person case (e.g., Dawes, 1980; however see Rubenstein et al., 1974; Stern, 1976, for early formulations that did not treat the commons as a PD game). When a PD game is played only once or is repeated with a definite ending time, a rational player has one—and only one—strategy that generates the highest immediate payoffs, assuming all players are using the same form of rationality. That strategy is to inform on the other players (called defection in the literature). Until recently, the dominant view has been that this one-shot, N-Person PD adequately models the nature of the situation faced in most commons settings. The research summarized by Kopelman et al. and Falk et al. in this volume shows that Hardin's predictions hold under a one-shot condition with no communication, but not necessarily in a world where the game is played repeatedly, where there is no predefined endpoint, or where communication is possible (see Axelrod, 1984).

Some researchers have argued that games other than PD, such as the "assurance game" or the "game of chicken," are more appropriate models for at least some of the situations facing users (Taylor, 1987). Unlike the PD game, which has a single equilibrium (and thus, each actor has a dominant strategy yielding a better individual outcome no matter what the other actor does), these games have multiple equilibria (and thus, neither actor has a dominant strategy), so both benefit from coordination.[7]

In a series of papers, Runge (1981, 1984a, 1984b) stressed that most users of a common-pool resource—at least in developing countries—live in the same villages where their families had lived for generations and intend to live in the same villages for generations to come. Given the level of poverty facing many villagers, their dependence on natural resources, and the randomness they all face in the availability of natural resources, Runge argued that it is implausible to assume that individuals have a dominant strategy of free riding. He suggested that users of common-pool resources in developing countries faced a repeated coordination game rather than a one-shot PD game. In such situations, all users would prefer to find ways of limiting their own use so long as others also committed themselves to stinting. Village institutions would provide mechanisms to enable users to arrive at agreements (within the village context) that would assure each user that others were conforming to the agreed-on set of rules. Thus, Runge and other scholars conceptualized the game as a coordination problem rather than a dilemma.

Anthropologists and human ecologists also challenged the concept of an inexorable tragedy of the commons. Dyson-Hudson and Smith (1978) reasoned that resources had characteristics that were valued by those living near them. Some of these attributes also affected whether individuals could defend private property or whether they needed to develop rules of access and use to regulate how resources would be owned by an entire local community. Similarly, Netting (1976), based on his extensive study of private and common property in the Swiss Alps, developed a clear set of resource characteristics that he argued would be associated with diverse forms of property. He predicted that when (1) the value of per-unit production was low, (2) the frequency and dependability of yield was low, (3) the possibility of improvement was low, (4) the area required for effective use was large, and (5) the size of the group needed to make capital investments was large, communal property would be developed by the users. Similarly, when the opposite conditions were present, Netting predicted that users would develop some form of private property (see also Netting, 1981). Netting provided substantial evidence to support his claims, also showing that common-property regimes developed under the above conditions had been sustained for centuries without overexploiting resources.

Other anthropologists argued that no single dimension was responsible for making some resources communal and other resources privately held and that there was no unidirectional tendency for resources to move over time from common property to private property. Leach (1954) documented long cycles of changes in social structure and property rights in Upper Burma, and Bauer (1977) documented short cycles of such changes in Ethiopia. McCay (1980, 1981) illustrated a wide diversity of local organizations developed by inshore fishers to keep access relatively open to those who lived and worked in a community. McCay describes the efforts by these fishers to try to organize themselves using forms of common property even when confronted with "modern" capitalist forms of organization.

Thus, by the mid-1980s, more and more questions were being raised about Hardin's model, the presumption that all commons situations were like a prisoners' dilemma, and the wisdom of policies based on these analyses. Scholars familiar with the qualitative case study literature in Africa, Latin America, Asia, and the United States were beginning to point out that the policy reforms that transformed resources from governance as common property by local communities into state governance were actually making things worse for the resource as well as for the users. The governments that took these actions frequently did not have enough trained personnel on the ground to monitor the resources. Thus, what had been de facto common property with some limitations on access and use patterns became de jure government property—but due to the lack of enforcement, it frequently became de facto open access. Corrupt public officials also faced opportunities to collect side payments from local resource users wishing to exploit resources that were officially government owned.

These questions and doubts were not discussed widely across scientific disciplines or communities, however, because each tended to use its own language and theory. As a result, very little bridging across disciplines and academic communities occurred before the mid-1980s. Scholars in one region of the world did not know about the research being undertaken by scholars in other parts of the world. Even scholars focusing on a single continent, such as Africa, who were studying forest resources were unaware of the findings of researchers studying pastoral resources or inshore fisheries on the same continent.

## Panel on Common Property Resource Management: A First Synthesis

In September 1983, the National Research Council appointed a Panel on the Study of Common Property Resource Management.[8] The panel recognized that one of its chief tasks was to create a framework whereby individuals from multiple disciplines could begin to communicate about the diverse property systems operating in different resource sectors. A framework was developed by Oakerson, drawing on many years of scholarship on institutions. The framework was used in a series of small meetings with scholars from diverse disciplines who each knew extremely well the patterns of user interactions around some common-pool resources. The challenge was finding a way that these scholars could communicate with one another and develop a common set of findings.

The panel organized a meeting in Annapolis, Maryland, in 1985 that provided a forum for exchange of ideas, synthesis, and growth. The Annapolis meeting was an unusual event for its era, given the diversity of disciplines, nations, and resource interests represented by the participants. The Oakerson framework was revised several times before and after the meeting and became the centerpiece of the final publication from the panel (Oakerson, 1986; National Research Council, 1986; see also Bromley et al., 1992).

In the last session at Annapolis, the panelists provided a cogent overview of lessons learned (Bromley, 1986; Ostrom, 1986; Peters, 1986). These included:

1. The need to define the performance of an institutional arrangement in terms of both environmental and human dimensions;

2. The importance of the initial situation as it affects emergence, performance, survival, and relative costs and benefits of institutional arrangements. Identifying correlations may be the best that social scientists could accomplish given the data available at the time;

3. The importance of the distinction between the characteristics of the resource (common-pool resource) and the regime that manages the resource (common property regime or some other kind of property regime). Analytical progress would be slow unless this distinction was taken seriously;

4. The need to compare and synthesize analyses of common-pool resources and common property regimes in various disciplines using a framework that enables scholars from different disciplinary backgrounds to communicate and compare findings;

5. The need, especially for international donors, to understand how various changes in property rights affect the distribution of income, wealth, and other resources that are important aspects of the creation and survival of institutional arrangements;

6. The need to understand how spatial and temporal heterogeneity in the resource endowment creates opportunities for some to benefit at the expense of others, thereby often exacerbating equity problems;

7. The need to compare the costs and benefits of various institutional arrangements for a given resource. Under some circumstances, common property regimes perform better than private property. This occurs when (a) the costs of creating and enforcing private property rights are high, (b) the economic value of the output produced from the resource is low, and (c) the benefits generated by the resources are distributed with high spatial uncertainty. Under these circumstances, a common property regime provides a way of reducing the risk of having no benefits at all in a given time period and thus may be preferable to private property (see Runge, 1986; Netting, 1976).

8. Resource users do not always choose to defect rather than cooperate. Individuals' decisions depend on their bargaining power, the initial endowment of resources, their shared values, and other factors.

The panelists also identified the following unanswered questions and areas for future research:

1. How do multiple levels of management interact and affect performance?
2. What is the effect of group size on the performance of institutional arrangements?
3. What are the roles of different mechanisms for dispute settlement?

These three questions identified an ambitious and scientifically difficult agenda. One of these unanswered questions (the effect of group size) has been addressed repeatedly in the research since 1985 and is discussed in Chapter 2 and several other chapters in this book. However, the question turns out to be deceptively simple. Different findings have been obtained depending on the context. The relationships among multiple levels of management are addressed in Chapters 8 and 9 and here, too, the results are complex. Less work has been done on diverse mechanisms for dispute settlement; this remains an important area for research where the tradition of work on commons could link to that on conflict resolution. This topic is reconsidered in Chapter 13.

A number of related activities followed the Annapolis conference. One was the publication of a series of book-length studies and edited volumes that led to a serious rethinking of the empirical foundations for the analysis of common-pool resources (see Berkes, 1986, 1989; Berkes et al.,1989; Blomquist, 1992; McCay and Acheson, 1987a, 1987b; Ostrom, 1990; Pinkerton, 1989; Tang, 1992). These studies were a serious challenge to the validity of Hardin's analysis and to the implication that government and private property were the "only" ways to manage common-pool resources. They demonstrated that under some conditions, local groups using a common property regime could manage their resources quite well. This challenge led to a move from seeing Hardin's formulation as a broad and accurate generalization to a special case that was observed only under certain circumstances. Furthermore, the rich case study literature illustrated a wide diversity of settings in which users dependent on common-pool resources have organized themselves to achieve much better outcomes than can be predicted by Hardin's model (Cordell, 1990; Ruddle and Johannes, 1985; Sengupta, 1991; Wade, 1994). This research changed the focus of the field from a search for the correct overall conception and the single right policy to a search for understanding of the conditions under which particular institutional forms serve user groups well in sustaining their resource bases over long periods of time. Conditional propositions of this sort have sometimes been formulated as "design principles" for resource institutions (Ostrom, 1990), a formulation that has stimulated considerable research interest since (see the discussion and synthesis of this literature by Agrawal, this volume:Chapter 2).

The Annapolis meeting also led to the development of several comparative databases designed to facilitate quantitative work related to the evolving theories. The first of these began at the Annapolis meeting as a draft coding form intended to capture most of the key variables contained in the cases. The form was revised on the basis of suggestions made at the meeting and further reworked by researchers at Indiana University. It was applied initially to a cross-national study of irrigation systems and inshore fisheries. In-depth case studies were evaluated for their completeness in regard to the variables in the database, and about 50 cases were coded for each of these two sectors (Schlager, 1994; Tang, 1992). This approach allowed substantial growth in understanding of the basic patterns of commons management (see, e.g., Bardhan and Dayton-Johnson, this volume: Chapter 3). The database was revised and updated to enable the coding of information on more than 100 irrigation systems in Nepal. The coded information from the case studies was supplemented by site visits to more than 80 of the systems to confirm initial coding and fill in missing information (see Lam, 1998). Another key database was developed by the International Forestry Resources and Institutions (IFRI) research program, and is used by collaborative research centers in Bolivia, Guatemala, India, Kenya, Madagascar, Mexico, Nepal, Tanzania, Uganda, and the United States. The purpose of this network of collaborating research centers is to apply the same core measurements to a series of cases within a country and to revisit locations regularly so it will be possible to study dynamic processes of common-pool resource management over time (see Gibson et al., 2000a).[9] Chapter 3 reviews some of the key research findings from more recently designed databases.

As the chapters that follow indicate, the present moment is not "the end of history" for research on commons. Rather, we seem to be at a point of rapid and exciting growth in work intended to aid our understanding of the dynamics of common-pool resources and the institutions that manage (and mismanage) them. New kinds of commons are being analyzed, new methodological tools and theoretical perspectives are being brought to bear, and ongoing work is increasingly synthetic and integrative. This effervescence in commons research is the motivation for this volume: A great deal has been learned and, based on that, research is moving forward at an exciting pace.

In the next section of this chapter we review the key concepts of commons research. The evolution of a clear conceptual framework has been an important part of commons research over the past decade. The growth in the field is being facilitated by clearer concepts and the concomitant recognition that similar ideas (albeit with different names) have emerged in several disciplines. As language and ideas are reconciled across disciplinary traditions, these relatively autonomous lines of work can cross-fertilize each other. So the discussion of conceptual developments is actually a continuation of our discussion of the history of the field and a prelude to the review of the current state of research.

## CONCEPTUAL DEVELOPMENTS AND KEY TERMS

An important outgrowth of the 1985 meeting has been a serious effort to untangle the various meanings of commons, common-pool resource, common property regimes, and related theoretical terms. As Bromley (1986) indicated in his synthesis at the Annapolis meeting, serious confusion had been introduced by using a property term—"common property"—to refer to a resource characterized by specific features. The term "common property" implies a kind of management arrangement created by humans rather than a characteristic of the resource itself. The preferred term for resources from which it is hard to exclude users is "common-pool" resource. The term "common-pool" focuses on the characteristics of the resource rather than on the human arrangements used to manage it. Such a resource could be left as open access without rules or could be managed by a government, as private property, or by a common property regime. The term "common property resource" had become so embedded in the language used in the economics and policy literatures that making this conceptual advance has been difficult. The confusion was embedded in the title of the NRC panel that organized the Annapolis conference, and it is still used in the title of the official newsletter of the association that emerged from this effort (*The Common Property Resource Digest*). After a somewhat heated debate, the word "resource" was dropped from the name of the IASCP itself so that the association's name includes the "property" term but not combined with the "resource" term. That both common-pool resource and common property resource can be abbreviated as CPR has added to the continued confusion. In this book, we do not use the CPR abbreviation at all to avoid further confusion.

Given this continued confusion, it is important that a clear set of definitions of key terms be presented in this initial chapter and used consistently throughout the book. In this chapter we focus on terms and concepts that now have gained relatively general agreement across disciplines. In Chapter 13, we turn to some of the newer conceptual developments in the field.

The term *commons* is used in everyday language to refer to a diversity of resources or facilities as well as to property institutions that involve some aspect of joint ownership or access. As mentioned, analytical advantages exist in separating the concept of the resource or good valued by humans from the concept of the rules that may be used to govern and manage the behavior and actions of humans using these resources.[10] In this view, a *common-pool resource* is a valued natural or human-made resource or facility that is available to more than one person and subject to degradation as a result of overuse. Common-pool resources are ones for which exclusion from the resource is costly and one person's use subtracts from what is available to others. The diversity of property rights regimes that can be used to regulate the use of common-pool resources is very large, including the broad categories of government ownership, private ownership, and ownership by a community.[11] When no property rights define who can use a common-pool resource and how its uses are regulated, a common-pool resource is under an open-access regime.

Human beings use common-pool resources by harvesting or extracting some of the finite flow of valued goods produced by them or by putting in unwanted byproducts, thus treating the resource as a sink.[12] In general, humans using resources of this type face at least two underlying incentive problems (Burger et al., 2001; Ostrom et al., 1994). The first is the problem of overuse, congestion, or even destruction because one person's use subtracts from the benefits available to others. The second is the free-rider problem that stems from the cost or difficulty of excluding some individuals from the benefits generated by the resource. The benefits of maintaining and enforcing rules of access and exclusion go to all users, regardless of whether they have paid a fair share of the costs. The institutions that humans devise to regulate the use of common-pool resources must somehow try to cope with these two basic incentive problems. They struggle with how to prevent overuse and how to ensure contributions to the mechanisms used to maintain both the resource and the institution itself.

## The Problem of Overuse

The first major characteristic of common-pool resources is the subtractability of resource units once extraction occurs. This characteristic is referred to by many other names, including jointness of consumption and rivalness of consumption.[13] All of these terms focus on the relationship that one person's use has on the availability of resource units for others. One person's harvest of fish, water, or timber subtracts from the amount left at any one time (and potentially, over time) for others. Because common-pool resources are subtractable, they can be easily congested, overharvested, degraded, and even destroyed. Many resources discussed in the theoretical literature on public goods are in fact common-pool resources because they have the attribute of subtractability, which classical public goods, such as world peace and scientific knowledge, do not have.

Some of the most challenging contemporary common-pool resource problems deal with the use of common-pool resources as sinks, which degrade through pollution. Common-pool sinks range in size from the global atmosphere, which is affected by the behavior of individuals in all countries of the world, to local watersheds and airsheds affected mainly by people at a single location. When a resource is a sink, the problem of overuse is putting too much of a contaminant into the resource as contrasted with the more familiar problem of taking too much out. Many watercourses suffer from both types of problems—too much water is extracted by each user, causing the costs of water for others to escalate, and too many pollutants are dumped into the resource, causing the quality of the water for others to decrease. Although the use of the common pool framework to understand sinks seems promising, this line of analysis is not as elaborate or as well studied as that examining resource extraction.

## The Free-Rider Problem

This problem was originally defined in its most extreme form—the impossibility of excluding beneficiaries once improvements to any set of resources had been made (Musgrave, 1959).[14] If the nature of certain resources made it truly impossible to solve the exclusion problem, however, institutions could not have any role in managing those resources. The contemporary view is that resources vary in the cost of excluding potential beneficiaries from deriving benefits from them. If it is not practical to exclude a user nor possible to force that user to contribute to the costs of developing and maintaining the resource, the noncontributing user is called a free rider. The cost of excluding potential users is often a function of technology. Prior to the invention of barbed wire fences, it was very expensive to exclude potential users from rangelands, but with barbed wire, it became more feasible to exclude those who did not have entry rights.

Thus, a core problem related to the use of common-pool resources is the cost of preventing access by potential users unless they agree to abide by a set of rules. In regard to a common-pool resource, users free ride when they harvest from or dump pollutants into the resource independently and take only their own costs and benefits into account. One "solves" the free-rider problem when rules are adopted and accepted that regulate individual actions so that social benefits and social costs are taken into account. The specific rules adopted in efforts to manage a common-pool resource sustainably are extremely numerous, but can be broadly classified into several general categories (Ostrom, 1999): boundary rules, position rules, authority rules, scope rules, aggregation rules, information rules, and payoff rules. Whether any particular rule configuration solves the free-rider problem in regard to a particular resource system depends on how well the rules address the biophysical structure of the resource, whether they are perceived by users as legitimate and are enforced, and whether they are understood by participants in a similar manner.

Analyzing the problem of exclusion and resulting free riding requires that a distinction be made between the system providing the resource itself (a river, a forest, or a fishery) and the resource units of value to humans (water, timber, or fish). After resource units have been extracted from the system, the cost of excluding potential beneficiaries from consuming the extracted units is often relatively low and the resource units may be considered to be private goods. That is, it may be hard to control who gets to go fishing but easy to control who gets the fish once they are caught. Effective markets for bottled water, fish, and timber are based on a low cost of excluding beneficiaries from the harvested units. A potential user can be easily prevented from acquiring them without paying the market price by the legal system and a strong set of norms providing enforcement to prevent theft. Ironically, these effective markets for harvested products are a major source of the incentives for users to overharvest. Harvesters obtain the full benefits from their overuse through the market for the resource units and suffer only a proportion of the costs they impose on others by overusing the system that provides the resource units.

Common-pool resources share the problem of difficult exclusion with another important policy problem—the provision of public goods such as international peace, knowledge, and living in a just society (Olson, 1965; Young, 1989). Once these goods are provided by someone—frequently a governmental agency—no one living within the scope of their provision can be easily excluded from enjoying the benefits. Although common-pool resources and public goods share this one characteristic, they differ in regard to subtractability: one person's use of a public good, such as the knowledge of a physical law, does not reduce the possibility for an infinite number of other persons to use the same knowledge.

As already noted, the key problem caused by high costs of exclusion for both common-pool resources and public goods is the free-rider problem. If exclusion is physically difficult and effective rules are not in place to limit who can use a resource and what can be withdrawn from it, then all harvesters face an incentive to increase their own harvesting rate without any concern for the impact of their actions on the costs for others (and eventually for themselves). Furthermore, the rules that govern a common-pool resource are themselves a public good because once they are provided, one person's use of the rules does not subtract from their availability for use by others. Thus, appropriation or harvesting from a common-pool resource has one structure of incentives that can lead to overuse. Providing rules to govern a common-pool resource has a second set of incentives that tempts participants to free ride on the time and effort required to craft effective rules because they will benefit from the adoption of such rules whether they contribute or not. The two sets of incentives work together to make the problem of avoiding overuse a real challenge. Contemporary scholars have stressed that there are actually many "games" involved in the governance and ongoing management of common-pool resources depending on many attributes of the resource and its users (see Ostrom et al., 1994).

## Institutional Attributes

Institutions are the rules that people develop to specify the "do's and don'ts" related to a particular situation. In regard to common-pool resources, rules define who has access to a resource; what can be harvested from, dumped into, or engineered within a resource; and who participates in key decisions about these issues and about transferring rights and duties to others. The stimulus for changes of institutional arrangements frequently has been fights over the distribution of resources (see Acheson and Knight, 2000; Knight, 1992; McCay, this volume:Chapter 11). Multiple types of institutional arrangements have been devised to try to reduce the problems of overuse and of free riding as well as distribution conflict.

As already noted, common-pool resources that do not have institutions governing their use are called open-access regimes. Institutions for governing use fit into three broad classes that are referred to as private property, common property, and government property. Each of these institutional types has a wide diversity of subtypes, and many hybrids exist as well. Something referred to as "government property," for example, may mean that a national government owns the property and that a national agency directly uses and manages that resource for its own purposes. Or, the resource may be "owned" by a national, state, or local government but users may have various rights to access, withdraw, manage, and determine who else is allowed to use the resource.[15] Use under a common-property regime may be restricted to members of a cooperative, an extended family, a formal corporation, a local community, or either a formally recognized or informally organized user group. A great variety of private-property regimes also have been devised to govern the use of common-pool resources (see Tietenberg, this volume:Chapter 6; see also Feeney et al., 1990).

## Additional Attributes of Common-Pool Resources

Costly exclusion and subtractability are the two defining attributes of common-pool resources.[16] A large number of other attributes are also important in shaping human resource use. Thus, developing a coherent theory of how institutions cope or do not cope effectively with the problems of overuse and free riding requires consideration of this diversity of attributes. Furthermore, some resource systems—such as groundwater basins or airsheds—provide only pure common-pool resources. Others, such as forests, yield some products that are subtractive (e.g., timber) and others that are nonsubtractive (e.g., flood control) (Gibson et al., 2000a). Thus, an analyst trying to understand how institutions affect behavior in regard to forest resources may need to understand which aspects of a forest are common-pool resources and which are public goods. (Subtractive and nonsubtractive products are related, however. For example, cutting timber can reduce a forest's ability to provide flood control.) We briefly describe three further attributes of resources that may have a major impact on the incentives that individuals face: renewability, scale, and cost of measurement.

*Renewable or Nonrenewable Common-Pool Resources*

Renewability relates to the rate at which resource units that are extracted (or used as a sink) replace themselves over time. The replacement rate over time can take any value between zero (nonrenewable) and one (instantly renewable). Mineral and oil resources are normally considered nonrenewable because once they are extracted from their source, no replacement is generated within a human time frame. Thus, the key problem faced in regulating nonrenewable resources is finding the optimal path toward efficient mining of the resource (Libecap, 1990).

On the other hand, biological species that are harvested for human use regenerate themselves in a cycle that varies from less than one year to decades, assuming the breeding stock and the breeding habitat are protected. Individuals who attempt to achieve sustainable use of such biological resources over time devise rules to limit the number of users; limit the technology, timing, quantity, or location of extraction; and protect the habitat of the species. The costs of designing, implementing, monitoring, and adapting these rules can vary substantially depending on the particular species characteristics, their habitat, the technology used, and the culture of the users. Resources that regenerate slowly are more challenging to manage because overharvesting may not be discovered until recovery of the resource is severely endangered. Fish that tend to cluster in groups are more likely to be destroyed with modern fishing technology because the marginal cost of searching for and harvesting the full extent of the fishery is much lower than for fish that spread out over a larger area (Clark, 1976, 1977).

Some human-made common-pool resources are renewed very rapidly once use has halted or been reduced. Broadcasting bandwidth, for example, is a common-pool resource because it is limited, one person's use is subtractive, and thus congestion can occur if too many users try to use the same bandwidth at the same time. The resource regenerates immediately, however, when usage declines, so subtractability exists across users, but not across time. Such commons cannot be destroyed permanently by overuse. The type of rules that are effective for regulating the use of radio bandwidth may thus be quite different from those needed to regulate the use of a biological species.

*Scale*

Major international problems, such as river and lake pollution, transmission of air pollutants across long distances, global climate change, threats to biodiversity, declines of ocean fisheries, and control of the use of outer space and the North and South Poles, have called attention to the attribute of scale among common-pool resources (Benedick, 1991; Buck, 1998; Gibson et al., 2000b; Haas et al., 1993; Young, 1989). Many important similarities exist between local and global common-pool resources even though there are obvious differences. Research has moved beyond studying resources at a single level (local or international) to comparing common-pool resources across levels and drawing lessons from one level to another (Keohane and Ostrom, 1995; Ostrom et al., 1999). One obvious difference between local and global resources is the sheer extent of the resource and thus the cost of monitoring use patterns at widely diverse locations. Global and local resources differ in two additional ways. The number of actors using, or having a say in decisions about, a global resource is usually larger than is the case for local resources, and these actors are usually much more heterogeneous. Both of these factors can affect the level of cooperation likely to be achieved in designing and complying with rules.

The literature on local common-pool resources suggests that a greater number of resource users does not necessarily impede cooperation (Ostrom, 1990), even though this may increase costs of devising, monitoring, and enforcing the rules. It also may make it necessary to design nested sets of institutions rather than a single layer. The literature on cooperation in international arenas, however, suggests that cooperation is less likely with a larger number of actors. These actors often include not only countries that are sovereign decision makers, but also a large number of nonstate actors that play important roles (Benedick, 1991; Mitchell, 1995; Vogel, 1986). The institutions granting these nonstate actors access to the political decision-making process also may play an important role in determining the potential for cooperation (Dolšak, 2001; International Human Dimensions Program, 1999; Weaver and Rockman, 1993; Young, 1997).

Heterogeneity of resource users may not have the same effects on local common-pool resources and on international resources. The literature on local common-pool resources suggests different, even opposing effects of heterogeneity among actors on cooperation. It has been argued that heterogeneity will induce cooperation (Olson, 1965) and that it will impede cooperation (Libecap, 1995). In empirical research, heterogeneity has been found to be a difficulty that users frequently are able to overcome so as to manage a common-pool resource (Lam, 1998; Varughese and Ostrom, 2001). This issue is discussed further by Bardhan and Dayton-Johnson (Chapter 3). However, studies at the international level, especially studies of international peace and provision of international public goods, suggest that heterogeneity induces cooperation (Martin, 1993, 1995). Although most scholars agree that heterogeneity of resource users makes a difference, considerably more work is needed to clarify this concept and its effects.[17]

It has become increasingly clear that global and local common-pool resources are not only analytically similar, but interrelated. The use of resources at the local level affects international and global resources, and vice versa. Thus devising the rules for using international and global resources requires a careful examination of local characteristics of resource use. For example, to devise a workable international regime for the use of global atmosphere as a sink for greenhouse gases, it is important to understand that different resource users emit various greenhouse gases for various reasons, that these uses cannot all be measured with the same degree of reliability, and that different resource users have drastically different capabilities to reduce their resource use. Many of these issues of linkage and interplay among institutions at different scales are discussed more fully in Chapters 8 and 9.

*Cost of Measurement*

To devise effective institutions that limit the use of common-pool resources so that they do not suffer congestion, overuse, or destruction, one needs to be able to measure the quantity and location of resource units. Common-pool resources vary substantially from one another in the reliability and cost of measuring current stocks and flows and predicting future conditions. Schlager and colleagues (1994) identify two physical attributes of resources that have a strong impact on the ease of measurement: the capability for storage and the mobility of resources. Storage (for example, a dam on a water distribution system) allows managers and users to measure the stock of a resource and to allocate its use over time in light of good information about what is currently available. Mobile resources, such as wildlife and undammed river water, are much harder to measure and account for than stable resources, such as forests and pasture lands. Again, the mobility of the resource makes measurement, and thus management, of wildlife much more difficult than stable resources.

## The Search for Effective Institutions

Devising better ways of governing resource systems will continue to be a major issue in the new century. Climate change, loss of biodiversity, ozone depletion, the widespread dispersal of persistent pollutants, and most other environmental problems involve the commons. Practitioners at international, national, regional, and local levels will continue to seek solutions and to debate the appropriate roles for government, private, and community ownership of natural resources. Meanwhile, considerable scientific uncertainty exists about how various property regimes and associated institutional forms affect resource sustainability.

The best available knowledge strongly suggests that the search for a single best strategy will be futile. The best tool for sustainable management of a common-pool resource depends on the characteristics of the resource and of the users. Substantial agreement is slowly evolving that multiple institutional strategies are needed given the wide diversity of threatened physical and biological resources. It requires substantial ingenuity to design institutions that cope effectively with the attributes of a particular resource given the larger macro-political institutions, culture, and economic environment in which that resource is embedded. With improved understanding, it may become possible to diagnose resource use situations well enough to separate promising institutional forms from those unlikely to achieve desired goals and thus provide useful scientific information to supplement ingenuity.

Analysis of the performance of a broad array of policy options at diverse levels of organization will be required to advance our knowledge. Analysis is proceeding from the early, rough classification of a few major categories of property rights regimes toward more refined typologies, from bivariate propositions about which institutional forms work better to more complex theories that take contextual differences into account, and from analyses at a single level of social organization to those that take into account linkages among institutional forms at different levels. An important advance was the idea that institutions face major design challenges (e.g., fit with resource characteristics, monitoring the resource and the users, enforcement of rules). This led to a search for robust "design principles" (Ostrom, 1990). Outcomes may be more dependent on the ability of institutions to meet design challenges than on institutional attributes such as the type of property rights they establish. We discuss these issues in more detail in Chapter 13.

Furthermore, recognition is growing that institutional performance may be assessed using multiple evaluative criteria, including efficiency, sustainability, and equity. The criterion of economic efficiency focuses on the relationship of total individual and social benefits to total individual and social costs. Even though it is often difficult to measure social benefits and costs, the conceptual unerpinning for efficiency analysis is clear. An institutional arrangement is considered economically efficient if no reallocation of resources will improve the welfare of some individuals affected by the resource without making someone else worse off. The criterion of sustainability can be applied to both the resource and the institutions governing the resource. In regard to the resource, sustainability refers to the continuance (or even improvement) of the resource system, facility, or stock that generates the flow of resource units. In regard to an institution, sustainability refers to the continued use of the institution over time with adaptation occurring in the day-to-day rules within the context of a stable constitution. Equity criteria are used to evaluate the distribution of costs and benefits either on the basis of the relationship between individuals' contributions to an effort and the benefits they derive or on the basis of their differential abilities to pay. Beyond efficiency, sustainability, and equity, criteria such as accountability and adaptability are frequently invoked. No institutional arrangement is likely to perform well on all evaluative criteria at all times. Thus, in practice, some tradeoff among performance criteria is usually involved. Economic efficiency has frequently dominated the policy debate, but concerns of equity and sustainability of the resource may be more important to those directly affected by policy proposals.

## STRUCTURE OF THE BOOK

An overview of a vibrant field of research can be organized in many ways. We have chosen to begin with chapters that review the most venerable traditions of research in the field and that at the same time display the diversity of methodological and theoretical tools that have been used to understand the commons. We hope this will give the reader a sense of the highly interdisciplinary and stimulating nature of the literature. We then move toward emerging topics in the commons literature, including the interplay between markets and other commons institutions and the problem of understanding the evolving relationships among local, national, and global institutions. Finally, we move to problems and approaches that are just on the horizon but that we believe will be central to any review of our understanding of the commons a decade hence. In our final chapter, we attempt to synthesize and suggest key problems for further research.

Chapters 2 through 12 provide reviews of key issues affecting the governance of common-pool resources. Generally, Chapters 2 through 9 summarize knowledge that has been developed in research over the past 15 years, while Chapters 10 through 12 give more emphasis to important issues that research has uncovered but that have not yet received detailed examination.

Chapters 2 through 5 are based on knowledge developed from quite different research methods. Agrawal (Chapter 2) examines the evidence regarding a number of empirical generalizations that have been proposed about the operation of institutions for managing common-pool resources. The chapter relies on evidence from structured qualitative case comparisons involving moderately large numbers of resource management institutions. Bardhan and Dayton-Johnson (Chapter 3) focus on the effects of heterogeneity among resource users, drawing evidence from quantitative analyses of irrigation systems. Kopelman, Weber, and Messick (Chapter 4) examine the effects of attributes of resource users, their groups, and the tasks they face by reviewing findings from experimental studies involving simulated common-pool resource users. Falk, Fehr, and Fischbacher (Chapter 5) use formal game theory to develop simple models that can generate empirically observed phenomena from a few behavioral assumptions. In addition to addressing important substantive issues in the theory of common-pool resource use, these chapters illustrate the variety of disciplines and research approaches that are contributing to knowledge in the field and the kinds of knowledge that can come from these disciplines and approaches.

Chapters 6 and 7 focus on what has been learned from policy experiments with two classes of property rights regimes: tradable environmental allowances and community property. Tietenberg (Chapter 6) examines the variety of tradable permits arrangements that have been used to govern air and water emissions and access rights in fisheries. He discusses both expectations from economic theory and results in practice, summarizes the factors associated with variations in outcomes, and discusses possible reasons for the observed outcomes. Rose (Chapter 7) considers tradable environmental allowances and common property as ideal types of property rights and offers a number of empirically based hypotheses about the conditions favoring success of each institutional type.

Chapters 8 and 9 address key issues of scale and linkage across institutions. Young (Chapter 8) presents a classification of cross-scale linkages and examines the evidence on their operation in land use and sea use. He offers conclusions about the strengths and weaknesses of larger and smaller scale units and the tradeoffs involved in vesting powers at the different scales. Berkes (Chapter 9) draws on the case literature to discuss conditions under which involvement by the state facilitates or impedes the operation of local institutions. He then discusses several institutional forms with the potential to improve cross-scale linkages.

Chapters 10 through 12 raise issues that have not as yet received the concerted research attention they deserve. Wilson (Chapter 10) discusses the history of scientific fisheries management to raise issues about the appropriate roles of standard science and local knowledge in resource management and about the effect of scientific uncertainty on the ability to use deterministic scientific models as a main management tool. McCay (Chapter 11) addresses several issues of process that have not received much research attention in the literature on common-pool resources, though they have received attention in other contexts. These include getting environmental issues on the agendas of decision-making bodies, the conflict management roles of institutions, problems of deliberative process in environmental institutions, and the uses of incremental change in resource management. Richerson, Boyd, and Paciotti (Chapter 12) discuss resource management institutions from the perspective of cultural evolutionary theory. They present a dual inheritance theory of culture that is applicable to institutions, discuss how important empirical regularities about commons institutions fit this theory, and identify a set of as-yet unexplored hypotheses that flow from the theory.

Finally, Chapter 13 provides an overview of the current state of knowledge about the potential of institutional design to help human groups avoid tragedies of the commons. It characterizes the development of common-pool resource management as a research field, summarizes some key substantive lessons that have been learned to date, and identifies the practical challenges of institutional design that have been uncovered by research. Finally, it suggests directions for future research, including further development of some ongoing lines of research and new attention to four critical but understudied issues: understanding the dynamics of resource management institutions, extending insights to more kinds of common-pool resources, understanding the effects of context on institutions, and understanding the operation of linkages across institutions.

## NOTES

1   In thinking about environmental concern, it has been useful to distinguish self-interest, concern with the welfare of other humans, and concern with other species, ecosystems, and the biosphere itself (Stern et al., 1993).

2   In a parallel argument, Axelrod (1997) suggests that game theory provides an *Escherichia coli* for the social sciences—an ideal experimental organism. We prefer the analogy to *Drosophila melanogaster*. *E. coli* has been studied primarily in the laboratory. *Drosophila* has been investigated both in the laboratory and in the field, and has been a key organism for making the link between the two (Dobzhansky et al., 1977; Rubin and Lewis, 2000). Thus it provides a closer parallel to the role the problem of the commons plays in the social sciences.

3   See Hardin's own discussion of the impact of his earlier article (Hardin, 1998).

4   The first bibliography on common-pool resources was started in 1985 by Martin (1989, 1992). In 1993 Hess developed a computerized database on common-pool resources and incorporated the earlier citations. She has continued building the bibliographic database through systematic searches (Hess, 1996a, 1996b, 1999). As of April 2001, 29,800 citations were in the Common-Pool Resources database. A searchable online version of this database is available at: http://www.indiana.edu/~iascp/Iforms/searchcpr.html.

5   This conference was cosponsored by the National Research Council, the U.S. Agency for International Development, the Ford Foundation, and the World Wildlife Fund. At about the same time as the NRC Panel on Common Property Resource Management was organized, Acheson and McCay organized two symposia and one workshop to bring together anthropologists from diverse subfields to examine the meaning of the concept "the commons" and to draw on the tools of sociocultural, economic, and ecological anthropology to examine basic questions of the commons (see McCay and Acheson, 1987b).

6   Hardin's argument is quite similar to the position held until recently by most evolutionary theorists: that selfish strategies would always obtain higher returns than reciprocal or cooperative strategies and drive out through competition any strategies other than selfish strategies. However, this view is losing its dominance. See Sober and Wilson (1998) and the discussion in Chapter 12.

7   A "game of chicken" can be illustrated with two drivers rapidly driving toward each other in a single lane. They both realize they will collide unless at least one swerves, so that they miss each other. Each prefers that the other swerves. The choice facing each is to go straight or swerve. If both go straight, they crash. The best joint outcome is for one to go straight and the other to swerve, but one player obtains more than the other in this outcome. The "assurance game" (also called "stag hunt") can be illustrated with two hunters following a stag. Catching the stag requires a joint effort of both, which yields the best joint outcomes. When a rabbit approaches the two hunters, they both face a temptation to catch a rabbit, which either can do alone, rather than chasing a stag with the uncertain help of the other. Going jointly for a stag is surely rational, but if the hunters have any reasons to doubt the effort of each other, then it is better to turn and start hunting a rabbit. For detailed discussion of the differences among these three types of games as applied to common-pool resources, see Ostrom et al. (1994).

8   The panel was composed of Daniel W. Bromley, David H. Feeny, Jere L. Gilles, William T.

Gladstone, Barbara J. Lausche, Margaret A. McKean, Ronald J. Oakerson, Elinor Ostrom, Pauline E. Peters, C. Ford Runge, and James T. Thomson.

9  It is hoped that the revisits can be scheduled at least every 5 years so as to observe changes in forest extent, biomass, and biodiversity as well as any demographic, economic, or institutional change that may have occurred (see Ostrom, 1998).

10  This is, of course, an analytical distinction. Behaviorally, an individual faces a resource and the institutions that are used to manage that resource (if any) at the same time, so the attributes that affect individual choice are derived from both the resource and the institutions. In examining theory and in proposing policies, the distinction is important because interventions are far more likely in regard to the institutional variables than in regard to the underlying attributes of the resource.

11  Given the wide diversity of rules used in practice, each of these categories includes very diverse institutions. The classification is a first cut and analysts will find it useful for some purposes. For others, one needs to know precisely the rules being used for controlling access and making other choices about the resource.

12  Schnaiberg (1980) discusses the use of the biophysical environment as a source or as a sink.

13  This attribute was posed initially by Samuelson (1954) as a way to divide the world of goods into two classes: private consumption goods and public consumption goods. Private goods are subtractable, public goods are not.

14  Musgrave, like Samuelson (1954), also used one attribute—exclusion—as a way of dividing the world into two types of goods: private and public. Having demonstrated that the market had desirable properties when used in relationship to private goods, a key theoretical debate among economists during the 1950s focused on the question of conditions leading to market failure. For some time, scholars tried to classify all goods, resources, and services into those that could be called "private goods" and were best provided by a market and those that could be called "public goods" and were best provided by a government. The recognition that there were multiple attributes of goods and resources that affect the incentives facing users came about gradually as the dichotomies posed by Samuelson and Musgrave proved to be theoretically inadequate to the task of predicting the effect of institutional arrangements (see Chamberlin, 1974; Ostrom and Ostrom, 1977; Taylor, 1987).

15  See Schlager and Ostrom (1992) for a discussion of the bundle of rights that may be involved in the use of common-pool resources.

16  As already noted, cost of exclusion is only partially an attribute of the resource. Although resource characteristics matter (e.g., exclusion is more difficult in an ocean fishery than in a lake), cost of exclusion also is affected by available technology and various other attributes of user groups and their institutions.

17  Keohane and Ostrom (1995), for example, focus on four types of heterogeneity: heterogeneity in capabilities, in preferences, in information and beliefs, and in institutions. In addition to these types, current debates on devising instruments for global climate change policy suggest that heterogeneity in the extent of the past use of the resource also plays an important role.

# REFERENCES

Acheson, J.M., and J. Knight
    2000    Distribution fights, coordination games, and lobster management. *Comparative Studies in Society and History* 42(1):209-238.
Arnold, J.E.M., and J.G. Campbell
    1986    Collective management of hill forests in Nepal: The community forestry development project. Pp. 425-454 in National Research Council, *Proceedings of the Conference on Common Property Resource Management.* Washington, DC: National Academy Press.
Axelrod, R.
    1984    *The Evolution of Cooperation.* New York: Basic Books.
    1997    *The Complexity of Cooperation.* Princeton, NJ: Princeton University Press.

Bauer, D.F.
    1977   *Household and Society in Ethiopia.* East Lansing, MI: African Studies Center, Michigan State University.
Baumol, W.J., and W.E. Oates
    1988   *The Theory of Environmental Policy.* 2d ed. Cambridge, Eng.: Cambridge University Press.
Benedick, R.E.
    1991   *Ozone Diplomacy: New Directions in Safeguarding the Planet.* Cambridge, MA: Harvard University Press.
Berkes, F.
    1986   Local-level management and the commons problem: A comparative study of Turkish coastal fisheries. *Marine Policy* 10:215-229.
Berkes, F., ed.
    1989   *Common Property Resources: Ecology and Community-Based Sustainable Development.* London: Belhaven Press.
Berkes, F., D. Feeny, B.J. McCay, and J.M. Acheson
    1989   The benefits of the commons. *Nature* 340:91-93.
Blomquist, A.
    1992   *Dividing the Waters: Governing Groundwater in Southern California.* San Francisco: ICS Press.
Bromley, D.W.
    1986   Closing comments at the conference on common property resource management. Pp. 591-597 in National Research Council, *Proceedings of the Conference on Common Property Resource Management.* Washington, DC: National Academy Press.
Bromley, D.W., D. Feeny, M.A. McKean, P. Peters, J.L. Gilles, R.J. Oakerson, C.F. Runge, and J.T. Thomson, eds.
    1992   *Making the Commons Work: Theory, Practice, and Policy.* San Francisco: ICS Press.
Buck, S.J.
    1998   *The Global Commons: An Introduction.* Washington, DC: Island Press.
Burger, J., E. Ostrom, R.B. Norgaard, D. Policansky, and B.D. Goldstein, eds.
    2001   *Protecting the Commons: A Framework for Resource Management in the Americas.* Washington, DC: Island Press.
Chamberlin, J.
    1974   Provision of collective goods as a function of group size. *American Political Science Review* 68:707-716.
Ciriacy-Wantrup, S.V., and R.C. Bishop
    1975   Common property as a concept in natural resources policy. *Natural Resources Journal* 15(4):713-727.
Clark, C.W.
    1976   *Mathematical Bioeconomics.* New York: Wiley Publishers.
    1977   The economics of over-exploitation. In *Managing the Commons*, G. Hardin and J. Baden, eds. San Francisco: W.H. Freeman.
Cordell, J., ed.
    1990   *A Sea of Small Boats.* Cambridge, MA: Cultural Survival.
Crutchfield, J.A.
    1964   The marine fisheries: The problem in international cooperation. *American Economic Review* 54:207-218.
Dawes, R.M.
    1980   Social dilemmas. *Annual Review of Psychology* 31:169-193.
Demsetz, H.
    1967   Toward a theory of property rights. *American Economic Review* 62:347-359.

Dobzhansky, T., F.J. Ayala, G.L. Stebbins, and J.W. Valente
   1977   *Evolution.* San Francisco: W.H. Freeman.
Dolšak, N.
   2001   Mitigating global climate change: Why are some countries more committed than others? *Policy Studies Journal,* 29(3):414-436.
Dyson-Hudson, R., and E.A. Smith
   1978   Human territoriality: An ecological reassessment. *American Anthropologist* 80:21-41.
Feeny, D., F. Berkes, B.J. McCay, and J.M. Acheson
   1990   The tragedy of the commons: Twenty-two years later. *Human Ecology* 18(1):1-19.
Gibson, C.
   2001   Forest resources: Institutions for local governance in Guatemala. Pp. 71-89 in *Protecting the Commons: A Framework for Resource Management in the Americas,* J. Burger, E. Ostrom, R.B. Norgaard, D. Policansky, and B.D. Goldstein, eds. Washington, DC: Island Press.
Gibson, C., M. McKean, and E. Ostrom, eds.
   2000a  *People and Forests: Communities, Institutions, and Governance.* Cambridge, MA: MIT Press.
Gibson, C., E. Ostrom, and T. Ahn
   2000b  The concept of scale and the human dimensions of global change: A survey. *Ecological Economics* 32(2):217-239.
Gordon, H.S.
   1954   The economic theory of a common-property resource: The fishery. *Journal of Political Economy* 62(April):124-142.
Grossi, P.
   1981   *An Alternative to Private Property. Collective Property in the Juridical Consciousness of the Nineteenth Century.* Chicago: University of Chicago Press.
Haas, P.M., R.O. Keohane, and M.A. Levy, eds.
   1993   *Institutions for the Earth: Sources of Effective International Environmental Protection.* Cambridge, MA: MIT Press.
Hardin, G.
   1968   The tragedy of the commons. *Science* 162:1243-1248.
   1998   Extensions of the tragedy of the commons. *Science* 280(May):682-683.
Hess, C.
   1996a  *Common Pool Resources and Collective Action: A Bibliography; Volume 3.* Bloomington, IN: Indiana University Workshop in Political Theory and Policy Analysis.
   1996b  *Forestry Resources and Institutions: A Bibliography.* Bloomington, IN: Indiana University Workshop in Political Theory and Policy Analysis.
   1999   *A Comprehensive Bibliography of Common Pool Resources (CD-ROM).* Bloomington, IN: Workshop in Political Theory and Policy Analysis.
Hoskins, W.G., and L.D. Stamp
   1963   *The Common Lands of England and Wales.* London: Collins.
International Human Dimensions Program
   1999   *Institutional Dimensions of Global Environmental Change.* Bonn, Ger.: International Human Dimensions Program.
Johnson, O.E.G.
   1972   Economic analysis, the legal framework and land tenure systems. *Journal of Law and Economics* 15:259-276.
Keohane, R.O., and E. Ostrom, eds.
   1995   *Local Commons and Global Interdependence: Heterogeneity and Cooperation in Two Domains.* London: Sage.

Knight, J.
  1992  *Institutions and Social Conflict.* New York: Cambridge University Press.
Lam, W.F.
  1998  *Governing Irrigation Systems in Nepal: Institutions, Infrastructure, and Collective Action.* San Francisco, CA: ICS Press.
Leach, E.
  1954  *Political Systems of Highland Burma.* London: Bell Publishers.
Libecap, G.
  1990  *Contracting for Property Rights.* New York: Cambridge University Press.
  1995  The conditions for successful collective action. Pp. 161-190 in *Local Commons and Global Interdependence: Heterogeneity and Cooperation in Two Domains*, R. Keohane and E. Ostrom, eds. London: Sage.
Lloyd, W.F.
  1977  [1833] On the checks to population. Pp. 8-15 in *Managing the Commons*, G. Hardin and J. Baden, eds. San Francisco: W.H. Freeman.
Maine, Sir Henry
  1871  *Village Communities in the East and West.* New York and London: Henry Holt and Company.
Malinowski, B.
  1926  *Crime and Punishment in Savage Society.* London: Kegan Paul, Trench and Trubner.
Martin, F.
  1989  *Common Pool Resources and Collective Action: A Bibliography; Volume 1.* Bloomington, IN: Indiana University Workshop in Political Theory and Policy Analysis.
  1992  *Common Pool Resources and Collective Action: A Bibliography; Volume 2.* Bloomington, IN: Indiana University Workshop in Political Theory and Policy Analysis.
Martin, L.
  1993  Credibility, costs, and institutions: Cooperation on economic sanctions. *World Politics* 45:406-432.
  1995  Heterogeneity, linkage and commons problems. Pp. 71-92 in *Local Commons and Global Interdependence*, R. Keohane and E. Ostrom, eds. London: Sage Publications.
McCay, B.J.
  1980  A fisherman's cooperative, *limited*: Indigenous resource management in a complex society. *Anthropological Quarterly* 53(January):29-38.
  1981  Optimal foragers or political actors? Ecological analyses of a New Jersey fishery. *American Ethnologist* 8(2):356-382.
  1995  Common and private concerns. *Advances in Human Ecology* 4:89-116.
  1996  Common and private concerns. Pp. 111-126 in *Rights to Nature: Ecological, Economic, Cultural, and Political Principles of Institutions for the Environment*, S. Hanna, C. Folke, and K.G. Mäler, eds. Washington, DC: Island Press.
McCay, B.J., and J.M. Acheson, eds.
  1987a Human ecology of the commons. Pp. 1-34 in *The Question of the Commons.* Tucson: University of Arizona Press.
  1987b *The Question of the Commons: The Culture and Ecology of Communal Resources.* Tucson: University of Arizona Press.
Mitchell, R.B.
  1995  Heterogeneities at two levels: States, non-state actors and international oil pollution. Pp. 223-254 in *Local Commons and Global Interdependence: Heterogeneity and Cooperation in Two Domains*, R. Keohane and E. Ostrom, eds. London: Sage.
Musgrave, R.A.
  1959  *The Theory of Public Finance: A Study in Public Economy.* New York: McGraw-Hill.

National Research Council
 1986 *Proceedings of the Conference on Common Property Resource Management.* Washington, DC: National Academy Press.
Netting, R.M.
 1976 What Alpine peasants have in common: Observations on communal tenure in a Swiss village. *Human Ecology* 4(2):135-146.
 1981 *Balancing on an Alp: Ecological Change and Continuity in a Swiss Mountain Community.* Cambridge, Eng.: Cambridge University Press.
Oakerson, R.J.
 1986 A model for the analysis of common property problems. Pp. 13-30 in *Proceedings of the Conference on Common Property Resource Management.* National Research Council. Washington, DC: National Academy Press.
Olson, M.
 1965 *The Logic of Collective Action: Public Goods and the Theory of Groups.* Cambridge, MA: Harvard University Press.
Ostrom, E.
 1986 Issues of definition and theory: Some conclusions and hypotheses. Pp. 599-614 in National Research Council, *Proceedings of the Conference on Common Property Resource Management.* Washington, DC: National Academy Press.
 1990 *Governing the Commons: The Evolution of Institutions for Collective Action.* New York: Cambridge University Press.
 1998 The international forestry resources and institutions research program: A methodology for relating human incentives and actions on forest cover and biodiversity. Pp. 1-28 in *Forest Biodiversity in North, Central, and South America, and the Caribbean: Research and Monitoring,* F. Dallmeier and J.A. Comiskey, eds. Man and the Biosphere Series, Vol. 21. Paris: UNESCO.
 1999 Coping with tragedies of the commons. *Annual Review of Political Science* 2:493-535.
Ostrom, E., J. Burger, C. Field, R.B. Norgaard, and D. Policansky
 1999 Revisiting the commons: Local lessons, global challenges. *Science* 284:278-282.
Ostrom, E., R. Gardner, and J. Walker
 1994 *Rules, Games, and Common-Pool Resources.* Ann Arbor: University of Michigan Press.
Ostrom, V., and E. Ostrom
 1977 Public goods and public choices. Pp. 7-49 in *Alternatives for Delivering Public Services: Toward Improved Performance,* E.S. Savas, ed. Boulder, CO: Westview Press.
Peters, P.E.
 1986 Concluding statement. Pp. 615-621 in National Research Council, *Proceedings of the Conference on Common Property Resource Management.* Washington, DC: National Academy Press.
Pinkerton, E., ed.
 1989 *Co-operative Management of Local Fisheries: New Directions for Improved Management and Community Development.* Vancouver: University of British Columbia Press.
Rappaport, R.A.
 1984 *Pigs for the Ancestors: Ritual in the Ecology of a New Guinea People.* New Haven, CT: Yale University Press.
Rose, C.
 1994 *Property and Persuasion: Essays on the History, Theory, and Rhetoric of Ownership.* Boulder, CO: Westview Press.
Rubenstein, F.D., G. Watzke, R.H. Doktor, and J. Dana
 1974 The effect of two incentive schemes upon the conservation of shared resources by five-person groups. *Organizational Behavior and Human Performance* 13:330-338.

Rubin, G.M., and E.B. Lewis
    2000    A brief history of *Drosophila's* contributions to genome research. *Science* 287:2216-2218.
Ruddle, K., and R.E. Johannes, eds.
    1985    *The Traditional Knowledge and Management of Coastal Systems in Asia and the Pacific.* Jakarta: UNESCO.
Runge, C.F.
    1981    Common property externalities: Isolation, assurance and resource depletion in a tradi- tional grazing context. *American Journal of Agricultural Economics* 63:595-606.
    1984a   Institutions and the free rider: The assurance problem in collective action. *Journal of Poli- tics* 46:154-181.
    1984b   Strategic interdependence in models of property rights. *American Journal of Agricultural Economics* 66:807-813.
    1986    Common property and collective action in economic development. Pp. 31-62 in National Research Council, *Proceedings of the Conference on Common Property Resource Man- agement.* Washington, DC: National Academy Press.
Samuelson, P.A.
    1954    The pure theory of public expenditure. *Review of Economics and Statistics* 36:387-389.
Schaefer, M.B.
    1957    Some considerations of population dynamics and economics in relation to the manage- ment of commercial marine fisheries. *Journal of the Fisheries Research Board of Canada* 14:669-681.
Schlager, E.
    1994    Fishers' institutional responses to common-pool resource dilemmas. Pp. 247-265 in *Rules, Games, and Common-Pool Resources*, E. Ostrom, R. Gardner, and J. Walker, eds. Ann Arbor: University of Michigan Press.
Schlager, E., W. Blomquist, and S.Y. Tang
    1994    Mobile flows, storage and self-organized institutions for governing common-pool re- sources. *Land Economics* 70(3):294-317.
Schlager, E., and E. Ostrom
    1992    Property rights and natural resources: A conceptual analysis. *Land Economics* 68(3):249- 262.
Schnaiberg, A.
    1980    *The Environment: From Surplus to Scarcity.* New York: Oxford University Press.
Sengupta, N.
    1991    *Managing Common Property: Irrigation in India and the Philippines.* New Delhi: Sage.
Smith, A.
    1977    [1804] *A Theory of Moral Sentiments.* New York: Oxford University Press.
Sober, E., and D.S. Wilson
    1998    *Unto Others: The Evolution and Psychology of Unselfish Behavior.* Cambridge, MA: Harvard University Press.
Stern, P.C.
    1976    Effect of incentives and education on resource conservation in a simulated commons di- lemma. *Journal of Personality and Social Psychology* 34:1285-1292.
Stern, P.C., T. Dietz, and L. Kalof
    1993    Value orientations, gender and environmental concern. *Environment and Behavior* 25:322- 348.
Tang, S.Y.
    1992    *Institutions and Collective Action: Self-governance in Irrigation.* San Francisco: ICS Press.
Taylor, M.
    1987    *The Possibility of Cooperation.* New York: Cambridge University Press.

Thompson, E.P.
   1975   *Whigs and Hunters.* London: Allen Lane.
Townsend, R., and J.A. Wilson
   1987   An economic view of the commons. Pp. 311-326 in *The Question of the Commons*, B.J. McCay and J.M. Acheson, eds. Tucson: University of Arizona Press.
Varughese, G., and E. Ostrom
   2001   The contested role of heterogeneity in collective action: Some evidence from community forestry in Nepal. *World Development* 29(5):747-765.
Vayda, A.P., and R.A. Rappaport
   1968   Ecology, cultural and noncultural. Pp. 477-497 in *Introduction to Cultural Anthropology*, J.A. Cliffton, ed. Boston: Houghton Mifflin.
Vogel, D.
   1986   *National Styles of Regulation: Environmental Policy in Great Britain and the United States.* Ithaca, NY: Cornell University Press.
Wade, R.
   1994   *Village Republics: Economic Conditions for Collective Action in South India.* San Francisco, CA: ICS Press.
Weaver, R.K., and B.A. Rockman, eds.
   1993   *Do Institutions Matter? Government Capabilities in the United States and Abroad.* Washington, DC: Brookings Institution.
Williams, G.C.
   1966   *Adaptation and Natural Selection: A Critique of Some Current Evolutionary Thought.* Princeton, NJ: Princeton University Press.
Young, O.
   1989   *International Cooperation. Building Regimes for Natural Resources and the Environment.* Ithaca, NY: Cornell University Press.
   1997   *Global Governance: Drawing Insights from the Environmental Experience.* Cambridge, MA: MIT Press.

# PART I

# RESOURCE USERS, RESOURCE SYSTEMS, AND BEHAVIOR IN THE DRAMA OF THE COMMONS

The chapters in Part I of the book illustrate some qualitatively different approaches to the question of how characteristics of the users of common-pool resources affect the way in which the drama of the commons unfolds and is resolved. These approaches include predictions derived in a top-down fashion from formal (game) theory (Chapter 5) and predictions induced in a more bottom-up fashion from observed behavioral regularities in controlled laboratory studies (Chapter 4) or in field studies (Chapter 3). The conclusions reached by these investigations add a level of complexity or contingency to the picture of behavior in commons situations drawn in this volume. In particular, they show the need to consider the effects of user characteristics (e.g., inequity aversion or economic heterogeneity) in conjunction with resource system characteristics (e.g., size, definition of boundaries) and task characteristics (e.g., the possibility of communication and sanctioning institutions). Chapter 2 makes a strong case for the need to study not just main effects but also the interactions between the classes of variables that have been identified as bearing a causal relationship with successful commons management and sustainability.

In their analyses of the effects of user characteristics, the chapters in Part I illustrate the relative strengths and complementary contributions that different academic disciplines bring to the table. Falk, Fehr, and Fischbacher (Chapter 5) demonstrate the power of the optimization framework of economic modeling that provides equilibrium predictions of behavior under a range of assumptions about user and task characteristics, which are operationalized as parameters in the optimization function. Falk et al. show that, in skillful hands, such models can account for a broad range of stylized facts obtained in carefully controlled laboratory experiments with a small number of parameters. Although elegant and

*37*

mathematically tractable, the top-down modeling of economics focuses solely on the *prediction* of behavior.

Theory building in other social sciences and particularly in psychology, on the other hand, is a more data-driven and less elegant enterprise, which has as its goal not only prediction but also *explanation* of observed behavior. Theory and research thus focus on the processes that give rise to observed behavior. It is this focus on explanation and process that leads psychologists to study a broader range of dependent measures (e.g., not just the magnitude of withdrawals from common-pool resources, but also users' justifications of such withdrawals as well as judgments of fairness). Kopelman, Weber, and Messick (Chapter 4) provide a very useful classification and road map to the effects of a large number of user and user group characteristics and situational variables that have been shown to affect behavior in commons dilemmas. Agrawal's critical analysis and synthesis of political science field studies of commons management regimes (Chapter 2) includes user group characteristics as well as resource system characteristics, institutional arrangements, and characteristics of the external environment. In all instances, he finds that the effect of these variables on sustainability of the commons is configural. The effect of even a basic characteristic such as user group size or heterogeneity depends on a range of other contextual and mediating factors. Closed-form solutions for models that would attempt to incorporate all of these variables into a prediction equation would be hard to come by. Nevertheless, awareness of these effects and causal explanations of their origin are of both theoretical and practical importance to researchers, resource users, and policy makers.

In their review of the effects of economic heterogeneity among water resource users on commons dilemma outcomes, Bardhan and Dayton-Johnson (Chapter 3) illustrate the utility of focusing on process-level explanations of behavior and the importance of checking the predictions of economic theory against observed regularities (in their case, regularities observed in large-$n$ field studies). The authors draw some careful distinctions between different processes by which economic heterogeneity might affect the resolution of common-pool resource management dilemmas (e.g., effects on incentives for cooperation versus effects on social norms and sanctions; effects on choice of institutions versus effects on their implementation and enforcement). Such distinctions allow them to reconcile apparently contradictory predictions about the effect of economic heterogeneity made by different theories. In their evaluation of the explanatory potential of these alternative process explanations for behavior observed in several international field studies, Bardhan and Dayton-Johnson show that some mechanisms (e.g., "Olson effects") may have greater theoretical plausibility than practical reality.

Agrawal (Chapter 2) explicitly addresses the relationship between theory-driven and data-driven research approaches. He argues that the two are not just synergistic in the insights they provide, but that they require each other at an operational level. In particular, given the large number of variables that have been identified as affecting sustainability in some form and their even more numerous potential interactions, a top-down approach driven by theory seems to be necessary to guide the design, execution, and analysis of field studies in ways that make maximal use of scarce empirical research resources.

In summary, the following four chapters provide us with a rich account of the way in which user and user group characteristics interact with resource system characteristics and affect the processes by which institutions are crafted, the types of institutions that emerge, the degree to which they are implemented successfully, and the way in which resulting conflicts are resolved. In combination, they show us that much has been learned over the past 15 years, with some substantive insights and—perhaps more importantly—significant methodological and meta-theoretical insights. They also provide us with a road map to yet unresolved questions of commons management design and with an appreciation that complex problems have highly contingent solutions which, in turn, require cross-disciplinary cooperation.

# 2

# Common Resources and Institutional Sustainability

*Arun Agrawal*

This chapter focuses on the large body of empirical work on common property. Its objective is delineate some of the most significant accomplishments of this literature, discuss some of its continuing deficiencies, and highlight shifts in research approaches and methods that can help address existing weaknesses. In an enduring achievement, scholars of common property have shown that markets or private property arrangements and state ownership or management do not exhaust the range of plausible institutional mechanisms to govern natural resource use. But the documentation and theoretical defense of this insight has rested chiefly on the analysis and examination of hundreds of separate case studies of successful common-pool resource governance. Each such study has generated different conclusions about what counts in "successful" resource management. The multiplicity of causal variables, and the lack of attention to how the observed effects of these variables depend on the state of the context, has created significant gaps in explanations of how common property institutions work. Addressing these gaps will require important shifts in how scholars of commons conduct their research.

Such a shift is important because common property institutions continue to frame how natural resources are governed in many countries throughout the world. In addition, national governments in nearly all developing countries have turned to local-level common property institutions in the past decade as a new policy thrust to decentralize the governance of the environment. This shift in policy is no more than a belated recognition that sustainable resource management can never be independent of sustainability of collective human institutions that frame resource governance, and that local users are often the ones with the greatest stakes in sustainability of resources and institutions. But until as late as the 1970s, writ-

*41*

ings about common property endowed their object of study with an antiquated flavor. Portrayals of the most famous example, the English Commons and their enclosures (Ault, 1952; Baker and Butlin, 1973; Thirsk, 1966; Yelling, 1977), suggested, if only by implication, that common property is a curious holdover from the past that was destined to disappear in the face of trends toward modernization.[1] To many observers, placing common property in the historical past seemed so obvious as to be natural.

The most sustained theoretical engagement with community and communal forms of life, occurring as it did toward the end of the nineteenth century, gave further credence to the idea that the disappearance of norms of community and forms of communal life is an integral if perhaps regrettable part of progress. Theorists of the time, among them Auguste Comte, Emile Durkheim, Karl Marx, Herbert Spencer, Ferdinand Tonnies, and Max Weber, wrote about the effects of industrialization on existing social arrangements, and gave theoretical expression to their observations. The dire tone many of them adopted as they theorized their concerns about the future of community came to constitute further evidence confirming an implicitly teleological theory of social change where communities and communal arrangements that governed social interactions inevitably would disappear over time.[2]

Similarly, the ethnographic work of many anthropologists sometimes described cooperative arrangements for managing rural resources, or resources owned by indigenous peoples. It implicitly implied that such arrangements lay outside modern life. If historical studies of community located common property in the past, contemporary work by anthropologists located the commons in nonmodern, nonwestern societies. Undoubtedly, sophisticated ethnographic analysis has contributed immensely to the current state of our knowledge about how common property institutions work. But it has also hinted despite itself, simply by virtue of its subject matter, that common property may be no more than the institutional debris of societal arrangements that somehow fall outside modernity.[3]

Therefore, it would be fair to say that for much of the twentieth century, the dominant theoretical lenses that have framed how social scientists view peasants and rural life have helped distort analytical vision so as to impart to community and communal forms of sociality only residual vigor, a transitional existence, and an exotic attraction. Economic and political power have been seen to rest on an urban-industrial social organization and the simultaneous eclipse of rural life. It should scarcely surprise that those writing about postcolonial states found them undertaking policies that would undermine rural communities by promoting fast-paced development and rapid urbanization (Bates, 1981). The effects of emerging and spreading market relations similarly were seen to assist the vast movement of history by promoting the pursuit of individual self-interest or contractual obligations, and destroying community ethic and customary rules.[4]

These background beliefs about peasants, communities, rural life, and their future have had quite specific implications for environmental conservation. As a number of analysts have pointed out, dominant beliefs structuring environmental policies until as late as the 1980s held that markets and states were the appropriate institutional means to address externalities stemming from the public goods nature of resources. Many scholars have held that only through a recourse to these institutional arrangements would it become possible to promote sustainable resource use.[5] Many still do.

However, discussions over what kind of institutional arrangements account for sustainable resource use have undergone a remarkable change since the mid-1980s. The shift has occurred in part as a response to developments in the field of noncooperative game theory (Falk et al., this volume:Chapter 5; Fudenberg and Maskin, 1986; Schotter, 1981; Sugden, 1984, 1989), but more directly as a result of the explosion of work on common property arrangements and common-pool resources (Berkes, 1989; McCay and Acheson, 1987; National Research Council, 1986; Ostrom, 1990). New understandings of resource institutions take common property as a viable mechanism to promote sustainable resource management. The work of scholars of common property has thus forced home a much-needed corrective to general policy prescriptions of privatization. This achievement cannot and should not be underrated. Scholars of common property, by shifting the focus of their investigations toward the analytical and structural elements that comprise successful management of the commons, have been in the vanguard of the bearers of the message that the commons and the community are an integral and indispensable part of contemporary efforts to conserve environmental resources. They have rewritten the text on how the environment should be governed.

Scholarship on common property spans many disciplines. Anthropologists, resource economists, environmentalists, historians, political scientists, and rural sociologists among others have contributed to the flood of writings on the subject. More recent empirical work on the commons draws significantly from theories of property rights and institutions.[6] It also uses many other approaches eclectically, including political ecology and ethnography. Using detailed historical and contemporary studies, writings on the commons have shown that resource users often create institutional arrangements and management regimes that help them allocate benefits equitably, over long time periods, and with only limited efficiency losses (Agrawal, 1999a; McKean, 1992a; Ostrom, 1992). Much of this research typically has focused on locally situated small user groups and communities.[7]

Of course, not all users of common-pool resources protect their resources successfully. Outcomes of experiences of commons management are highly variable. Documentation of the performance of regimes of local resource management provides us with many cases of successful local management of common-pool resources. In light of this knowledge, scholars and policy makers are less likely to propose central state intervention, markets, or privatization of property rights over resources as a matter of course. Rather, many scholars examine the conditions under which communal arrangements compare favorably with private or state ownership, even on efficiency criteria, but especially where equity and sustainability are concerned. Other scholars of commons and some institutional theorists question the familiar trichotomy of private, communal, and state ownership and instead focus more directly on underlying rights and powers of access, use, management, exclusion, and transferability that are conferred through rules governing resources.[8] The work initiated and carried out by scholars of common property parallels important developments in the world of policy making and resource management. Governments in more than 50 countries, according to a recent survey on national forestry policies (Food and Agriculture Organization, 1999), claim to be following new initiatives that would devolve some control over resources to local users.

In synthesizing the extensive empirical work that has occurred over the past two decades, this chapter draws on rich descriptions of particular cases, comparative studies, and insights from works on social scientific methods to suggest how it might be possible to develop plausible causal mechanisms to link outcomes with causal variables. An enormous experimental and game theoretic literature also has begun to inform our understanding of how humans act under different incentive structures (see Falk et al., this volume:Chapter 5; Kopelman et al., this volume:Chapter 4). But the most valuable resources for this chapter are studies whose conclusions are based on explicit comparisons using relatively large samples of cases (Baland and Platteau, 1996; Ostrom et al., 1994; Pinkerton, 1989; Pinkerton and Weinstein, 1995; Sengupta, 1991; Tang, 1992).

The exact definitions of terms such as efficiency, equity, or sustainability that characterize outcomes related to common-pool resource management are beyond the scope of this chapter.[9] However, it might be useful to indicate that by "sustainability on the commons," I primarily have in mind the durability of institutions that frame the governance of common-pool resources. Such a general view of sustainability is justified because few studies of the commons provide rigorous measures of their dependent variables. To use a strict definition of sustainability, therefore, is likely to make comparisons across studies difficult. At the same time, it must be admitted that most writings on the commons implicitly define successful institutions as those that last over time, constrain users to safeguard the resource, and produce fair outcomes.[10]

The next section begins by focusing on three comprehensive attempts to produce theoretically informed generalizations about the conditions under which groups of self-organized users are successful in managing their commons dilemmas.[11] These studies are Wade ([1988] 1994), Ostrom (1990), and Baland and Platteau (1996).[12] I examine the robustness of their conclusions by comparing them with findings that a larger set of studies of the commons has identified.

Many of the conclusions of scholars of common property closely match theoretical generalizations in the literature on collective action and institutional analysis.[13] But just as institutional analysts and theorists of collective action provide inferences that are sometimes in tension, scholars of common property also highlight outcomes and causal connections that often run counter to each other. One significant reason for divergent conclusions of empirical studies of commons is that most of them are based on the case study method. The multiplicity of research designs, sampling techniques, and data collection methods present within each study can be welcomed on the grounds that a hundred flowers should bloom; it also means that careful specification of the contextual and historical factors relevant to findings, systematic tests of findings, and comparisons of postulated causal connections have been relatively few. In analyzing the mostly case study-based empirical literature on the commons, the following section focuses on some of the typical problems of method that plague many studies of self-organized resource management institutions. I suggest that studies of the commons need to be especially attentive to areas in which case analysis is deficient, explicitly highlight the objectives of their studies, and explain the advantages of adopting a case study approach. The subsequent section proposes possible complementary methods and areas of emphasis for further research on common property.[14]

The main argument of the paper is that existing studies of sustainable institutions around common-pool resources suffer from two types of problems. The first is substantive—many scholars of commons have focused narrowly on institutions around common-pool resources. Such a focus on institutions is understandable in light of the objective of showing that common property arrangements can result in efficient use, equitable allocation, and sustainable conservation. But it comes at a cost. The cost is the lack of careful analysis of the contextual factors that frame all institutions and that affect the extent to which some institutions are more likely to be effective than others. The same institutional rules can have different effects on resource governance depending on variations in the biophysical, social, economic, and cultural contexts. Because existing studies of commons are relatively negligent in examining how aspects of the resource system, some aspects of user group membership, and the external social, physical, and institutional environments affect institutional durability and long-term management at the local level, we need new work that considers these questions explicitly (but see Lam, 1998; Ostrom, 1999; Ostrom et al., 1994; and Tang, 1992).

The second problem relates to methods and is more fundamental. Given the large number of factors, perhaps as many as 35 of them (see the next section), that have been highlighted as being critical to the organization, adaptability, and sustainability of common property, it is fair to suggest that existing work has not yet fully developed a theory of what makes for sustainable common-pool resource management. Systematic tests of the relative importance of factors important to sustainability, equity, or efficiency of commons are relatively uncommon (but see Bardhan and Dayton-Johnson, this volume:Chapter 3 and Lam, 1998).

Also uncommon are studies that connect the different variables they identify in causal chains or propose plausible causal mechanisms. Problems of incomplete model specification and omitted variables in hypothesis testing are widespread in the literature on common property. These problems of method often characterize even those writings that claim to address problems of substance.[15] Therefore, it is likely that many conclusions from case studies of common-pool resource management and even from comparative studies of the commons are relevant primarily for the sample under consideration, rather than applying more generally.

Of course, there are good reasons for the existence of these problems in studies of sustainability on the commons. Some of these reasons have to do with difficulties of data availability and collection, regional and area expertise of those who study the commons, disciplinary allegiances, and the tendency in single case studies to select instances of successful common-pool resource management. But these reasons do not obviate the need for a more viable and compelling theory of common-pool resource management. Such a theory is even more important today because of the increasing number of policy experiments in commons management that are under way. These policy experiments, and their vast human and territorial coverage, make it imperative that scholars of the commons squarely confront two critical questions: (1) Which of the lessons learned from current studies are sufficiently reliable to help diagnose institutional malfunctioning?; and (2) How can studies of common property contribute reliably to greater equity and justice in the implementation of revised institutional arrangements?

## ANALYSES OF SUSTAINABLE MANAGEMENT
## OF COMMON-POOL RESOURCES

Of the significant number of comparative studies on the commons, I have chosen the book-length studies by Wade ([1988] 1994), Ostrom (1990), and Baland and Platteau (1996). Two of them, by Wade and Ostrom, appeared more than a decade ago, and can be seen as the advance guard of a veritable flood of new writings on the commons that have put an end to the notion that common property is a historical curiosity. The main positive lessons I derive by comparing these authors are how they show that under some combinations of frequently occurring conditions, members of small groups can design institutional arrangements that help sustainable management of resources. They go further and identify the specific conditions that are most likely to promote local self-management of resources. Not only that, they use theoretical insights to defend and explain the empirical regularities they find.

It would be fair to say that each of the three books is a careful and rigorous conversation between theory and empirical investigation because of their attention to theoretical developments at the time of writing, their effort to relate theory to the cases they examine, and their contributions to common property theory.

They all use a large body of empirical materials to test the validity of the theoretical insights they garner. Although the three books embody very different approaches to empirical comparative research, and rely on very different kinds of data, their concern for being empirically relevant and holding theory accountable to data is evident. For this paper, one of the most appealing aspects of their argument is that after wide-ranging discussion and consideration of many factors, each author arrives at a summary set of conditions and conclusions he or she believes to be critical to sustainability of commons institutions. Together, their conclusions form a viable starting point for the analysis of the ensemble of factors that account for sustainable institutional arrangements to manage the commons. But a discussion of their conclusions and some of the implications of their work also demonstrates that their propositions about sustainability on the commons need to be supplemented.

Because there is no single widely accepted theory of what makes common property institutions sustainable, it is important to point out that differences of method are significant among these three authors. Wade relies primarily on data he collected from South Indian villages in a single district. His sample is not representative of irrigation institutions in the region, but at least we can presume that the data collection in each case is consistent. To test her theory, Ostrom uses detailed case studies that other scholars generated. The independent production of the research she samples means that all her cases may not have consistently collected data. But she examines each case using the same set of independent and dependent variables. Baland and Platteau are more relaxed in the methodological constraints they impose on themselves. To motivate their empirical analysis, they use a wide-ranging review of the economic literature on property rights and the inability of this literature to generate unambiguous conclusions about whether private property is superior to regulated common property. But to examine the validity of their conclusions, they use information from different sets of cases. In an important sense, the "model specification" is incomplete in each test (King et al., 1994).

Wade's (1994) important work on commonly managed irrigation systems in South India uses data on 31 villages to examine when it is that corporate institutions arise in these villages and what accounts for their success in resolving commons dilemmas.[16] His arguments about the origins of commons institutions point, in brief, toward environmental risks as being a crucial factor. But he also provides a highly nuanced and thoughtful set of reasons about successful management of commons. According to Wade, effective rules of restraint on access and use are unlikely to last when there are many users, when the boundaries of the common-pool resource are unclear, when users live in groups scattered over a large area, when detection of rule breakers is difficult, and so on (Wade, 1988: 215).[17] Wade specifies his conclusions in greater detail by classifying different variables under the headings of resources, technology, user group, noticeability, relationship between resources and user group, and relationship between users and the state (1988:215-216).[18] The full set of conditions that Wade considers important for sustainable governance are listed in Box 2-1.

In all, Wade finds 14 conditions to be important in facilitating successful management of the commons he investigates.[19] Most of his conditions are general statements about the local context, user groups, and the resource system, but some of them are about the relationship between users and resources. Only one of his conditions pertains to external relationships of the group or of other local factors.

---

**BOX 2-1**
**Facilitating Conditions Identified by Wade**

(1)  Resource system characteristics
    (i)    Small size
    (ii)   Well-defined boundaries
(2)  Group characteristics
    (i)    Small size
    (ii)   Clearly defined boundaries
    (iii)  Past successful experiences—social capital
    (iv)  Interdependence among group members
(1 and 2) Relationship between resource system characteristics and group characteristics
    (i)    Overlap between user group residential location and resource location
    (ii)   High levels of dependence by group members on resource system
(3)  Institutional arrangements
    (i)    Locally devised access and management rules
    (ii)   Ease in enforcement of rules
    (iii)  Graduated sanctions
(1 and 3) Relationship between resource system and institutional arrangements
    (i)    Match restrictions on harvests to regeneration of resources
(4)  External environment
    (i)    Technology: Low-cost exclusion technology
    (ii)   State:
        (a)  Central governments should not undermine local authority

SOURCE: Wade (1988).

Studies appearing since Wade's work on irrigation institutions have added to his list of factors that facilitate institutional success, but some factors have received mention regularly. Among these are small group size, well-defined bounds on resources and user group membership, ease in monitoring and enforcement, and closeness between the location of users and the resource. Consider, for example, the eight design principles that Ostrom (1990) lists in her defining work on community-level governance of resources. She crafts these principles on the basis of lessons from a sample of 14 cases where users attempted, with varying degrees of success, to create, adapt, and sustain institutions to manage the commons. A design principle for Ostrom is "an essential element or condition that helps to account for the success of these institutions in sustaining the [common-pool resources] and gaining the compliance of generation after generation of appropriators to the rules in use" (1990:90). She emphasizes that these principles do not provide a blueprint to be imposed on resource management regimes. Seven of the principles are present in a significant manner in all the robust commons institutions she analyzes. The eighth covers more complexly organized cases such as federated systems.

Although Ostrom lists eight principles, on closer examination the number of conditions turns out to be larger. For example, her first design principle refers to clearly defined boundaries of the common-pool resource and of membership in a group, and is in fact listed as two separate conditions by Wade. Her second principle, similarly, is an amalgam of two elements: a match between levels of restrictions and local conditions, and between appropriation and provision rules. Ostrom thus should be seen as considering 10, not 8, general principles as facilitating better performance of commons institutions over time (see Box 2-2).

A second aspect of the design principles, again something that parallels Wade's facilitating conditions, is that most of them are expressed as general features of long-lived, successful commons management rather than as relationships between characteristics of the constituent analytical units or as factors that depend for their efficacy on the presence (or absence) of other variables. Thus, principle seven suggests that users are more likely to manage their commons sustainably when their rights to devise institutions are not challenged by external government authorities. This is a general principle that is supposed to characterize all commons situations. The principle says that whenever external governments do not interfere, users are more likely to manage sustainably. In contrast, principle two suggests that restrictions on harvests of resource units should be related to local conditions (rather than saying that the lower [or higher] the level of withdrawal, the more [or less] likely would be success in management). Thus, it is possible to imagine certain resource and user group characteristics for which withdrawal levels should be high, and where setting them at a low level may lead to difficulties in management. For example, when supplements to resource stock are regular and high, and user group members depend on resources significantly, setting low harvesting levels will likely lead to unnecessary rule infractions. Thus

---

**BOX 2-2**
**Ostrom's Design Principles**

(1)  Resource system characteristics
     (i)   Well-defined boundaries
(2)  Group characteristics
     (i)   Clearly defined boundaries
(1 and 2) Relationship between resource system characteristics and group characteristics
     None presented as important
(3)  Institutional arrangements
     (i)   Locally devised access and management rules
     (ii)  Ease in enforcement of rules
     (iii) Graduated sanctions
     (iv)  Availability of low-cost adjudication
     (v)   Accountability of monitors and other officials to users
(1 and 3) Relationship between resource system and institutional arrangements
     (i)   Match restrictions on harvests to regeneration of resources
(4)  External environment
     (i)   Technology: None presented as important
     (ii)  State:
           (a)  Central governments should not undermine local authority
           (b)  Nested levels of appropriation, provision, enforcement, governance

SOURCE: Ostrom (1990).

---

principle two covers a wider range of variations across cases, but at the cost of some ambiguity. In contrast, principle seven is more definite, but it is easy to imagine situations where it is likely not to hold.

Finally, most of Ostrom's principles focus primarily on local institutions, or on relationships within this context. Only two of them, about legal recognition of institutions by higher level authorities and about nested institutions, can be seen to express the relationship of a given group with other groups or authorities.

Baland and Platteau (1996), in their comprehensive and synthetic review of a large number of studies on the commons, follow a similar strategy as does Ostrom (1990). Beginning with an examination of competing theoretical claims by scholars of different types of property regimes, they suggest that the core argument in favor of privatization "rests on the comparison between an idealized fully efficient private property system and the anarchical situations created by open access" (Baland and Platteau, 1996:175). Echoing earlier scholarship on the commons, they emphasize the distinction between open access and common property arrangements and suggest that when private property regimes are compared with regulated common property systems (and when information is perfect and there are no transaction costs), then *"regulated common property and private property are equivalent from the standpoint of the efficiency of resource use"* (Baland and Platteau, 1996:175, emphasis in original).[20] Furthermore, they argue, the privatization of common-pool resources or their appropriation and regulation by central authorities tends to eliminate the implicit entitlements and personalized relationships that are characteristic of common property arrangements. These steps, therefore, are likely to impair efficiency, and even more likely to disadvantage traditional users whose rights of use seldom get recognized under privatization or expropriation by the state.[21]

Their review of the existing literature from property rights and economic theory leads them to assert that "none of the property rights regimes appears intrinsically efficient" and that the reasons for which common property arrangements are criticized for their inefficiency can also haunt privatization measures. Where agents are not fully aware of ecological processes, or are unable to protect their resources against intruders, or their opportunity costs of degrading the environment are low,[22] state intervention may be needed to support both private *and* common property (Baland and Platteau, 1996:178). In the absence of clear theoretical predictions regarding the superiority of one property regime over another, they argue in favor of attention to specific histories of concrete societies, and explicit incorporation of cultural and political factors[23] into analysis. Only then might it be possible to know when people cooperate, and when inveterate opportunists dominate and make collective action impossible.

After a wide-ranging review of empirical studies of common-pool resource management, and focusing on several variables that existing research has suggested as crucial to community-level institutions, Baland and Platteau arrive at conclusions that significantly overlap with those of Wade and Ostrom. Small size of a user group, a location close to the resource, homogeneity among group members, effective enforcement mechanisms, and past experiences of cooperation are some of the themes they emphasize as significant to achieve cooperation (Baland and Platteau, 1996:343-345). In addition, they highlight the importance of external aid and strong leadership.[24]

As is true for Ostrom, several of the factors they list are actually a joining together of multiple conditions. For example, their third point incorporates what Wade and Ostrom would count as four different conditions: the relationship between the location of the users and the resources on which they rely, the ability of users to create their own rules, the ease with which rules are understood by members of the user group and are enforced, and whether rules of allocation are considered fair. Some of their other conditions also signify more than one variable. Therefore, instead of 8 conditions, Baland and Platteau should be seen as identifying 12 conditions (see Box 2-3).

**BOX 2-3**
**Conclusions Presented by Baland and Platteau**
**as Facilitating Successful Governance**
**of the Commons**

(1) Resource system characteristics
    None presented as important
(2) Group characteristics
    (i) Small size
    (ii) Shared norms
    (iii) Past successful experiences—social capital
    (iv) Appropriate leadership—young, familiar with changing external
         environments, connected to local traditional elite
    (v) Interdependence among group members
    (vi) Heterogeneity of endowments, homogeneity of identities and
         interests
(1 and 2) Relationship between resource system characteristics and
group characteristics
    (i) Overlap between user group residential location and resource
        location
    (ii) Fairness in allocation of benefits from common resources
(3) Institutional arrangements
    (i) Rules are simple and easy to understand
    (ii) Locally devised access and management rules
    (iii) Ease in enforcement of rules
    (iv) Accountability of monitors and other officials to users
(1 and 3) Relationship between resource system and institutional arrangements
    None presented as important
(4) External environment
    (i) Technology: None presented as important
    (ii) State:
        (a) Supportive external sanctioning institutions
        (b) Appropriate levels of external aid to compensate local us-
            ers for conservation activities

SOURCE: Baland and Platteau (1996).

The conclusions that Baland and Platteau reach typically are stated as general statements about users, resources, and institutions rather than about relationships between characteristics of these constituent analytical units. Only one of their conclusions is relational: contiguous residential location of group members and of the resource system. Finally, in comparison to Wade and Ostrom, Baland

and Platteau pay somewhat greater attention to external forces, such as in their discussions of external aid, enforcement, and leadership with broad experience.

Box 2-4 summarizes the different conditions that Wade, Ostrom, and Baland and Platteau have identified as important in promoting sustainable use of common-pool resources. Even a quick examination of the conditions listed in Box 2-4 makes evident some of the patterns in the conclusions of these three landmark studies.[25] The examples they consider have ample variation on the causal and dependent variables, and they use this variation to identify a set of conditions that facilitate greater success on the commons. Whereas Ostrom focuses primarily on the specifics of institutional arrangements in accounting for successful governance of the commons, Wade and Baland and Platteau cast a wider net, and incorporate noninstitutional variables in their conclusions. The regularities in successful management that they discover pertain to one of four sets of variables: (1) characteristics of resources, (2) nature of groups that depend on resources, (3) particulars of institutional regimes through which resources are managed, and (4) the nature of the relationship between a group and external forces and authorities such as markets, states, and technology.[26]

Characteristics of resources can include, for example, features such as well-defined boundaries of the resource, riskiness and unpredictability of resource flows, and mobility of the resource. Characteristics of groups, among other aspects, relate to size, levels of wealth and income, different types of heterogeneity, power relations among subgroups, and experience. Particulars of institutional regimes have an enormous range of possibilities, but some of the critical identified aspects of institutional arrangements concern monitoring and sanctions, adjudication, and accountability. Finally, a number of characteristics pertain to the relationships of the locally situated groups, resource systems, and institutional arrangements with the external environment in the form of demographic changes, technology, markets, and the state.

The analysis of the information in Box 2-4 reveals several significant obstacles to the identification of a universal set of factors that are critical to successful governance of common-pool resources. Of these, three relate to substantive issues and two stem from conundrums of method. The missing substantive concerns of these three scholars are examined at greater length in the next section, which widens the net I cast to examine additional important research on common property institutions. Unfortunately, attempts to redress substantive issues tend to exacerbate problems of method that I explain later in the chapter. We have to contend with the possibility that attempts to create lists of critical enabling conditions that apply universally founder at an epistemological level. Lists of factors can be only a starting point in the search for a compelling theorization of how these factors are related to each other and to outcomes. Instead of focusing on lists of factors that apply to all commons institutions, it is likely more fruitful to focus on configurations of conditions that contribute to sustainability. The identification of such configurations requires sharp analytical insights. Such insights

---

**BOX 2-4**
**Synthesis of Facilitating Conditions Identified by Wade,**
**Ostrom, and Baland and Platteau**

(1) Resource system characteristics
    (i)   Small size (RW)
    (ii)  Well-defined boundaries (RW, EO)
(2) Group Characteristics
    (i)   Small size (RW, B&P)
    (ii)  Clearly defined boundaries (RW, EO)
    (iii) Shared norms (B&P)
    (iv) Past successful experiences—social capital (RW, B&P)
    (v)  Appropriate leadership—young, familiar with changing external environments, connected to local traditional elite (B&P)
    (vi) Interdependence among group members (RW, B&P)
    (vii) Heterogeneity of endowments, homogeneity of identities and interests (B&P)
(1 and 2) Relationship between resource system characteristics and group characteristics
    (i)   Overlap between user group residential location and resource location (RW, B&P)
    (ii)  High levels of dependence by group members on resource system (RW)

---

are most likely to follow from comparative research that is either based on carefully selected cases, or uses statistical techniques to analyze data from multiple cases after ensuring that the selection of cases conforms to theoretical specification of causal connections.

## SUPPLEMENTING THE SET OF SUBSTANTIVE FACTORS

The set of factors identified by Wade, Ostrom, and Baland and Platteau is relatively deficient in considering resource characteristics. Only two aspects of resource systems find explicit mention by the three authors. Baland and Platteau do not include aspects of resources in their final conclusions at all.

The limited attention to resource characteristics is unfortunate. Even if we leave aside the climatic and edaphic variables that may have an impact on levels of regeneration and possibility of use, there are grounds to believe that other

(iii) Fairness in allocation of benefits from common resources (B&P)

(3) Institutional arrangements
- (i) Rules are simple and easy to understand (B&P)
- (ii) Locally devised access and management rules (RW, EO, B&P)
- (iii) Ease in enforcement of rules (RW, EO, B&P)
- (iv) Graduated sanctions (RW, EO)
- (v) Availability of low-cost adjudication (EO)
- (vi) Accountability of monitors and other officials to users (EO, B&P)

(1 and 3) Relationship between resource system and institutional arrangements
- (i) Match restrictions on harvests to regeneration of resources (RW, EO)

(4) External environment
- (i) Technology: Low-cost exclusion technology (RW)
- (ii) State:
  - (a) Central governments should not undermine local authority (RW, EO)
  - (b) Supportive external sanctioning institutions (B&P)
  - (c) Appropriate levels of external aid to compensate local users for conservation activities (B&P)
  - (d) Nested levels of appropriation, provision, enforcement, governance (EO)

SOURCES: RW, Wade (1988); EO, Ostrom (1990); B&P, Baland and Platteau (1996).

aspects of a resource may be relevant to how and whether users are able to sustain effective institutions.[27] For example, it is easy to see that extensive movements of many forms of wildlife can make them less suited to local management alone (Moseley, 1999; Naughton-Treves and Sanderson, 1995).[28] This aspect of common-pool resources is different from Wade's argument about small size in that the issue is one of mobility of the resource, and volatility and unpredictability in the flow of benefits from a resource; it is not just about size.

In a carefully argued paper on resource characteristics, Blomquist et al. (1994) focus on two physical features of resource systems: stationarity and storage. Stationarity refers to whether a resource is mobile and storage concerns the extent to which it is possible to "collect and hold resources" (1994:309).[29] Stationarity and storage, if considered as dichotomous variables, lead to a four-fold typology of common-pool resources. Resources such as wildlife are mobile and cannot be stored, and groundwater basins and lakes have stationary water resources that can be stored. Shellfish and grazing lands are stationary but their degree of storage is limited, and conversely, irrigation canals with reservoirs have water resources that can be stored, but are mobile. Sheep flocks and cattle herds owned and/or managed as common property also would fall in this last category.

After examining the impact of these two physical characteristics of resources on externalities, Blomquist and colleagues conclude that these two factors have an impact on management because of their relationship to information. Greater mobility of resources and difficulties of storage make it more difficult for users to adhere to institutional solutions to common-pool resource dilemmas because of their impact on the reliability and costs of information needed for such solutions.[30] This point also can be seen as a question about the extent to which resource availability is predictable, something noted by Naughton-Treves and Sanderson (1995) as well, and how unpredictability affects the abilities of users to allocate available resources or undertake activities that would augment supply (see also Wilson, this volume:Chapter 10).[31]

A second broad area to which the analyses by Wade, Ostrom, and Baland and Platteau pay only limited attention is the external social, institutional, and physical environment.[32] Thus none of them explicitly remark on demographic issues in their conclusions, and they put equally little emphasis on market-related demands that may make local demand pressures relatively trivial. But variations in levels of population and changes in demographic pressures, whether as a result of local changes or through migration, are surely significant in influencing the ability of users to follow existing rules and norms for resource management. Indeed, there is an enormous literature that focuses on questions of population and market pressures on resource use and asserts the importance of these two complex factors.[33]

Writings on the role of population in resource management have a long history and an impressive theoretical pedigree (Ehrlich, 1968:15-16; Malthus, 1798, 1803, rpt. 1960). Much recent scholarship links environmental degradation in a relatively straightforward fashion with population growth (Abernathy, 1993; Durning, 1989; Fischer, 1993; Hardin, 1993; Low and Heinen, 1993; Pimental et al., 1994). On the whole it is clear that the debate is highly polarized. Some scholars assert that population pressures have an enormous effect (Ehrlich and Ehrlich, 1991; Myers, 1991; Wilson, 1992), and a smaller but vocal group suggests the impact to be far more limited (Lappé and Shurman, 1989; Leach and Mearns, 1996; Simon, 1990; Tiffen et al., 1994; Varughese and Ostrom, 1998).

The story is somewhat similar where markets are concerned, except that the terms of the debate are less polarized and there is wider agreement that increasing integration with markets usually has an adverse impact on the management of common-pool resources, especially when roads begin to integrate distant resource systems and their users with other users and markets (Chomitz, 1995; Young, 1994). As local economies become better connected to larger markets and common property systems confront cash exchanges, subsistence users are likely to increase harvesting levels because they can now exploit resources for cash income as well (Carrier, 1987; Colchester, 1994:86-87; Stocks, 1987:119-120).

It is important to note that apart from potentially higher returns, there are additional reasons why common property arrangements may be undermined by market pressures. Market integration introduces new ways of resolving the risks that common property institutions are often designed to address. Pooling of resources that becomes possible under common property regimes helps those who are subject to such regimes. It helps by allowing them to reduce risks they would face were they to exploit the same resources individually.[34] Mobility over space and through time (storage) are comparable mechanisms to address production fluctuations, but markets and exchange compete with them by encouraging individuals to specialize in different kinds of economic activities. By specializing in different occupations and exchanging their surplus output, individual producers can alleviate the need for migration (with or without their means of production) and storage. In addition, markets also form alternative arenas for the provision of credit and generation of prestige in ways that can undermine the importance of other local institutions.

Analogous to market articulation is the question of technological means available to exploit the commons. Sudden emergence of new technological innovations that transform the cost-benefit ratios of harvesting benefits from commons are likely to undermine the sustainability of institutions. Sufficient time may be necessary before users are able to adapt to the new technologies. Furthermore, technological change is capable of disrupting not just the extent to which existing mechanisms of coordination around mobility, storage, and exchange can continue to serve their members, but the very nature of the political and economic calculation that goes into the invention and definition of common property. Recall how the invention of barbed wire permitted cheap fencing, and helped convert rangelands in the U.S. west into an excludable resource.[35]

The arrival of markets and new technologies, and the changes they might prompt in existing resource management regimes, is not a bloodless or innocent process (Oates, 1999). Typically, new demand pressures originating from markets and technological changes are likely to create different incentives about the products to be harvested, technologies of harvest, and rates of harvest. They are also likely to change local power relations as different subgroups within a group using a common-pool resource gain different types of access and maneuver to ensure their gains (Fernandes et al., 1988; Jessup and Peluso, 1986; Peluso, 1992). And in many cases, as new market actors gain access to a particular common-pool resource, they may seek alliances with state actors in efforts to privatize commons or defend the primacy of their claims (Ascher and Healy, 1990; Azhar, 1993). Indeed, state officials themselves can become involved in the privatization of commons and the selling of products from resources that were earlier under common property arrangements (Rangarajan, 1996; Sivaramakrishnan, 1999; Skaria, 1999).

These specific arguments about changes in resource use and management institutions under the influence of markets are in line with more general perceptions about the transformative role and potential of capital and market forces.[36] But clearly, differences in market and population pressures need greater attention in any examination of the factors that affect sustainability of commons institutions. It is important not only to attend to different levels of these pressures, but also to the effect of changes and rates of changes in them.

As the ultimate guarantor of property rights arrangements, the role of the state and overarching structures of administration have been decisive under many historical circumstances in governing common-pool resources. It is true that many communities and local user groups have the right to craft and implement new institutional arrangements. But unspecified rights and the settlement of major disputes often cannot be addressed without state intervention (Rangan, 1997). Although the three authors are more attentive to the potential role of central governments in local commons than they are to issues of population and market pressures, the nature of local-state relations requires more careful exploration.[37] As an increasing number of governments decentralize control over diverse natural resources to local user groups, questions about the reasons behind such loosening of control and the effects of differences in organization of authority across levels of governance become extremely important. A large number of studies have attempted to explore these issues, either by focusing on decentralization of resource management in general (Ascher, 1995; Poffenberger, 1990) or by examining the role of resource management-related laws and national policies (Ascher and Healy, 1990; Lynch and Talbott, 1995; Repetto and Gillis, 1988). But as yet we do not have a systematic examination or clear understanding of variations in these relationships and how these variations affect the nature and outcome of common-pool resource management.

One reason scholars of commons have focused so little on external factors like markets, technology, states, and population pressures lies simply in the nature of their intellectual enterprise. Because their efforts have aimed at showing the importance of local groups, institutions, and resource-system related factors, they have focused relatively little on those factors that have received attention from many other streams of scholarship. But it seems that in focusing on the locality and the importance of local factors, they have ignored how what is local is often created in conjunction with the external and the nonlocal environment. The almost exclusive focus on the local has made the work on common property vulnerable to the same criticisms that apply to the work of those anthropologists who saw their field sites as miniature worlds in themselves, changing only in response to political or economic influences from outside.[38] The attention to the locality in preference to the context within which localities are shaped has thus prevented the emergence of a better understanding of how factors such as population, market demand, and state policies interact with local institutional arrangements and resource systems.

My argument in favor of attention to markets, demography, and the state addresses the nature and importance of contextual factors only to a partial degree. In research, the context can be defined as the encompassing variables that remain constant for a given study, but not across studies. Furthermore, the state of contextual variables affects the impact of variables that are being studied explicitly. It is likely impossible to define a priori the ensemble of factors that constitute the context because contextual factors for a given study depend on the questions it seeks to answer. However, studies of commons that examine institutional sustainability can afford to ignore the nature of markets and market-related changes, population and demographic changes, and the state and its policies only when these remain constant. For many single-time period, single-location case studies, inattention to these critical contextual variables may be justifiable. But where studies seek to develop more general arguments, attention to context and how contextual factors relate to specified causal arguments become extremely important. Even within a case study, it may be possible to examine how formerly constant ("slow") variables change, driving and interacting with other ("fast") variables. In such a situation, sustainability itself can be thought of as a dynamically maintained system condition rather than a static equilibrium (National Research Council, 1999).

But even where the locality itself is concerned, and even where some important features of groups that manage commons are concerned, there are important gaps in our understanding. Take three aspects of groups as an illustration: size, heterogeneity, and poverty.

According to an enormous literature on the commons and collective action, sparked in part by Olson's seminal work (1965), smaller groups are more likely to engage in successful collective action. This conclusion is supported by Baland and Platteau (1999:773), who reiterate Olson: "The smaller the group the stronger its ability to perform collectively." But other scholars have remarked on the ambiguities in Olson's argument and suggested that the relationship between group size and collective action is not very straightforward. For example, Marwell and Oliver (1993:38) emphatically claim, "a significant body of empirical research…finds that the size of a group is positively related to its level of collective action."[39] Agrawal and Goyal (2001) use two analytical features of common-pool resources—imperfect exclusion and lumpiness of third-party monitoring[40]—to hypothesize a curvilinear relationship between group size and successful collective action. They test their hypothesis using a sample of 28 cases from the Kumaon Himalaya. The current state of knowledge is perhaps best summarized by Ostrom (1997), who says that the impact of group size on collective action is usually mediated by many other variables. These variables include the production technology of the collective good, its degree of excludability, jointness of supply, and the level of heterogeneity in the group (Hardin, 1982:44-49). After more than 30 years of research on group size and collective action, there is still a need to tease out more carefully the relationship between group size and successful collective action.

Cumulation of knowledge into a coherent and empirically supported theory has proved even more difficult in relation to group heterogeneity. It can be argued fairly that most resources are managed by groups divided along multiple axes, among them ethnicity, gender, religion, wealth, and caste (Agrawal and Gibson, 1999). The nature of heterogeneities within groups can have multiple and contradictory effects.[41] Wade and Baland and Platteau highlight the importance of greater interdependence among group members as a basis for building institutions that would promote sustainable resource management. In addition, Baland and Platteau also provide an initial assessment of the nature of heterogeneities by classifying them into three types and hypothesizing that heterogeneities of endowments have a positive effect on resource management whereas heterogeneities of identity and interests create obstacles to collective action. Their first point, about heterogeneities of endowments enhancing the possibilities of collective action, is similar to that made by Olson (1965). But the categories into which they classify heterogeneities are not mutually exclusive. For example, heterogeneities of interests may lead to different types of economic specialization and different levels of endowments, which could in turn lead to mutually beneficial exchange.[42] Further, empirical evidence on how heterogeneities affect collective action is still highly ambiguous (Baland and Platteau, 1999; Bardhan and Dayton-Johnson, this volume:Chapter 3; Kanbur, 1992; Quiggin, 1993; Varughese and Ostrom, 1998). It is possible, thus, even in groups that have high levels of heterogeneities of interest, to ensure collective action if some subgroups can coercively enforce conservationist institutions (Agrawal, 1999a; Jodha, 1986; Peluso, 1993; but see also Libecap, 1989, 1990). On the other hand, the role of intragroup heterogeneities in distribution may be more amenable to definition. Significant research on the effects of development projects and on commons suggests that better off group members often are likely to gain a larger share of benefits from a resource (see, for example, Agrawal, 2001). This is not to say that collective action always exacerbates intragroup inequalities; rather it is simply to point out that inequalities within a group are not necessarily reduced because group members are willing to cooperate in the accomplishment of a collective goal.

Another locality-related factor that is critical to outcomes, and on which much research has been carried out without the emergence of a consensus, is the relation of poverty of users to their levels of exploitation of common-pool resources. Whether poverty leads to a greater reliance on the commons (Jodha, 1986) and their degradation, or whether increasing levels of wealth, at least initially, lead to greater use of commons by users is a question on whose answer contours of many commons-related policies would hinge. But to a significant degree, government interventions in this arena are based on limited information and even less reliable analysis.

For each of the three factors—size, heterogeneity, and poverty—the extent to which existing research has settled the question of the direction of their effect on the sustainability of commons institutions is ambiguous at best. Whether the relationship between sustainability and these variables is negative, positive, or curvilinear seems subject to a range of other contextual and mediating factors, not all of which are clearly understood. Box 2-5 constitutes an effort to supplement the set of variables presented in Box 2-4. The additional factors presented in the box are the ones that are not followed by the initials of a particular author. Although the factors in Box 2-5 are among those that many scholars of commons would consider most important for achieving institutional sustainability on the commons, they do not form an exhaustive set of factors that affect common-pool resource management. Nor is it likely that an undisputed exhaustive set of variables can be created.[43]

Box 2-5 highlights the most significant variables that scholars of commons have identified as being critical to the sustainable functioning of commons institutions.[44] That the set of enabling conditions presented in Box 2-5 is reasonably comprehensive can be tested by examining it in relation to the independent study and conclusions of another scholar of the commons (McKean, 1992b:275-276). McKean examines the historical experience of communities in managing Japanese forests and identifies nine conditions, which she arranges as six conclusions at the end of her essay. Her nine conditions are follows: (1) co-owners of the commons should have some autonomy of management; (2) distribution of rights to shares in commons should be carefully outlined in terms of (2a) equality, (2b) economic efficiency, and (2c) product specificity; (3) rich and poor subgroups among a community of users should both support the commons institutions; (4) there should be low incentives to harvest heavily from the commons; (5) rules should be easily enforceable; and (6) careful monitoring and sanctioning (6a) should be undertaken by the group itself and (6b) should incorporate graduated sanctions. Each of these conditions is included in Box 2-5. Admittedly, the language in which the conditions are expressed in Box 2-5 is not always the same as McKean's. For example, McKean's point that rich and poor subgroups should both support commons institutions is represented in different ways by two separate variables in Box 2-5: shared norms and interdependence among users.

The box makes clear that policy innovations can influence and change the state of only some of the different variables that scholars of commons consider to be important in sustainable management of resources. Current policy experiments, aiming to improve the local management of common-pool resources, need to be especially attentive to the shared conceptual lessons that studies of the commons have generated. Among these would be fairness in the allocation of benefits from the commons; local autonomy to craft, implement, and enforce institutional arrangements that users believe to be critical in managing their resources; low-cost mechanisms for adjudication of disputes and accountability of office holders to users; and local incentives to develop substitutes.

It may be argued that some of the factors listed in Box 2-5 are important to

---

**BOX 2-5**
**Critical Enabling Conditions for**
**Sustainability on the Commons**

(1)  Resource system characteristics
    (i)   Small size (RW)
    (ii)  Well-defined boundaries (RW, EO)
    (iii) Low levels of mobility
    (iv) Possibilities of storage of benefits from the resource
    (v)  Predictability
(2)  Group characteristics
    (i)   Small size (RW, B&P)
    (ii)  Clearly defined boundaries (RW, EO)
    (iii) Shared norms (B&P)
    (iv) Past successful experiences—social capital (RW, B&P)
    (v)  Appropriate leadership—young, familiar with changing external environments, connected to local traditional elite (B&P)
    (vi) Interdependence among group members (RW, B&P)
    (vii) Heterogeneity of endowments, homogeneity of identities and interests (B&P)
    (viii)Low levels of poverty
(1 and 2) Relationship between resource system characteristics and group characteristics
    (i)   Overlap between user group residential location and resource location (RW, B&P)
    (ii)  High levels of dependence by group members on resource system (RW)
    (iii) Fairness in allocation of benefits from common resources (B&P)
    (iv) Low levels of user demand

---

explain the emergence of commons institutions, not their sustainable management. For example, Ostrom (1999) examines a large literature to cull four attributes of resources and seven attributes of users that she suggests are important to the emergence of self-organization among users of a resource. Some of these—feasible improvement of the resource and low discount rate—are absent from Box 2-5. But other attributes she lists are present in Box 2-5, including predictability of benefit flow from the resource, dependence of users on the resource, and successful experience in other arenas of self-organization. Indeed, at least one of the factors that she counts as being important for emergence of commons institutions is also one of her design principles (recognition by external authorities of the ability of users to create their own access and harvesting rules). The

(v)   Gradual change in levels of demand
(3)   Institutional arrangements
    (i)    Rules are simple and easy to understand (B&P)
    (ii)   Locally devised access and management rules (RW, EO, B&P)
    (iii)  Ease in enforcement of rules (RW, EO, B&P)
    (iv)   Graduated sanctions (RW, EO)
    (v)    Availability of low-cost adjudication (EO)
    (vi)   Accountability of monitors and other officials to users (EO, B&P)
(1 and 3) Relationship between resource system and institutional arrangements
    (i)    Match restrictions on harvests to regeneration of resources (RW, EO)
(4)   External environment
    (i)    Technology:
        (a)   Low-cost exclusion technology (RW)
        (b)   Time for adaptation to new technologies related to the commons
    (ii)   Low levels of articulation with external markets
    (iii)  Gradual change in articulation with external markets
    (iv)   State:
        (a)   Central governments should not undermine local authority (RW, EO)
        (b)   Supportive external sanctioning institutions (B&P)
        (c)   Appropriate levels of external aid to compensate local users for conservation activities (B&P)
        (d)   Nested levels of appropriation, provision, enforcement, governance (EO)

SOURCES: RW, Wade (1988); EO, Ostrom (1990); B&P, Baland and Platteau (1996).

overlap between conditions that facilitate emergence and those that facilitate continued successful functioning of institutions points to the close and complex relationship between origins and continued existence, without any suggestion that the two can be explained by an identical set of facilitating conditions.

## ADDRESSING PROBLEMS OF METHOD

The factors presented in Box 2-5 above, relating to resource characteristics, group features, institutional arrangements, and the external environment refer to the substantive aspects of the careful analyses that scholars of common property have conducted. Continued successful research on the commons will depend on the ability of those interested in the commons to resolve some important methodological obstacles that this list of factors raises.

One important problem that is evident from the factors specified in Box 2-5 is a consequence of the fact that most of the conditions cited as facilitating successful use of common-pool resources are general: They are expected to pertain to all common-pool resources and institutions, rather than being related to or dependent on some aspect of the situation.[45] As an illustration, consider the first two conditions in Box 2-5 under the broad class of resource system characteristics: small size and well-defined boundaries. According to Wade, relatively small resource systems are likely to be managed better under common property arrangements, and according to both Ostrom and Wade, resources that have well-defined boundaries are likely better managed as common property. Although these conditions are couched as general statements about all commons, it is in principle possible, and perhaps more defensible, to think of the question of resource size or boundary definition as a contingent one, where the effects of one variable depend on the state of another variable.[46]

It may be possible, thus, to suggest that boundaries of resources should be well defined when flows of benefits are predictable and groups relying on them stationary, but when there are large variations in flows of benefits, and/or the group relying on a resource system is mobile, then resource boundaries should be fuzzy to accommodate variations in group needs and resource flows (see also McCarthy et al., 1999). The effects of resource size, it can be similarly argued, are also contingent on the state of other variables, rather than always flowing in the same direction. Instead of accepting that small resource systems are likely to have a positive relationship with institutional sustainability, for example, it may be more defensible to hypothesize that "size of the resource system should vary with group size, and for larger resources, authority relations within a group should be organized in a nested fashion."

Attempts to identify such conjunctural relationships are critically important for the commons literature because many of the causal relationships in commons situations may be contingent relationships where the impact of a particular variable is likely to depend on the state attained by a different causal factor, or on the relationship of the variable with some contextual factors (Rose, this volume:Chapter 7). As another example, consider the question of fairness in allocation of benefits from the commons. Typically, intuition as well as much of the scholarship on the commons suggests that fairer allocation of benefits is likely to lead to more sustainable institutional arrangements. In a social context characterized by highly hierarchical social and political organization, however, institutional arrangements specifying asymmetric distribution of benefits may be more sustainable.

But the most significant issues of method stem from the sheer number of conditions that seem relevant to the successful management of common-pool resources.[47] Wade, Ostrom, and Baland and Platteau jointly identify 36 important conditions. On the whole there are relatively few areas of common emphasis among them. If one compares across their list of conditions, interprets them carefully, and eliminates the common conditions, 24 different conditions are still to be found (as shown in Box 2-4). Because these authors argue from theoretical foundations, the conditions they find empirically critical in their sample also can be defended on broader grounds. Thus it is difficult to eliminate a priori any of the conditions they consider important.

The discussion of substantive conclusions of Wade, Ostrom, and Baland and Platteau in the previous section reveals that even the 24 factors they have identified do not exhaust the full set of conditions that may be important in common-pool resource management.[48] Once we take into account additional factors identified in the vast literature on the local governance of common-pool resources as being important in sustainably managing those resources, it is reasonable to suppose that the total number of factors that affect successful management of commons is greater than 30, and may be closer to 40. Box 2-5 lists a total of 33 factors, and there is some reason to believe that this a relatively comprehensive list of factors that potentially affect common-pool resource management. Not all of these factors are independent of each other. Some of them are empirically correlated as, for example, group size and resource size, or shared norms, interdependence among group members, and fairness in allocation rules, or ease of enforcement and supportive external sanctioning institutions. We do not, however, have any reliable way of assessing the degree of correlation among these and other variables that have emerged as important in the discussion.

Furthermore, because the effects of some variables may depend on the state of other variables and interactional effects among variables may also affect outcomes, any careful analysis of sustainability on the commons needs to incorporate interaction effects among many of the variables under consideration. As soon as we concede the possibility that between 30 and 40 variables affect the management of common-pool resources, and that some of these variables may have important interactional effects, we confront severe additional analytical problems.

When a large number of variables exist, the absence of careful research design that controls for factors that are not the subject of investigation makes it almost impossible to be sure that the observed differences in outcomes are indeed a result of hypothesized causes. Consider an example. One can select between large group size or high levels of mobility as the relevant causal variables that adversely affect successful management only if the selected cases are matched on other critical variables, and differ (significantly) in relation to group size and mobility. If the researcher does not explicitly take into account the relevant variables that might affect success, then the number of selected cases must be (much) larger than the number of variables. But there are no studies of common-pool resources that develop a research design by explicitly taking into account the different variables considered critical to successful management. In an important sense, then, many of the existing works on the management of common-pool resources, especially those conducted as case studies or those that base their conclusions on a very small number of cases, suffer from the problem of not specifying carefully the causal model they are testing. In the absence of such specification, qualitative studies of the commons are potentially subject to significant problems of method. Two of the most important of these problems are those stemming from "omitted variable bias" and the problem of endogeneity (King et al., 1994:168-182, 185-195). These biases resulting from deficiencies of method have the potential to produce an emphasis on causal factors that may not be relevant, ignoring of other factors that may be relevant, and the generation of spurious correlations.

An incorrect emphasis on some causal variables also may result from the underlying problem of multiple causation, where different causal factors or combinations of causal factors may have similar impacts on outcomes (Ragin, 1987). Thus unpredictable benefit flows and unfair allocation both may have adverse effects on durability of institutions. But in a particular case, it is possible that although benefit flows are unpredictable, they have a much smaller effect on outcomes compared to "unfair allocation of benefits," and that the researcher has ignored the nature of allocation. In such a situation, the conclusions from the study would be flawed in that they would under- or overemphasize variables inappropriately. This issue is especially acute for commons researchers because conclusions from much case study analysis are couched in terms of directional effects of independent variables: positive or negative. "Unpredictable benefit flow," it can be argued, undermines the sustainability of commons institutions. But in a case study it may be difficult to discover how particular independent variables are related to each other, or the strength of their relationship to observed outcomes. In an important sense, single-case analyses, especially when they cover a single time period, limit conclusions about cause-effect relationships to bivariate statements when actual relationships are likely to be more contingent, or continuous.

The large number of variables potentially affecting the sustainability of institutions that govern common resources, thus, has important theoretical implications for future research. The most important implication is perhaps for research design. Because the requirement of a random or representative selection of cases is typically very hard to satisfy where common-pool resources are concerned (even when the universe of cases is narrowed geographically), purposive sampling easily becomes the theoretically defensible strategy for selecting cases whether the objective is statistical analysis or structured comparative case analysis. In purposive sampling, the selected cases will be chosen for the variation they represent on theoretically significant variables. This strategy can be defended both because it is easier to implement than an effort to select a representative sample, and because it requires explicit consideration of theoretically relevant variables (Bennett and George, 2001; Stern and Druckman, 2000).[49]

There is no general theory of purposive sampling apart from the common-sense consideration that selected cases should represent variation on theoretically significant causal factors, and that the investigator should ensure that the selected cases contain at least some variation in terms of observed outcomes.[50] Therefore two factors are likely to be critical in research design: awareness of the variables that are theoretically relevant, and particular knowledge of the case(s) to be researched so that the theoretically relevant variables can be operationalized. For example, when constructing a research design where the variables of interest have to do with mechanisms of monitoring and sanctioning, it would be important for the researcher to be familiar with the different forms of monitoring that groups can use. The presence or absence of a guard may only be indicative of the presence or absence of third-party monitoring, and may reveal nothing about whether the group being studied has adopted monitoring mechanisms. Other forms of monitoring could include mutual monitoring, and rotational monitoring where households in a group jointly share the tasks related to monitoring and enforcement.

The information presented in Box 2-5, organized into four major categories, can therefore be useful in the creation of a research design and for case selection. Given a particular context, it can help in the selection of the variables that need closest attention in the selection of cases. For example, if the cases to be selected lie in the same ecological zone and represent the same resource type, then variables related to resource characteristics may not be very important for case selection. The obvious tradeoff for this reduction in the number of variables is that the research will provide little or no insight into the effect of differing levels of predictability on institutional sustainability. If the research objective were to understand the effects of unpredictability, then it would be imperative to select cases where resource output varied from highly predictable to unpredictable.

However, a large-N study of commons institutions that incorporated more than 30 independent variables and their interactions would require impossibly large samples and entail astronomically high costs. Researchers conducting such studies are likely to face complex problems in interpreting the data and stating their results, even if they could collect information on thousands of cases. Even if it were possible to create purposive samples of cases that accommodated variation on more than 30 causal factors and their interactions, the problems related to contingent and multiple causation will not fade away. The problems of contingent and multiple causation make it necessary for researchers of the commons to also postulate causal relationships among the critical theoretical variables they have identified, and then conduct structured studies that examine the postulated causal links among variables.

Larger sample sizes and statistical analyses also do not constitute a global answer to the problem of many independent variables for another reason. As argued earlier in the chapter, the set of variables that constitutes the context is potentially infinite. Multiplying the number of cases may simultaneously imply an increase in the number of contextual variables that affect outcomes in a specific selected case. Because conclusions from empirical analysis cannot conceivably control causal variables in the same manner as a laboratory study can, it is far more important to understand more carefully the causal relations in a given study than just to test for the robust correlations that a statistical analysis generates. It is when the causal argument is well specified, the research design is carefully crafted, and the sample of cases is rigorously selected that we are most likely to be able to make definitive conclusions about factors leading to sustainability of institutional arrangements around common-pool resources.

A two-pronged approach to advance the research program related to institutional solutions to commons dilemmas, then, seems advisable. On the one hand, scholars of commons need to deploy theoretically motivated comparative case analyses to identify the most important causal mechanisms and narrow the range of relevant theoretical variables and their interactions. On the other hand, they also need to conduct large-$n$ studies to identify the strength of causal relations. Only then would it be possible to advance our understanding of how institutional sustainability can be achieved on the commons.

Once again, the list of factors in Box 2-5 can serve as a starting point for postulating such causal links. For example, a significant body of research on the commons suggests that the nature of monitoring and enforcement is a crucial variable in determining whether existing institutional arrangements to manage the commons will endure. This is to be expected because common property institutions typically are aimed to constrain resource use, and therefore are likely to require enforcement. A complex causal chain to test this finding carefully might be constructed out of the following three hypotheses that connect some of the factors listed in Box 2-5 in causal chains (see Box 2-6):

1. Small size of the resource and the group, low levels of mobility of the resource, and low levels of articulation with markets promote high levels of interdependence among group members;
2. Interdependence, social capital, and low levels of poverty promote well-defined boundaries for the group and the resource; and
3. Well-defined boundaries, ease of enforcement, and recognition of group rights by external governments lead to sustainable institutional performance.

Other variables may be causally related to social capital, ease of enforcement, or recognition of group rights, and such relationships among different variables can be elaborated on in turn. The effect of institutional arrangements related to monitoring and enforcement may be dwarfed by variations in population density or unpredictability of benefit flows. But it still may be possible to investigate some of the causal links listed with a relatively small number of case studies because each comparative study may be used to throw light on only one or two causal chains. The investigation of such causal chains, especially with attention to contextual variables on which particular causal effects may be dependent, therefore, continues to be necessary in commons research.

**BOX 2-6**
**An Illustrative Set of Causal Links**
**in Commons Research**

Durable institutions     = f(boundary definition, enforcement, govern-
                      ment recognition) + error

Boundary definition      = f(group interdependence, poverty, social
                      capital) + error

Group interdependence = f(group size, resource size, mobility, market
                      pressures) + error

The above equations would lead to:

Durable Institutions     = f(group size, resource size, mobility, market
                      pressures, group interdependence, poverty,
                      social capital, enforcement, government
                      recognition) + error

Consider another example. Common property theorists have argued that high levels of dependence on resources in a subsistence-oriented economy are likely to be associated with better governance of common resources. Once again, a chain of causal relationships might be stated as follows (see Box 2-7):

1. Low levels of articulation with the market, high population pressures, low availability of substitutes, and relatively less developed technology promote high dependence on common resources;

2. High dependence on common resources and low possibilities of migration lead users to devise strong constraints on resource use, including strong enforcement mechanisms; and

3. Strong enforcement mechanisms and predictability in flow of benefits leads to sustainable institutional arrangements for governing common resources.

Boxes 2-6 and 2-7 hint at some of the problems of method highlighted in this section. They show that different analysts, depending on the context, may choose to highlight very different causal variables to explain the same phenomenon. They also show how multiple causation is a real-world phenomenon that most commons scholars need to confront explicitly. Finally, Boxes 2-6 and 2-7 show that the factors presented in Box 2-5, when considered by an analyst in the empirical context of his or her research, can help construct causal links and thereby help in research design and case selection.

---

**BOX 2-7**
**Another Illustrative Set of Causal Links in**
**Commons Research**

Durable institutions   = f(strong enforcement, predictable benefit
                                flow) + error

Strong enforcement   = f(high dependence on resource, low migra-
                                tion levels) + error

High dependence   = f(market pressures, population pressures,
                                migration levels, technology levels) + error

                                The above equations would lead to

Durable institutions   = f(technology levels, migration levels, popu-
                                lation pressures, market pressures, strong
                                enforcement, predictable benefit flows) +
                                error

---

To examine such causal links as presented for illustrative purposes in Boxes 2-6 and 2-7, it may not be necessary to launch fresh case studies. Given the large number of studies of commons dilemmas that exist already, it is likely possible to draw on their empirical contents and compare them systematically for understanding the operations of specific causal mechanisms. Postulating causal links among the listed variables also can help reduce the total number of variables on which data need to be collected, and thereby make large-N studies more practical. But it should be obvious that to investigate the full ensemble of relationships depicted in Boxes 2-6 and 2-7, it will be necessary to undertake analyses that draw information from a large number of studies that contain data on each of the identified variables. A large number of studies is also important because more than one empirical measure might be needed to assess some of the theoretical variables listed in the box.

## CONCLUSION

This chapter synthesizes the findings of the empirical literature on the governance of common-pool resources in an effort to identify the contributions and weaknesses of writings on the commons. The chapter suggests that scholarship on common property has made a valuable and distinctive contribution to a better understanding of resource management by focusing on the analytical and structural elements that underpin successful governance. The vigorous flood of recent writings on the commons has presented a wealth of empirical material on how communities and states around the world are using common property institutions to facilitate better governance of natural resources. According to this body of scholarship, robust institutional performance around common-pool resources is positively related to policy choices that encourage fairness in the allocation of benefits from the commons; grant autonomy to users for crafting, implementing, and enforcing institutional arrangements that they identify as being critical in managing resources; institutionalize low-cost mechanisms for adjudication of disputes; promote accountability of office holders to users; and create local-level incentives to develop substitutes. These policy choices are then likely to spur local institutional innovation where users develop clear criteria for group membership, match harvesting rules to the regenerative capacities of the resources they own, and articulate better with state-level institutions. In diverse contexts, other causal stories may turn out to be more compelling.

My examination of the empirical work on the commons ultimately is aimed at analyzing the theoretical underpinnings and methodological assumptions of this field of research. After reviewing several landmark studies and a large number of additional writings, I adopt the position that writings on the commons have come of age, and that it is now necessary to undertake comparative and statistical work that is undergirded by careful research design and rigorous sample selection. Only then can existing understanding of common property institutions and their role in resource management be advanced further.

The chief criticisms I highlight relate to the very large number of factors that commons scholars have postulated as being critical to successful management of natural resources, and the fact that the effect of many of these factors depends on the state assumed by other factors. Directly in tension with this finding is the way research on the commons is conducted. The case study approach remains the preferred mode of analysis of most commons scholars. Even the best known studies of the commons usually have no more than 15 to 30 cases in their sample. When the number of causal factors is higher than that, it is obvious that the case study approach to understanding how commons institutions work is inadequate, especially when authors of case studies focus on one or two factors as determining success. It is especially urgent to devise a way out of this methodological bind because the practical importance of commons research has never been greater: In the past decade, governments in nearly every developing country have turned to decentralized community-level institutions to localize their environmental policies and make them more effective.

One way out of the bind would be to undertake multiple case studies, each using the same methods and variables to ensure comparability. This would, however, be an enormously expensive affair in terms of time, finances, and keeping one's involvement in the case at bay. Few such ambitious projects have been attempted.[51] The paper instead identifies the need for new research that would rely on more careful research design and case selection. It advocates studies that explicitly (1) postulate causal links that can be investigated through structured case comparisons, and (2) use a large number of cases that are purposively selected on the basis of causal variables.

The current stage of research on common property arrangements makes such systematic studies more possible. One means for conducting such causal tests would be to use some of the more careful case studies that already have been completed and that contain information on the critical variables related to resource systems, user groups, institutional arrangements, and external environment that I identify and present in Box 2-5 (Tang, 1992; Schlager, 1990). It is unlikely that the cases for such an enterprise could be selected randomly. But the objective of random selection of cases is unrealistic perhaps in any case. Even an intentional selection of cases that ensures variation on independent variables will allow causal inferences and relatively low levels of bias. What is exciting about studies of commons is that the collective scholarship on local institutions has made it possible for us to approach the construction of a coherent, empirically relevant theory of the commons.

## NOTES

1  Even Netting's sterling study (1981) of the commons in Switzerland possesses the implicit assumption that as resources become more scarce (perhaps because of increasing population pressures, or for any other reasons), common property arrangements will be replaced with more precise and efficient forms of management that private property facilitates.

2  For a review of some of the writings around the turn of the past century, see Agrawal (1999b).

3  Ethnographic writings that can be located in an ancestral relationship to the current scholarship on the commons form a very large set. For some illustrative and magisterial works, see Alexander (1977, 1982), Berreman (1963), Brush (1977), Cole and Wolf (1974), Dahl (1976), and Netting (1972, 1981).

4  The view that community relations are undermined by the intrusion of state policies and market forces formed the basis of much familiar research in the middle of the 1970s (Dunn and Robertson, 1974; O'Brien, 1975; Scott, 1976). Earlier work, especially by Polanyi (1957), had an immense influence on progressive writings on community and market interactions.

5  For a review of some of this literature, see Leach and Mearns (1996) and Ostrom (1990).

6  A vast literature on institutions and property rights proves relevant for the study of common property. Some illustrative starting points for pursuing an interest may be Bates (1989), Eggertsson (1990), Hechter et al., (1990), Knight and Sened (1995), Libecap (1989), North (1980, 1990), and Rose (1994). Some of the early foundations of this literature can be traced back to Commons [1924 (1968)], two influential articles of Coase (1937, 1960), and contributions by scholars such as Alchian and Demsetz (1972), Cheung (1970), and Demsetz (1964). A review of some of this literature is ably presented in Ensminger's (1992) introduction.

7  To say that groups and resources under consideration are situated locally is not to deny the often-intimate connections that exist between external forces and what is considered to be local. In any case, the influence of research on common property is also visible in larger arenas, such as international relations (Keohane and Ostrom, 1995).

8  See Schlager and Ostrom (1992) for a discussion of types of rights and the nature of incentives related to resource use and management that their different combinations create.

9   But see Chapter 1 of this volume for some brief reflections on definitional issues.

10  See, for example, Ostrom (1990:89). Baland and Platteau (1996:285) highlight the difficulties inherent in deciding on parameters of successful management when they say, "It is perhaps too simplistic to view the experiences of common-property management in terms of outright failure or success. It is likely that a good number of these experiences are only partially successful." They do not, however, define precisely what they mean by success.

11  See Blomquist and Ostrom (1985) for a distinction between "commons situations" that are potentially subject to problems of crowding and depletion, and "commons dilemmas" in which private actions of users of commons have costs that cannot be overcome without collective organization.

12  There are other valuable comparative studies of commons management as well that readers can examine at greater length than has been possible in this paper. Pinkerton and Weinstein (1995) and Steins (1999) focus on fisheries; Arnold and Stewart (1991) are concerned mainly with land-based resources in India, while Raintree (1987) examines tenure-related issues in agroforestry more widely; Peters (1994) and Lane (1998) examine livelihood importance of common grazing resources in Africa; Sengupta (1991) compares 12 cases of community irrigation management in India and the Philippines; and Redford and Padoch (1992) and Sandbukt (1995) analyze different institutional regimes around forest commons. Some general overview studies about designing sustainable institutions are also available in Hanna and Munasinghe (1995). The interested reader will find these additional texts well worth pursuing.

13  Hardin (1982), Hechter (1987), Sandler (1992), and Lichbach (1996) provide useful reviews of the collective action literature.

14  Those who already have an extensive familiarity with the common property literature might wish to skip directly to the discussion that is of greater interest.

15  See, for example, Steins and Edwards (1999), who attempt to examine how context affects the incentives of users of a resource, but derive their conclusions from a single case study related to a single resource type.

16  For some comparisons, Wade also uses data on 10 villages that have no irrigation.

17  These empirical observations of Wade are also corroborated in theoretical terms by Ostrom et al. (1994:319), who suggest that when individuals do not trust each other, cannot communicate effectively, and cannot develop agreements, then outcomes are likely to match theoretical predictions of noncooperative behavior among fully rational individuals playing finitely repeated, complete information, common-pool resource games.

18  Wade relies in part on Ostrom's (1985) list of variables that facilitate collective action on the common. Wade interprets "noticeability" as "ease of detection of rule breakers" and considers it to be a function of resource size, group size, and overlap between the location of the resource and the residential location of the group. In Table 2-1, "ease in enforcement" is the variable that stands for Wade's use of "ease of detection."

19  Wade states that he has a set of 13 conditions, but the first condition identified by Wade is in effect 2 different conditions: small size and clearly defined boundaries of the common-pool resource.

20  Note that this particular result is a formal expression of Coase's insight (1960) about the irrelevance of property rights arrangements in the absence of transaction costs. See also Lueck (1994), who examines conditions under which common property can generate greater wealth than private property.

21  See also Maggs and Hoddinott (1999) for a study of how intrahousehold allocation of resources is affected by changes in common property regimes.

22  Baland and Platteau see poverty as the force that often drives users to overexploit environmental resources. But because the rich can consume at even higher levels with scant attention to the environment—witness pronouncements by American President George W. Bush about not honoring campaign promises to restrict carbon emissions because of the potential effects on energy costs—it seems appropriate to recast their argument in terms of opportunity costs.

23  See the important work of Greif (1994a) on how cultural beliefs are an integral part of institu-

tions and affect the evolution and persistence of different societal organizations. In another paper, Greif (1994b) examines the relationship of political institutions to economic growth. A more discursive discussion of political and social relations in the context of common-pool resources is presented by Cleaver (2000) and McCay and Jentoft (1998).

24   The full list of factors they cite is summarized in Box 2-4. Their factors are the ones that are followed by "B&P."

25   For a review of experimental and game theoretic evidence on the same issues, see Kopelman et al. (this volume:Chapter 4) and Falk, et al. (this volume:Chapter 5).

26   To a significant extent, my choice of these four broad categories to classify the conditions identified by Wade, Ostrom, and Baland and Platteau is motivated by the work carried out by Ostrom and her colleagues at the Workshop in Political Theory and Policy Analysis at Indiana University since the mid-1980s on fisheries, forests, irrigation, and pastoral resources. For attempts to establish relationships among these different sets of variables, see discussions of the Institutional Analysis and Development framework (Ostrom et al., 1994) developed by scholars at Indiana. See also Oakerson (1992) and Edwards and Steins (1998).

27   An excellent example of a study that relates characteristics of resource systems to the viability of institutions to manage resources is Netting (1981), who focuses on scarcity and value of resources and the relationship of these two factors to whether common property institutions will endure. See also Thompson and Wisen (1994) for a similar case study from Mexico. Another study that examines common property arrangements, but focuses on environmental risks, is Nugent and Sanchez (1998).

28   The same argument would hold for some forms of humanly created products—for example, greenhouse gases or industrial pollutants—that create externalities across many groups and jurisdictions.

29   As a reviewer of this paper pointed out, movement of resources such as wildlife, and collecting and holding resources such as irrigation water, can be seen as mobility in space and time, and both aim to address fluctuations inherent in the production functions associated with the output from a resource system. Also, in one sense, markets provide individual producers with mobility across functionally specialized tasks.

30   Indeed, as Ostrom points out, the impact of all the independent variables on sustainability of commons institutions can be depicted in terms of a cost-benefit calculus related to individual decision making.

31   See also Bardhan (1993) on the role of scarcity, and Scherr et al. (1995).

32   Although this paper does not focus on cultural contextual factors that may affect how local conservation and resource use processes unfold, such factors may also, in some instances, have important effects (Uphoff and Langholz, 1998).

33   For a review of some of the writings on this subject, and for a test of the relative importance of population pressures, market pressures, and enforcement institutions on the condition of resources, see Agrawal and Yadama (1997). Regev et al. (1998) examine how market-related and technological changes may affect rates of harvest and resource use.

34   In the absence of transaction costs related to exchange and political gains to be had from cornering the supply of scarce resources, no benefits could be derived from pooling. For a more familiar example of the redundancy of pooling institutions in the absence of transaction costs, think of insurance organizations. None of them would be necessary were pooling of individual-level risks to become pointless.

35   Hechter (1987) discusses how new technology in the cable television industry determined excludability.

36   The issue is not whether markets and capital availability have an effect today in comparison to the past. It is one of the degree or intensity with which market forces and capital availability have an impact at different time periods in specific places. Even if processes of globalization make the presence of money and capital more widespread, they do not accomplish it in any homogeneous fashion.

37  Three studies that examine some of the complexities of state-local relationships are Gibson (1999), Ribot (1999), and Richards (1997). For analyses that focus directly on decentralization in the context of common-pool resources, see Agrawal and Ostrom (forthcoming, 2001) and Agrawal and Ribot (1999).

38  In contrast to the current fascination of many anthropologists with history and globalization, much anthropological writing in the 1960s and 1970s saw its object of ethnographic analysis as an ensemble of timeless relations. For some critical reviews on this subject, see Dirks et al. (1994), Donham (1990), Mathur (2000), Roseberry (1989), Sahlins (1999), and Wolf (1982).

39  Marwell and Oliver's conclusion is for public goods rather than common goods, but even for common goods, Esman and Uphoff (1984) find that larger local organizations were associated with greater success in rural development initiatives. The extent to which groups might have grown in size as they experienced success is not clear, but in any case their finding suggests that larger groups might function more effectively even if smaller groups are more successful in initiating collective action. I am grateful to Ruth Meinzen-Dick for drawing this reference to my attention.

40  Lumpiness of monitoring refers to the situation in which a specialist guard is hired to enforce common property arrangements. In this situation, the guard needs to be paid a salary for fixed periods such as a few months or a year, rather than just for an hour or a day in the year. The exact relationship Agrawal and Goyal (2001) identify suggests that in the Kumaon Himalaya context, user groups larger than 100 households and smaller than 30 households have difficulties in finding the levels of surplus needed to ensure adequate monitoring.

41  In the introduction to their recent discussion of inequality, Bowles and Gintis (1998:4) state, "economic theory has proven, one hears, that any but cosmetic modifications of capitalism in the direction of equality and democratic control will exact a heavy toll of reduced economic performance. Yet economic theory suggests no such thing. On the contrary, there are compelling economic arguments and ample empirical support for the proposition that there exist changes in the rules of the economic game that can foster both greater economic equality and improved economic performance…inequality is often an impediment to productivity."

42  For a concrete example of heterogeneities of interests leading to heterogeneous endowments and mutually beneficial exchanges, see Agrawal (1999a). More generally, exchanges between pastoralists and agriculturalists depend on distinct and heterogeneous interests and endowments in relation to land-based resources.

43  Elster (1992:14) writing about the study of local justice, suggests that "it is a very messy business, and that it may be impossible to identify a set of necessary and sufficient conditions that constitute a theory of local justice." His diagnosis for local justice may be equally applicable to the study of commons, as also his prescription: Instead of making a choice between theory and description, focus on "identifiable causal patterns" (Elster, 1992:16).

44  For a discussion of what accounts for the emergence of common property institutions, see McCay (this volume:Chapter 11).

45  Commenting on a similar tendency in political analysis, Ostrom (1998:16) recognizes that "political systems are complexly organized, and that we will rarely be able to state that one variable is always positively or negatively related to a dependent variable."

46  This issue of the effects of a given variable being very different depending on the state of another variable is not addressed by the ceteris paribus clause that is implicit in all the conditions stated by these authors. Depending on the state of a related variable, the effects of another variable may even run counter to the suggested direction. Thus, Turner (1999) shows how clearer definition of boundaries and strengthening of exclusionary powers in the context of high levels of variability and mobility can lead to increased conflict. Such conflicts can endure over long time periods if those who are excluded cannot find alternative occupational opportunities. Agrawal (1999a) uses the example of the raika shepherds in western Rajasthan to make a related argument about the marginalization of

mobile shepherds through firmer delineation of boundaries to resources and exclusionary powers of communities.

47 A somewhat different but also very critical question of method is whether conclusions derived from one level of analysis or at a particular spatial/temporal level apply to other levels. Do inferences that are valid at the local level also apply to more macro-level phenomena? Although I do not address this question, both Berkes (this volume:Chapter 9) and Young (this volume:Chapter 8) examine it carefully.

48 Indeed, it should be clear that my discussion of potentially missing variables was aimed not just to highlight deficiencies of substance in these careful analyses, but even more to focus on a general problem of method that characterizes most studies of common property, and that these studies avoid to the extent possible.

49 For discussions of problems of bias that result from sampling on the dependent variable, see King et al. (1994) and Collier and Mahoney (1996).

50 Although cases should not be selected so as to include some instances of success and some of failure, because this is likely to introduce bias in sample selection (King et al., 1994), it should be kept in mind that if there is no variation in outcomes, then even if the selected cases vary on the factors that are deemed causally significant, the research will reveal little about the differing effects of hypothesized causes because outcomes are invariant in the selected sample.

51 The International Forestry Resources and Institutions Program at the Workshop in Political Theory and Policy Analysis, Indiana University, is in the middle of such an ambitious project. Members are just initiating analysis that may address some of the substantive and methodological criticisms voiced in this paper (see the collection of studies in Gibson et al., 2000). Even in this project, however, case selection can sometimes depend on availability of funding, an individual researcher's interests, and the ease of establishing collaborative partnerships with research institutions in different countries.

# REFERENCES

Abernathy, V.
1993 *Population Politics: The Choices that Shape Our Future.* New York: Plenum Press/Insight Books.

Agrawal, A.
1999a *Greener Pastures: Politics, Markets, and Community among a Migrant Pastoral People.* Durham, NC: Duke University Press.
1999b Community: Tracing the outlines of an enchanting concept. Pp. 92-108 in *A New Moral Economy for India's Forests? Discourses of Community and Participation*, Roger Jeffrey and Nandini Sundar, eds., New Delhi: Sage Publications.
2001 State formation in community spaces? Decentralization of control over forests in the Kumaon Himalaya, India. *Journal of Asian Studies* 60(1):1-32.

Agrawal, A., and C. Gibson
1999 Community and conservation: Beyond enchantment and disenchantment. *World Development* 27(4):629-649.

Agrawal, A., and S. Goyal
2001 Group size and collective action: Third party monitoring in common-pool resources. *Comparative Political Studies* 34(1):63-93.

Agrawal, A., and E. Ostrom
2001 Collective action, property rights, and decentralization in resource use in India and Nepal. *Politics and Society.*

Agrawal, A., and J. Ribot
1999 Accountability in decentralization: A framework with south Asian and west African cases. *Journal of Developing Areas* 33(Summer):473-502.

Agrawal, A., and G. Yadama
  1997   How do local institutions mediate market and population pressures on resources? Forest Panchayats in Kumaon, India. *Development and Change* 28(3):437-466.
Alchian, A., and H. Demsetz
  1972   Production, information costs, and economic organization. *American Economic Review* 62(December):777-795.
Alexander, P.
  1977   South Sri Lanka sea tenure. *Ethnology* 16:231-255.
  1982   *Sri Lankan Fishermen: Rural Capitalism and Peasant Society.* Canberra: Australian National University.
Arnold, J.E.M., and W.C. Stewart
  1991   *Common Property Resource Management in India.* Oxford, Eng.: Oxford Forestry Institute, University of Oxford.
Ascher, W.
  1995   *Communities and Sustainable Forestry in Developing Countries.* San Francisco: ICS Press.
Ascher, W., and R. Healy
  1990   *Natural Resource Policymaking in Developing Countries: Environment, Economic Growth, and Income Distribution.* Durham, NC: Duke University Press.
Ault, W.O.
  1952   *Open Field Farming in Medieval England: The Self-Directing Activities of Village Communities in Medieval England.* Boston: Boston University Press.
Azhar, R.
  1993   Commons, regulation, and rent-seeking behavior: The dilemma of Pakistan's *Guzara* forests. *Economic Development and Cultural Change* 42(1):115-128.
Baker, A., and R. Butlin
  1973   *Studies of Field Systems in the British Isles.* Cambridge, Eng.: Cambridge University Press.
Baland, J., and J. Platteau
  1996   *Halting Degradation of Natural Resources: Is There a Role for Rural Communities?* Oxford, Eng.: Clarendon Press.
  1999   The ambiguous impact of inequality on local resource management. *World Development* 27:773-788.
Bardhan, P.
  1993   Analytics of the institutions of informal cooperation in rural development. *World Development* 21(4):633-639.
Bates, R.
  1981   *Markets and States in Tropical Africa: The Political Basis of Agricultural Policies.* Berkeley: University of California Press.
  1989   *Beyond the Miracle of the Market: The Political Economy of Agrarian Development in Kenya.* Cambridge, Eng.: Cambridge University Press.
Bennett, A., and A. George
  2001   *Case Studies and Theory Development.* Cambridge, MA.: MIT Press.
Berkes, F., ed.
  1989   *Common Property Resources: Ecology and Community-Based Sustainable Development.* London: Belhaven Press.
Berreman, G.D.
  1963   *Hindus of the Himalayas: Ethnography and Change.* Berkeley: University of California Press.
Blomquist, W., and E. Ostrom
  1985   Institutional capacity and the resolution of a commons dilemma. *Policies Studies Review* 5(2):383-393.

Blomquist, W., E. Schlager, S. Yan Tang, and E. Ostrom
    1994    Regularities from the field and possible explanations. Pp. 301-316 in *Rules, Games, and Common-Pool Resources*, E. Ostrom, R. Gardner, and J. Walker, eds. Ann Arbor: University of Michigan Press.
Bowles, S., and H. Gintis
    1998    Effective redistribution: New rules of markets, states, and communities. Pp. 3-71 in *Recasting Egalitarianism: New Rules for Communities, States, and Markets*, E.O. Wright, ed., London:Verso.
Brush, S.
    1977    *Mountain, Field, and Family: The Economy and Human Ecology of an Andean Valley.* Philadelphia: University of Pennsylvania Press.
Carrier, J.
    1987    Marine tenure and conservation in Papua New Guinea: Problems in interpretation. Pp. 142-170 in *The Question of the Commons: The Culture and Ecology of Communal Resources*, B.J. McCay and J.M. Acheson, eds., Tucson: University of Arizona Press.
Cheung, S.N.S.
    1970    The structure of a contract and the theory of a Non-Exclusive Resource. *Journal of Law and Economics* 13(1, April):49-70.
Chomitz, K.
    1995    Roads, Land, Markets and Deforestation: A Spatial Model of Land Use in Belize. Unpublished paper presented at the First Open Meeting of the Human Dimensions of Global Environmental Change Community, Duke University, Durham, NC, June 1-3.
Cleaver, F.
    2000    Moral ecological rationality, institutions, and the management of common property resources. *Development and Change* 31:361-383.
Coase, R.
    1937    The nature of the firm. *Economica* 4(3):386-405.
    1960    The problem of social cost. *Journal of Law and Economics* 3:1-44.
Colchester, M.
    1994    Sustaining the forests: The community-based approach in South and South-east Asia. *Development and Change* 25(1):69-100.
Cole, J.W., and E.R. Wolf
    1974    *The Hidden Frontier: Ecology and Ethnicity in an Alpine Valley.* New York: Academic Press.
Collier, D., and J. Mahoney
    1996    Insights and pitfalls: Selection bias in quantitative research. *World Politics* 49(1):56-91.
Commons, J.R.
    [1924] 1968 *Legal Foundations of Capitalism.* Madison: University of Wisconsin Press.
Dahl, G.
    1976    *Suffering Grass: Subsistence and Society of Waso Borana.* Stockholm: University of Stockholm Press.
Demsetz, H.
    1964    The exchange and enforcement of property rights. *The Journal of Law and Economics* 3(1, October):1-44.
Dirks, N., G. Eley, and S. Ortner, eds.
    1994    *Culture/Power/History: A Reader in Contemporary Social Theory.* Princeton, NJ: Princeton University Press.
Donham, D.
    1990    *History, Power, Ideology: Central Issues in Marxism and Anthropology.* Cambridge, Eng.: Cambridge University Press.

Dunn, J., and A.F. Robertson
    1974   *Dependence and Opportunity: Political Change in Ahafo.* New York and London: Cambridge University Press.
Durning, A.
    1989   *Poverty and the Environment: Reversing the Downward Spiral.* Washington, DC: Worldwatch Institute.
Edwards, V.M., and N.A. Steins
    1998   Developing an analytical framework for multiple-use commons. *Journal of Theoretical Politics* 10(3):347-383.
Eggertsson, T.
    1990   *Economic Behavior and Institutions.* Cambridge, Eng.: Cambridge University Press.
Ehrlich, P.
    1968   *The Population Bomb.* New York: Ballantine.
Ehrlich, P., and A. Ehrlich
    1991   *The Population Explosion.* New York: Touchstone, Simon and Schuster Inc.
Elster, J.
    1992   *Local Justice: How Institutions Allocate Scarce Goods and Necessary Burdens.* New York: Russell Sage Foundation.
Ensminger, J.
    1992   *Making a Market: The Institutional Transformation of an African Society.* Cambridge, Eng.: Cambridge University Press.
Esman, M., and N. Uphoff
    1984   *Local Organizations: Intermediaries in Rural Development.* Ithaca, NY: Cornell University Press.
Fernandes, W., G. Menon, and P. Viegas
    1988   *Forests, Environment, and Tribal Economy.* New Delhi: Indian Social Institute.
Fischer, G.
    1993   The Population Explosion: Where is it Leading? *Population and Environment* 15(2):139-153.
Food and Agriculture Organization (FAO)
    1999   Status and Progress in the Implementation of National Forest Programmes: Outcomes of an FAO Worldwide Survey. Mimeo. Rome: FAO.
Fudenberg, D., and E. Maskin
    1986   The folk theorem in repeated games with discounting and imperfect information. *Econometrica* 54(3):533-554.
Gibson, C.
    1999   *Politicians and Poachers: The Political Economy of Wildlife Policy.* New York: Cambridge University Press.
Gibson, C., M.A. McKean, and E. Ostrom, eds.
    2000   *People and Forests: Communities, Institutions, and Governance.* Cambridge, MA: MIT Press.
Greif, A.
    1994a  Cultural beliefs and the organization of society: A historical and theoretical reflection on collectivist and individualist societies. *The Journal of Political Economy* 102(5):912-950.
    1994b  On the political foundations of the late medieval commercial revolution: Genoa during the twelfth and thirteenth centuries. *Journal of Economic History* 54(2):271-287.
Hanna, S., and M. Munasinghe, eds.
    1995   *Property Rights in a Social and Ecological Context: Case Studies and Design Applications.* Washington, DC: The Beijer International Institute of Ecological Economics and The World Bank.

Hardin, G.
    1993    *Living Within Limits.* New York: Oxford University Press.
Hardin, R.
    1982    *Collective Action.* Baltimore, MD: Johns Hopkins University Press.
Hechter, M.
    1987    *Principles of Group Solidarity.* Berkeley: University of California Press.
Hechter, M., K. Opp, and R. Wippler, eds.
    1990    *Social Institutions: Their Emergence, Maintenance, and Effects.* New York: Aldine de
            Gruyter.
Jessup, T.C., and N.L. Peluso
    1986    Minor forest products as common property resources in East Kalimantan, Indonesia. Pp.
            501-531 in National Research Council, *Proceedings of the Conference on Common Prop-
            erty Resource Management.* Washington, DC: National Academy Press.
Jodha, N.S.
    1986    Common property resources and rural poor in dry regions of India. *Economic and Political
            Weekly* 21(27):1169-1182.
Kanbur, R.
    1992    *Heterogeneity, Distribution, and Cooperation in Common Property Resource Manage-
            ment.* Policy Research Working Papers, WPS 844. Washington DC: World Bank.
Keohane, R., and E. Ostrom, eds.
    1995    *Local Commons and Global Interdependence: Heterogeneity and Cooperation in Two
            Domains.* Thousand Oaks, CA: Sage Publications.
King, G., R. Keohane, and S. Verba
    1994    *Designing Social Inquiry: Scientific Inference in Qualitative Research.* Princeton, NJ:
            Princeton University Press.
Knight, J., and I. Sened, eds.
    1995    *Explaining Social Institutions.* Ann Arbor: University of Michigan Press.
Lam, W.F.
    1998    *Governing Irrigation Systems in Nepal: Institutions, Infrastructure, and Collective Action.*
            San Francisco, CA: ICS Press.
Lane, C., ed.
    1998    *Custodians of the Commons: Pastoral Land Tenure in East and West Africa.* London:
            Earthscan.
Lappé, F.M., and R. Shurman
    1989    *Taking Population Seriously.* London: Earthscan.
Leach, M., and R. Mearns, eds.
    1996    *The Lie of the Land: Challenging Received Wisdom on the African Environment.* Oxford,
            Eng., and Portsmouth, NH: James Currey and Heinemann.
Libecap, G.
    1989    Distributional issues in contracting for property rights. *Journal of Institutional and Theo-
            retical Economics* 145:6-24.
    1990    *Contracting for Property Rights.* New York: Cambridge University Press.
Lichbach, M.
    1996    *The Cooperator's Dilemma.* Ann Arbor: University of Michigan Press.
Low, B., and J. Heinen
    1993    Population, resources and environment: Implications of human behavioral ecology for
            conservation. *Population and Environment* 15(1):7-41.
Lueck, D.
    1994    Common property as an egalitarian share contract. *Journal of Economic Behavior and
            Organization* 25(1):93-108.

Lynch, O.J., and K. Talbott
  1995  *Balancing Acts: Community-Based Forest Management and National Law in Asia and the Pacific.* Washington, DC: World Resources Institute.
Maggs, P., and J. Hoddinott
  1999  The impact of changes in common property resource management in intrahousehold allocation. *Journal of Public Economics* 72:317-324.
Malthus, T.
  1960  *On Population (First Essay on Population, 1798, and Second Essay on Population, 1803).* New York: Random House.
Marwell, G., and P. Oliver
  1993  *The Critical Mass in Collective Action: A Micro-Social Theory.* Cambridge, Eng.: Cambridge University Press.
Mathur, S.
  2000  History and anthropology in south Asia: Rethinking the archive. *Annual Reviews of Anthropology* 29:89-106.
McCarthy, N., B. Swallow, M. Kirk, and P. Hazell
  1999  *Property Rights, Risk, and Livestock Development in Africa.* Washington DC: International Livestock Research Institute and International Food Policy Research Institute.
McCay, B.J., and J. Acheson, eds.
  1987  *The Question of the Commons: The Culture and Ecology of Communal Resources.* Tucson: University of Arizona Press.
McCay, B.J., and S. Jentoft
  1998  Market or community failure? Critical perspectives on common property research. *Human Organization* 57(1):21-29.
McKean, M.
  1992a Management of traditional common lands *(Iriaichi)* in Japan. Pp. 63-98 in *Making the Commons Work: Theory, Practice, and Policy*, D.W. Bromley, ed. San Francisco: ICS Press.
  1992b Success on the commons: A comparative examination of institutions for common property resource management. *Journal of Theoretical Politics* 4(3):247-281.
Moseley, C.
  1999  New Ideas, Old Institutions: Environment, Community, and State in the Pacific Northwest. Unpublished PhD. dissertation, Department of Political Science, Yale University.
Myers, N.
  1991  The world's forests and human populations: The environmental interconnections. Pp. 237-251 in *Resources, Environment, and Population: Present Knowledge, Future Options*, K. Davis and M. Bernstam, eds. New York: Oxford University Press.
National Research Council
  1986  *Proceedings of the Conference on Common Property Resource Management.* Washington, DC: National Academy Press.
  1999  *Our Common Journey: A Transition Toward Sustainability.* Washington, DC: National Academy Press.
Naughton-Treves, L., and S. Sanderson
  1995  Property, politics and wildlife conservation. *World Development* 23(8):1265-1275.
Netting, R.M.
  1972  Of men and meadows: Strategies of alpine land use. *Anthropological Quarterly* 45:132-144.
  1981  *Balancing on an Alp.* Cambridge, Eng.: Cambridge University Press.
North, D.C.
  1980  *Structure and Change in Economic History.* New York: Norton.

1990    *Institutions, Institutional Change and Economic Performance*. Cambridge, Eng.: Cambridge University Press.

Nugent, J.B., and N. Sanchez
1998    Common property rights as an endogenous response to risk. *American Journal of Agricultural Economics* 80(3):651-658.

Oakerson, R.J.
1992    Analyzing the commons: A framework. Pp. 41-59 in *Making the Commons Work: Theory, Practice and Policy*, D. Bromley, ed. San Francisco: ICS Press.

Oates, J.F.
1999    *Myth and Reality in the Rain Forest: How Conservation Strategies are Failing in West Africa*. Berkeley: University of California Press.

O'Brien, Donal C.
1975    *Saints and Politicians: Essays in the Organization of a Senegalese Peasant Society*. New York and London: Cambridge University Press.

Olson, M.
1965    *The Logic of Collective Action: Public Goods and the Theory of Groups*. Cambridge, MA: Harvard University Press.

Ostrom, E.
1990    *Governing the Commons: The Evolution of Institutions for Collective Action*. Cambridge, Eng.: Cambridge University Press.
1992    *Crafting Institutions for Self-Governing Irrigation Systems*. San Francisco: ICS Press.
1995    The Rudiments of a Revised Theory of the Origins, Survival, and Performance of Institutions for Collective Action. Working Paper No. W85-32, Workshop in Political Theory and Policy Analysis, Indiana University, Bloomington.
1997    Self-Governance of Common-Pool Resources. W97-2, Workshop in Political Theory and Policy Analysis, Indiana University, Bloomington.
1998    A behavioral approach to the rational choice theory of collective action. *American Political Science Review* 92(1):1-22.
1999    Self Governance and Forest Resources. Occasional Paper No. 20, Center for International Forestry Research, Bogor, Indonesia. Available: [http://www.cifor.cgiar.org/publications/pdf_files/OccPapers/OP-20.pdf [accessed Sept. 3, 2001].

Ostrom, E., R. Gardner, and J. Walker
1994    *Rules, Games, and Common-Pool Resources*. Ann Arbor: University of Michigan Press.

Peluso, N.L.
1992    *Rich Forests, Poor People*. Berkeley: University of California Press.
1993    Coercing conservation: The politics of state resource control. *Global Environmental Change* 3(2):199-217.

Peters, P.
1994    *Dividing the Commons: Politics, Policy, and Culture in Botswana*. Charlottesville: University Press of Virginia.

Pimental, D., R. Harman, M. Pacenza, J. Pecarsky, and M. Pimental
1994    Natural resources and an optimal human population. *Population and Environment* 15(5):347-369.

Pinkerton, E., ed.
1989    *Cooperative Management of Local Fisheries: New Directions for Improved Management and Community Development*. Vancouver: University of British Columbia Press.

Pinkerton, E., and M. Weinstein
1995    *Sustainability through Community-Based Management*. Vancouver, BC: The David Suzuki Foundation.

Poffenberger, M., ed.
  1990   *Keepers of the Forest: Land Management Alternatives in Southeast Asia.* West Hartford, CT: Kumarian.
Polanyi, K.
  1957   *The Great Transformation: The Political and Economic Origins of Our Time.* Boston: Beacon.
Quiggin, J.
  1993   Common property, equality, and development. *World Development* 21:1123-1138.
Ragin, C.C.
  1987   *The Comparative Method: Moving beyond Qualitative and Quantitative Strategies.* Berkeley: University of California Press.
Raintree, J.B., ed.
  1987   *Land, Trees and Tenure: Proceedings of an International Workshop on Tenure Issues in Agroforestry.* Nairobi, Kenya: ICRAF and the Land Tenure Center.
Rangan, H.
  1997   Property vs. control. The state and forest management in the Indian Himalaya. *Development and Change* 28(1):71-94.
Rangarajan, M.
  1996   *Fencing the Forest: Conservation and Ecological Change in India's Central Provinces, 1860-1914.* New Delhi: Oxford University Press.
Redford, K.H., and C. Padoch, eds.
  1992   *Conservation of Neotropical Forests: Working from Traditional Resource Use.* New York: Columbia University Press.
Regev, U., A.P. Gutierrez, S.J. Schreiber, and D. Zilberman
  1998   Biological and economic foundations of renewable resource exploitation. *Ecological Economics* 26:227-242.
Repetto, R., and M. Gillis, eds.
  1988   *Public Policies and the Misuse of Forest Resources.* Cambridge, Eng.: Cambridge University Press.
Ribot, J.C.
  1999   Decentralization, participation, and accountability in Sahelian forestry: Legal instruments of political-administrative control. *Africa* 69(1):23-65.
Richards, M.
  1997   Common property resource institutions and forest management in Latin America. *Development and Change* 28(1):95-117.
Rose, C.
  1994   *Property and Persuasion: Essays on the History, Theory, and Rhetoric of Ownership.* Boulder, CO: Westview.
Roseberry, W.
  1989   *Anthropologies and Histories: Essays in Culture, History, and Political Economy.* New Brunswick, NJ: Rutgers University Press.
Sahlins, M.
  1999   What is anthropological enlightenment? Some lessons of the twentieth century. *Annual Review of Anthropology* 28:i-xxiii.
Sandbukt, Ø., ed.
  1995   *Management of Tropical Forests: Towards an Integrated Perspective.* Oslo, Norway: Center for Development and the Environment, University of Oslo.
Sandler, T.
  1992   *Collective Action: Theory and Applications.* Ann Arbor: University of Michigan Press.

Scherr, S.J., L. Buck, R. Meinzen-Dick, and L.A. Jackson
    1995   *Designing Policy Research on Local Organizations in Natural Resource Management.*
           EPTD Workshop Summary Paper 2. Washington, DC: International Food Policy Research
           Institute.
Schlager, E.
    1990   Model Specification and Policy Analysis: The Governance of Coastal Fisheries. Unpub-
           lished PhD dissertation, Department of Political Science, Indiana University.
Schlager, E., and E. Ostrom
    1992   Property rights regimes and natural resources: A conceptual analysis. *Land Economics*
           68(3):249-262.
Schotter, A.
    1981   *The Economic Theory of Social Institutions.* New York: Cambridge University Press.
Scott, J.C.
    1976   *The Moral Economy of the Peasant: Rebellion and Subsistence in Southeast Asia.* New
           Haven, CT:Yale University Press.
Sengupta, N.
    1991   *Managing Common Property: Irrigation in India and the Philippines.* London: Sage Pub-
           lications.
Simon, J.
    1990   *Population Matters: People, Resources, Environment and Integration.* New Brunswick,
           NJ: Transaction Publishers.
Sivaramakrishnan, K.
    1999   *Modern Forests: Statemaking and Environmental Change in Colonial Eastern India.*
           Stanford, CA: Stanford University Press.
Skaria, A.
    1999   *Hybrid Histories: Forests, Frontiers, and Wildness in Western India.* New Delhi: Oxford
           University Press.
Steins, N.A.
    1999   All Hands on Deck: An Interactive Perspective on Complex Common-Pool Resource Man-
           agement Based on Case Studies in the Coastal Waters of the Isle of Wight (UK), Con-
           nemara (Ireland), and the Dutch Walden Sea. Unpublished Ph.D. thesis, Wageningen
           University, The Netherlands.
Steins, N.A., and V.M. Edwards
    1999   Collective action in common-pool resource management: The contribution of a social
           constructivist perspective to existing theory. *Society and Natural Resources* 12:539-557.
Stern, P.C., and D. Druckman
    2000   Evaluating interventions in history: The case of international conflict resolution. Pp. 38-89
           in National Research Council, *International Conflict Resolution after the Cold War*, P.C.
           Stern and D. Druckman, eds. Washington, DC: National Academy Press.
Stocks, A.
    1987   Resource management in an Amazon Varzea lake ecosystem: The Cocamilla case. Pp.
           108-120 in *The Question of the Commons: The Culture and Ecology of Communal Re-
           sources*, B.J. McCay and J.M. Acheson, eds. Tucson, AZ: University of Arizona Press.
Sugden, R.
    1984   Reciprocity: The supply of public goods through voluntary contributions. *Economic Jour-
           nal* 94:772-787.
    1989   Spontaneous order. *Journal of Economic Perspectives* 13(4):85-97.
Tang, S.Y.
    1992   *Institutions and Collective Action: Self Governance in Irrigation Systems.* San Francisco:
           ICS Press.

Thirsk, J.
1966    The origins of the common fields. *Past and Present* 33(April):142-147.
Thompson, G.D., and P.N. Wisen
1994    Ejido reforms in Mexico: Conceptual issues and potential outcomes. *Land Economics* 70(4):448-465.
Tiffen, M., M. Mortimore, and F. Gichuki
1994    *More People, Less Erosion: Environmental Recovery in Kenya.* Chichester, Eng. John Wiley & Sons.
Turner, M.D.
1999    Conflict, environmental change, and social institutions in dryland Africa: Limitations of the community resource management approach. *Society and Natural Resources* 12(7):643-658.
Uphoff, N., and J. Langholz
1998    Incentives for avoiding the tragedy of the commons. *Environmental Conservation* 25(3):251-261.
Varughese, G., and E. Ostrom
1998    The Contested Role of Heterogeneity. Unpublished paper, Workshop in Political Theory and Policy Analysis, Indiana University.
Wade, R.
[1988] 1994 *Village Republics: Economic Conditions for Collective Action in South India.* San Francisco, CA: ICS Press.
Wilson, E.O.
1992    *The Diversity of Life.* New York: W.W. Norton.
Wolf, E.
1982    *Europe and the People Without History.* Berkeley: University of California Press.
Yelling, J.A.
1977    *Common Field and Enclosure in England, 1450-1850.* London: MacMillan.
Young, K.R.
1994    Roads and the environmental degradation of tropical montane forests. *Conservation Biology* 8(4):972-976.

# 3

# Unequal Irrigators: Heterogeneity and Commons Management in Large-Scale Multivariate Research

*Pranab Bardhan and Jeff Dayton-Johnson*

C ommon-pool resources play a decisive role in determining the livelihood of the rural poor (Jodha, 1986, 1990) and local environmental conditions. The past 15 years of research have clearly demonstrated the importance of institutional form to the performance of commons-using communities; the notion that such situations always should be viewed within the framework of the "Tragedy of the Commons" has been decisively dispelled. Nevertheless, much remains to be understood and synthesized. Among the unsettled questions is the following: What is the impact of heterogeneity among the users of a community-based natural resource? Many field studies of the commons have addressed this question, although generally only tangentially. Nevertheless, the only consensus that emerges from the multidisciplinary empirical literature is that the relationship between heterogeneity and commons use and management is complicated. Recent theoretical research in economics has clarified some of the complicated mechanisms that link inequality and commons outcomes, and we will consider much of the case-study literature in light of this economic work. This chapter identifies the most important types of heterogeneity, the commons outcomes they might affect, and the mechanisms that link the two.[1]

The principal task of this chapter is to review large-scale surveys of locally managed irrigation systems as an empirical illustration of the relationship between heterogeneity and success in managing the commons. Until recently, most

The authors extend appreciation to the editors, Elinor Ostrom in particular, to the other authors in this volume, to Ruth Meinzen-Dick, and to an anonymous referee. Thanks are also due to Emannuel Bon and other participants in a workshop at the Eighth Biennial Conference of the International Association for the Study of Common Property (IASCP) at Indiana University in June 2000.

studies of community-based irrigation (like most studies of common-pool resource systems generally) focused on one or two systems. We have learned a great deal from these case studies, but they do not permit generalizations about the relationship between heterogeneity and commons management. If one village with a high level of inequality also has a relatively successful management regime for its irrigation system, it is difficult to deduce from that a general relationship between heterogeneity and management. In statistical terms, the case studies do not have the degrees of freedom necessary to discern relationships among the institutions of governance, various dimensions of performance, and the structural characteristics of resource-using communities. More recently, a small number of studies have sought to complement the case study approach with information culled from relatively large numbers of resource-using systems. This chapter is unique in this volume in that it gives pride of place to this large-scale multivariate research, rather than other forms of empirical inquiry (e.g., laboratory experiments or anthropological case studies). More specifically it synthesizes lessons learned from a subset of these studies focused on irrigation systems.

The empirical context for our chapter, thus, is the poor hydraulic economy: peasant water users in conditions of low-income rural sectors. The unit of analysis is the resource-using group, of which heterogeneity is a characteristic.[2]

## HETEROGENEITY

Irrigators, or users of some other common-pool resource, may be heterogeneous in economic, social, cultural, or other dimensions. There are many relevant types of economic inequality alone. Variants of economic heterogeneity include: inequality in wealth or income among the members of a resource-using group; inequalities in the sacrifices community members make in cooperating with commons-management regimes; inequalities in the benefits they derive from such regimes; and inequalities in outside earnings opportunities ("exit options"). There are other kinds of disparities that may have economic consequences, and those in turn affect cooperation. For example, locational differences, to the extent that they are not already reflected in landholding or wealth differences, might not be taken into account adequately if one considers only wealth inequality. Head-end and tail-end farmers in irrigation systems face different incentives to cooperate (Bardhan, 1984; Ostrom, 1994), as do fishers with access to more or less productive fishing spots (Berkes, 1986). For irrigators, long-run locational advantages and disadvantages will be capitalized into land values if land markets work reasonably well. Thus, the head-end/tail-end inequality is another version of wealth inequality.[3] Of course, in many parts of the world, land markets notoriously do not work reasonably well. Even if head-end/tail-end differences are captured perfectly in land values, such locational differences provide strategic opportunities that are not normally available simply as a result of wealth differences. Head-end farmers, poor or not, get the water first. Similarly, differences in ability or efficiency in resource extraction will affect cooperative behavior (Johnson and Libecap, 1982). These differences in many cases will be correlated closely with wealth: Fishers with more gear will have lower unit costs of harvesting. Differences in what economists call *rate of time preference*—essentially differences among resource users in the degree to which they consider the future in their current extraction activities (Ostrom, 1990:passim)—will lead to differential impatience among commons users in making short-run sacrifices for resource conservation.

Ethnic heterogeneity such as differences in language or caste among irrigators also will affect cooperative behavior.[4] An irrigating community may be socially heterogeneous if its users come from various villages. Of course, in many cases, ethnic or social heterogeneity will be correlated with economic heterogeneity, as certain castes or ethnic groups are also more likely to be richer or poorer than other groups. Nevertheless, these noneconomic types of heterogeneity potentially have effects independent of the economic heterogeneity with which they are correlated.

Other types of inequality or heterogeneity are measured by state variables like trust or social cohesion—the absence of which Baland and Platteau (1995) called *cultural heterogeneity*. Generally, shared values or interpretations of social problems—cultural homogeneity—can facilitate cooperation in the use of the commons. It is conceivable that cultural homogeneity and pronounced economic heterogeneity coexist in a stable relationship. For example, highly unequal agrarian societies might sometimes exhibit widespread adherence to a hierarchical ideology that facilitates monitoring and enforcement of cooperative agreements.[5]

Cultural heterogeneity exists, then, when there is more than one community of interpretation or community of shared values among the members of a group. This can overlap with ethnic or social or locational heterogeneity, but need not. The experimental social-psychology literature reviewed by Kopelman et al. (this volume:Chapter 4) demonstrates the critical importance of the relative shares of "prosocial" and "proself" individuals in a group. This division could exist in the absence of other types of differentiation.[6] A related type of difference arises in the game-theoretic context of Falk et al. (this volume:Chapter 5) that allows for the possibility that some players have a preference for reciprocity or equity.

The sources of heterogeneity considered in this chapter do not pretend to exhaust the possibilities for differentiation. One type of heterogeneity not considered here is different uses of the resource. In the western United States, this conflict pits residential water users against agricultural ones. In agrarian economies, an ever-present, and arguably the most important, dimension of heterogeneity is gender. For water users, Meinzen-Dick and Jackson (1996) argue that differentiation by gender and by type of use overlap substantially: In many cases, women need water for cleaning and cooking, while men need it to irrigate cash crops.[7]

Why is heterogeneity important? Generally speaking, if we can discern empirical regularities that link inequality to better or worse commons outcomes, then this has consequences for asset redistribution policies like land reform, and for poverty alleviation and development programs that target communities based on the level of inequality. For example, when policy makers contemplate turning over the management of public irrigation assets to local communities, special types of assistance could be called for where irrigators are especially heterogeneous.

How does heterogeneity affect commons outcomes? The answer depends on which "commons outcomes" we mean. There are many of those: the success with which a community of resource users conserves a resource system (whether through a formal regulatory regime, or through social norms that prevail even in a community with no explicit resource-using rules); the success with which a community crafts rules for managing the commons (what Ostrom, 1990, referred to as the problem of *institutional supply*); the success with which a community monitors and enforces its regulatory regime; the success with which a community resolves conflicts and modifies the regulatory regime in response to changes in social and environmental conditions. Broadly, theoretical and case study research has tended to diverge into two camps: those studies that find a positive role for heterogeneity, and those that point out a negative role. For the moment, let us restrict attention to economic inequality—which, as we have argued, is actually quite broad and inclusive—and look more closely at the inequality-is-good and inequality-is-bad schools of thought.

That inequality may favor provision of collective goods can justifiably be called the "Olson effect." Olson (1965:34), in a classic hypothesis, explained the effect this way:

> In smaller groups marked by considerable degrees of inequality—that is, in groups of members of unequal "size" or extent of interest in the collective good—there is the greatest likelihood that a collective good will be provided; for the greater the interest in the collective good of any single member, the greater the likelihood that that member will get such a significant proportion of the total benefit from the collective good that he will gain from seeing that the good is provided, even if he has to pay all of the cost himself.

Restraint in resource exploitation and cooperation with maintenance efforts (e.g., fire prevention measures in community forests or canal cleaning in irrigation systems) are approximately public goods: One villager's actions provide benefits to most or all other members of the community. In such settings a dominant player might "internalize" a sufficiently large share of the collective good he or she provides. Thus Olson's hypothesis suggests that inequality is beneficial to successful commons management.[8]

The Olson effect makes sense if, for example, each farmer in a community-managed irrigation system is responsible for cleaning the portion of the common canal network that passes through his or her land *and* if the amount of canal passing through one's land is proportional to one's landholding size. Then an irrigator's payoff from canal cleaning is also proportional to his or her landholding wealth. This is the case in Leach's (1961) account of collective duties known as *rajakariya* or "king's work" in the preindependence Ceylonese village of Pul Eliya. Large landowners might provide canal-cleaning efforts even if no other irrigators follow suit. In such a case, the smaller, noncooperative irrigators free ride on the effort of the large player.[9]

Olson effects are also likely if large fixed costs are involved in setting up a commons management regime. These costs might be material, such as the building of fences around pasturelands, or the construction of irrigation canals. Such startup costs also involve the organizational effort to collectively mobilize a community of resource users. Vaidyanathan (1986) illustrates the historical importance of local elites in promoting the emergence of irrigation management regimes in India, China, and Japan. Ruttan and Borgerhoff Mulder's (1999) model of pastoralists illustrates the conditions under which the wealthier players choose to coerce the poorer players into conserving. Powerful elites in Vaidyanathan's history are successful in part because they centralize decision-making power as much as they command material wealth. Heterogeneity in *decision-making power*, considered by McCay (this volume:Chapter 11), is another relevant dimension of inequality.

Large startup costs of this type are an example of *nonconvexities* in the production technology.[10] Roughly speaking, benefits from collective action are a nonconvex function of the effort provided to produce those benefits if there is a threshold level of aggregate effort that must be supplied before *any* benefits are realized. As effort increases beyond the threshold, however, benefits to the group begin to increase. Irrigation, for example, provides no benefit until the expense of building a dam or a canal (or both), or drilling a tubewell, has been undertaken; but thereafter added effort systematically increases crop yields. In this setting, wealthier farmers may be able to mobilize the capital necessary to build the dam or install the tubewell. Nonconvexities also exist if there is a threshold stock of the resource (e.g., fish or pasture) below which regeneration is impossible. Baland and Platteau (1997) confirm the theoretical possibility of this Olson effect when there are such nonconvexities. Widening inequality in this setting can lead to discrete jumps in cooperative actions (e.g., maintenance effort or restraint in resource use) by the wealthier players. But they show that this result depends critically on assumptions about the characteristics of the resource-using technology.

Not everyone agrees, of course, that inequality is good for successful management of the commons. The case study literature in particular is replete with examples of the harmful effects of inequality. This is not necessarily inconsistent with Olson. It is easy to imagine irrigation systems wherein the canal length passing through one's parcel is not proportional to one's parcel size, or commons where there are no significant nonconvexities. In such cases, the Olson effect need not hold. Indeed, the (quite heterogeneous, incidentally) field work seems to speak with one voice, and that voice says that inequality is harmful. Consider a handful of Indian irrigation examples. Jayaraman's (1981) study of surface-water irrigation projects in Gujarat notes the importance of a relatively egalitarian structure to farmers' coming together to form a water users' association. Similarly, Easter and Palanisami's (1986) study of 10 tank irrigation groups in Tamil Nadu shows that the smaller the variation in farm size among farmers, the more likely that water users' associations will form.[11] Varughese and Ostrom (2001) also find a modest negative correlation between the level of wealth disparity and collective activity in forest use in 18 Nepali villages.

This ambiguous relationship between inequality and successful commons management is borne out by more recent theoretical work in economics. Dayton-Johnson and Bardhan (in press) verify the Olson effect in a model of the commons, but show, nevertheless, that the relationship between inequality and conservation is U-shaped. They assume two things. A linear harvesting technology means that a given increase in harvesting effort always leads to the same increase in resources harvested, until the resource is completely depleted: sending one more boat into the fishery always increases catch by, say, 1,000 tons as long as there are still fish in the sea. (Although a linear production technology might be considered a restrictive assumption, it permits fairly clear results. A more general assumption regarding production technology, as will be seen, complicates the results considerably.) Second, there are no formal rules constraining commons users. With these assumptions, Dayton-Johnson and Bardhan find that communities with more equally distributed wealth exhibit higher rates of resource conservation than more unequal ones. Resource harvesters with wealth below a threshold level will not conserve, regardless of what others do. Beyond that threshold, however, a resource user will conserve *conditional* on the conservation of others on the commons. If sufficiently many resource users have wealth below the threshold—a consequence of inequality—then conservation will break down.

Bardhan et al. (2000) construct a general model of collective goods with strictly concave production functions to determine the effect of wealth inequality on provision of the collective good. Their results are generally indeterminate. On the one hand, extreme inequality favors collective-good provision, as the dominant player has an incentive to provide the collective good, even if others free ride on his effort. This is one version of the "Olson effect," already described. On the other hand, with a concave technology, wealth equality promotes higher levels of collective-good provision. The net effect of inequality on collective-good provision depends on the relative magnitude of these two effects.

Bardhan et al. furthermore, show that market imperfections (for example, in land, credit, or insurance)—a pervasive feature of poor agrarian economies—complicate the Olson effect. In particular, the effect of inequality in the presence of market imperfections depends on the characteristics of the collective good in question. The authors introduce a distinction between common-pool resource products and public goods. When the positive spillovers from collective-good provision, whether the result of restraint in harvesting effort or the addition of abatement technology, for example, outweigh the negative externalities of provision, such as the classic congestion externalities of the commons, the collective good is a *public good*. When the opposite is true—the negative spillovers exceed the positive ones—the collective good is a *common-pool resource*. (When the positive externality of a good's provision exactly cancels the negative one, incidentally, it is a private good.) Bardhan et al. consider only spillovers that affect other users of the resource.

To return to the effect of inequality on collective-good provision in the presence of market imperfections: In the commons case, if the goods traded in imperfect markets (credit, agricultural land) are complementary to the commonly provided resources (water, grazing land) in households' production, then the higher the level of inequality, the greater the overall efficiency (measured by the gross rate of return on the collective good). (Water and land, for example, are said to be *complementary* if increased use of water raises land productivity, and vice versa. The definition generalizes to any two factors of production.) This is consistent with Olson effects, which predict better commons outcomes where inequality is greater. If those productive factors are not complementary, this result does not hold. Moreover, in the case of public goods (as opposed to common-pool resources), overall efficiency *falls* as inequality rises when complementary goods are transacted in imperfect markets.

Economic inequality might influence commons outcomes via differences in costs of resource harvesting. Although it is likely that if there is any difference in costs, richer commons users will enjoy lower input costs, inequality in costs is conceptually distinct from inequality in wealth or income. Aggarwal and Narayan (1999) provide a two-stage model of groundwater use that incorporates differences in cost among water users. They motivate this asymmetry with the observation that in poor agrarian economies, agents face different costs of securing credit to install extraction capacity. They demonstrate a U-shaped relationship between cost inequality and the resource stock: Starting from low levels of cost inequality, increasing inequality first reduces, then increases, water-use efficiency.[12]

Both economic and social heterogeneities may be especially salient in precluding the collective action needed to establish local institutions for managing the commons in the first place, that is, in the problem of *institutional supply*. Social heterogeneity increases the cost of negotiation and bargaining inherent in the process of crafting institutions; economic inequality, combined with other constraints, severely limits the possible bargaining outcomes available to commons users. Johnson and Libecap (1982), for example, formulate a model based on their observation of the South Texas shrimp fishery, where fishers are differentiated by their productivity. They find that both fisher-specific quotas and bilateral payments among fishers (which amount to the same thing) are impractical to administer: Presumably such schemes are too difficult to monitor and enforce. The only option, therefore, is a system of uniform quotas. The more productive the fisher, however, the larger the restriction implied by this regime. Thus more productive fishers might stand to lose under a cooperative arrangement, and hence will oppose it. (This logic is further developed in Kanbur, 1991, and Baland and Platteau, 1998.)

In a related vein, Quiggin (1993) hypothesizes that common property arises as a legal mode where there are certain scale economies in production (as in Baland and Platteau's 1997 nonconvexity case, summarized earlier). Wealthier agents will gain less from economies of scale from collectively owned assets and consequently could seek such a high share of benefits from the collective organization that the group fails to form.[13] In both the Johnson-Libecap and Quiggin stories, the feasible set of institutional arrangements is restricted in some way; under certain configurations of parameters, this restriction makes it impossible to craft an arrangement that satisfies both rich and poor commons users. Even though especially resourceful harvesters might successfully overcome this obstacle, in general, inequality makes it more difficult to self-organize.

An important complication is the presence of *exit options*. If resource users have relatively lucrative earnings opportunities outside the commons, this can affect their individual incentives, as well as the power of social cohesion to promote cooperative behavior. In Dayton-Johnson and Bardhan's (in press) model, the harvester can stay and conserve, stay and degrade, or degrade and then leave. They demonstrate that the effect of these exit options on conservation is predictably complicated, but depends in part on whether the relationship between wealth and exit options is concave. If a resource harvester's exit option is a concave function of wealth, then the value of the exit option increases as the harvester's wealth level rises, but, by definition, at a decreasing rate.[14] If a resource harvester's wealth were to double, his or her exit-option value might rise considerably; if that person's wealth were to double *again*, the exit-option value would increase by less than it did with the first doubling. If the exit-option function is concave, then, as Dayton-Johnson and Bardhan show, increases in inequality, starting from relatively equal wealth distributions, will reduce conservation. In that case, the relatively poorer harvester will optimally choose not to conserve: As his or her wealth declines, the gain from conserving (which is a linear function of wealth) falls off more rapidly than the gain from exercising his or her exit option (which is a concave function of wealth). The poor harvester thus derives a higher return from staying on the commons and degrading it than he or she does from exiting. If on the other hand the exit option is a convex function of the wealth level (which would be true, for example, if villagers faced borrowing constraints in credit markets), then increased inequality might either enhance or damage the prospects for conservation: The effect is indeterminate.

Conservation in the Dayton-Johnson/Bardhan model occurs in a completely noncooperative setting, that is, where no management regime necessarily exists. In the presence of a management regime, exit might help resource conservation, if the outmigrants are those most likely to discount the future substantially. Nevertheless, exit might hamper collective action in this situation. For example, if the amount of labor available to maintain irrigation infrastructure falls, the infrastructure's condition could decay if the community is not wealthy enough to hire guards and workers. Furthermore, in the presence of exit options, communities may have fewer mechanisms to enforce cooperation in a footloose population. An open question is whether conservation is reduced appreciably if the poor do not conserve. The poor harvester may have such a small effect on the resource that his or her lack of adherence to the rules has a negligible effect, as with the pastoralists studied by Ruttan and Borgerhoff Mulder (1999). Baland and Platteau (1999), similarly, argue that with inequality, increasing participation by the wealthy can compensate for the poor's lack of participation; the net effect on conservation depends on the extraction technology involved.

There is empirical evidence that exit options weaken the prospects for cooperation. Baland and Platteau (1996) illustrate this phenomenon with reference to conflicts between artisanal and industrial fishers in fisheries around the world. The former group is tied by their technology to a very circumscribed fishing ground, while the latter are highly mobile. In Mali and Mauritania, large (usually absentee) livestock herd owners have been much less interested than small herders in local arrangements for rangeland management to prevent overgrazing and desertification (Shanmugaratnam et al., 1992). Freudenberger (1991) describes the deforestation of a forest ecosystem in Senegal by the local unit of a nationwide agricultural entity known as the Mouride. A relatively low-intensity pattern of resource use by nearby peasant producers and pastoralists gave way to intensive cash-crop (groundnut) production. After the soil's rapid exhaustion by groundnut farming, the Mouride's national decision-making body could open up new territory elsewhere, unlike traditional users who were more interested in the long-term viability of the local forest. Shanmugaratnam (1996) notes that after the privatization of some village grazing areas in Western Rajasthan, large landowners, now able to produce a large part of their fodder needs on their private land or to buy supplementary fodder in the market, tend to be uninterested in the sustainable management of the remaining commons.

In many of the cases cited, the richer or larger commons users were prone to defect. This need not always be the case. Other authors have reported that the poorer or smaller users exercise exit options. Bergeret and Ribot (1990), in a study similar to that of Freudenberger, describe deforestation in a larger area and over a longer time frame, also in the Senegalese Sahel. Trees are harvested by Fulani refugees from Guinea, who are more likely to be landless than other peasants, in order to produce charcoal for the rapidly growing urban market. A qualitatively similar situation has been described in southern Burkina Faso, where immigrants are more prone to use destructive gathering techniques in communal forests (Laurent et al., 1994). (Because the poor in these examples are also immigrants, it may be more accurate to infer that they have already exercised their exit options from a previous place. Because of their lack of connection to their new locales, they pursue environmentally unsound practices.)

How do noneconomic sources of heterogeneity affect economic outcomes? In a completely different public goods domain, Alesina et al. (1999) find that ethnic diversity is associated with lower public goods funding across U.S. municipalities because different ethnic groups have different preferences over the type of public good (such as language of school instruction). In the kind of rural societies considered in this chapter, ethnic heterogeneity works through social norms and sanctions. The effectiveness of social sanctions weakens as they cross ethnic reference groups. In this vein, Miguel (2000) constructs a theoretical model where the defining characteristics of ethnic groups are the ability to impose social sanctions within the community against deviant individuals and the ability to coordinate on efficient equilibria in settings of multiple equilibria. With data from the activities of primary school committees in rural western Kenya, Miguel then shows that higher levels of ethnic diversity are associated with significantly lower parent participation in parent meetings, worse attendance at school committee meetings, and sharply lower teacher attendance and motivation.[15]

If social groups (not solely ethnic groups) are defined as those whose boundaries coincide with the effective monitoring and enforcement of shared social norms, then this provides a workable concept of social heterogeneity for irrigation communities. Indeed, this is one way of understanding the notion cited earlier of cultural homogeneity, a variant of what many authors have called social capital or social cohesion.[16] Irrigation systems whose boundaries obey hydrological rather than social boundaries will comprise irrigators from many villages. Irrigation organizations that cross village boundaries can rely less on social sanctions and norms to enforce cooperative behavior than those that comprise a single village.

## LARGE-*N* STUDIES

The foregoing discussion demonstrates the richness of the theoretical and empirical literature on the commons. A careful reading of the empirical literature demonstrates that case studies (whether by anthropologists, political scientists, sociologists, engineers, or the odd economist) prevail. Larger scale surveys of several resource-using systems that would permit statistical analysis of the empirical regularities present on the commons are still relatively rare. Ostrom et al. (1994) attempt to remedy this shortage by systematically combining the results of the voluminous case study literature on irrigation systems, community forests, and fisheries. Although useful, such "meta-evaluations" are not substitutes for survey research of large groups of resource-using communities. Even careful compilation of case studies cannot address biases in the selection of studied systems. In this section we synthesize results from a handful of studies of farmer-managed irrigation systems that seek to fill this gap.[17] All of the studies mentioned herein share the objective of establishing empirical regularities among structural characteristics, institutions of governance, and various measures of performance; we will focus our attention on one particular characteristic, namely heterogeneity, and its links to institutions and performance. The principal studies considered in this section are Lam's (1998) Nepali study, Dayton-Johnson's (1999, 2000a, 2000b) Mexican study, Bardhan's (2000) Indian study, and Tang's (1991, 1992, 1994) meta-evaluation of the case study literature. We will also have occasion to refer to Fujita et al.'s (2000) study of 46 surface-water systems recently transferred to their users by the Philippine public irrigation authorities. Theirs is clearly in the vein of the other studies considered here, but its usefulness for our chapter is limited because it does not explicitly consider the effects of economic inequality on irrigation performance. Khwaja's (2000) study of infrastructure investment in Pakistan includes information on inequality, although the research is not limited to irrigation projects.[18]

Lam (1998) analyzes data from a data set covering 127 irrigation systems throughout Nepal, as well as nearly 25 further systems surveyed by the author. Unlike the Mexican and Indian studies, the sample of irrigation communities is not randomly drawn; nevertheless, the data set is regionally representative within Nepal. For the basic 127-system data set, 104 systems were farmer managed, while the remaining systems were managed by the public Department of Irrigation. The mean system area was 399 hectares, and the mean number of appropriators was 585, implying that the average irrigated area per appropriator was just over two-thirds of a hectare. Bardhan (2000) analyzes data from a survey of 48 irrigation units known as *ayacuts*, each in a different village within six districts in the south Indian state of Tamil Nadu. Half of the *ayacuts* were members of larger canal systems, and half were members of more traditional tank systems. All were ostensibly under the control of the government, but most *ayacuts* had traditional and informal community management regimes. The average number of households per irrigation source was 53; the *ayacut* area per household was just slightly more than a third of a hectare.

Dayton-Johnson (1999) describes a field study of 54 farmer-managed irrigation systems known as *unidades de riego* in the central Mexican state of Guanajuato. All of these *unidades* were autonomous from state control, and all were based on surface-water irrigation derived from reservoirs. The average number of irrigating households in the systems was 123; the average command area was 449 hectares; and the average land holding per household was 3.3 hectares. Tang (1991, 1992, 1994) has applied the meta-evaluation approach to the inventory of case studies of irrigation. He aggregates information from 47 irrigation systems in 15 countries. Twenty-nine are farmer managed, 14 are agency managed, and 4 are other types of systems. Khwaja's (2000) study of a variety of infrastructural projects in the Himalayas of northern Pakistan considers various forms of inequality with great care. His 123 externally funded projects include investment in irrigation, but also in roads and other forms of infrastructure.

These studies define certain commons outcomes, allowing estimation of the impact of heterogeneity. Broadly, there are two types of outcomes: institutions and performance. Both may vary systematically with the degree of heterogeneity. With respect to institutions, Tang (1994:231) describes two types of "rules-in-use": *boundary* rules ("the requirements one must fulfill before appropriating water") and *authority* rules (the procedure and basis for withdrawal, including fixed shares or rotating turns). Bardhan (2000) and Dayton-Johnson (2000a) consider cost-sharing rules for mobilizing canal-cleaning and maintenance efforts, as well as water allocation rules that define households' claims on irrigation resources.

Performance is measured in various ways. An obvious dimension is the degree to which irrigators adhere to the rules established above: *rule conformance*. Bardhan (2000) measures whether water allocation rules are frequently violated by one group; Tang (1994) codes more generally whether the irrigation rules are followed. Alternatively, one could measure not rule conformance per se, but rather the level of infrastructure maintenance. Bardhan (2000) uses a categorical-variable index of maintenance of distributaries and field channels. Dayton-Johnson (2000b) uses disaggregated variables and estimates statistical models of three dimensions of maintenance: the degree of definition of canal side slopes, state of repair of field intakes, and degree of control of leakage around the canals. Lam (1998) uses the overall physical condition of the system. Another class of performance variables measures the adequacy of water delivery; Lam aggregates information on adequacy of water delivery at various points in the system, equity among users, and reliability of water supply at the tail end. An imperfect indicator of the success of irrigation is crop yields, considered by Dayton-Johnson (1999) and by Lam (1998), who aggregates information on output per hectare, and cropping intensity at the head and tail ends of the system. Lam subjects the three dimensions of his performance measure (condition, water delivery, and productivity) to confirmatory factor analysis, finding, among other things, that the dimensions are not highly correlated. Fujita et al. (2000) propose a four-dimensional concept of irrigation system performance, based on the existence of rules for maintenance, coordination in rice-cropping schedules, practice of water rotation, and organized monitoring of rules. They perform a principal components analysis on the four measures to derive appropriate weights for an index of performance. Finally, Bardhan (2000) also considers the absence of water-related conflicts as a measure of performance.

All of the studies considered here are multivariate analyses, but the list of independent variables differs considerably from study to study. To some extent, then, we are comparing estimated coefficients that are not strictly comparable: This should be borne in mind when considering the results reviewed in the following paragraphs.

*Income inequality.* What then, are the effects of heterogeneity? Consider first inequality in incomes. Tang (1991) finds that "a low variance of the average annual family income among irrigators tends to be associated with a high degree of rule conformance and good maintenance." Tang (1992:72-73) identifies 27 cases where the degree of income variance can be gleaned from published research. He finds that in systems where income variance is high, 17 percent exhibit high levels of rule conformance and maintenance; where income variance is moderate, 75 percent exhibit high levels of these performance measures; and where variance is low, 89 percent exhibit high rates of performance. (Tang cautions against inferring too much from these results, because the degree of income variance could not be identified in a significant fraction of the case studies he compiled.) Lam's (1998) regression analysis shows that income inequality (measured by a zero-one variable indicating either "low/medium" or "high" variance in average annual family income) is significantly and negatively related to water delivery performance. Income inequality is also significantly and negatively related to productivity in the Nepali systems, but it is not significantly related to physical condition of the system. Khwaja (2000) finds a U-shaped relationship between inequality in one form of income—project returns—and project maintenance in his Pakistan study. Starting from a low level of inequality in project returns among beneficiaries, increased inequality tends to reduce project maintenance. Beyond a certain level of project-return inequality, however, maintenance improves as inequality widens.

*Wealth inequality.* Wealth inequality is likely quite highly correlated with income inequality, and its effects are similar here. Bardhan (2000) and Dayton-Johnson (2000a, 2000b) compute the Gini coefficient based on irrigated land holding for their Indian and Mexican studies. The Gini coefficient is related to performance: The relationship, where it is significant, is negative in the Indian study. Bardhan finds that landholding inequality is significantly and negatively associated with canal maintenance in the Tamil Nadu systems. For Bardhan's indicator of intravillage conflict over water, he finds evidence of a U-shaped relationship between the Gini coefficient and this indicator of performance.[19] That is, at low and high levels of inequality, there is little intravillage conflict, but for inequality in the middle range, conflicts are more likely. Bardhan finds no statistically significant effect of inequality on rule conformance. For the Mexican study, the full effect of landholding inequality on maintenance (accounting for the indirect effect on the choice of rules) is negative, but complicated.[20] Khwaja (2000) once again finds a U-shaped relationship between landholding inequality and project maintenance in his Pakistan study: Starting at perfect equality, increasing inequality reduces maintenance, while at high inequality levels, increasing inequality raises maintenance.

*Head-enders and tail-enders.* Another source of inequality is the asymmetry between those at the head and tail ends of the canal network. As noted, this is probably only imperfectly correlated with inequalities in landholding wealth given that land markets do not function very well. Tang (1992:60-63,73-74) considers the impact on rule conformance and maintenance of the presence of "disadvantaged groups." In most cases, this refers to tail-enders, although in a few instances it refers to groups against which system rules systematically discriminate. In a simple bivariate comparison, most systems without disadvantaged groups exhibited high rule conformance and maintenance, while fewer than a third of those with disadvantaged groups exhibited these high performance levels.

One predictor of conflicts between irrigators at the extremities of the network is the presence of modern headworks to divert water from its source. Lam (1998) finds that the presence of modern headworks was negatively and significantly associated with *all* of his indicators of performance. A strict engineering view would predict that modern headworks would improve performance: Lam interprets his results as confirming that headworks increase the bargaining power of those at the head end of the network. For Lam, water delivery and cropping intensity at the tail end are dependent variables. The Philippine study by Fujita et al. (2000), however, considers disparities in water availability between the head and tail ends as independent variables; they can do this because they measure availability before the systems were transferred from the government to their users, and they measure performance after transfer. They find that disparities wherein head-enders have relatively abundant availability and tail-enders relatively scarce availability are significantly associated with poorer performance.[21]

***Exit options.*** Another dimension of economic inequality already mentioned is differential earnings opportunities not fundamentally tied to the commons. Exit options can be empirically detected in several ways. Bardhan (2000) includes an indicator of linkage (e.g., by bus or telephone) to urban centers in the south Indian study. This linkage variable is negatively and significantly related to system maintenance, suggesting that the proximity to the city makes it harder to enforce rules for cleaning canals and the like. This is verified in Bardhan's statistical model of rule conformance, where the linkage effect is negative. Linkage is similarly positively and significantly related to the presence of intravillage water conflicts. Note that this does not reveal the degree to which exit options are distributed unequally, however. The Indian results merely verify that cooperative behavior is more difficult to sustain in the presence of outside opportunities. The Philippine study by Fujita et al. (2000) attempt to measure asymmetry by the ratio of nonfarm to farm households within the territory of the irrigation community. (Nonfarm employment is the exit option in this case.) A higher proportion of nonfarm households indicates asymmetry in the way people gain their livelihoods, an asymmetry that might weaken the enforcement power of informal sanctions within the farming subset. The nonfarm household ratio is significantly and negatively associated with their measure of performance. (In a separate series of regressions where the components of their performance index are separately estimated as dependent variables, the nonfarm household ratio negatively affects the probability of observing a water rotation scheme, and the probability of observing organized monitoring.) Bardhan finds in the Indian study some (weak) evidence that when farmers have access to alternative sources of irrigation, they tend to violate the water rules more frequently: This is an exit option outside the commons, but not outside of agriculture.

***Ethnic and social heterogeneity.*** The boundary between economic and non-economic heterogeneity is a fuzzy one. Differential exit options appear to be economic because they are like unequal assets; nevertheless, the effect of different exit options, we suspect, operates through the weakening of social norms and sanctions. Less explicitly economic forms of heterogeneity have significant effects in the studies considered here. Bardhan (2000) controls for whether at least three-quarters of surveyed farmers in an *ayacut* are members of the same caste. This kind of caste *homogeneity* is strongly associated with the absence of intravillage conflict, but it is not significantly associated with rule conformance. (Bardhan did not include caste homogeneity in his statistical models of maintenance.) Khwaja (2000) computes a "fragmentation index" that is the average of ethnic, political, and religious fragmentation indices for the communities in his Pakistan study. Each fragmentation index is the probability, in a given community, that two randomly selected individuals belong to different groups. Khwaja's fragmentation index bears a negative relationship with project maintenance.

A measure of social heterogeneity is whether or not irrigators come from more than one community. Dayton-Johnson (2000b) includes in his statistical models the number of *ejidos* (Mexican agrarian reform communities) from which *unidad* members are drawn: It is consistently and negatively associated with infrastructure maintenance. This provides strong support that when enforcement crosses *ejido* boundaries, it is less effective. Nevertheless, the Philippine study of Fujita et al. (2000) includes a similar variable—the number of villages represented in the irrigation system—and it is not significant. Baker's (1997) study of 39 *kuhl* irrigation systems in Himachal Pradesh considers the effect of *differentiation*, which is "high when a kuhl irrigates more than one village, the irrigators of the kuhl are composed of multiple castes, and land distribution is relatively unequal" (1997:204). Baker proposes that, in the presence of high differentiation, increased opportunities for nonfarm employment can place intolerable stress on traditional *kuhl* management regimes.[22] Tang's (1992:68-72) evidence on the effect of "social and cultural divisions" is ambiguous and marked by small numbers of observations. The effect is mediated by the institutional nature of the irrigation system. Where it is community managed, sociocultural heterogeneity does not preclude good performance; where the system is agency managed, this heterogeneity is associated uniformly with poor performance. Possibly overlooked is the selection bias in community-managed systems: If a group is not able to organize itself, or if conflicts become too severe, the system will not exist when the researcher goes into the field. The same cannot be said of agency-managed systems. Thus, we are left with a sample of community-managed systems that have survived a process that requires high levels of cooperation.

***Choosing rules.*** Heterogeneities, finally, can affect performance indirectly by means of their effect on the choice of rules. Dayton-Johnson (2000a) finds that wealth inequality increases the probability of observing proportional water allocation (as opposed to equal shares for all farmers). This is consistent with richer landowners pressing for proportionally more of the irrigation supply. The Mexican study also finds that proportional water allocation is associated with poorer maintenance. Inequality thus may lead to a particular set of rules under which irrigation systems do not perform as well.[23] Bardhan's (2000) south Indian study also provides significant evidence that when the water allocation rules are crafted by the village elite, the latter violate the rules less frequently; otherwise the elite violate the rules more frequently. (Overall, the elite break the rules more often than the nonelite, but by definition, the former is a smaller group than the latter.) It is also observed that when an average farmer believes the water rules have been crafted jointly (i.e., with collective participation, as opposed to rule crafting only by the village elite, or by government), he is more likely to have positive comments about the water allocation system and about rule compliance by other farmers. Joint rule crafting of this kind is associated with the highest level of rule compliance in the Indian case. Bardhan also estimates the likelihood that villages have adopted proportional cost-sharing rules—that is, rules that specify that the labor costs of maintaining irrigation infrastructure are shared proportionally to (irrigated and nonirrigated) landholding wealth.[24] This rule is in general positively associated with cooperative outcomes; adoption of this rule is, in turn, significantly and positively associated with landholding inequality. This might be an indication of social pressure for a redistributive adjustment of the cost-sharing rule to take account of wealth disparities. This points to an important and more general observation noted by Varughese and Ostrom (2001:762). They find that many groups "overcome stressful heterogeneities by crafting innovative institutional arrangements well-matched to their local circumstances." In their Nepal study, forest users created diverse forms of memberships with different rights and duties to cope with heterogeneity, particularly when there are substantial benefits to be obtained through collective action.

## CONCLUSIONS

Table 3-1 summarizes the evidence from the large-$n$ studies reviewed in the previous section. The evidence is tentative, but sufficient to permit us to hazard a few conclusions. First, there is a confirmation of the case study literature's conclusion that heterogeneity, however it might be measured, has a negative impact on cooperation on the commons in these irrigation cases. Heterogeneity tends to have a discernable negative effect, or no effect at all. Second, the evidence is consistent with the hypothesis that heterogeneity weakens the effect of social norms and sanctions to enforce cooperative behavior and collective agreements. Support for this conclusion derives from the negative effect on performance in multivillage, multicaste irrigation systems. Third, however, controlling for this social heterogeneity, there is an independent, and largely negative, effect of economic heterogeneity per se: This is borne out by the significant effect of Gini coefficients, for example, in the Mexican and Indian studies. This conclusion underscores the importance of economic mechanisms in the theoretical literature, starting with Olson (1965). These economic mechanisms are based on the differential incentives to cooperate created by the distribution of wealth or income, distinct from social norms. Moreover, although economic theory cannot predict whether "Olson effects"—a positive impact of inequality—will predominate, the empirical evidence for irrigators is that they do not. This finding also underscores the value of the multivariate analysis approach adopted in the studies summarized here: Such an approach allows one to isolate the effect of particular structural characteristics (like wealth inequality) while controlling for the effect of others (like social heterogeneity). Fourth, and finally, there is evidence that heterogeneity affects system performance both directly and indirectly via its effect on the institutions adopted by an irrigating community. Inequality might affect the degree to which irrigators follow the rules, but it also affects the type of rules chosen, and not all rules are equally conducive to good performance. Quantifying the magnitude of these direct and indirect effects requires the adoption of the multivariate approach used in these studies.

A question raised by the studies summarized here is the degree to which these results based on irrigation systems can be generalized to other types of commons. Blomquist et al. (1994) consider a typology of common-pool resources that situates irrigation with respect to other types of commons. Two physical dimensions that matter are *stationarity* of the resource ("the resource units... remain spatially confined prior to harvest" [1994:308]) and the possibility of *storage*. This two-way typology generates four classes of physical resources, and irrigation systems are found in three of the four categories: groundwater-based systems are stationary and storage is available; reservoir-based canal systems are nonstationary but storage is possible; river-diversion systems are nonstationary and storage is impossible. Many of the systems considered in this chapter are canal-based systems, with a minority of run-of-the-river systems. (In Bardhan's 2000 South Indian study, half of the systems are not served by canals.) Most other commons do not share this combination of characteristics: Forests, rangelands, and community threshing grounds are stationary without recourse to storage. Migratory species are nonstationary with no possibility for storage. In the presence of distinct structural characteristics, further analysis is necessary before extending these irrigation-based findings to other settings.

One distinction between a flowing resource (like water) and a standing resource (like forest) is that gravity makes locational (head-tail) heterogeneity more salient in the former (although if land markets work, one presumes that in the long run locational advantage gets capitalized into wealth inequality). Another difference is that in forests, part of the collective action is in replanting and regeneration efforts, which are less important in canal irrigation. Third, issues of intertemporal conservation are somewhat less salient in canal irrigation, although

TABLE 3-1 Heterogeneity and Commons Outcomes: Summary of Empirical Studies

| | South India[a] | Central Mexico[b] | Philippines[c] | North Pakistan[d] | Nepal[e] | Meta-Analysis[f] |
|---|---|---|---|---|---|---|
| Income inequality | | | | U-shaped association between project-income inequality and project maintenance | High variance of family incomes associated with lower water delivery performance and lower productivity | Low variance of family incomes associated with better rule conformance and maintenance |
| Wealth inequality | Higher landholding inequality associated with lower canal maintenance; U-shaped association between landholding inequality and intravillage conflict | Landholding inequality positively associated with maintenance | | U-shaped association between landholding inequality and project maintenance | | |
| Head-ender/ tail-ender | | | Head-end/tail-end disparities in water availability associated with lower levels of performance index | | Presence of modern headworks negatively associated with water delivery and cropping intensity at tail end | Presence of "disadvantaged groups" associated with lower rule conformance and maintenance |

| | | | | | |
|---|---|---|---|---|---|
| Exit options | Linkage to urban centers associated with poorer maintenance and rule conformance; alternative irrigation availability associated with poorer rule conformance | | Higher proportion of nonfarm households associated with lower levels of performance index | | |
| Ethnic/social heterogeneity | Caste homogeneity associated with lower incidence of intravillage conflict | Higher number of ejidos represented among system members associated with lower maintenance | No effect of number of villages represented in system | Ethnic/religious/ political "fragmentation" associated with poorer maintenance | Ambiguous effect of "social, cultural divisions" |
| Rule choice | Jointly chosen rules more frequently followed; inequality positively associated with proportional cost sharing | Wealth inequality positively associated with proportional water allocation | | | |

SOURCES:
[a]Bardhan, 2000.
[b]Dayton-Johnson, 1999, 2000a, 2000b.
[c]Fujita et al., 2000.
[d]Khawja, 2000.
[e]Lam, 1998.
[f]Tang, 1991, 1992, 1994.

they are important in groundwater irrigation. This refers not only to dynamic conservation of the resource, but intertemporal externalities of harvesters' behavior, whereby my extraction this period affects your payoff next period.[25]

One can reasonably question whether surface-water irrigation systems are common-pool resources at all. Groundwater-based irrigation, of course, draws from a resource (an aquifer) that is subject to regeneration as well as the risk of depletion, as is the case with pastures, forests, and fish. To the extent that canal systems based on reservoirs or river diversions bear a formal resemblance to such common pools, it lies in the collectively maintained infrastructure: water source, canals, and water-control devices. Cleaning canals and repairing equipment is formally similar to replanting, recharge, and regeneration. Moreover, one person's use of irrigation water reduces the availability for others, just as in other types of common-pool resource systems. Nevertheless, human intervention in natural systems is arguably much more invasive in irrigation than in, say, fishing or fuelwood collection. (These and other aspects of "irrigation exceptionalism" are considered in passing in Rose, this volume:Chapter 7.)

To a large extent, of course, the problems of successful commons management are not necessarily based on the characteristics of the natural resource itself—as the earlier, tragedy-of-the-commons tradition would have it—but rather the more prosaic problem of getting people to cooperate. Thus the problem is particularly closely related to those of producer and worker cooperatives. Mobilizing cooperative effort is especially problematic at the level of institutional supply, but also in the running of the institution.

Another social—rather than natural—phenomenon deserving increased attention is the effect of market failure. Market failure is said to exist when the market for a good or service fails to be efficient or, in the extreme, fails to exist at all. Such market failures in credit, insurance, and land are endemic in agrarian economies, and interact with the problem of cooperation. Optimal regulatory regimes are not difficult to describe in theory, but real-world market failures constrain the set of feasible arrangements. These constraints may be such that commons users are unable to negotiate any kind of cooperation whatsoever; or, they reach an accord that nevertheless leads to environmental degradation. Another such market failure that is frequently invoked but not quite justified is the impossibility of side payments—or equivalently, the absence of secondary markets for the common-pool resource.

Our empirical reference point limits the generalizability of our results to global-scale commons, where actors are states and international agencies rather than peasant households. Our conclusions regarding the effect of heterogeneity could be implausible in the setting of international climate-change agreements, for example; there, the presence of disproportionately powerful actors might enhance the prospects for cooperation.[26]

This survey illustrates the utility of the large-scale multivariate analysis of resource-using communities. Similar syntheses could be compiled for other types of commons (fisheries, forests, rangelands), and for other structural characteristics of communities (group size).[27] Despite the impressive advances of research chronicled in this volume, the lacunae in our knowledge are still great, as is the potential contribution to commons users' welfare of policy makers' judicious application of that knowledge.

## NOTES

1  Ours is not the first attempt to survey this literature; see also Baland and Platteau (1999).

2  The effects of many other group characteristics are considered systematically in Agrawal (this volume:Chapter 2).

3  Ostrom and Gardner (1993) recount the experience of an irrigation system in Nepal, where the richer farmers are located at the tail-end; this system is better maintained than those where the tail-enders are poorer.

4  Bardhan (1997) analyzes the economic aspects of ethnic conflict.

5  Fafchamps (1992) explores the emergence of such patron-client relationships in agrarian economies using the theory of repeated games.

6  See Cárdenas (1999) and Henrich et al. (2000) for field-based laboratory experiments among actual commons users.

7  See also Zwarteveen (1997) for a discussion of gender in the context of irrigation management transfer.

8  This result is generalized in a pure public goods model by Bergstrom et al. (1986). (Parenthetically, Bergstrom et al. sought to dispel the earlier conventional wisdom in economics, namely, that changes in the distribution of wealth would *not* affect the overall level of public goods provided in society.)

9  Alternatively, the *rajakariya* case might be an example of the proportionality between costs and benefits that Ostrom (1990) claims is exhibited by successful commons management regimes. In one variant of this story, proportionality should neutralize the effects, good or bad, of inequality. Dayton-Johnson (2000a) provides a simple game-theoretic illustration of the proportionality principle.

10  The "lumpiness" in third-party monitoring to which Agrawal (this volume:Chapter 2) refers is another example of nonconvexity.

11  See Bardhan (1995) for further examples from the case study literature on farmer-managed irrigation systems in Asia.

12  Baland and Platteau (in press) note that the effect of increased wealth inequality on inequality in costs is likely to be hard to predict. On the issue of cost inequalities on the commons, see also the paper by Hackett (1992).

13  Parenthetically, these mechanisms might be viewed as variants of the macroeconomic redistributive-pressure mechanism modeled by Persson and Tabellini (1994). Inequality in their model leads to pressure from below to redistribute income; this in turn leads to a tax on capital that lowers investment and growth. In the arena of institutional supply on the commons, in contrast, the wealthy require more of the notional gains from cooperation than the poor are willing to accept and commons management regimes fail to emerge.

14  Graphically, the exit-option function (measured on the vertical axis) rises sharply from the origin, but gradually levels off as wealth increases (on the horizontal axis).

15  Contrary to Miguel's definition, ethnicity might be conceived alternatively as an identity associated with shared norms, but not necessarily with sanctioning. As such, ethnic homogeneity or heterogeneity would affect commons management more through shared understandings than through sanctioning behavior.

16   Dayton-Johnson (2001) constructs a game-theoretic model of social cohesion, which he differentiates from "community": The latter is based on shared values and shared interpretation of social reality, a stronger condition than social cohesion. These concepts must be viewed in the present context as exogenous in the first instance. As Agrawal (1999:103) notes, "The aspect of community that stands for shared understanding is precisely what external interventions can do very little about. States, NGOs [nongovernmental organizations], bureaucratic authorities, aid agencies, and policy makers cannot directly create community-as-shared-understanding."

17   Recent empirical research on producer cooperatives in developing countries can be interpreted as part of the same research agenda. See the recent studies by Banerjee et al. (2001) and Seabright (1997). Similarly, empirical studies of people's propensity to join voluntary organizations in Paraguay (Molinas, 1998) and rural Tanzania (La Ferrara, 1999) demonstrate a negative effect of economic inequality.

18   A much earlier quantitative study of irrigation systems, not considered in this chapter, was carried out by de los Reyes (1980).

19   The estimated coefficient on the Gini variable is negative and significant, while the estimated coefficient on the square of the Gini is positive and significant.

20   The estimated coefficient on the square of the Gini term was *positive* and significant in two of the three models in Dayton-Johnson (2000b), suggesting a strongly positive effect of inequality on maintenance. Nevertheless, inequality was significantly related to proportional water allocation; that rule, in turn, was associated with *lower* levels of maintenance. The full effect, direct plus indirect, of inequality was negative.

21   Bardhan (2000) includes a variable indicating whether an *ayacut* is located at the tail end of a larger system; an entire village of tail-enders, however, does not behave differently from other villages, all else equal.

22   Baker's argument is more nuanced than that stated here: He claims that the effect of exit options is mediated not only by differentiation, but by *reliance* on the water source. Where reliance is high and differentiation is low, management regimes can withstand increased exit options. Nevertheless, our reading of his argument is that where differentiation is high, regardless of the level of reliance, the stress on the institutions of governance is critical.

23   In the Mexican study, landholding inequality is essentially not endogenous because the distribution of landholding was frozen by the agrarian reform. Otherwise, it would be more difficult to determine the indirect effect of inequality on performance via the choice of rules.

24   To recapitulate, landholding inequality is associated with proportional *cost sharing* in the Indian study, and with proportional *water allocation* in the Mexican study. Proportional cost sharing is, in turn, associated with better performance in India, while proportional water allocation is associated with *poorer* maintenance in Mexico.

25   This is the principal spillover exploited in the model of Dayton-Johnson and Bardhan (in press).

26   These problems are considered in Young (this volume:Chapter 8).

27   As an example of this kind of research beyond the realm of unequal irrigators, Agrawal and Goyal (1999) analyze the question of group size based on data from 28 forest councils from the Indian Himalaya. Their appealing result is that there is a U-shaped relationship between group size and effective monitoring, rather than the classic monotonic result.

# REFERENCES

Aggarwal, R., and T. Narayan
   1999   Does Inequality Lead to Greater Efficiency in The Use of Local Commons? The Role of Sunk Costs and Strategic Investments. Unpublished paper, Department of Agricultural and Resource Economics, University of Maryland, College Park.

Agrawal, A.
  1999   Community-in-conservation: Tracing the outlines of an enchanting concept. Pp. 92-108 in
         *A new moral economy for India's forests? Discourses of community and participation*, R.
         Jeffery and N. Sundar, eds. New Delhi: Sage Publications.
Agrawal, A., and S. Goyal
  1999   Group Size and Collective Action: Third-Party Monitoring in Common-Pool Resources.
         Leitner Working Paper No. 1999-09. New Haven, CT: The Leitner Program in Interna-
         tional Political Economy, Yale University.
Alesina, A., R. Baqir, and W. Easterly
  1999   Public goods and ethnic divisions. *The Quarterly Journal of Economics* 114:1243-1284.
Baker, J.M.
  1997   Common property resource theory and the *kuhl* irrigation systems of Himachal Pradesh,
         India. *Human Organization* 56:199-208.
Baland, J.M., and J.P. Platteau
  1995   Does Heterogeneity Hinder Collective Action? Cahiers de la Faculté des sciences écono-
         miques et sociales de Namur, Serie Recherche No. 146, Collection "Développement."
  1996   *Halting Degradation of Natural Resources: Is There a Role for Rural Communities?* Ox-
         ford, Eng.: Oxford University Press.
  1997   Wealth inequality and efficiency in the commons, i: The unregulated case. *Oxford Eco-
         nomic Papers* 49:451-482.
  1998   Wealth inequality and efficiency in the commons, ii: The regulated case. *Oxford Economic
         Papers* 50:1-22.
  1999   The ambiguous impact of inequality on local resource management. *World Development*
         27:773-788.
  in     Institutions and the efficient management of environmental resources. In *Handbook of
  press  Environmental Economics*, K.G. Mäler and J. Vincent, eds. Amsterdam: Elsevier.
Banerjee, A., D. Mookherjee, K. Munshi, and D. Ray
  2001   Inequality, control rights, and rent seeking: Sugar cooperatives in Maharashtra. *Journal of
         Political Economy* 109:138-190.
Bardhan, P.K.
  1984   *Land, Labor and Rural Poverty: Essays in Development Economics.* New York: Columbia
         University Press.
  1995   Rational fools and cooperation in a poor hydraulic economy. In *Choice, Welfare, And
         Development: A Festschrift in Honour of Amartya K. Sen*, K. Basu, P. Pattanaik, and K.
         Suzumura, eds. Oxford, Eng.: Clarendon Press.
  1997   Method in the madness? A political-economy analysis of the ethnic conflicts in less devel-
         oped countries. *World Development* 25:1381-1398.
  2000   Irrigation and cooperation: An empirical analysis of 48 irrigation communities in South
         India. *Economic Development and Cultural Change* 48:847-865.
Bardhan, P.K., M. Ghatak, and A. Karaivanov
  2000   Inequality, Market Imperfections, and Collective Action Problems. Unpublished paper,
         Department of Economics, University of California, Berkeley, and Department of Eco-
         nomics, University of Chicago.
Bergeret, A., and J.C. Ribot
  1990   *L'arbre Nourricier en Pays Sahelien.* Paris: Éditions de la Maison des Sciences de
         L'homme.
Bergstrom, T.C., L. Blume, and H. Varian
  1986   On the private provision of public goods. *Journal of Public Economics* 29:25-49.

Berkes, F.
    1986    Marine inshore fishery management in Turkey. Pp. 63-83 in National Research Council,
            *Proceedings of the Conference on Common Property Resource Management.* Washing-
            ton, DC: National Academy Press.
Blomquist, W., E. Schlager, and S.Y. Tang
    1994    Regularities from the field and possible explanations. Pp. 301-316 in *Rules, Games, And
            Common-Pool Resources*, E. Ostrom, R. Gardner, and J. Walker, eds. Ann Arbor: Univer-
            sity of Michigan Press.
Cárdenas, J.C.
    1999    Real Wealth and Experimental Cooperation: Evidence from Field Experiments. Unpub-
            lished paper, Department of Environmental and Resource Economics, University of Mas-
            sachusetts, Amherst.
Dayton-Johnson, J.
    1999    Irrigation organization in Mexican *unidades de riego*: Results of a field study. *Irrigation
            and Drainage Systems* 13:55-74.
    2000a   Choosing rules to govern the commons: A model with evidence from Mexico. *Journal of
            Economic Behavior and Organization* 42:19-41.
    2000b   The determinants of collective action on the local commons: A model with evidence from
            Mexico. *Journal of Development Economics* 62:181-208.
    2001    Social Capital, Social Cohesion, Community: A Microeconomic Analysis. Mimeo. Dal-
            housie University.
Dayton-Johnson, J., and P.K. Bardhan
    in      Inequality and conservation on the local commons: A theoretical exercise. *Economic Jour-
    press   nal.*
de los Reyes, R.P.
    1980    47 Communal Gravity Systems: Organizational Profiles. Quezon City: Institute of Philip-
            pine Culture.
Easter, K.W., and K. Palanisami
    1986    Tank Irrigation in India and Thailand: An Example of Common Property Resource Man-
            agement. Minneapolis: Department of Agricultural and Applied Economics Staff, Univer-
            sity of Minnesota.
Fafchamps, M.
    1992    Solidarity networks in preindustrial societies: Rational peasants with a moral economy.
            *Economic Development and Cultural Change* 41:147-175.
Freudenberger, K.S.
    1991    *Mbegué: The Disingenuous Destruction of a Sahelian Rainforest.* Paper No. 29. London:
            International Institute for Environment and Development.
Fujita, M., Y. Hayami, and M. Kikuchi
    2000    The Conditions of Collective Action for Local Commons Management: The Case of Irri-
            gation in the Philippines. Unpublished manuscript, Takushoku University, Foundation for
            Advanced Studies on International Development, and Chiba University.
Hackett, S.C.
    1992    Heterogeneity and the provision of governance for common-pool resources. *Journal of
            Theoretical Politics* 4:325-342.
Henrich, J., R. Boyd, S. Bowles, H. Gintis, E. Fehr, and R. MacElreath
    2000    Cooperation, Reciprocity and Punishment: Experiments in 15 Small-Scale Societies.
            Mimeo, MacArthur Foundation.
Jayaraman, T.K.
    1981    Farmers' organizations in surface irrigation project: Two empirical studies from Gujarat.
            *Economic and Political Weekly* 16:A89-A98.
Jodha, N.S.

1986 Common property resources and rural poor in dry regions of India. *Economic and Political Weekly* 21:1169-1181.
1990 Rural common property resources: Contributions and crisis. *Economic and Political Weekly* 25:A65-A78.

Johnson, R.N., and G.D. Libecap
1982 Contracting problems and regulation: The case of the fishery. *American Economic Review* 72:1005-1022.

Kanbur, R.
1991 Heterogeneity, Distribution and Cooperation in Common Property Resource Management. Background paper for the 1992 *World Development Report*, World Bank.

Khwaja, A.I.
2000 Leadership, Rights and Project Complexity: Determinants of Collective Action in the Maintenance of Infrastructural Projects in the Himalayas. Unpublished paper, Harvard University.

La Ferrara, E.
1999 Inequality and Participation: Theory and Evidence from Rural Tanzania. Unpublished paper, Innocenzo Gasparini Institute for Economic Research (IGIER), Università Bocconi.

Lam, W.F.
1998 *Governing Irrigation Systems in Nepal: Institutions, Infrastructure, and Collective Action.* Oakland, CA: Institute for Contemporary Studies Press.

Laurent, P.J., P. Mathieu, and M. Totté
1994 Populations et Environnement Rural au Burkina Faso. Paris and Louvain-la-Neuve, Belg.: Université Catholique du Louvain/L'Harmattan (Les Cahiers du CIDEP, no. 20).

Leach, E.R.
1961 *Pul Eliya: A village in Ceylon.* Cambridge, Eng.: Cambridge University Press.

Meinzen-Dick, R., and L.A. Jackson
1996 Multiple Uses, Multiple Users of Water Resources. Unpublished paper presented at Voices from the Commons, the Sixth Conference of the International Association for the Study of Common Property, University of California, Berkeley, June 5-8.

Miguel, E.
2000 The Political Economy Of Education and Health in Kenya. Unpublished Ph.D. dissertation, Harvard University.

Molinas, J.R.
1998 The impact of inequality, gender, external assistance and social capital on local-level cooperation. *World Development* 26:413-431.

Olson, M.
1965 *The Logic of Collective Action: Public Goods and the Theory of Groups.* Harvard Economic Studies 124. Cambridge, MA: Harvard University Press.

Ostrom, E.
1990 *Governing the Commons: The Evolution of Institutions for Collective Action.* New York: Cambridge University Press.
1994 Constituting social capital and collective action. *Journal of Theoretical Politics* 6:527-562.

Ostrom, E., and R. Gardner
1993 Coping with asymmetries in the commons: Self-governing irrigation systems can work. *Journal of Economic Perspectives* 7:93-112.

Ostrom, E., R. Gardner, and J. Walker, eds.
1994 *Rules, Games, and Common-Pool Resources.* Ann Arbor: University of Michigan Press.

Persson, T., and G. Tabellini
1994 Is inequality harmful for growth? *American Economic Review* 84:600-621.

Quiggin, J.

 1993   Common property, equality, and development. *World Development* 21:1123-1138.
Ruttan, L., and M. Borgerhoff Mulder
 1999   Are East African pastoralists truly conservationists? *Current Anthropology* 40:621-652.
Seabright, P.
 1997   Is cooperation habit-forming? In *The Environment and Emerging Development Issues*,
        vol. 2, P. Dasgupta and K.-G. Mäler, eds. Oxford, Eng.: Clarendon Press.
Shanmugaratnam, N.
 1996   Nationalization, privatization and the dilemmas of common property management in West-
        ern Rajasthan. *Journal of Development Studies* 33:163-187.
Shanmugaratnam, N., T. Vedeld, A. Mossige, and M. Bovin
 1992   Resource Management and Pastoral Institution-Building in the West African Sahel. Dis-
        cussion Paper No. 175. Washington, DC: World Bank.
Tang, S.Y.
 1991   Institutional arrangements and the management of common-pool resources. *Public Ad-
        ministration Review* 51:42-51.
 1992   *Institutions and Collective Action: Self-Governance in Irrigation.* San Francisco: ICS
        Press.
 1994   Institutions and performance in irrigation systems, in *Rules, Games, and Common-Pool
        Resources*, E. Ostrom, R. Gardner, and J. Walker eds. Ann Arbor: University of Michigan
        Press.
Vaidyanathan, A.
 1986   Water control institutions and agriculture: A comparative perspective. *Indian Economic
        Review* 20:25-83.
Varughese, G., and E. Ostrom
 2001   The contested role of heterogeneity in collective action: Some evidence from community
        forestry in Nepal. *World Development* 29(5):747-765.
Zwarteveen, M.
 1997   Water: From basic need to commodity. *World Development* 25:1335-1349.

# 4

# Factors Influencing Cooperation in Commons Dilemmas: A Review of Experimental Psychological Research

*Shirli Kopelman, J. Mark Weber, and David M. Messick*

This chapter reviews recent experiments on psychological factors that influence cooperation in commons dilemmas. Commons dilemmas are social dilemmas in which noncooperation between individual people leads to the deterioration and possible collapse of a resource (Hardin, 1968; Van Lange et al., 1992a). Hardin's parable about herdsmen who share a common pasture—each has an incentive to raise the number of sheep grazing, but if each herdsman does so they risk ruining the pasture—illustrates the prototypical commons dilemma. From an economic perspective, commons dilemmas are one class of social interactions in which equilibrium outcomes are (Pareto) inefficient. Such inefficient equilibria are not confined to resource and environmental situations, but arise in other domains as diverse as industrial organization, public finance, and macroeconomic policy.

Formally, all social dilemmas can be defined by three characteristics (Dawes, 1980; Messick and Brewer, 1983; Yamagishi, 1986): (1) a noncooperative choice is always more profitable to the individual than a cooperative choice, regardless

We would like to thank the National Science Foundation for funding this ambitious project, and Elke Weber and Paul Stern for shepherding our paper and this project to completion. We would like to thank the three blind reviewers and an external coordinator for valuable comments that helped us frame the final draft of this chapter. We would also like to thank colleagues in our field, as well as the other authors and editors of this volume, for their comments on the early drafts of this chapter. We are grateful for their encouragement and the many ways this chapter has improved because of their input. We also want to express special appreciation to the "practitioner participants" of the International Association for the Study of Common Property 2000 Conference who assured us that the experimental work being done in behavioral labs around the world is relevant to their work and sheds explanatory light on their efforts in the field.

of the choices made by others; (2) a noncooperative choice is always harmful to others compared to a cooperative choice; and (3) the aggregate amount of harm done to others by a noncooperative choice is greater than the profit to the individual. The commons dilemmas (also called resource dilemmas) are a subset of social dilemmas that have traditionally been defined as situations in which collective noncooperation leads to a serious threat of depletion of future resources (Hardin, 1968; Van Lange et al., 1992a). They can be categorized as "social traps" because behavior that is personally gratifying in the short term can lead to long-term collective costs (Cross and Guyer, 1980; Platt, 1973). Although we focus on commons dilemmas, we also draw on relevant research on other types of social dilemmas such as the prisoners' dilemma and the problem of public goods.

The first part of this chapter places recent research in a historical perspective lays out our framework and provides basic definitions. The second part provides a critical review of the recent literature within a categorical framework we developed. The third part concludes by linking the issues raised in our review to the other chapters in this volume.

## INTRODUCTION

### Historical Roots of Experimental Research on Commons Dilemmas

The modern history of social psychological research on common property management, commons dilemmas, resource dilemmas, or social dilemmas—as the field is variously labeled—began in the 1950s. In their path-breaking book, *Theory of Games and Economic Behavior* (1944), Von Neumann and Morgenstern introduced a specific class of models that outlined a theory of individual decision making (with the axiomatization of preferences and utilities) and proposed a theory of social interdependence for both zero-sum and nonzero-sum games. Although economists had been studying departures from competitive equilibrium since the turn of the century, this book spurred a flurry of empirical investigations that explored decision making and utility functions. By the late 1950s, the general ideas of game theory had been introduced to social psychologists in a formal manner by Luce and Raiffa (1957) and in terms of psychological theory by Thibaut and Kelley (1959).

The 1960s saw the proliferation of experiments on two-person games, largely prisoners' dilemma games, and, more importantly, on the generalization of the prisoners' dilemma idea to applied multiperson situations. Two of the important publications of this time, Olson's (1965) *The Logic of Collective Action* and Hardin's (1968) celebrated article "The Tragedy of the Commons," highlighted the issues for the scientific community. During this period, the interests of experimental psychologists and experimental economists diverged. Economists continued to focus on rules and institutions, as well as payoff structures (for an excellent account of the early development of experimental economics, see Davis and

Holt, 1993; Roth, 1995). Psychologists became interested in psychological factors such as individual differences (Kelley and Stahelski, 1970; Messick and McClintock, 1968), the effects on behavior of changing the payoffs (Kelley and Grezlak, 1972), and the effects of communication (Dawes, et al., 1977).

More generally, throughout the 1970s and 1980s, psychologists examined factors that influence cooperation across the range of social dilemmas, including commons dilemmas, prisoners' dilemmas, and public goods tasks (for a broader review of social dilemmas in the social psychological research, see Dawes, 1980; Komorita and Parks, 1994; Messick and Brewer, 1983). Much of the early work on prisoners' dilemmas was criticized on the grounds that it was atheoretical and that it had little to say about extra-laboratory affairs (Pruitt and Kimmel, 1977).

One interesting theme that has emerged from the more recent research we reviewed is the extent to which people are, or are not, other-regarding (how, if at all, people take others' welfare into account). The nature in which they are, or they become, other-regarding has become a central research question. Although the hypothesis that people have preferences for the welfare of others is at least as old as Adam Smith's *Theory of Moral Sentiments*,[1] psychologists have found this question pivotal for understanding choice behavior in interdependent situations. Early efforts in the latter half of the 20th century were made by Sawyer (1966), who tried to measure altruism, by Conrath and Deci (1969), who were estimating a "bivariate" utility function, and by Messick and McClintock (1968), who used a type of random utility model to assess social motives for allocating distributive outcomes in situations of social interdependence. In the Messick and McClintock model, each preference (maximize own outcome in absolute terms, maximize own outcome in relative terms, and maximize joint outcomes of both self and other) had sizable nonzero probabilities. In the 1970s researchers in economics (e.g., Scott, 1972) and in the behavioral sciences (e.g., MacCrimmon and Messick, 1976) began to explore preference structures that could produce behavior that appeared to be altruistic, selfish, and competitive at the same time.

In the 1980s, Messick and Sentis (1985) introduced the concept of a "social utility function" that was later expanded by Lowenstein et al., (1989). A social utility function posits additive preferences for one's own outcomes and preferences for the difference between one's outcome and that of others. Both studies found that the latter function takes its maximum when payoffs to self and other are equal, supporting the assumption made by Falk et al. (this volume:Chapter 5). Their economic model further generalizes the social utility component to comparisons with more than one other person.

## Our Framework

This chapter focuses on experimental work published in major peer-reviewed journals in psychology. In passing, we note experimental work in economics that bears on variables of interests to psychologists. We included studies that manipulated factors that influence cooperation in commons dilemmas and sorted these factors according to the aspect of the type of manipulation involved.

We identified nine classes of independent variables that influence cooperation in commons dilemmas: social motives, gender, payoff structure, uncertainty, power and status, group size, communication, causes, and frames. We organized these classes to first distinguish between individual differences (stable personality traits) and situational factors (the environment). Situational factors were further differentiated into those related to the task structure itself (the decision structure and the social structure) and those related to the perception of the task (see Figure 4-1).

In the psychological literature, the main types of individual differences that have been studied are social motives and gender. The decision structure of the

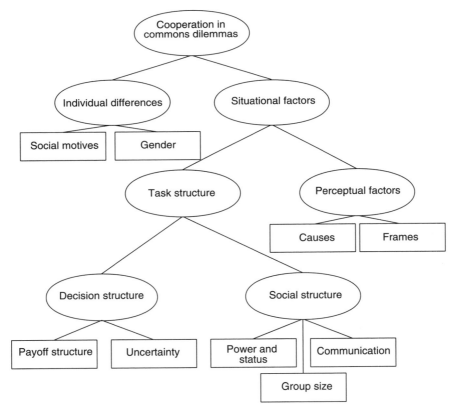

FIGURE 4-1 Elements influencing cooperation in commons dilemmas.

task includes factors like the payoff structure and the amount and type of uncertainty involved in the resource. The social structure includes factors such as the power and status of the individuals or organizations involved, the size of the group, and the ability of people to communicate with one another. Perceptual factors include perceived causes of shortages, or the way cooperation is framed.

## An Experimental Primer

Psychologists generally use an experimental approach to test hypotheses in a laboratory environment. They use scientific and statistical methods that control for extraneous influences and thereby reveal causal relationships between the variables studied. Some participants are assigned to perform a task in a control condition, while others are assigned to an experimental condition. The only difference between these two conditions is an experimental manipulation. As a result, if the two groups have statistically different outcomes (dependent variable[s]), these can be attributed to the experimental manipulation (independent variable[s]). Random assignment of participants to the experimental and control groups enables scientists to identify causal factors.

Imagine you just entered an experimental lab as a participant in a study. You are told that you will be participating in a decision-making task. You and several other people will be playing a game that simulates harvesting decisions by commercial fishermen over a period of 10 seasons. You receive some background information and are asked to make harvesting decisions over several rounds (each round representing a consecutive fishing season). You may be told that it is in your interest to maximize profits, but if the level of fish drops below a certain level, the reproduction rate will drop and there may be less fish to go around. You may or may not receive feedback about simultaneous decisions of other participants, about the size of the resource pool, about the replenishment rate, and other variables. As a participant you are not aware of the factors being studied, nor do you know whether you are in a control or experimental group.

If a researcher wants to study the influence of communication on cooperation in a commons dilemma, then the information you and the other participants receive will be identical. However, in the experimental condition and not in the control condition, the fishermen may be allowed to communicate after five rounds (i.e., five seasons). Indeed, a well-documented finding reveals that experimental groups that are allowed to communicate consistently cooperate more than groups in which no communication is allowed (for a review see Dawes, 1980; Kerr and Kaufman-Gilliland, 1994; Messick and Brewer, 1983). Research described later in this chapter attempts to identify what aspects of communication are critical for developing cooperation.

The strength of the experimental method is its ability to test causal relationships between isolated variables in a controlled environment. Achieving such control over interacting variables is not generally possible in the field. However, the degree of control has also, at times, been construed as a limitation. Despite the common assumption that lab research offers poor external validity (i.e., ability to generalize findings outside the lab), recent empirical work suggests that lab research reliably yields findings comparable in both nature and effect size to those of field research across multiple domains of inquiry (Anderson et al., 1999).

Although a lab environment is by design artificial in that it isolates behavior from many of the large number of simultaneous and interacting influences that affect behavior in the field, it need not ignore context. Often an experimental design simultaneously tests the influence of two independent variables (e.g., trust and communication) so that the influence of one on the other can be evaluated. For example, a recent study on the prisoners' dilemma suggests that in simple tasks, there is no difference between face-to-face communication and e-mail communication, while in complex settings, face-to-face communication elicits more cooperation than e-mail communication (Frohlich and Oppenheimer, 1998). The interaction between the type of communication and the type of task informs us that without examining both factors, it is difficult to predict cooperation.

## REVIEW OF RECENT FINDINGS
## IN THE EXPERIMENTAL LITERATURE

We begin this section by discussing the effects of differences among people, namely social motives and gender.

### Individual Differences

*Social Motives*

Social motives have been conceptualized as stable individual characteristics. Based on experiments using the prisoners' dilemma, Kelley and Stahelski (1970:89) concluded that "two types of persons (cooperative versus competitive personalities) exist in the world whose dispositions are so stable and their interaction so 'programmed' by these dispositions that (a) they do not influence each other at the dispositional level, and (b) they do not influence each other's world views."

Although in theory, an infinite number of social motives (sometimes referred to as social value orientations) can be distinguished (McClintock, 1976, 1978), a common theoretical classification identifies four major motivational orientations (McClintock, 1972): (1) individualism—the motivation to maximize one's own gains; (2) competition—the motivation to maximize relative gains, the difference between one's outcome and that of the other; (3) cooperation—the motivation to maximize joint gain; and (4) altruism—the motivation to maximize other parties' gains. Individualism and competition motives often are referred to as "proself" motives, whereas cooperation and altruism are referred to as "prosocial" motives.

Social motives are measured using a series of decomposed games—each game requires a decision regarding points to be allocated to oneself and a contingent sum to be allocated to some other person—with fixed choices that represent the three most empirically frequent types: individualistic, competitive, and cooperative social motives (Kuhlman and Marshello, 1975). Because the task used to evaluate social motives is an internally consistent measure (Liebrand and Van Run, 1985) with high test-retest reliability (Kuhlman et al., 1986), it provides a dependable tool for measuring social motives.

In the context of resource dilemmas, consistent findings demonstrate that proself individuals harvest significantly more than people with prosocial motives (Kramer et al., 1986; Parks, 1994; Roch and Samuelson, 1997). Similarly, in scenarios that mirror "real-life" social dilemmas such as traffic congestion, prosocial individuals exhibit a greater preference to commute by public transportation rather than private car, and are more concerned with collective outcomes vis-à-vis the environment than proself individuals (Van Vugt et al., 1995; Van Vugt et al., 1996).

The "Might versus Morality Effect" provides a clear example of how social motives influence not only choice behavior but also the interpretation of behavior. Liebrand et al. (1986) examined the relationship between social motives and interpretations of cooperative and competitive behavior. They found that people with individualist social motives tend to interpret behavior along the might dimension (what works), whereas cooperators tend to view cooperation and competition as varying on the moral dimension (what is good or bad). Moreover, prosocials view rationality in social dilemmas from the perspective of the collective (community, group-level), whereas proself people may view it more from a perspective of individual rationality (egocentrically). Van Lange et al. (1990:36) argue that "if one accepts the idea that a perceiver's own goal or predisposition affects his/her choice and also indicates the perspective (collective or individualistic) taken on rationality, it follows that attributions to intelligence should be determined by the combination of the target's choice and the subject's own choice. Thus, social motives may relate not only to differences in choice behavior but also to different perceptions of rationality and intelligence.

Van Lange and colleagues (1990) confirmed that cooperators make larger distinctions between cooperative and noncooperative people than do competitors when making attributions about their behavior on a scale that measures "concern for others." Both cooperators and defectors (noncooperative people) agreed that cooperation is more related to concern for others than noncooperation. In three N-person prisoners' dilemma games (varying in the extent to which fear and greed could be the cause of noncooperation), they compared causal attributions made by cooperative versus noncooperative people. Following each game, participants were asked to make causal interpretations of cooperative and noncooperative choices performed by two imaginary target people (one was a cooperative person, the other was noncooperative). Their findings suggested that cooperators (participants who made cooperative choices in the prisoners' dilemma) were more likely than defectors to attribute cooperation to intelligence, whereas defectors were more likely than cooperators to attribute noncooperation to intelligence.

Van Lange and Liebrand (1991) specifically tested whether individual differences in social motives influence perceptions of rationality in social dilemmas. They manipulated the perception of another person in terms of intelligence in a public goods dilemma. The findings supported their prediction that prosocial individuals expected more cooperation from an intelligent than an unintelligent person, while competitors expected significantly more cooperative behavior from an unintelligent other than an intelligent one.

Van Lange and Kuhlman (1994) evaluated whether social motives influence how information about others is interpreted. In this experiment, people with different social motives made different interpretations of a commons dilemma. Impressions of honesty or intelligence, as well as fairness and self-interest, fell in line with the might versus morality perspective. Cooperative individuals assigned greater weight to honesty than did individualist and competitive participants, while individualists and competitors placed greater weight on intelligence than prosocial participants. Similarly, Samuelson (1993) found systematic differences between cooperators and noncooperators in the importance they assign to dimensions of fairness and self-interest in resource dilemmas. Cooperators assigned greater weight to a fairness dimension, whereas noncooperators assigned greater weight to a self-interest dimension.

Another dimension that may relate to social motives is culture. People from collectivist cultures—cultures that view the self as interdependent with others—behave cooperatively with members of their own group and competitively with members of an out-group, whereas people from individualist cultures—cultures in which the self is perceived as an independent entity—focus less on the social environment and are more task oriented, focusing on their individual goals (Hofstede, 1980; Leung, 1997; Schwartz, 1994; Triandis, 1989). The relationship between culture and social motives is not as straightforward. In a study using an intergroup prisoners' dilemma, Probst and colleagues (1999) found that cultural values of individualism versus collectivism and social motives measured superficially similar constructs. However, the correlations between these measures were low and the authors caution against assuming overlap. Gaerling (1999) found that social motives are related to some cultural values but not to others. Prosocial individuals scored significantly higher on measures of universalism (a cultural value that relates to equality, social justice, and solidarity) but not on benevolence (a cultural value that relates to inner harmony, friendship, good relations, being liked, and security). Because culture is a complex group-level phenomenon, it may not map on directly to measures of individual differences such as social motives. Researchers are only now beginning to focus on the influence of culture on social dilemmas (Kopelman and Brett, in press).

The main conclusions that may be drawn from the research on social motives is that prosocials who tend to view rationality in collective terms are more likely to cooperate in commons dilemmas than proselves who tend to view rationality in individual terms. Prosocials tend to think of cooperation as moral and of competition as immoral, while proselves tend to think of competition as effective and cooperation as less so. Both prosocials and proselves think that their own preferred strategy is more intelligent.

### Gender

Not much research has focused on gender in resource dilemmas. There seems to be a weak but reliable relationship between gender and social motives such that the percentage of prosocials (cooperators) is slightly higher among women than men, while that of proselves (individualists and competitors) is higher among men (e.g., Van Lange et al., 1997). A recent meta-analysis on gender and negotiator competitiveness also found a slight tendency for women to appear more cooperative than men in negotiations (Walters et al., 1998). Some experiments on gender differences and social dilemmas have been conducted using the public goods paradigm, but findings are contradictory.

Gender may influence cooperation because men and women respond differently to one another in group interactions and discussions (Stockard et al., 1988), because they differ in understanding and reacting to each other's actions (Cadsby and Maynes, 1998), or because they respond differently to certain types of resources (Sell et al., 1993). In one study, when participating in four-person same-sex groups, men contributed to a public good at higher rates than women (Brown Kruse and Hummels, 1993). In contrast, another study found all-female groups were more cooperative than either all-male groups or mixed-gender groups (Nowell and Tinkler, 1994). Similarly, Stockard et al. (1988) found that in mixed groups, women were more likely to cooperate than men, especially when discussion among group members was permitted. Yet another study found that women initially contributed significantly more than men, but that the difference disappeared with subsequent trials (Cadsby and Maynes, 1998). Sell and colleagues (1993) found no influence of group gender composition on contributions to a public good, nor did they find a gender effect when money was the resource; however, when the resource was changed to time with an expert, men cooperated significantly more than women.

These mixed findings suggest that gender may have an influence on cooperation in social dilemmas, but its effect may be small and variable. It may be that group diversity is more relevant than the specific gender composition. Research on minority opinions (Nemeth, 1986) and intragroup diversity (Gruenfeld et al., 1996; Williams and O'Reilly, 1998) in decision making suggests that divergence of opinion about the task—task conflict, in contrast to relationship conflict (Jehn, 1995)—leads to better decisions and thus also could influence the development of norms for cooperation in social dilemmas.

## Decision Structure of the Task

*Payoff Structure*

Historically, experimental research on social dilemmas of all kinds has demonstrated significant effects attributable to changes in the "payoff structure" underlying a situation. What are the payoffs associated with cooperation or defection? What are the risks associated with different choices? The influence of payoff structures has been demonstrated not only in the laboratory, but also in the field (Van Lange et al., 1992b). Although emphasis most often has been placed on the monetary payoff structure in experimental games, the present review considers a broader array of structural factors that affect individuals' choices. Central to popular and psychological understandings of behavior is the notion that behaviors generally are more likely to be exhibited when rewarded, and less likely to be exhibited when punished. The central question in any given situation is what combination or form of rewards and punishments (sanctions) will yield optimal or desirable results. A number of recent studies have offered new insights that may be applied productively to the development of better commons management techniques.

Gachter and Fehr (1999) moved beyond the familiar experimental manipulation of material economic rewards or punishments to examine the effect of social rewards on people's willingness to contribute to public goods. They were specifically interested in whether social rewards alone could overcome free-rider problems. First, the investigators conducted a questionnaire study. The questionnaire results confirmed that participants "expect [to] receive more approval if they contribute more, and less approval if others contribute more. In addition, they expect higher marginal approval gains if others contribute more" (p. 346). In the main study, participants faced a public goods dilemma in one of four conditions: (1) an anonymous condition in which participants never knew who they were playing with; (2) a "social exchange" condition in which participants had an opportunity to interact after the game; (3) a "group identity" condition in which participants met one another before playing, but knew they would not see one another afterwards; and (4) a combination of conditions 2 and 3 in which participants met ahead of time, and had a chance to interact afterwards. Neither social familiarity (condition 3) nor the opportunity to receive social rewards in the form of expressions of appreciation after the fact (condition 2) improved the level of cooperation relative to the baseline anonymous condition. However, the combination of the two (condition 4) resulted in significantly higher levels of contribution.

Gachter and Fehr (1999:361-362) conclude that "social approval has a rather weak and insignificant positive effect on participation in collective actions if subjects are complete strangers. Yet, if the social distance between subjects is somewhat reduced by allowing the creation of a group identity and of forming weak social ties, approval incentives give rise to a large and significant reduction in free-riding." They go on to suggest that group identity effects may act as a facilitating "lubricant" for social exchange. It is important to note that there remained, even in the combined condition 4, a minority of participants who seemed unmotivated by social approval and willing to exploit the end-game round. A consistent finding in the gaming literature is that cooperation drops off as the end of the interaction draws near. Although many real-world commons dilemmas are related to resources that parties want to last indefinitely, a similar effect is likely to arise when a given party or parties sees an end to their interest in the commons, and therefore, the relationships that attend its management. Nonetheless, consistent with findings described elsewhere in this chapter, the effectiveness of social rewards in reducing free riding and increasing cooperation is enhanced by reductions in social distance and the facilitation of group identity.

Bell et al. (1989) offer a unique solution to the problem of overconsumption: Let consumers steal from one another. The investigators ran an experiment with a 3 (probability of punishment for stealing) × 3 (probability of punishment for overconsumption) design. The levels of probability for each factor were zero percent (control), 25 percent (low), and 75 percent (high). The punishment in both cases was a loss of points. In each round of play, participants could harvest from the common resource pool, or they could steal from the other players. The results suggest that increasing the probability of punishment for a behavior has a significant deterrence effect; there were main effects for punishment of both behaviors. However, "punishment of one behavior increased the occurrence of the selfish alternative" (p. 1483). If the probability of punishment for overconsumption increased, so did the likelihood of stealing from neighbors. If the probability of punishment for stealing from neighbors increased, so did the likelihood of overconsumption. "To summarize, in the commons simulation, punishment for overconsumption reduced overconsumption, helped preserve the commons, but increased stealing. Punishment of stealing deterred stealing, promoted depletion of the commons and increased oveconsumption" (p. 1495).

Of course, in the real world more than one kind and level of reinforcer is operational at any given time. "Poaching wildlife, for example, may involve perceived rewards of food and hides, perceived thrill of the hunting experience, risk of being caught and punished, potential inconvenience, as well as depletion of the resource, among other consequences" (Bell et al., 1989:1491). Understanding the interplay of such factors is clearly a complex task that is, at least to some extent, unique to any given context.

The Bell et al. (1989) findings also should be read with an understanding that their experimental framework made stealing a highly public act. Although there are real-world analogues (e.g., parking in a handicapped parking spot), the majority of resource theft is done under the assumption that detection is improbable. Although their experiment fixed the probabilities of punishment regardless of an offense's public nature, whether the potential for secret theft under the same probability conditions would yield different behaviors is an open empirical question. Certainly, given the findings reported earlier on the motivating influence of social approval or disapproval (Gachter and Fehr, 1999), one could reasonably anticipate greater willingness to offend if offered the opportunity to do so more discreetly.

In another interesting commons study, Martichuski and Bell (1991) crossed three levels of reinforcement (reward, punishment, or no reinforcement) with three different game structures (territoriality, "golden rule" moral suasion, and a basic structure). Rewards were affirmations for making commons-sustaining harvest choices (i.e., "Good choice, player X"), and punishments for commons-depleting harvest choices were simply the inverse (i.e., "Bad choice, player X"). The territorial structure involved splitting the larger pool so individuals essentially managed their own access to a personal resource pool. The golden rule moral suasion structure involved an initial suggestion that when participants made harvesting decisions, they could make "a lot of points" by making their decisions "exactly the way that [they] would want other people to make their choices." The basic structure was a straightforward commons dilemma (Edney and Harper, 1978).

Those in the privatization (i.e., "territorial") condition were more effective in preserving the commons than those in the moral suasion condition, who were in turn more effective than those in the basic structure condition. Reward and punishment improved the life of the commons in the moral suasion and basic structure conditions, but had no appreciable impact on the privatized condition. Furthermore, reward and punishment had equivalent effects. Martichuski and Bell (1991:1367) suggest that "it seems that a privatized resource maximizes individual harvests while preserving the slowly regenerating resource, and that rewards and punishments do not add to these maxima." This raises a number of interesting questions. For example, would an elaborate system of metering and rationing (with limits or tiered pricing) be a simpler and more effective mechanism for managing certain resources (e.g., water) than elaborate reward and punishment systems? Where it is difficult to effect a system akin to privatization, moral suasion combined with a reinforcement system seems to be a strategy worthy of consideration.

This final point is particularly interesting in light of the rather weak manipulations of this study. The statement "Good move" flashing on one's computer screen is hardly a powerful reward. There is, however, at least one problem from our perspective with the moral suasion condition: It appears to confound what the morally right thing to do is (golden rule) with maximizing personal utility ("Here is a way to make a lot of points..."). This is problematic given that, unlike the typical understanding of social dilemmas, the manipulation seems to suggest that participants' short-term gains can be improved by considering community issues. Further testing of these findings in a context where moral suasion is less confounded, and in which more powerful and realistic rewards and punishments are utilized, could be both interesting and worthwhile.

The value, necessity, and effectiveness of sanctioning systems can vary across cultures. Yamagishi (1988:271) found that American participants in a public goods experiment "cooperated more strongly than Japanese subjects when no sanctioning existed." The sanctions were monetary and were double the amount a person contributed to a "punishment fund." Yamagishi (1988:271) explains his finding in terms of Taylor's (1976) argument that the existence of "a strong external system of sanctioning destroys the basis for voluntary cooperation." Therefore, the existence of such a system "exacerbates the conditions which are claimed to provide its justification and for which it is supposed to be the remedy." He suggests that Japan's more collectivist culture and the culture's tendency toward mutual monitoring and sanctioning result in a decrease of trust in the absence of such control mechanisms relative to America's more individualistic society. This was further supported by questionnaire findings that indicated a lower level of interpersonal trust among Japanese participants than their American counterparts. This finding poses at least two challenges for those interested in commons management. The first is to give careful consideration to cultural factors when making statements about commons dilemma strategies. The second is to consider the long-term consequences of sanctioning systems and authorities on trust and general cooperative tendencies in communities. This is a difficult balance.

Although we focus on experimental and not on applied commons dilemma research, it is important to note that there have been numerous studies on the effects of reward/punishment strategies outside the lab. In this vein, Van Vugt and Samuelson (1999) conducted a field experiment on structural solutions that promote water conservation. They made explicit use of the social dilemma framework to test the effect of personal metering during a naturally occurring resource crisis—a water shortage. They found that conservation efforts were greater among metered (versus unmetered) households when people perceived the water shortage as severe. They suggest "it is time to move beyond the simplified taxonomy (of individual versus structural solutions) to investigate the dynamic interrelationship between structural changes and individuals' psychological and behavioral responses within their new interdependence structure" (p. 743).

In conclusion, sanctioning systems offer potential benefits to the management of common resources. On the other hand, sanctioning systems may undermine intrinsic motivations for cooperation and other generally helpful factors for community life such as interpersonal trust.

*Uncertainty*

Environmental uncertainty increases the difficulty of solving social dilemmas. For example, in many environmental problems the size of the resource pool and its replenishment rate may not be known, or estimates may be contested. For a discussion of the institutional response to uncertainty in complex adaptive systems such as commons dilemmas see Wilson (this volume:Chapter 10). Other authors in this volume confirm that uncertainty of one kind or another can complicate both the exercise of sustaining a common resource (Agrawal, this volume:Chapter 2) and the possibility of one emerging (McCay, this volume:Chapter 11). In the experimental literature, too, the influence of environmental uncertainty on cooperation has emerged as a focal issue.

Ignorance of crucial parameters tends to reduce cooperation in commons dilemmas. In the face of increasing levels of environmental uncertainty about the pool size, people request more for themselves, expect others also will request more, overestimate the size of the resource pool, and display more variability in their harvesting efforts (Budescu et al., 1990, 1992, 1995). These experiments establish that pool size uncertainty affects behavior in both symmetric and asymmetric payoff structures. The effects of pool size uncertainty were corroborated by Hine and Gifford (1996) in an experiment that extended the experimental manipulation of uncertainty to situations of regeneration rate uncertainty; both types of environmental uncertainty led to greater probability of overharvesting. These findings were also supported by Gustafsson et al. (1999a; 1999b).

Why does increased variability about the potential size of the resource or uncertainty regarding the replenishment lead to increased overuse? One explanation is that increased variability of the pool size makes people think that others' requests also will be more variable. Budescu et al. (1990) suggest that, depending on whether an individual is risk seeking or risk averse, environmental uncertainty may respectively lead to either increased or decreased requests from the commons. They found that risk-seeking people requested more from the resource pool than risk-averse people.

Work by Roch and Samuelson (1997) supports the hypothesis that different types of people perceive environmental uncertainty differently. Specifically, social motives moderated the effect of environmental uncertainty on harvesting behavior. These authors found that individualists and competitors (proselves) increased their harvesting under situations of uncertainty. In contrast, prosocial individuals (cooperators and altruists) held their harvest constant, or harvested less.

Another possible explanation for increased harvesting in the face of environmental uncertainty relates to the finding that in situations of uncertainty, people overestimate the size of the pool. As uncertainty about the common resource increases, both the mean estimate and their associated standard deviations increase (Budescu et al., 1990). On one hand, people may believe that the pool is larger because it potentially can be larger. However, this may be a justification for their overharvesting behavior. Uncertainty about pool size may provide a stable external justification for greed: "It's not my greed, I simply assumed the pool was larger – who knew?" Like the diffusion of social responsibility in large groups (Darley and Latané, 1968; Fleishman, 1980),[2] uncertainty also may act to diffuse personal accountability.

Increased harvesting from a common resource under circumstances of uncertainty occurs both in situations of simultaneous protocol of play (Budescu et al.,

1990) and sequential protocol of play (Budescu et al., 1992; Rapoport et al., 1993). The "protocol of play" refers to the temporal order in which people harvest from a shared resource pool (Budescu et al., 1997). Using a simultaneous protocol, people make their harvesting decisions simultaneously and often anonymously. Under a sequential protocol, there is a prespecified order and each person knows his or her position in the sequence and the sum of previous harvests (i.e., current size of resource). In the sequential protocol of play, an additional effect results such that an inverse relationship characterizes the player's position and the size of the request—the first player is likely to make the largest harvest.

An interesting variant of the sequential protocol is the positional protocol, where there is uncertainty about the resource size for subsequent players. In this case, first movers cannot depend on those who come later to adapt to larger initial harvests because the magnitude of the early harvests will not be known. The positional protocol permits three hypotheses about decision making. First, because sequential pool size information is unavailable, there should be no position effect—the results should look like the simultaneous protocol. Alternatively, if players all expect the position effect to exist, then they will act in accordance with it and create the effect and the results should look like the sequential protocol. Finally, the ambiguity and uncertainty about how to approach harvesting, even for the early players, will result in some harvesters thinking the appropriate model is the simultaneous protocol and others thinking the appropriate model is the sequential protocol. If this were to occur, the results should fall somewhere between the two "pure" benchmarks of simultaneous versus sequential protocols. Budescu et al. (1995), Budescu et al. (1997), and others have confirmed this latter hypothesis.

Van Dijk et al. (1999) have questioned the dominant view that environmental uncertainty leads to defection. They have found that environmental uncertainty is not necessarily detrimental to collective interest. In a complex experimental setting, they show that cooperation in social dilemmas depends on the type of dilemma (public goods or common resource dilemma), the asymmetry of position in the group (e.g., high-position members have more resources in a public goods dilemma and are allowed to harvest more in a common resource dilemma), and the type of uncertainty faced by a group. The authors found that groups dismiss uncertain information and base their decisions on environmental information that is certain.

In conclusion, uncertainty tends to reduce cooperation in commons dilemmas, although not always. Although uncertainty is not easily resolved by facts because scientific findings about the size of the resource and its replenishment rate are often controversial, it is important to note the potentially negative influences that uncertainty has on cooperation in commons dilemmas.

## Social Structure of the Task

In the past 10 years, research on various elements of the social context of commons-related decisions has yielded a number of important clarifications to earlier findings and charted worthwhile new territory. Although there is still some debate, it is fairly clear today that groups that interact repeatedly have higher cooperation rates in social dilemmas than groups that are rebuilt every time (Keser and van Winden, 2000). This line of research highlights a potential difference in cooperation between interactions with "strangers" and interactions in familiar social contexts. There is an array of research on issues relating to social structure. In this section, we will focus on three broad categories of research: (1) power, status, and leadership; (2) group size; and (3) understanding the role of communication and communication-related factors in commons settings.

### Power and Status

Issues of power and status have long been a subject of focal interest for social scientists (e.g., Weber, 1924). In recent years, work by Pfeffer (1981) and others has reinvigorated efforts to better understand the ubiquitous role of power in governing and influencing human behavior. This lens now is being focused on social dilemma settings.

It is not uncommon for individuals to violate the expectations of others in ways that hurt other members of their group. Social dilemmas in general, and commons dilemmas in particular, offer a fertile context for this kind of betrayal of expectations. Someone is expected to contribute to a public good, or exercise restraint in harvesting a common resource, and fails to do so—causing negative outcomes for everyone else. In such circumstances, it is typical for the offending party to offer a justification for offending behavior. (A justification is defined as accepting responsibility for an act, but denying that it was wrong. It is distinct from an excuse, in which the offending party agrees that an act was wrong, but denies responsibility for it.)

A group of researchers examined the impact of power and status on the judgments people make about justifications that are offered in a common resource dilemma setting (Massey et al., 1997). Justifications are significant in common resource dilemmas; they are assertions that behaviors that seem a violation of the rules or norms that govern a resource—or the spirit behind them—are not violations at all. Broad acceptance of a justification can redefine fundamental understandings and rules of behavior.

A series of three experiments yielded four interesting findings (Massey et al., 1997). First, and perhaps least surprising, an offending act was judged to be less proper if the justification was invalid than if it was valid. (The validity of justifications was determined through extensive pretesting with a random sample of a similar population.) Second, when an offending individual had higher status than other group members (i.e., a Ph.D. in resource management), it positively impacted others' judgments of the offending act's propriety if the individual's justification was also valid or at least ambiguous in terms of validity. The augmenting effect was greatest when the justification was ambiguous in terms of its validity. Strikingly, however, an offending individual's higher status was a liability if the justification was invalid. Third, an offending individual's greater level of power had a positive impact on others' public judgments of the offending act's propriety, but not on their private judgments. Finally, if an offending individual had both high status and greater power, the combination resulted in a positive impact on even others' private judgments about the act's propriety.

Clearly, the power and status of actors in a commons dilemma context can have a significant effect on how both individuals and their actions are perceived. Further study of such variables is certainly merited. Because a justification constitutes a denial that an act was wrong, one of the interesting implications of these findings is that those with status and power may be in a privileged position when it comes to defining propriety concerning a common resource and its management.

Mannix (1991) compared the resource distribution strategies of organizational groups as a function of discount rate—of what the value of resources would be over time. Her high discount rate condition was assigned a value of 12 percent, while the low discount rate condition was assigned a value of 2 percent. Groups in the high discount rate condition were more likely to adopt coalition strategies that involved fewer group members than groups in the low discount rate condition. This strategy resulted in lower individual and group outcomes. The low discount rate groups, by contrast, actually achieved growth in their resource pool over time. Why the increased competitiveness and destructive behavior among those facing a high discount rate? Mannix offers a few hypotheses. First, she suggests that the rapid devaluation of the resource pool might have led group members to treat every round "as if it were the last" (1991:388). Second, she suggests that the rapid discounting of resource value might have seemed startling relative to anchoring on initial harvesting values, and that group members quickly shifted to short-term strategies to compensate. Finally, she suggests that deep discounting also could affect the value of relationships: "one defector in a high discount condition may generate more fear and defensive behavior than the same defector in a more stable environment" (1991:389). This study raises a number of largely unresolved questions regarding the effects of participants' valuations of future resources on their harvesting decisions. Nonetheless, Mannix's finding that perceived rapid devaluation can lead to increased competition and the formation of excluding coalitions is a noteworthy and instructive caution to those who manage resources.

In addition to the discounting of resource value, and perhaps uncertainty, power imbalances within groups that draw on a common resource can increase the likelihood of coalition formation (Mannix, 1993:2). Mannix argues that when imbalances exist, individual group members have a harder time focusing on mutual gains, and instead focus on protecting their own interests. Coalitions can have significant negative effects on a group's overall outcomes because they can deprive individuals and subgroups of access to the resources they require to succeed or survive. Consistent with her hypotheses, Mannix found that, relative to groups with equalized power relations, groups with power imbalances: (1) made less efficient use of available resources; (2) were more likely to begin the exercise distributing resources to a subset of the group; (3) included fewer people in resource utilization across multiple rounds; and (4) took more effort to reach agreements on resource distributions. Power imbalance was manipulated by assigning different profit percentages to divisions in a decentralized organization (equal versus unequal). In addition, members of groups with power imbalances were more likely to see the group as competitive, be motivated by individual gains, and retaliate against those who omitted them from a coalition. Evidently it also was easier for groups with power imbalances to form small coalitions rather than large ones.

Mannix (1993:16) concludes that power imbalance can be detrimental to group outcomes, noting that "power imbalance appears to encourage competition and a focus on individual outcomes resulting in less integrative agreements." She does, however, offer a possible prescription for better functioning groups: "One of the ways to balance power is to assemble group members from the same position in the hierarchy who have various sources of expertise that are all necessary to the functioning of the group. This way, although the group members would still have their own interests and goals, they might not be as threatened by the positions of other group members" (pp. 18-19).

Wade-Benzoni et al. (1996) offer some important insight into both asymmetric power distributions between people in a commons dilemma and the role of egocentrism (the tendency to see the world only from one's own point of view) in commons management. In an elaborate study that simulated a real-world fish-stock dilemma, they found that levels of egocentrism affect individuals' and groups' perceptions of fairness in asymmetric dilemmas. Next, and more importantly, they found that overharvesting behavior was positively correlated with levels of egocentrism. These two findings naturally lead to the question of whether anything can be done to decrease egocentric biases in dilemma settings. By examining egocentrism before and after discussion, the investigators learned that discussion appeared to decrease egocentric biases. This suggests that the reduction of egocentrism may be one of the reasons why communication has a positive effect on cooperation in social dilemmas in general (see section on communication later in this review). In keeping with Mannix's (1993) conclusions, the study's results suggest that overharvesting tendencies are greater in asymmetric than in symmetric dilemmas. Finally, overharvesting behavior also is related to participants' beliefs about what other participants are likely to do.

Also related to the study of coalitions and power distribution is research on voting institutions. Walker et al. (2000) found that voting substantially increases the efficiency of the outcomes in commons dilemma games. Voting can act as a communication signal when no communication is possible. "The very act of making a proposal and voting on a set of proposals signals limited information to all involved. In particular, it appears to generate information that enables a learning process to occur" (p. 231). This learning extends to subsequent situations and enables people to coordinate their activities even in rounds where no proposals are made.

In 1991, the California water shortage offered Tyler and Degoey (1995) a natural commons dilemma to study. With complete survey data from 400 people directly affected by the shortage, they were able to pose a number of interesting questions about authorities and leadership in relation to the management of a common resource dilemma. Their results replicated earlier experimental findings that people confronted with a severe resource shortage willingly endow authorities with additional control over the resource (e.g., Messick et al., 1983). They also found that the legitimacy of such authorities was determined in large part by the authorities' commitment to fair allocation and decision-making procedures (procedural justice). Perhaps most interesting was their finding that respondents' social identifications with their community moderated the relationship between authorities' use of fair procedures and the support of the authorities. Those who felt pride in their community and perceived procedures to be fair expressed particularly strong support for the regulating authorities. In fact, people who took pride in their community cared even less about their personal outcomes. Taken as a whole, Tyler and Degoey (1995:482) suggest that authorities' effectiveness is "primarily linked to the nature of their social bonds with community members." Social identification with community is an important variable that should not be overlooked in future studies of resource dilemmas.

A number of recent findings speak to contingency issues related to leadership and administration in social dilemma settings. Wit and Wilke (1990), for example, examined the role of who presented rewards and punishments in a social dilemma, and to whom they were presented. The experimental procedures placed participants in the role of chemical company managers concerned with making waste storage versus waste treatment decisions. The former choice was in participants' short-term financial interests, while the latter choice was better for the community and promised greater long-term value. For 124 undergraduates they found no difference between the effectiveness of rewards or punishments on their choices, regardless of whether they were presented by the government or by their parent companies. In contrast, for 239 managers, rewards supplied by the parent company were highly effective, while those supplied by government were actually counterproductive. This finding suggests an interesting consideration for those attempting to manage dilemmas in the real world: What source of sanctioning is most likely to be embraced and respected by the people who make the important decisions?

A large existing literature has explored the conditions under which group members opt to appoint a leader to aid them in achieving their goals in a commons dilemma (e.g., Messick et al., 1983; Samuelson and Messick, 1986). It indicates that groups will opt for a leader when they have failed to manage a resource efficiently and inequalities in harvesting outcomes emerge and that followers will endorse leaders when they are successful in maintaining the common resource (Wilke et al.,1986; Wit and Wilke, 1988; Wit et al., 1989). Studies on public goods also point out that leaders are not autocratic decision makers but rather need some form of legitimacy in order to be effective in persuading members to cooperate (Van Vugt and De Cremer, 1999).

Wit and Wilke (1988) examined the role of leaders' allocation decisions in determining whether or not their leadership is endorsed. Their experiment varied both the outcomes the leader allocated to himself or herself (leader overpayment, leader equal payment, leader underpayment) and his or her allocation to subordinates (participant overpayment, participant equal payment, participant underpayment). They found that leader "endorsement was weakest when the leader overpaid himself or herself" (p. 151) and when the participant making the evaluation had been underpaid relative to other group members. Three more specific findings are also worth noting. First, the leader received his or her greatest endorsement when all allocations were equal. Second, when the leader paid himself or herself less than his or her fair share, participants seemed to take little notice of differences between themselves and other subordinates. Third, when participants were overpaid, they took little notice of how the leader and the other subordinates were paid.

*Group Size*

Earlier research established the much-replicated tendency of small groups to achieve more cooperative outcomes than larger groups (e.g., Dawes, 1980). One recent study offers an interesting insight into a mechanism that may partly explain this tendency: self-efficacy. Self-efficacy is an individual's belief that he or she is competent and capable of taking effective action to achieve a given outcome (Bandura, 1986). In a series of experiments, Kerr (1989) demonstrated that even when group size was objectively irrelevant to the impact a participant could have on an outcome, members of small groups felt more "self-efficacious" than members of larger groups. In the last experiment in this series, the effect of group size on assessments of "collective" efficacy—the perception that one's group can succeed at a given task—was measured. A largely parallel effect to the self-efficacy results was found. When the provision point (proportion of group members demonstrating contributing behavior necessary to achieve the public good) was high (67 percent), group size had no significant impact on assessments of collective efficacy. However, when the provision point was low (33 percent), smaller groups were perceived to be more efficacious than large groups. Kerr (1989:307) observes that "The striking thing is that this belief persisted even when exactly the opposite was objectively true."

Despite Kerr's consistent finding across three studies that smaller group size resulted in judgments of greater self- and collective efficacy to attain a public good, only in the last study were there significant group size effects on actual cooperative behavior. Kerr hypothesized that reductions in group size may increase assessments of the efficacy of others' cooperative behaviors, and therefore encourage free riding. Kerr's experimental paradigm may have encouraged free riding relative to other settings "by minimizing interaction and identifiability" (p. 310).

Kerr refers to his findings as "illusions of efficacy," which he attributes to "familiar judgmental heuristics, involving an overgeneralization of experience in groups of varying sizes" (p. 287). It would be interesting to test whether segmenting an affected population and highlighting subgroup goals or restraints encourages cooperative behaviors in commons dilemmas. For example, one might highlight water consumption behavior in a given apartment building or neighborhood rather than simply highlighting a statewide need for restraint. Other work suggests that small groups are more motivated to divide resources equally than are members of large groups (Allison et al., 1992). This tendency might make it easier for members of smaller groups to make appropriate harvesting decisions.

In contrast, recent studies in economics contradict the widely held view that a group's ability to provide an optimal level of a pure public good is inversely related to group size. Isaac et al. (1994) investigated free-riding behavior in public goods provision and found that groups of sizes 40 and 100 actually provided the public good more efficiently than groups of sizes 4 and 10. To overcome methodological problems that may be associated with studying large groups, they make two methodological modifications: (1) decision-making rounds last several days rather than a few minutes, and (2) rewards are based on extra-credit points rather than cash. The high level of cooperation in large groups is inconsistent with the standard Nash model, but can be explained by alternative approaches such as that of Ledyard (1993), who proposed an equilibrium model in which individuals get some satisfaction (a warm glow) from participating in a cooperative group.

An experiment that introduced a market mechanism for managing the commons provides a somewhat different perspective on group size (Blount White, 1994). Each participant played the role of a corporation that drew on a finite water supply. As it became apparent that the common resource was dwindling at a dangerous pace, half of the participant groups were given the option of buying out other participants. In the "transfer payment" condition, each participant could set a price for his or her right to consume water from the supply, and the other participants could make contributions to buy a seller out. Once a participant was bought out by the others, that participant closed up shop. Therefore, the buyouts could reduce the number of participants drawing on the water supply—effectively reducing group size. Note that participants were not buying a right to a fixed quota of consumption, but simply a reduction in the number of enterprises drawing on the common resource.

Blount White (1994) initially hypothesized that the act of paying compensation to remove a participant from the commons would make the true costs of overconsumption more salient for the remaining participants, and thereby reduce the speed with which they exhausted the remaining water supply. Interestingly, not only did the water supply of groups with the transfer payment option last no longer than the water supply of groups without the transfer payment option, but those with the option consumed significantly more in later rounds than those without the option. Thus, "the market-based intervention hastened depletion" (p. 443). The transfer payment option actually motivated greater self-interest, rather than greater attention to conservation. Why? In debriefing, participants commonly "cited the strategy of trying to take out as much as possible for oneself and then trying to get bought out" (p. 443). Blount White suggested that "when participants pay compensation they may not cognitively interpret it as a cost of consumption but as the purchase of the right to consume more" (p. 453). She concluded, "a self-regulated, market-based approach is not necessarily effective at controlling detrimental social choice patterns" (p. 454). Of course, any number of additional tests of this conclusion would be merited, but the finding is nonetheless interesting and relevant to real-world commons management.

*Communication*

Among the most consistent findings in the experimental social dilemma literature is that a period of discussion among participants yields positive cooperative effects. In the face of an impressive and systematic research program on the effect of communication on cooperation, all but two explanations of this phenomenon had been dismissed as insufficient explanations of the communication effect (Dawes et al., 1990). Those two explanations were: (1) Group discussion enhances group identity or solidarity, and (2) group discussion elicits commitments to cooperate. Still greater clarity regarding the causal mechanism at work was necessary to move forward and more effectively develop optimizing strategies for real-world dilemmas. It is precisely this kind of research enterprise—teasing apart the mechanisms driving an effect—for which experimental laboratory methods seem uniquely well suited.

Kerr and Kaufman-Gilliland (1994) competitively tested the group identity versus commitment explanations in a step-level public goods task. In an elegant 8 $\times$ 2 $\times$ 2 factorial design, they manipulated the self-efficacy of participants' cooperation, the presence or absence of discussion, and the anonymity or public nature of cooperation decisions after discussion. They found a clear pattern of results consistent with the "elicitation of commitments" explanation. "Regardless of how inefficacious a cooperative act was for providing the public good, those who had previously discussed the public-good cooperated at a rate about 30 percent higher than those who had not participated in such a group discussion" (p. 521). While groups that engaged in discussion demonstrated a stronger, more positive sense of group identification, and group identification accounted for some variance beyond that accounted for by discussion condition, it clearly was not a sufficient explanation for the communication effect. Discussion resulted in commitments, and, on average, people followed through with their commitments. These results are also consistent with the finding that, in a public goods dilemma, "a pledge with a certain degree of commitment may facilitate cooperative behavior" (Chen and Komorita, 1994).

Bouas and Komorita (1996) further confirmed Kerr and his colleague's finding that group identity enhancement is an insufficient explanation for the effect of group discussion. However, the structure of their study led them to a somewhat different conclusion about what constituted a sufficient explanation. Whereas Kerr and Kaufman-Gilliland's (1994) study tested the effects of a universal consensus (commitment), Bouas and Komorita (1996) found that a more generalized perception of a degree of consensus was also sufficient to elicit the communication effect. For those managing real-world resources, this stream of research suggests that finding ways to elicit commitments and maximize perceptions of cooperative consensus might be worthwhile.

A natural follow-up question flows from these studies: Why do people follow through on their commitments? Do they fear social sanctions (social norm), or are they internally motivated (internalized or personal norm)? One of the interesting findings of Kerr and Kaufman-Gilliland's original study (1994) was that the anonymity of actual contribution decisions had no effect on the decisions. People honored their commitments even if there was no chance of getting caught cheating. Kerr and his colleagues (Kerr et al., 1997) followed up with a more rigorous test of whether anonymity would moderate the effects of group discussion. Although it was possible for participants in the original study to believe the experimenter might know whether they cheated or not, this follow-up study made it seem impossible for the experimenter to determine whether or not participants honored the commitments they made. In the anonymous condition, the videotape of each session was purportedly mangled and dangled in its damaged state before participants' eyes before they had to make their decisions. The results of this study suggest that the functioning norm in such situations is governed predominantly by self-monitoring. It appears that for most people, the norm against violating their stated commitments is an "internal personal one" as opposed to a social one. This suggests that, paired with dialogue, a society's ability to instill well-internalized personal commitment norms among its citizens may be more effective in managing resource dilemmas in the long run than sanctioning systems. However, as Kerr and his colleagues make sure to point out, not everyone strictly adheres to such an internalized norm. Thirty-two percent of their participants failed to do so. This may simply underscore the value of developing better paradigms for moral education. However, it perhaps further reinforces the importance of finding the right kind of sanctioning system to deal with those inclined to act selfishly and imprudently. This stream of research implies that further empirical study of promising and committing in groups and ways to encourage trustworthiness in those inclined to renege on commitments would be worthwhile pursuits.

Our increasingly electronic age is changing the kinds of communication that may occur in commons settings. Commons dilemmas often involve actors from a variety of institutions who are dispersed geographically, and thus e-mail communication may be commonly used to discuss and negotiate the use of a common resource. Comparing the efficacy of e-mail versus face-to-face communications is of both theoretical and practical interest. As mentioned earlier, research on the prisoners' dilemma suggests that in simple tasks there is no difference between face-to-face communication and e-mail communication, while in complex settings, face-to-face communication elicits more cooperation than e-mail communication (Frohlich and Oppenheimer, 1998). The investigators also examined whether one form of communication had better outcomes for cooperation in later rounds when no further communication was allowed. They found no differences in the "staying power" of the communication effect on cooperation as a function of communication channel. These results raise important issues that are just as relevant in commons resource management. They suggest that there are subtleties worth exploring in the communication effect as a function of communication channel. Furthermore, the study may have implications for researchers. For pragmatic and economic reasons, many researchers have adopted experimental techniques that offer e-mail (usually to a fictitious other) as the communication channel open to participants in lab experiments. The reported study raises a caution for such researchers regarding the generalizability of effect sizes as a function of computer-based versus "live" methods.

Communication can vary not only in terms of the medium that is used but also with respect to directionality. One question that has been raised is whether the unidirectional flow of information can also yield a positive effect on cooperation. Using prisoners' dilemma game and dictator game paradigms, Bohnet and Frey (1999) concluded that two-way communication is not always required to yield "solidarity" (cooperation). They found that one-way identification alone was sufficient for participants to personalize an anonymous stranger, reduce social distance, and positively affect participants' behavior. (Mutual identification and two-way communication generally still had more powerful effects.) The authors cite their study as supportive of Schelling's (1968) claim that "the more we know, the more we care." For the management of resource dilemmas, these findings suggest that actions diminishing social distance between "harvesters" and those who stand to suffer first or most from the depletion of a resource may have advantageous consequences.

## Perceptual Factors

In this section we review recent studies that have questioned the effects of manipulating perceived causes and cognitive frames on cooperation in resource dilemmas. The general methodological structure of these studies is to hold constant the basic economic structure of the decision problem (or to manipulate it systematically) and to systematically change the reasons why things are as they are—the framing, verbal description, or context of the problem. The goal is to determine if these noneconomic and noninstitutional variations influence cooperation in the social dilemmas and if so, how.

### Causes

Hoffman and Spitzer (1985) were perhaps the first researchers to show that the reason given for people's priority position with regard to access to a shared resource made a difference in how much of the resource they claimed for themselves. When the researchers told their participants that they had "earned the right" to go first, to be the "controller," people took more of the resource than when they were told they had been "designated" as the controller by the experimenter. This study was followed by Samuelson and Allison (1994), who systematically varied, among other things, the reasons participants were given for having been assigned a priority position with regard to a resource presumably shared with five other participants. All participants were told they had been assigned to be the first of the six-member group to extract resources from a common pool. However, four groups of participants were given different descriptions about how they achieved this position. The underlying idea of the experiment was that a legitimate method for assigning a privileged position would lead the people to believe they were justified to take more than an equal share of the resources, whereas an illegitimate or questionable procedure would not support such justification. The better the "fit" between the means of getting the privilege and the justification, the more likely it is that people will depart from a "share equally" rule that allocation tasks evoke (Messick, 1993).

According to Samuelson and Allison (1994), this fit is maximal when the process resulting in the first position is a good example of a fair mechanism, which is to say when it is a good prototype of a selection process that leads to a "first come, first served" rule. Two such mechanisms, they propose, are flipping a coin and excelling on an achievement test. Roughly a quarter of their participants were told that they got first position by means of a coin toss, and a quarter were told that they got first position because they answered the most questions correctly on a test of general knowledge. Two other equally random, but less prototypical ways were used to putatively assign the first position for the other participants. One quarter were told they had gotten the most answers correct on an achievement test, but they had seen that one of the six tests was much easier than the other five. The lucky person would get first place, not the person who knew the most. As a test this was unfair, but as a random device it was fair because tests were assigned randomly (subjects were told). In any case, it was not a prototypical process. Neither was the fourth mechanism, which involved calculating the distance of a participant's birthday from a randomly selected day of the year. Although participants rated this process as fair, they also rated it as unprototypical.

The study results showed that participants given the two prototypical justifications for their privileged position took nearly 50 percent more of the shared resource than those given the less prototypical justifications. Moreover, the importance of the justification depended on the details of the decision problem. When overuse resulted in zero payoffs for everyone, the effect of the justification was nonexistent; when people were allowed to keep whatever they had taken, the participants with prototypical justifications took nearly twice as much as those with unusual justifications.

Causal attributions are also important with regard to scarcity or abundance of the resource pool. Why there is a lot or a little has been shown to make a difference in how people treat the resource. In a field study of water use during the 1976-77 drought in California, Talarowski (1982) found that people who stayed within their water allocation limits tended to believe the drought was caused by a natural shortage. Those who exceeded their allocation, however, expressed the view that the shortage was people-induced. In this type of study, it is impossible to say whether the beliefs cause the behavior or the behavior causes the beliefs, or whether both are being caused by some other factor.

Rutte et al. (1987) tried to provide an experimental answer to this question. In their study, participants were told that they would be the fifth person of a six-person group to harvest from a shared pool. All subjects saw the harvests of the previous four (bogus) group members. Collectively, these first four members took 20 points (Dutch guilders—the experiment was conducted in the Netherlands). Half of the subjects were told that the pool initially contained 35 points (leaving 15 for the last two members to share) and half were told that it contained 25 (leaving just 5 for the last two members to share). Half of the people in these two conditions were told that all group members knew the size of the pool from the beginning, and the other half were told that the first four were ignorant of the pool size. When everyone knew the pool size, the shortage or abundance would be attributable to the others, whereas it would be attributable to luck when the first four did not know.

When all group members knew the pool size, the behavior of the first four tends to establish a norm, either a norm of generosity (when there are 35 points) or a norm of greed (when there are 25). Thus the prediction was that when the group was seen as the cause, the participants would be more greedy (when the pool had 25 points), and less greedy (when it had 35) than the participants in groups whose first four members did not know the pool size. The data confirmed this pattern. People-caused shortages reflect a lack of restraint, whereas nature-caused shortages need not.

Samuelson (1991) showed that causal attributions were important in preferences for structural solutions to commons crises. Groups were given a chance to collectively manage an experimental resource pool and were given feedback that they had not done well in maintaining the pool. Roughly half of the people were told that most groups did well and that the task was rather easy, inducing an attribution that the people in the group were greedy. The other half were told that the task was a difficult one and that most groups did not do well, inducing the attribution that poor performance was due to the difficult environment. They were then told that they would be given a chance to do the task for a second time. At this point the subjects were told that they could do the task in the same way they had done it in the past or, if they wished, they could elect a leader who would make a group harvest on each trial and allocate the resources to the members. Samuelson (1991) found that nearly twice as many subjects favored having a leader when they thought that the reason for the prior failure was task difficulty (57 percent favored having a leader) than when they thought it was personal greed (30 percent favored the leader), suggesting that preferences for "solutions" depend on perceptions of causes.

There is one other point about causes that needs to be made in this section, which is that people will only try to solve social dilemmas if they think it is their responsibility to do so, and if they place causal agency on themselves. A study by Guagnano et al. (1994) showed that the ascription of personal responsibility was highly correlated with reported willingness to pay for a variety of environmental goods. This work suggests that people need to see themselves as appropriate causal agents in order to contribute at a higher level to the solution of environmental dilemmas.

In comparison with individual differences and both the decision and social task structure variables, perceptual factors may be easier to manipulate in real-world dilemmas. The scope of causal attribution and cognitive frames, however, goes beyond the "spin" given to the dilemma by the media or by another social institution. Causal attributions—how people explain a certain situation—influence how much of a resource people claim for themselves. This is evident with respect to the priority position regarding access to a shared resource, scarcity or abundance of the resource pool, and preferences for structural solutions to commons crises.

## Frames

Framing, in the study of decision making, concerns the ways in which outcomes, options, and actions are described. Interest in framing can be traced to "prospect theory," the seminal work of Kahneman and Tversky (1979), which showed that people respond differently to decision problems in which the same outcomes are described either as gains or as losses. These authors introduced the concept of loss aversion, which refers to the empirical observation that people evaluate the loss of a given amount more seriously than they evaluate a gain of the same (absolute) amount in risky choices. Moreover, risk attitudes may change as a function of outcome framing. Kahneman and Tversky (1979) proposed that people tend to be risk averse with gains and risk seeking with respect to losses. Monetary outcomes can be framed by changing the reference point from which they are evaluated. A salary of $60,000 could be described as $10,000 more than the average for an industry (a positive frame), or $10,000 less than the mean salary of people with a comparable education (a negative frame).

In the study of social dilemmas, the idea of outcome framing seemed to correspond to the distinction between public goods dilemmas and common-pool dilemmas. In public goods problems, people must make a contribution or give money and hence experience a loss; in common-pool problems, people will make harvests from a resource and hence experience a gain (e.g., Brewer and Kramer, 1986). Thus there seemed to be a one-to-one correspondence between social dilemmas and outcome framing, and many of the early experiments on framing were based on this correspondence. These early studies found inconsistent and puzzling results (see Aquino et at., 1992; Brewer and Kramer, 1986; De Dreu et al., 1992; Fleishman, 1988; McDaniel and Sistrunk, 1991). In these early studies, it was not always clear whether the predictions being made were based on the loss aversion concept or on the assumed difference in risk attitudes for gains and losses.

A recent study of this type (Sonnemans et al., 1998) makes it clear that there is no simple way to apply prospect theory to social dilemmas. Prospect theory requires the specification of a clear reference point for the evaluation of prospects, and social dilemmas are complicated decision situations with a multitude of potential reference points. Moreover, these authors found that although there were no initial differences in cooperation between two versions of a game—one in which people gave money to create a public good and one in which people restrained themselves from taking to create the good—differences did emerge as the participants gained experience with the task. The authors argue that these results require a dynamic theory that can highlight the learning that takes place in the two different environments as participants explore the consequences of their choices.

Although there is little doubt that framing effects occur, there is no consensus on the underlying cause or causes. Indeed, there may be many ways to frame social dilemmas and to influence rates of cooperation, and that fact may be the most important result of this line of experimentation. The following experiments illustrate some of these framing manipulations and their consequences.

De Dreu and McCusker (1997) pursued the loss aversion concept by creating payoff matrices for a two-person prisoners' dilemma game that expressed payoffs either in terms of gains or in terms of losses. They then argued that framing outcomes as gains or losses changes the relative utility or preference between cooperating and defecting in different ways depending on the person's social value orientation. On the assumption that choice frequencies are a direct function of the difference in payoff magnitudes, these authors argued that the incentive to cooperate should be greater in a loss frame than in a gain frame for cooperatively oriented people (who are trying to maximize the sum of the payoffs for the two parties). However, for individualists (trying to maximize their own payoff) and for competitors (trying to maximize the difference between what they get and the other's payoff), the incentive to defect is stronger in loss frames than in gain frames. Thus, they argue, framing can make some people more cooperative and others less so, depending on their utilities. These authors report a series of three experiments that provide impressive support for their hypothesis. Cooperative subjects cooperated more in loss-framed games than in gain-framed ones, while the reverse tended to be true for individualists and competitors. De Dreu and McCusker (1997) also reviewed more than a dozen previously conducted experiments to marshal suggestive evidence that the instructions in these studies determined if loss frames influenced cooperation and, if so, how.

Not all framing has to do with losses and gains. Batson and Moran (1999) conducted a prisoners' dilemma experiment in which the game was described as either a "Business Transaction Study" or a "Social Exchange Study." The instructions for the former consisted of business examples, while the instructions for the latter referred to noneconomic social exchange. The idea was that the description of the task could trigger different means of evaluating strategies for interacting in it. As expected, people made more cooperative choices when the task was framed as a social exchange study than as a business transaction study. These authors also demonstrated that when empathy was created for the other participant in the experiment, the level of cooperation was increased regardless of the frame.

Frames also can be implied by institutions, as has been shown by Elliot et al. (1998). In this experiment, subjects read a series of news briefs, either about entrepreneurial business strategies or about cooperative business strategies. They were also asked to generate examples of successful business strategies that were, respectively, entrepreneurial or cooperative. Then, in the context of doing another experiment, they were given the chance to engage in a public goods social dilemma for a series of six trials. Unlike the Batson and Moran (1999) experiment, here there was no direct labeling of the game, but the labels had been primed in the first part of the study. The entrepreneur-framed people cooperated in about 39 percent of the trials, whereas the cooperative-framed people cooperated in 75 percent of the trials.

Larrick and Blount (1997) have reported a related finding. They noted that the underlying structure of an ultimatum bargaining game and a sequential social dilemma were identical. Yet typically, social dilemma studies produce more cooperation than is reported with ultimatum bargaining games. In a clever series of studies, Larrick and Blount (1997) were able to show that the differences in cooperation rates were attributable to procedural frames: differences in the ways the actions were described. Specifically, second movers in ultimatum bargaining games are told they may "accept or reject" the offer left by the first mover, while in sequential social dilemmas, the second movers are told they can "claim" what is left by the first mover. It is of interest that the connotations of the verb "to claim" not only affect the second mover, who is more likely to accept whatever is left, but also the first mover, who is more likely to leave more than in the accept or reject frame.

Van Dijk and Wilke (1997) have argued that the framing of property rights or the implied ownership of common or personal resources can influence cooperation. These authors contrasted a commons dilemma framework with a public goods dilemma framework. In the resource dilemma, participants were told either that they could harvest up to 20 units from a common pool of 80 (there were four people in a group) or that they could harvest as many units as they wished from their own pool of 20. In the public goods version, they were told either that they could contribute up to 20 units of their own property, or that they could contribute up to 20 units from a common pool of 80. In this experiment, the framing of the pool as one's own or as a common pool had an impact in the resource dilemma. People took more when taking from their own pool than when taking from the common pool. In the latter case, the authors speculate people were concerned about the others' fate; in the former there was less need to think about the others. However, in the public goods context, the authors argue, because the goal of the contribution is to create a shared result, people will think about the others regardless of whether the contributions come from a private or public pool. Thus the authors did not expect nor did they find a framing difference in the public goods situation.

Van Dijk and Wilke (2000) took this a step further than their previous article and suggested that what is really happening with framing manipulations is that the decisions people are being asked to make induce the people to focus on one aspect or variable of the decision problem. For instance, one difference between cooperation in resource dilemmas and public goods dilemmas is that the decision in the former is how much to take, while the decision in the latter is how much to give. The correspondence between the two dilemmas, however, in terms of measures of cooperation, is how much one leaves and how much one gives. The choice of the verb, either giving or keeping in public goods games, and taking or leaving in resource games, may frame the decision independently of the consequences of the choice. Taking and keeping refer to what one will have oneself, and leaving and giving refer to the collective component.

It may be that the actual decision (take, keep, leave, or give) causes one to focus on a quantity that determines one's strategy. For instance, in giving in public goods dilemmas, there is a tendency for people with different endowments to give equal proportions of their endowments. Perhaps this is not the result of the public goods dilemmas but rather because people are focusing on what is necessary to meet the criterion rather than what they have left. Likewise, in resource dilemmas, people typically focus on achieving equal final outcomes. Perhaps this is because they are induced to focus on what they get, rather than what they leave. To test this hypothesis, resource and public goods dilemmas were created in which the participants were either focused on what they ended up with (take and keep) or on what they contributed (give and leave). Van Dijk and Wilke (2000) then calculated whether the person seemed more to be trying to achieve proportionality or equal final outcomes. The results indicated that a large part of the difference between the two types of games could be accounted for by decision-induced focusing, by the quantity on which one was induced to focus.

Most of the studies we have discussed in this section directly manipulated the decision frame in one way or another. One study that indirectly manipulated the frame was reported by Tenbrunsel and Messick (1999). This investigation into the effects of economic sanctions on cooperation in a hypothetical pollution decision suggested that economic sanctions, the possibility of being fined for violating an agreement to reduce emissions, had at least two effects on decision makers. First, they may transform what previously had been considered an ethical issue, whether we have a duty to reduce emissions or not, into a business issue, whether it pays to reduce emissions or not. Second, they change the cost/benefit calculation to make cheating less profitable. However, the authors argued that the cost/benefit analysis would be done only for those people who saw the problem as a business problem. If the decision is seen as an ethical one, then the right thing to do is clear—do not cheat.

So if economic sanctions are introduced that are weak, if the fines are small and the probability of detection is remote, the result may be an increase in cheating. The sanctions will induce more people to think of the problem as a business problem and to find, as a result of the cost/benefit analysis, that cheating is profitable. However, if the sanctions are strong, they should have a deterrent effect on cheating, but only for people who frame the decision as a business decision. Cheating should remain rare among people who frame the decision as an ethical one. The results of the experiments reported by Tenbrunsel and Messick (1999) supported these expectations. Cheating was more likely with weak sanctions than with no sanctions, and the sanctions made more people think of the decision as a business decision than an ethical decision. However, when the sanctions were strong, cheating was reduced, but only for people who viewed the decision as a business problem.

It is clear that cooperation in social dilemmas can be influenced strongly by framing effects, and it seems equally clear that these effects can be of a variety of types—such as framing outcomes as gains or losses, framing games as entrepreneurial or social exchange, or framing choices as taking, keeping, leaving, or giving. Outside a lab environment, however, one must contend with intervening variables such as the challenges of alternative frames of reference advocated by people with competing interests.

## CONCLUSIONS AND SYNTHESIS

The research reviewed in this chapter demonstrates the breadth of experimental work done on commons dilemmas and the advances that have been made in this area over the past decade. Relative to earlier research in psychology, recent work has been more theoretically grounded and more sensitive to field implementation. In this section we link theoretical and empirical findings from the disciplines discussed in other chapters of this volume with the topics we have surveyed.

### Emergence of Other-Regarding Behavior

Many of the experimental findings that we have reviewed are consistent with the general economic model proposed by Falk et al. (this volume:Chapter 5). Their theory suggests that people evaluate their outcomes, at least in part, by comparing them to the outcomes of others with a general preference, all else equal, for equality. The research we reviewed suggests that differences in other-regarding behavior are sometimes viewed as individual differences and other times as situational attributes.

The individual differences approach assumes that people have stable preferences for what they consider fair distributions of outcomes irrespective of the specific person involved. For example, social motives in the social psychological literature are found to be stable individual differences that persist over time (Kuhlman et al., 1986). Other-regarding behavior by cooperative individuals has, until recently, been considered by economists as "anomalous" and "sub optimal" in that it departs from the assumption of rational, self-interested behavior that underlies economic theory (Thaler, 1992). However, a recent experiment by Clark (1998) finds that people who choose such "sub optimal" strategies do not depart from economically rational behavior because of heuristic errors in their decision-making process. Indeed, research surveyed in this chapter suggests that they follow a "collective" versus an "individual" level of rationality.

Situations matter too. In athletic situations the important outcome is usually the score difference—who wins and who loses. In some judicial matters the court decides in favor of one party. Both are competitive situations and all types of people understand this and change their motives and objectives accordingly. One situational or social factor that may influence people's preferences is the extent to which others are seen as cooperative. People evaluate the behavior of others before deciding on their own preferences for a given situation. If others are willing to exercise self-restraint, then so am I. In this case, social mechanisms, such as the norm of reciprocity (Cialdini, 1993; Gouldner, 1960) or social history between people, may come into play and influence interpersonal exchange behavior (e.g., Gallucci and Perugini, 2000; Ortmann et al., 2000).

Conditional preferences can, as Falk and colleagues (this volume:Chapter 5) perceptively note, convert common-pool resource problems into coordination problems. They have shown that their model may provide a sufficient explanation for some communication effects, sanctioning effects, willingness to do what others did, and other departures from strict, self-interested rationality. The model they offer has the powerful virtue of parsimony at the level of "stylized" facts and there is nothing that we reviewed that would constitute a refutation of their ideas. What our review suggests is that preferences may be more complex than just the "inequity aversion" process that Falk et al. propose. For instance, the perception of the causal texture of problems may influence willingness to cooperate or the way that choices are framed or described. We believe the model offered by Falk and colleagues is a valuable first step in the direction of creating a theory of individual human choice that is sufficiently rich to accommodate the wide variety of results that we have described.

The experimental research we have reviewed also confirms the conclusions of Richerson and colleagues (this volume:Chapter 12) in that we find that people do cooperate with strangers, that cooperation is contingent on many things, and that institutions, and cues that imply institutions, do matter. There is little doubt that important aspects of human sociability are part of our evolutionary nature. Most trivially, although it may not be in a woman's best interest to assume the risks of bearing children, we are not the offspring of women who chose not to take this risk. And, just as we have evolved rules for cooperation, institutions that govern the form and pattern of the cooperation also need to evolve. Furthermore, there must be a "fit" between the individual psychology of cooperation and the institutions that foster and regulate it. We may be "wired" to cooperate in small egalitarian family and communal groups, but we must also find ways in large hierarchical groups of strangers to "work around" our evolutionary tendencies to make stable, efficient, and sustainable shared resources.

McCay (this volume:Chapter 11) offers a thoughtful model for the emergence of self-organized cooperation. When do people mobilize themselves to coordinate a common resource? McCay proposes that people must recognize a serious problem, determine the attendant cause and effect relations, and answer the question "is the problem too far gone?" Parts of her model are supported by research reviewed in this chapter. For example, "is the problem too far gone?"

relates directly to the question of efficacy: Can we make a difference? The literature on self-efficacy that we have reviewed indicates that McCay is absolutely correct to see an affirmative answer to her question as an important determinant of whether or not people mobilize.

McCay also argues that communication and persuasion are important for mobilizing people. We would add that experimental lab research on communication suggests that the elicitation of commitments from the parties involved is likely to have the greatest impact. Similarly, experimental work on the nature of decision structures and power may be of use in further specifying what parts of the macro-institutional structures identified are of greatest interest in understanding mobilization. It may be complemented by a model of "structural change in resource dilemmas" that was proposed based on earlier studies in the experimental literature (Samuelson and Messick, 1995).

## Social Heterogeneity

A question that has sparked opposing theoretical perspectives in the broader literature on commons dilemmas is whether socioeconomic heterogeneity leads to cooperation or hinders it. Bardhan and Dayton-Johnson (this volume:Chapter 3), who focus on economic heterogeneity in large-scale studies of locally managed irrigation systems, find support for the latter—heterogeneity hinders cooperation. As Bardhan and Dayton-Johnson note, other types of heterogeneity (social, ethnic, and cultural) may also play an important role. Some research we surveyed on gender composition of groups points out that such group heterogeneity can influence cooperation, although the direction of influence demands further specification of relevant contingencies.

One way to narrow the gap between laboratory and local common-pool resource dilemmas is by actually conducting experiments in the field. An excellent example is an experiment conducted by Cardenas (2000:4) that focused on the influence of economic heterogeneity: "[I]nstead of introducing these effects [economic heterogeneity] artificially through experimental institutions or incentives, and instead of attempting to avoid these factors to enter the experimental design as noise, we accounted for such information that people may bring into the field lab, and analyzed it against the experimental behavior and outcomes." Rather than bringing participants to an experimental lab, this study took the experimental lab to a community (several villages in Colombia). Similar to other findings reported by Bardhan and Dayton-Johnson (this volume:Chapter 3), economic heterogeneity decreased cooperation.

## The Scale of the Dilemma

Social heterogeneity may be especially salient in cross-national dilemmas where members of different cultures come together to solve commons dilemmas.

These may translate not only into differences in cultural values and norms at the group level, but as Young (this volume:Chapter 8) points out, international regimes also operate in social settings that feature a substantial amount of institutional heterogeneity. Decisions at this international level are complicated further by the tensions involved in shifting vertically to national levels of authority. Young describes how implementation of such agreements may vary due to differences in competence, compatibility, and capacity of national governments. The experimental literature would point out another hurdle: The chore of implementing international agreements often becomes fragmented among different subgroups, potentially turning the resource dilemma structurally from an intragroup to an intergroup conflict. Changing the paradigm to an intergroup dilemma changes the incentives and behavior of people in social dilemmas (Bornstein, 1992). Changes along levels of analysis become especially relevant when designing experiments because variables influencing cooperation may not have the same effect when evaluated in small-scale versus large-scale commons situations.

A recent chapter by Biel (2000) discusses similarities and differences between factors promoting cooperation when evaluated (1) in a laboratory environment; (2) in small-scale communal property regimes; and (3) in large-scale societal dilemmas. For example, social norms of reciprocity and commitment may not play as key a role in large-scale dilemmas where the social group is intangible and face-to-face communication is unlikely. On the other hand, environmental uncertainty is likely to play a much larger role because the resources involved in large-scale dilemmas are often less visible (e.g., air pollution) and less quantifiable (e.g., oceans). When evaluating differences across scales, it is important to note whether the characteristics of the resource and/or the complexity of institutional arrangements may account for these differences.

Rose (this volume:Chapter 7) offers a significant real-world example that fleshes out the different structural solutions that may be effective in large-scale dilemmas versus smaller scale common property regimes. As she points out, real-world commons dilemmas occur in complex, dynamic systems in which disagreement over the truth of "facts" must be expected. Some level of uncertainty is the norm. Small communities have developed complex rules and norms that protect the resource as well as the interests of the local community by providing barriers of entry. Developing similar mechanisms in large-scale market regimes is challenging in that instituting a system of tradable environmental allowances that create a level of certainty around the rights that such allowances convey is not a trivial task. Will they be durable rights? Will the volume of entitlement associated with each allowance remain constant? In facing this challenge it is both valuable to understand the predictable ways uncertainty affects individual actors, and to appreciate the positive impact reductions in uncertainty can have on cooperation in commons dilemmas.

## Environmental Uncertainty

Both lab and field studies have pointed to the importance of reducing uncertainty to promote cooperation on both individual and organizational levels. Research we reviewed highlights how environmental uncertainty increases harvesting behavior by individual decision makers. Wilson (this volume:Chapter 10) points out that better institutions for managing the commons can be designed, but that this requires a paradigmatic shift in the way that environmental uncertainty is approached. From the perspective of institutional design, the goal is to create the circumstances under which the average user views restraint as rational. Wilson suggests that the reductionist scientific approach, which has dominated the field, needs to incorporate complex, dynamic, and adaptive processes (like oceans and weather patterns). In such "complex adaptive systems," cause and effect relationships are weakened and predictability decreases.

## A Final Word

A dynamic dialogue between experimentalists and field researchers can yield fruitful results for both. Qualitative research is key to developing rich models that can be subjected to experimental testing and controlled decomposition, which can in turn offer insight for future theoretical model development and field-based interventions. Agrawal's review (this volume:Chapter 2) of the traditional, largely case-based literature on common-pool resources points at a substantial overlap between lab and field studies both in terms of the choice of variables studied and their implications. Readers of his review should find striking parallels with the findings reported in this chapter on issues ranging from group size to sanctions and the significance of communication and a sense of efficacy. Agrawal (this volume:Chapter 2) identifies the importance of employing a "careful research design that controls for factors that are not the subject of investigation" (p. 65). This is exactly what the experimental approach has to offer. The strength of the experimental method is that by isolating variables, it enables social scientists to pit theoretical concepts against one another and establish causal linkages.

## NOTES

1  In this first book (published in the middle of the 18th century, a decade before his more famous book on the wealth of nations, his hypothesis is made clear early on: "However selfish man may be supposed, there are evidently some principles of his nature, which interest him in the fortune of others, and render their happiness necessary to him, though he derives nothing from it..." (Werhane, 1991:25).

2  A person is less likely to respond to an emergency situation when there are many bystanders than when that person thinks he or she is the only witness.

# REFERENCES

Allison, S.T., L.R. McQueen, and L.M. Schaerfl
  1992   Social decision making processes and the equal partitionment of shared resources. *Journal of Experimental Social Psychology* 28(1):23-42.
Anderson, C.A., J.J. Lindsay, and B.J. Bushman
  1999   Research in the psychological laboratory: Truth or triviality. *Current Directions in Psychological Science* 8(1):3-9.
Aquino, K., V. Steisel, and A. Kay
  1992   The effects of resource distribution, voice, and decision framing on the provision of public goods. *Journal of Conflict Resolution* 36(4):665-687.
Bandura, A.
  1986   *Social Foundation of Thought and Action: A Social Cognitive Theory.* Englewood Cliffs, NJ: Prentice-Hall.
Batson, C.D., and T. Moran
  1999   Empathy-induced altruism in a prisoner's dilemma. *European Journal of Social Psychology* 29(7):909-924.
Bell, P.A., T.R. Petersen, and J.E. Hautaluoma
  1989   The effect of punishment probability on overconsumption and stealing in a simulated commons. *Journal of Applied Social Psychology* 19(17, Pt. 1):1483-1495.
Biel, A.
  2000   Factors promoting cooperation in the laboratory, in common pool resource dilemmas, and in large-scale dilemmas: similarities and differences. Pp. 25-41 in *Cooperation in Modern Society*, M. Van Vugt, M. Snyder, T. Tyler, and A. Biel, eds. London: Routledge.
Blount White, S.B.
  1994   Testing an economic approach to resource dilemmas. *Organizational Behavior and Human Decision Processes* 58(3):428-456.
Bohnet, I., and B.S. Frey
  1999   The sound of silence in Prisoner's Dilemma and Dictator Games. *Journal of Economic Behavior and Organization* 38(1):43-57.
Bornstein, G.
  1992   The free-rider problem in intergroup conflicts over step-level and continuous public goods. *Journal of Personality and Social Psychology* 62(4):597-606.
Bouas, K.S., and S.S. Komorita
  1996   Group discussion and cooperation in social dilemmas. *Personality and Social Psychology Bulletin* 22(11):1144-1150.
Brewer, M.B., and R.M. Kramer
  1986   Choice behavior in social dilemmas: Effects of social identity, group size, and decision framing. *Journal of Personality and Social Psychology* 50(3):543-549.
Brown Kruse, J., and D. Hummels
  1993   Gender effects in laboratory public goods contribution: Do individuals put their money where their mouth is? *Journal of Economic Behavior and Organization* 22(3):255-67.
Budescu, D.V., A. Rapoport, and R. Suleiman
  1990   Resource dilemmas with environmental uncertainty and asymmetric players. *European Journal of Social Psychology* 20(6):475-487.
  1992   Simultaneous vs. sequential requests in resource dilemmas with incomplete information. *Acta Psychologica* 80:297-310.
Budescu, D.V., R. Suleiman, and A. Rapoport
  1995   Positional and group size effects in resource dilemmas with uncertain resources. *Organizational Behavior and Human Decision Processes* 61(3):225-238.

Budescu, D.V., W. Tung Au, and X.P. Chen
    1997    Effects of protocol of play and social orientation on behavior in sequential resource dilem-
            mas. *Organizational Behavior and Human Decision Processes* 69(3):179-193.
Cadsby, C.B., and E. Maynes
    1998    Gender and free riding in a threshold public goods game: Experimental evidence. *Journal
            of Economic Behavior and Organization* 34(4):603-620.
Cárdenas, J.C.
    2000    Real Wealth and Experimental Cooperation: Evidence from Field Experiments. Unpub-
            lished paper presented at the International Association for the Study of Common Property
            2000 conference, Bloomington, IN, June 1-4.
Chen, X., and S.S. Komorita
    1994    The effects of communication and commitment in a public goods social dilemma. *Organi-
            zational Behavior and Human Decision Processes* 60(3):367-386.
Cialdini, R.B.
    1993    *Influence: Science and Practice.* New York: Harper Collins College Publishers.
Clark, J.
    1998    Fairness preferences and optimization skills: Are they substitutes? An experimental inves-
            tigation. *Journal of Economic Behavior and Organization* 34(4):541-557.
Conrath, D.W., and E.L. Deci
    1969    The determination and scaling of a bivariate utility function. *Behavioral Science* 14(4):316-
            327.
Cross, J.G., and M.J. Guyer
    1980    *Social Traps.* Ann Arbor: University of Michigan Press.
Darley, J.M., and B. Latané
    1968    Bystander intervention in emergencies: Diffusion of responsibility. *Journal of Personality
            and Social Psychology* 8:377-383.
Davis, D.D., and C.A. Holt
    1993    *Experimental Economics.* Princeton: Princeton University Press.
Dawes, R.M.
    1980    Social dilemmas. *Annual Review of Psychology* 31:169-193.
Dawes, R.M., J. McTavish, and H. Shaklee
    1977    Behavior, communication, and assumptions about other people's behavior in a commons
            dilemma situation, *Journal of Personality and Social Psychology* 33(1).
Dawes, R.M., A.J.C. van de Kragt, and J.M. Orbell
    1990    Cooperation for the benefit of us—Not me, or my conscience. Pp. 97-110 in *Beyond Self-
            Interest*, J.J. Mansbridge, ed. Chicago: University of Chicago Press.
De Dreu, C.K., B.J. Emans, and E. Van de Vliert
    1992    Frames of reference and cooperative social decision-making. *European Journal of Social
            Psychology* 22(3):297-302.
De Dreu, C.K.W., and C. McCusker
    1997    Gain-loss frames and cooperation in two-person social dilemmas: A transformational
            analysis. *Journal of Personality and Social Psychology* 72(5):1093-1106.
Edney, J.J., and C.S. Harper
    1978    Heroism in a resource crisis: A simulation study. *Environmental Managment* 2:523-527.
Elliott, C.S., D.M. Hayward, and S. Canon
    1998    Institutional framing: Some experimental evidence. *Journal of Economic Behavior and
            Organization* 35(4):455-464.
Fleishman, J.A.
    1980    Collective action as helping behavior: Effects of responsibility diffusion on contributions
            to a public good. *Journal of Personality and Social Psychology* 38(4):629-637.
    1988    The effects of decision framing and others' behavior on cooperation in a social dilemma.
            *Journal of Conflict Resolution* 32(1):162-180.

Frohlich, N., and J. Oppenheimer
    1998    Some consequences of e-mail vs. face-to-face communication in experiment. *Journal of Economic Behavior and Organization* 35(3):389-403.
Gachter, S., and E. Fehr
    1999    Collective action as a social exchange. *Journal of Economic Behavior and Organization* 39(4):341-369.
Gaerling, T.
    1999    Value priorities, social value orientations and cooperation in social dilemmas. *British Journal of Social Psychology* 38(4):397-408.
Gallucci, M., and M. Perugini
    2000    An experimental test of a game - Theoretical model of reciprocity. *Journal of Behavioral Decision Making* 13:367-389.
Gouldner, A.W.
    1960    The norm of reciprocity: A preliminary statement. *American Sociological Review* 25: 161-179.
Gruenfeld, D.H., E.A. Mannix, K.Y. Williams, and M.A. Neale
    1996    Group composition and decision making: How member familiarity and information distribution affect process and performance. *Organizational Behavior and Human Decision Processes* 67(1):1-15.
Guagnano, G.A., T. Dietz, and P.C. Stern
    1994    Willingness to pay for public goods: A test of the contribution model. *Psychological Science* 5:411-415.
Gustafsson, M., A. Biel, and T. Gaerling
    1999a   Outcome-desirability bias in resource management problems. *Thinking and Reasoning* 5(4):327-337.
    1999b   Overharvesting of resources of unknown size. *Acta Psychologica* 103:47-64.
Hardin, G.
    1968    The tragedy of the commons. *Science* 162:1243-1248.
Hine, D.W., and R. Gifford
    1996    Individual restraint and group efficiency in commons dilemmas: The effects of two types of environmental uncertainty. *Journal of Applied Social Psychology* 26(11):993-1009.
Hoffman, E., and M.L. Spitzer
    1985    Entitlements, rights, and fairness: An experimental examination of subjects' concepts of distributive justice. *Journal of Legal Studies* 14:259-297.
Hofstede, G.
    1980    *Culture's Consequences: International Differences in Work-Related Values.* Newbury Park, CA: Sage.
Isaac, M.R., J.M. Walker, and A.W. Williams
    1994    Group size and the voluntary provision of public goods. *Journal of Public Economics* 54: 1-36.
Jehn, K.A.
    1995    A multimethod examination of the benefits and detriments of intragroup conflict. *Administrative Science Quarterly* 40:256-282.
Kahneman, D., and A. Tversky
    1979    Prospect theory: An analysis of decision under risk. *Econometrica* 47:263-291.
Kelley, H.H., and J. Grezlak
    1972    Conflict between individual and common interest in an N-person relationship. *Journal of Personality and Social Psychology* 21(2):190-197.
Kelley, H.H., and A.J. Stahelski
    1970    Social interaction basis of cooperators' and competitors' beliefs about others. *Journal of Personality and Social Psychology* 16(1):66-91.

Kerr, N.L.
    1989    Illusions of efficacy: The effects of group size on perceived efficacy in social dilemmas. *Journal of Experimental Social Psychology* 25(4):287-313.
Kerr, N.L., J. Garst, D.A. Lewandowski, and S.E. Harris
    1997    That still, small voice: Commitment to cooperate as an internalized versus a social norm. *Personality and Social Psychology Bulletin* 23(12):1300-1311.
Kerr, N.L., and C.M. Kaufman-Gilliland
    1994    Communication, commitment, and cooperation in social dilemma. *Journal of Personality and Social Psychology* 66(3):513-529.
Keser, C., and F. van Winden
    2000    Conditional cooperation and voluntary contributions to public goods. *Scandinavian Journal of Economics* 102:29-39.
Komorita, S., and C.D. Parks
    1994    *Social Dilemmas.* Madison, WI: Brown and Benchmark.
Kopelman, S., and J.M. Brett
    in      Culture and social dilemmas. In M. Gelfand and J.M. Brett (Eds.), *Negotiation and Cul-*
    press   *tures: Research Perspectives.* Stanford, CA: Stanford University Press.
Kramer, R.M., C.G. McClintock, and D.M. Messick
    1986    Social values and cooperative response to a simulated resource conservation crisis. *Journal of Personality* 54(3):576-592.
Kuhlman, D.M., C. Camac, and D.A. Cunha
    1986    Individual differences in social orientation. Pp. 151-176 in *Experimental Social Dilemmas*, D.M.H. Wilke, and C. Rutte, eds. New York: Verlag Peter Lang.
Kuhlman, D.M., and A.F. Marshello
    1975    Individual differences in game motivation as moderators of preprogrammed strategy effects in prisoner's dilemma. *Journal of Personality and Social Psychology* 32(5):922-931.
Larrick, R.P., and S. Blount
    1997    The claiming effect: Why players are more generous in social dilemmas than in ultimatum games. *Journal of Personality and Social Psychology* 72(4):810-825.
Ledyard, J.O.
    1993    Is there a problem with public goods provision? Pp. 111-194 in *The Handbook of Experimental Economics*, J. Kagel and A.Roth, eds. Princeton: Princeton University Press.
Leung, K.
    1997    Negotiation and reward allocations across cultures. Pp. 640-675 in *New Perspectives on International Industrial/Organizational Psychology*, P.C. Earley and M. Erez, eds. San Francisco: New Lexington Press.
Liebrand, W.B., R.W. Jansen, V.M. Rijken, and C.J. Suhre
    1986    Might over morality: Social values and the perception of other players in experimental games. *Journal of Experimental Social Psychology* 22(3):203-215.
Liebrand, W.B., and G.J. Van Run
    1985    The effects of social motives on behavior in social dilemmas in two cultures. *Journal of Experimental Social Psychology* 21(1):86-102.
Lowenstein, G.F., L. Thompson, and M.H. Bazerman
    1989    Social utility and decision making in interpersonal contexts. *Journal of Personality and Social Psychology* 57(3):426-441.
Luce, R.D., and H. Raiffa
    1957    *Games and Decisions: Introduction and Critical Survey.* New York: Wiley.
MacCrimmon, K.R., and D.M. Messick
    1976    A framework for social motives. *Behavioral Science* 21(2):86-100.
Mannix, E.A.
    1991    Resource dilemmas and discount rates in decision making groups. *Journal of Experimental Social Psychology* 27(4):379-391.

1993    Organizations as resource dilemmas: The effects of power balance on coalition formation in small groups. *Organizational Behavior and Human Decision Processes* 55(1):1-22.

Martichuski, D.K., and P.A. Bell
1991    Reward, punishment, privatization, and moral suasion in a commons dilemma. *Journal of Applied Social Psychology* 21(16):1356-1369.

Massey, K., S. Freeman, and M. Zelditch
1997    Status, power, and accounts. *Social Psychology Quarterly* 60(3):238-251.

McClintock, C.G.
1972    Social motivation: A set of propositions. *Behavioral Science* 17(5):438-455.
1976    Social motivations in settings of outcome interdependence. Pp. 49-77 in *Negotiations: Social Psychology Perspective*, D. Druckman, ed. Beverly Hills: Sage.
1978    Social values: Their definition, measurement, and development. *Journal of Research and Development in Education* 12:121-137.

McDaniel, W.C., and F. Sistrunk
1991    Management dilemmas and decisions: Impact of framing and anticipated responses. *Journal of Conflict Resolution* 35(1):21-42.

Messick, D.M.
1993    Equality as a decision heuristic. Pp. 11-31 in *Psychological Perspectives on Justice: Theory and Applications*, B.A.B.J. Mellers, ed. Cambridge Series on Judgment and Decision Making. New York: Cambridge University Press.

Messick, D.M., and M.B. Brewer
1983    Solving social dilemmas: A review. Pp. 11-44 in *Review of Personality and Social Psychology*, Vol. 4, L. Wheeler and P. Shaver, eds. Beverly Hills: Sage.

Messick, D.M., and C.G. McClintock
1968    Motivational bases of choice in experimental games. *Journal of Experimental Social Psychology* 4(1):1-25.

Messick, D.M., and K.P. Sentis
1985    Estimating social and nonsocial utility functions from ordinal data. *European Journal of Social Psychology* 15(4):389-399.

Messick, D.M., H. Wilke, M.B. Brewer, R.M. Kramer, P.E. Zemke, and L. Lui
1983    Individual adaptations and structural change as solutions to social dilemmas. *Journal of Personality and Social Psychology* 44(2):294-309.

Nemeth, C.J.
1986    Differential contributions of majority and minority influence. *Psychological Review* 93(1):23-32.

Nowell, C., and S. Tinkler
1994    The influence of gender on the provision of a public good. *Journal of Economic Behavior and Organization* 25(1):25-36.

Olson, M.
1965    *The Logic of Collective Action.* Cambridge, MA: Harvard University Press.

Ortmann, A., J. Fitzgerald, and C. Boeing
2000    Trust, reciprocity, and social history: A re-examination. *Experimental Economics* 3:81-100.

Parks, C.D.
1994    The predictive ability of social values in resource dilemmas and public goods games. *Personality and Social Psychology Bulletin* 20(4):431-438.

Pfeffer, J.
1981    *Power in Organizations.* Marchfield, MA: Pitman.

Platt, J.
1973    Social traps. *American Psychologist* 28(8):641-651.

Probst, T., P.J. Carnevale, and H.C. Triandis

1999    Cultural values in intergroup and single-group social dilemmas. *Organizational Behavior and Human Decision Processes* 77(3):171-191.

Pruitt, D.G., and M. Kimmel
1977    Twenty years of experimental gaming: Critique, synthesis, and suggestions for the future. *Annual Review of Psychology* 28:363-392.

Rapoport, A., D.V. Budescu, and R. Suleiman
1993    Sequential requests from randomly distributed shared resources. *Journal of Mathematical Psychology* 37(2):241-265.

Roch, S.G., and C.D. Samuelson
1997    Effects of environmental uncertainty and social value orientation in resource dilemmas. *Organizational Behavior and Human Decision Processes* 70(3):221-235.

Roth, A.E.
1995    A brief history of experimental economics. Pp. 3-109 in *The Handbook of Experimental Economics*, J.H. Kagel and A. E. Roth, eds. Princeton, NJ: Princeton University Press.

Rutte, C.G., H.A. Wilke, and D.M. Messick
1987    Scarcity or abundance caused by people or the environment as determinants of behavior in the resource dilemma. *Journal of Experimental Social Psychology* 23(3):208-216.

Samuelson, C.D.
1991    Perceived task difficulty, causal attributions, and preferences for structural change in resource dilemmas. *Personality and Social Psychology Bulletin* 17(2):181-187.

1993    A multiattribute evaluation approach to structural change in resource dilemmas. *Organizational Behavior and Human Decision Processes* 55(2):298-324.

Samuelson, C.D., and S.T. Allison
1994    Cognitive factors affecting the use of social decision heuristics in resource-sharing tasks. *Organizational Behavior and Human Decision Processes* 58(1):1-27.

Samuelson, C.D., and D.M. Messick
1986    Inequities in access to and use of shared resources in social dilemmas. *Journal of Personality and Social Psychology* 51(5):960-967.

1995    Let's make some new rules: Social factors that make freedom unattractive. In *Negotiation as a Social Process*, R. Kramer and D.M. Messick eds., Thousand Oaks, CA: Sage.

Sawyer, J.
1966    The Altruism Scale: A measure of co-operative, individualistic, and competitive interpersonal orientation. *American Journal of Sociology* 71(4):407-416.

Schelling, T.C.
1968    The life you save may be your own. In *Problems in Public Expenditure Analysis*, S. Chase, ed., Washington, DC: Brookings Institution.

Schwartz, S.H.
1994    Beyond individualism/collectivism: New cultural dimensions of values. Pp. 85-117 in *Individualism and Collectivism*, U. Kim, H.C. Triandis, and G. Yoon, eds., London:Sage.

Scott, R.H.
1972    Avarice, altruism, and second party preferences. *Quarterly Journal of Economics* 86(1): 1-18.

Sell, J., W.I. Griffith, and R.K. Wilson
1993    Are women more cooperative than men in social dilemmas? *Social Psychology Quarterly* 56(3):211-222.

Sonnemans, J., A. Schram, and T. Offerman
1998    Public good provision and public bad prevention: The effect of framing. *Journal of Economic Behavior and Organization* 34(1):143-161.

Stockard, J., A.J. Van de Kragt, and P.J. Dodge
1988    Gender roles and behavior in social dilemmas: Are there sex differences in cooperation and in its justification? *Social Psychology Quarterly* 51(2):154-163.

Talarowski, F.S.
 1982  Attitudes Toward and Perceptions of Water Conservation in a Southern California Community. Unpublished dissertation, University of California, Santa Barbara.
Taylor, M.
 1976  *Anarchy and Cooperation.* New York: Wiley.
Tenbrunsel, A.E., and D.M. Messick
 1999  Sanctioning systems, decision frames, and cooperation. *Administrative Science Quarterly* 44(4):684-707.
Thaler, R.H.
 1992  *The Winner's Curse.* New York: Free Press.
Thibaut, J.W., and H.H. Kelley
 1959  *The Social Psychology of Groups.* New York: Wiley.
Triandis, H.C.
 1989  Cross-cultural studies of individualism and collectivism. Pp. 41-133 in *Nebraska Symposium on Motivation,* J. Berman, ed., Lincoln:University of Nebraska Press.
Tyler, T.R., and P. Degoey
 1995  Collective restraint in social dilemmas: Procedural justice and social identification effects on support for authorities. *Journal of Personality and Social Psychology* 69(3):482-497.
van Dijk, E., and H. Wilke
 1997  Is it mine or is it ours? Framing property rights and decision making in social dilemmas. *Organizational Behavior and Human Decision Processes* 71(2):195-209.
 2000  Decision-induced focusing in social dilemmas: Give-some, keep-some, take-some, and leave-some dilemmas. *Journal of Personality and Social Psychology* 78(1):92-104.
van Dijk, E., H. Wilke, M. Wilke, and L. Metman
 1999  What information do we use in social dilemmas? Environmental uncertainty and the employment of coordination rules. *Journal of Experimental Social Psychology* 35(2):109-135.
Van Lange, P.A., and W.B. Liebrand
 1991  Social value orientation and intelligence: A test of the Goal Prescribes Rationality Principle. *European Journal of Social Psychology* 21(4):273-292.
Van Lange, P.A., W.B. Liebrand, and D.M. Kuhlman
 1990  Causal attribution of choice behavior in three N-Person Prisoner's Dilemmas. *Journal of Experimental Social Psychology* 26(1):34-48.
Van Lange, P.A.M., E.M.N. De Bruin, W. Otten, and J.A. Joireman
 1997  Development of prosocial, individualistic, and competitive orientations: Theory and preliminary evidence. *Journal of Personality and Social Psychology* 73(4):733-746.
Van Lange, P.A.M., and D.M. Kuhlman
 1994  Social value orientations and impressions of partner's honesty and intelligence: A test of the might versus morality effect. *Journal of Personality and Social Psychology* 67(1):126-141.
Van Lange, P.A.M., W.B.G. Liebrand, D.M. Messick, and H.A.M. Wilke
 1992a Introduction and literature. In *Social Dilemmas: Theoretical Issues and Research Findings,* W. Leibrand, D. Messick, and H. Wilke, eds., Oxford: Pergamon Press.
 1992b Social dilemmas: The state of the art. In *Social Dilemmas: Theoretical Issues and Research Findings,* W. Leibrand, D. Messick, and H. Wilke, eds., Oxford: Pergamon Press.
Van Vugt, M., and D. De Cremer
 1999  Leadership in social dilemmas: The effects of group identification on collective actions to provide public goods. *Journal of Personality and Social Psychology* 76(4):587-599.
Van Vugt, M., R.M. Meertens, and P.A.M. Van Lange
 1995  Car versus public transportation? The role of social value orientations in a real-life social dilemma. *Journal of Applied Social Psychology* 25(3):258-278.

Van Vugt, M., and C.D. Samuelson
    1999    The impact of personal metering in the management of a natural resource crisis: A social
            dilemma analysis. *Personality and Social Psychology Bulletin* 25(6):731-745.
Van Vugt, M., P.A.M. Van Lange, and R.M. Meertens
    1996    Commuting by car or public transportation? A social dilemma analysis of travel mode
            judgments. *European Journal of Social Psychology* 26(3):373-395.
Von Neumann, J., and O. Morgenstern
    1944    *Theory of Games and Economic Behavior.* New York:Wiley.
Wade-Benzoni, K.A., A.E. Tenbrunsel, and M.H. Bazerman
    1996    Egocentric interpretations of fairness in asymmetric, environmental social dilemmas: Ex-
            plaining harvesting behavior and the role of communication. *Organizational Behavior and
            Human Decision Processes* 67(2):111-126.
Walker, J.M., R. Gardner, A. Herr, and E. Ostrom
    2000    Collective choice in the commons: Experimental results on proposed allocation rules and
            votes. *The Economic Journal* 110:212-234.
Walters, A.E., A.F. Stuhlmacher, and L.L. Meyer
    1998    Gender and negotiator competitiveness: A meta-analysis. *Organizational Behavior and
            Human Decision Processes* 76(1):1-29.
Weber, M.
    1978    *Economy and Society: An Outline of Interpretive Sociology*, E. Fischoff et al., trans. 1978
            ed. Berkeley: University of California Press (first published in 1924).
Werhane, P.H.
    1991    *Adam Smith and His Legacy for Modern Capitalism.* New York: Oxford University Press.
Wilke, H.A.M., W. Liebrand, and K. De Boer
    1986    Differential access in a social dilemma situation. *British Journal of Social Psychology* 25:
            57-65.
Williams, K.Y., and C.A. O'Reilly, eds.
    1998    *Demography and Diversity in Organizations: A Review of 40 Years of Research.* Research
            in Organizational Behavior, Vol. 20. Greenwich, CT: JAI Press.
Wit, A., and H. Wilke
    1988    Subordinates' endorsement of an allocating leader in a commons dilemma: An equity
            theoretical approach. *Journal of Economic Psychology* 9(2):151-168.
    1990    The presentation of rewards and punishments in a simulated social dilemma. *Social
            Behaviour* 5(4):231-245.
Wit, A.P., H.A. Wilke, and E. Van Dijk
    1989    Attribution of leadership in a resource management situation. *European Journal of Social
            Psychology* 19(4):327-338.
Yamagishi, T.
    1986    The structural goal/expectation theory of cooperation in social dilemmas. Pp. 51-87 in
            *Advances in Group Processes*, Vol. 3, E. Lawler, ed. Greenwich, CT:JAI Press.
    1988    The provision of a sanctioning system in the United States and Japan. *Social Psychology
            Quarterly* 51(3):265-271.

# 5

# Appropriating the Commons:
# A Theoretical Explanation

*Armin Falk, Ernst Fehr, and Urs Fischbacher*

I n his classic account of social dilemma situations, Hardin (1968) develops his pessimistic view of the "tragedy of the commons." Given the incentive structure of social dilemmas, he predicts inefficient excess appropriation of common-pool resources. Hardin's view has been challenged by the insights of numerous field studies reported in the seminal book by Ostrom (1990). In this book the metaphor of a tragedy is replaced by the emphasis that people are able to govern the commons. Ostrom shows that in many situations people are able to cooperate and improve their joint outcomes. Moreover, the reported field studies point to the importance of behavioral factors, institutions, and motivations. However, although it has been shown that these factors collectively influence behavior, it is of course nearly impossible to isolate the impact of individual factors.

This is why we need controlled laboratory experiments: Only in an experiment is it possible to study the role of each factor in isolation. In carefully varying the institutional environment, the experimenter is able to disentangle the importance of different institutions and motivations. The regularities discovered in the lab then can be used to better understand the behavior in the field. In this paper we concentrate on three such empirical regularities, which are reported in Walker et al., (1990), and Ostrom et al. (1992).[1] They first study a baseline situation that captures the central feature of all common-pool resource problems: Because of negative externalities, individually rational decisions and socially optimal outcomes do not coincide. In a next step, the baseline treatment is enriched with two institutional features, the possibility of informal sanctions and the possibility to communicate. The empirical findings can be summarized as follows: In the baseline common-pool resource experiment, aggregate behavior is best described by the Nash equilibrium of selfish money maximizers. People excessively appropriate the common-pool resource and thereby give rise to the "tragedy" predicted by Hardin (1968). Giving subjects the possibility to sanction each other, however, strongly improves the prospects for cooperative behavior. The reason is that many people sanction defectors. This is surprising because sanctioning is costly and therefore not consistent with the assumptions that provide the basis for Hardin's pessimistic view, that is, that subjects are selfish and rational. A similar observation holds for the communication environment. Allowing for communication also increases cooperative behavior. The resulting efficiency improvement is again inconsistent with the behavioral assumptions underlying Hardin's analysis because communication does not alter the material incentives.

Taken together, therefore, we have the following puzzle: In a sparse institutional environment, people tend to overharvest common-pool resources. In this sense the pessimistic predictions by Gordon (1954) and Hardin (1968), which are based on the assumptions of selfish preferences, are supported. At the same time, however, we find the efficiency-enhancing effect of informal sanctions and communication. This is in clear contradiction to the standard rational choice view, because why should a rational and selfish individual sacrifice money in order to sanction the behavior of another subject? And why should a money-maximizing subject reduce his or her appropriation level following some cheap talk? The question is more general: Why is the rational choice conception correct in one setting and wrong in another?

In this paper we suggest an integrated theoretical framework that is capable of explaining this puzzle. We argue that the reported regularities are compatible with a model of human behavior that extends the standard rational choice approach and incorporates *preferences for reciprocity and equity.* The basic behavioral principle that is formalized in our model is that a substantial fraction of the subjects act conditionally on what other subjects do. If others are nice or cooperative, they act cooperatively as well, but if others are hostile, they retaliate.[2] Our model also accounts for the fact that there are selfish subjects who behave in the way predicted by standard rational choice theory. We formally show that the interaction of these two diverse motivations (reciprocity and selfishness) and the institutional setup is responsible for the observed experimental outcomes. In the absence of an institution that externally enforces efficient appropriation levels, the selfish players are pivotal for the aggregate outcome. However, if there is an institutional setup that enables people to impose informal sanctions or allows for communication, the reciprocal subjects discipline selfish players and thus shape the aggregate outcome. Moreover, our model shows that when the members of a group have a preference for reciprocity or equity, the common-pool resource problem is transformed into a coordination game with efficient and inefficient equilibria. If subjects are given the opportunity to communicate, they can, therefore, ensure that the equilibrium with the efficient appropriation level is reached.

In the presence of a preference for reciprocity and equity, communication is a coordination device that helps subjects to coordinate their behavior on the low—

appropriation equilibrium. Thus, if the institutional setup allows for sanctions or for communication, there is less appropriation in common-pool resource problems and higher voluntary contributions in public goods situations. Even though it is our main purpose in this chapter to show that the approach is able to account for the seemingly contradictory evidence of common-pool resource experiments, we believe the developed arguments are very general and likely to extend beyond the lab.

In the next section, we briefly outline the basic structure of our approach and recently developed fairness models. Then we apply our model to the standard common-pool resource game and discuss the theoretical predictions in light of empirical findings. We also provide propositions for a common-pool resource game with sanctioning opportunities as well as a discussion on the role of communication in the presence of reciprocal preferences. The subsequent section contains a comparison of common-pool resource results to those arrived in public goods games. The final section provides the conclusion.

## THEORETICAL MODELS OF RECIPROCITY AND FAIRNESS

A large body of evidence indicates that fairness and reciprocity are powerful determinants of human behavior (for an overview, see e.g., Fehr and Gächter, 2000b). As a response to this evidence, various theories of reciprocity and fairness have been developed (Rabin, 1993; Levine, 1998; Bolton and Ockenfels, 2000; Fehr and Schmidt, 1999; Falk and Fischbacher, 1999; Dufwenberg and Kirchsteiger, 1998; Charness and Rabin, 2000). These models assume that—in addition to their material self-interest—people also have a concern for fair outcomes or fair treatments. The impressive feature of several of these models is that they are capable of correctly predicting experimental outcomes in a wide variety of experimental games. Common to all of these models is the premise that the players' utility depends not only on their own payoff but also on *the payoff(s) of the other player(s)*. This assumption stands in sharp contrast to the standard economic model according to which subjects' utility is based solely on their own absolute payoff.

Some of the models mentioned are based on the notion that people care for fair outcomes (Bolton and Ockenfels, 2000; Fehr and Schmidt, 1999). Other models are based on the assumption that people evaluate the fairness of others' action in terms of the kindness of the intentions that triggered the action (Rabin, 1993; Dufwenberg and Kirchsteiger, 1999). Intention-based fairness models capture an important aspect of what has been called procedural fairness by some authors (e.g., Lind and Tyler, 1988). A third class of models combines outcome-based *and* intention-based notions of fairness (Falk and Fischbacher, 1998; Charness and Rabin, 2000). The experimental evidence (see, e.g., Blount, 1995; Falk et al., 2000a) indicates that subjects do not only sanction because they want to achieve fair outcomes but that the motive to punish unfair intentions is also a major determinant of sanctioning behavior. Therefore, approaches that rely on distributional concerns *and* on rewarding and sanctioning of intentions (as, e.g., in Falk and Fischbacher, 1998) best capture the experimental regularities.

All mentioned fairness theories are rational choice theories in the sense that they allow for interdependent preferences but assume rational individuals. This assumption may be criticized because often people act not fully rational but boundedly rational (e.g., Selten, 1998; Dietz and Stern, 1995). Although we are generally sympathetic with this view, we would like to point out that so far there is no formal model of bounded rationality that is able to predict the experimental results presented in this chapter in a rigorous way.

In the games analyzed in this chapter, the Falk and Fischbacher model and the Fehr and Schmidt model yield similar predictions. We therefore restrict our attention to the latter model because it is relatively easy to apply in our context. The Bolton and Ockenfels model also yields similar predictions in the baseline common-pool resource environment, but predicts a wrong punishment pattern in the common-pool resource game with punishment opportunities. The reason is that in their model, each player does not evaluate fairness toward each other player (as in the Falk and Fischbacher and the Fehr and Schmidt models), but rather toward the group average. This basically means that people are indifferent between punishing defectors or punishing cooperators, a prediction that is at odds with the experimental data. Finally, the Dufwenberg and Kirchsteiger model and the Charness and Rabin model are extremely complicated and often do not generate precise predictions. Both models often predict many equilibria. Finally, we do not apply altruism models (see, e.g., Palfrey and Prisbrey, 1997) because these models are not compatible with sanctions, nor with the fact that people cooperate conditionally.

In the Fehr and Schmidt model, fairness is modeled as "inequity aversion." An individual is inequity averse if he or she dislikes outcomes that are perceived as inequitable. This definition raises, of course, the difficult question of how individuals measure or perceive the fairness of outcomes. Fairness judgments inevitably are based on a kind of neutral reference outcome. The reference outcome that is used to evaluate a given situation is itself the product of complicated social comparison processes. In social psychology (Adams, 1963; Festinger, 1954; Homans, 1961) and sociology (Davis, 1959; Pollis, 1968; Runciman, 1966), the relevance of social comparison processes has been emphasized for a long time. One key insight of this literature is that relative material payoffs affect people's well-being and behavior. As we will see, without the assumption that relative payoffs matter at least to some people, it is difficult, if not impossible, to make sense of the empirical regularities observed in common-pool resource experiments. There is, moreover, direct empirical evidence suggesting the importance of relative payoffs. The results in Agell and Lundborg (1995) and Bewley (1998), for example, indicate that relative payoff considerations constitute an important constraint for the internal wage structure of firms. In addition, Clark and Oswald

(1996) show that comparison incomes have a significant impact on overall job satisfaction. Strong evidence for the importance of relative payoffs also is provided by Loewenstein et al. (1989). These authors asked subjects to ordinally rank outcomes that differ in the distribution of payoffs between the subject and a comparison person. On the basis of these ordinal rankings, the authors estimate how relative material payoffs enter the person's utility function. The results show that subjects exhibit a strong and robust aversion against disadvantageous inequality: For a given own income $x_i$ subjects rank outcomes in which a comparison person earns more than $x_i$ substantially lower than an outcome with equal material payoffs. Many subjects also exhibit an aversion against advantageous inequality, although this effect seems to be significantly weaker than the aversion against disadvantageous inequality.

The determination of the relevant reference group and the relevant reference outcome for a given class of individuals ultimately is an empirical question. The social context, the saliency of particular agents, and the social proximity among individuals are all likely to influence reference groups and outcomes.[3] Because in the following discussion we restrict attention to individual behavior in economic experiments, we have to make assumptions about reference groups and outcomes that are likely to prevail in this context. In the laboratory it is usually much simpler to define what is perceived as an equitable allocation by the subjects. The subjects enter the laboratory as equals, they don't know anything about each other, and they are allocated to different roles in the experiment at random. Thus, it is natural to assume that the reference group is simply the set of subjects playing against each other and that the reference point, that is, the equitable outcome, is given by the egalitarian outcome.

So far we have stressed the importance of the concern for relative payoffs. This does not mean, however, that the absolute payoff should be viewed as a quantité negliable. Moreover, we do not claim that all people share a (similar) concern for an equitable share. In fact, many experiments have demonstrated the heterogeneity of subjects and the importance of absolute payoffs. A discussion on the heterogeneity of individual preferences is given, for example, in Parks (1994), Van Lange et al. (1997), and Kopelman et al. (this volume:Chapter 4). In this literature different types are discussed, such as cooperators, individualists, and competitors. In a similar vein, we assume that there are selfish people who care only for their material payoff and fair-minded people who reward fair and punish unfair behavior. As we will see, the interaction between these two types explains much of the observed data.

To be precise, in the Fehr and Schmidt model it is assumed that in addition to purely selfish subjects, there are subjects who dislike inequitable outcomes. They experience inequity if they are worse off in material terms than the other players in the experiment and they also feel inequity if they are better off. Moreover, it is assumed that, in general, subjects suffer more from inequity that is to their material disadvantage than from inequity that is to their material advantage. Formally, consider a set of $n$ players indexed by $i \in \{1,\dots, n\}$ and let $\pi = (\pi_1,\dots,\pi_n)$ denote the vector of monetary payoffs. The utility function of player $i$ is given by

$$U_i = \pi_i - \frac{\alpha_i}{n-1} \sum_{j,\pi_j > \pi_i} (\pi_j - \pi_i) - \frac{\beta_i}{n-1} \sum_{j,\pi_i > \pi_j} (\pi_i - \pi_j) \qquad (5\text{-}1)$$

where $\alpha_i \geq \beta_i \geq 0$ and $\beta_i < 1$.

The first term in Equation 5-1, $\pi_i$, is the material payoff of player $i$. The second term in Equation 5-1 measures the utility loss from disadvantageous inequality, and the third term measures the loss from advantageous inequality. Figure 5-1 illustrates the utility of player $i$ as a function of $x_j$ for a given income $x_i$. Given his own monetary payoff $x_i$, player $i$'s utility function obtains a maximum at $x_j = x_i$. The utility loss from disadvantageous inequality $(x_j > x_i)$ is larger than the utility loss if player $i$ is better off than player $j$ $(x_j < x_i)$.

To evaluate the implications of this utility function, let us start with the two player case. For simplicity the model assumes that the utility function is linear in inequality aversion as well as in $x_i$. Furthermore, the assumption $\alpha_i \geq \beta_i$ captures the idea that a player suffers more from inequality that is to his disadvantage. The paper mentioned by Loewenstein et al. (1989) provides strong evidence that this assumption is, in general, valid. Note that $\alpha_i \geq \beta_i$ essentially means that a subject is loss averse in social comparisons: Negative deviations from the reference outcome count more than positive deviations. The model also assumes that $0 \leq$

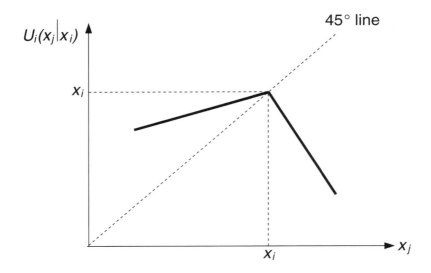

FIGURE 5-1 Preferences of inequity aversion.

$\beta_i < 1$. $\beta_i \geq 0$ means that the model rules out the existence of subjects who like to be better off than others. To interpret the restriction $\beta_i < 1$, suppose that player $i$ has a higher monetary payoff than player $j$. In this case $\beta_i = 0.5$ implies that player $i$ is just indifferent between keeping $1 to himself and giving this dollar to player $j$. If $\beta = 1$, then player $i$ is prepared to throw away $1 in order to reduce his advantage relative to player $j$ which seems very implausible. This is why we do not consider the case $\beta_i \geq 1$. On the other hand, there is no justification to put an upper bound on $\alpha_i$. To see this, suppose that player $i$ has a lower monetary payoff than player $j$. In this case player $i$ is prepared to give up $1 of his own monetary payoff if this reduces the payoff of his opponent by $(1 + \alpha_i) / \alpha_i$ dollars. For example, if $\alpha_i = 4$, then player $i$ is willing to give up $1 if this reduces the payoff of his opponent by $1.25.

If there are $n > 2$ players, player $i$ compares his income to all other $n - 1$ players. In this case the disutility from inequality has been normalized by dividing the second and third term by $n - 1$. This normalization is necessary to make sure that the relative impact of inequality aversion on player $i$'s total payoff is independent of the number of players. Furthermore, we assume for simplicity that the disutility from inequality is self-centered in the sense that player $i$ compares himself to each of the other players, but does not care per se about inequalities within the group of his opponents.

## THEORETICAL PREDICTIONS

In the following text we discuss the impact of inequity aversion in typical common-pool resource games. The first game we analyze is a standard common-pool resource game without communication and sanctioning opportunities. We proceed by analyzing games that add the possibilities of costly sanctioning and communication, respectively. For all games we first derive the standard economic prediction, that is, the Nash equilibrium assuming that everybody is selfish and rational. We contrast this prediction with experimental results and the prediction derived by our fairness model. In presenting the experimental results, we restrict our attention to behavior of subjects in the final period because in that period, nonselfish behavior cannot be rationalized by the expectation of rewards in future periods. Furthermore, in the final period, we have more confidence that the players fully understand the game being played. The reason we do not analyze one-shot data (as, e.g., in Rutte and Wilke, 1985) is simple: To our knowledge there are no one-shot experiments where the same common-pool resource game has been studied in various environments. Only the repeated game data by Walker and colleagues (1990) allows this type of analysis because they studied the same game in various institutional setups. Of course the final period of a repeated interaction may be different in some way from a pure one-shot game. It has been argued, for example, that people might not sanction if they interact only once. This conjecture, however, clearly is refuted by recent experimental evidence showing that even in a pure one-shot game, many people engage in costly sanctions and punish defectors (Falk et al. 2000b).

## The Standard Common-Pool Resource Game

In a standard common-pool resource game, each player is endowed with an endowment $e$. All $n$ players in the group decide independently and simultaneously how much they want to appropriate from a common-pool resource. Individual $i$'s appropriation decision is denoted by $x_i$. The appropriation decision causes a cost $c$ per unit of appropriation but also yields a revenue. Although the cost is assumed to be independent of the decisions of the other group members, the revenue depends on the appropriation decisions of all players. More specifically, the total revenue of all players from the common-pool resource is given by $f(\Sigma x_j)$ where $\Sigma x_j$ is the amount of total appropriation. For low levels of total appropriation $f(\Sigma x_j)$ is increasing in $\Sigma x_j$, but beyond a certain level $f(\Sigma x_j)$ is decreasing in $\Sigma x_j$. An individual subject $i$ receives a fraction of $f(\Sigma x_j)$ according to the individual's share in total appropriation $\dfrac{x_i}{\Sigma x_j}$. Thus the total material payoff of $i$ is given by:

$$\pi_i = e - cx_i + \left[\frac{x_i}{\Sigma x_j}\right] f\left(\Sigma x_j\right)$$

In the experiments of Walker and colleagues (1990), $e = 10$ (or 25) and $c = 5$. The total revenue is given by $f(\Sigma x_j) = 23\,\Sigma x_j - .25(\Sigma x_j)^2$. Thus in this experiment, material payoffs are:

$$\pi_i = 10 - 5x_i + \left[\frac{x_i}{\Sigma x_j}\right]\left[23\Sigma x_j - .25\left(\Sigma x_j\right)^2\right]$$

Intuitively this is a social dilemma problem because individual $i$'s appropriation decision $x_i$ does not only affect player $i$'s payoff, but also that of all other players. Beyond a certain level of total appropriations, an increase in the appropriation of player $i$ lowers the other players' revenue from the common-pool resource. Because selfish players are concerned only with their own well-being, they do not care about the negative externalities they impose on others. As we will discuss, this leads to the typical inefficiencies that are characteristic for this type of dilemma games.

The above payoff function from Walker and colleagues can be transformed into $\pi_i = 10 + 18x_i - 0.25x_i \Sigma x_j$ or more abstractly as $\pi_i =: e + \alpha x_i - bx_i \Sigma x_j$. As we will see this notation will be useful in the following discussion.

In this common-pool resource game, the standard economic prediction (assuming completely selfish and rational subjects) is as stated in Proposition 1:[4]

*Proposition 1 (Selfish Nash Equilibrium)*

> *If all players have purely selfish preferences, the unique Nash equilibrium is symmetric and individual appropriation levels are given by* $x_i^* = \dfrac{a}{b(n+1)}$.

In the following we denote this equilibrium as *SNE* (Selfish Nash Equilibrium) and the corresponding individual appropriation levels as $x_{SNE}$. As can be seen from Proposition 1, the individual contribution is independent of the endowment and it is decreasing in the number of players. In the specification of Ostrom et al., groups of eight players participated in the experiment. Thus in their experiment the predicted individual contribution amounts to $x_i^* = \dfrac{18}{0.25(8+1)} = 8$. Given the group size, total appropriation is 64. Compared to the social optimum of 36, this equilibrium yields substantial inefficiencies.[5] The point is that in their decisions, subjects ignore the negative externality imposed on the other players. Because players are assumed to care only about their own material payoff, they simply don't care about such externalities.

How does the presence of inequity-averse or reciprocally motivated subjects alter the standard economic prediction? To answer this question, we will discuss two propositions. The first proposition considers symmetric equilibria whereas the second deals with asymmetric equilibria.

It is useful to start our discussion of the properties of symmetric equilibria with the nature of the *best response function* of an inequity-averse subject. The best response function indicates the optimal appropriation response of an inequity-averse player to the average appropriation of all other players. Figure 5-2 shows the best response of an inequity-averse subject (with positive $\alpha$ and $\beta$) and compares it to that of a selfish subject. The thin line represents the optimal appropriation of a selfish subject given the average appropriation level of the other group members.[6] As can be seen in Figure 5-2, a selfish player appropriates less, the more the other group members appropriate. At the point where the best response function intersects the diagonal, the *SNE* prevails. At this point the average appropriation level of the other $n-1$ players is 8 from the viewpoint of each individual player. Moreover, it is in the self-interest of each player to respond to this average appropriation of the $n-1$ other players with an own appropriation level of 8. Now look at the bold line in Figure 5-2. This line shows the best response behavior of an inequity averse subject. Four aspects of this function are important to emphasize.

First, in the area above the diagonal, that is, where the other players appropriate *less* than in the *SNE*, the best response curve of an inequity-averse player lies *below* that of a selfish player. This means that if the other players are "nice" in the sense that they appropriate less than what is in their material self-interest, an inequity-averse subject also appropriates less. Because inequity-averse players dislike being in a too favorable position, they do not exploit the kindness of the

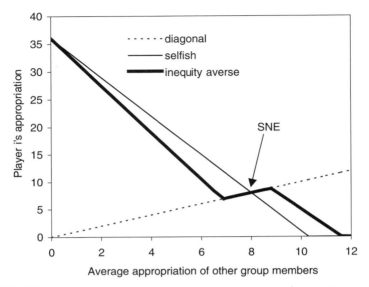

FIGURE 5-2  Best response behavior in a standard common-pool resource game (alpha = 4, beta = 0.6).

other players but instead voluntarily sacrifice some of their resources in favor of the other players.

Second, there is an area below the diagonal. In this area the other group members appropriate *more* than in the *SNE*. The best response behavior of an inequity-averse subject dictates to appropriate more than is compatible with pure self-interest in this case. Here, the intuition is that because the other players appropriate more than in the *SNE*, the inequity-averse player takes revenge by imposing negative externalities on the other players. The desire to take revenge results from the fact that the large appropriation levels of the others cause disadvantageous inequality for the inequity-averse subject. Because appropriating in this area reduces the payoff of the others more than their own payoff, an inequity-averse player can reduce the payoff differences. The selfish player, on the other hand, does not care about payoff differences and therefore appropriates less in this situation.

Third, a part of the inequity-averse player's best response lies right *on* the diagonal. This is the area in which symmetric equilibria may exist. There may be equilibria in which subjects appropriate less than in the *SNE* as well as equilibria in which they appropriate more. Of particular interest are the equilibria to the left of the *SNE* because in this direction efficiency is increasing (up to the optimal appropriation level of 36). Whether such equilibria do exist depends on the distribution of the parameters $\alpha$ and $\beta$ (see our discussion on Proposition 2).

Fourth, notice that in case the others do not appropriate at all the best responses of selfish and inequity-averse players coincide. At first glance, this seems counterintuitive, because in a certain sense appropriating nothing is the most friendly choice of the other group members. However, the coincidence of the two best response functions at that point is quite sensible. The reason is that if the other group members do not appropriate at all, the appropriation decisions of a player do not affect the other players' payoffs at all. This is because the other players' share of the total revenue $\dfrac{x_i}{\sum x_j}$ is zero. So why should an inequity averse-player not choose the money-maximizing appropriation level of 36 units? Remember that the utility function specified in equation 5-1 combines a concern for absolute income and for payoff differences. In case the other players do appropriate nothing, utility is equal to $U_i = \pi_i - \beta_i(\pi_i - \pi_{-i}) = \pi_i(1 - \beta_i) + \beta_i \pi_{-i}$, where $\pi_{-i}$ is the individual payoff of each of the $n - 1$ other players who appropriate zero. Because $\pi_{-i}$ is equal to $e$, it does not depend on the choice of player $i$, and because $\beta_i < 1$, it is clear that even for a highly inequity-averse subject, money-maximizing behavior and utility-maximizing behavior coincide.

Given the best response behavior of inequity-averse subjects, the existence conditions and the nature of symmetric equilibria are described in the next proposition. Note that in this proposition, $\min(\beta_i)$ denotes the smallest $\beta_i$ among all $n$ players and $\min(\alpha_i)$ denotes the smallest $\alpha_i$.

*Proposition 2 (Symmetric Equilibria with Inequity-Averse Subjects)*

*There is a symmetric equilibrium in which each subject chooses*

$$x_i^* = \hat{x} \; \textit{if} \; \hat{x} \; \textit{is in the interval} \left[ \frac{a(1 - \min(\beta_i))}{b(1 + n(1 - \min(\beta_i)))}, \frac{a(1 + \min(\alpha_i))}{b(1 + n(1 + \min(\alpha_i)))} \right].$$

The intuition of Proposition 2 is as follows: If both the smallest $\alpha_i$ and the smallest $\beta_i$ are equal to zero, the only equilibrium is the *SNE*, that is, $\hat{x} = \dfrac{a}{b(n+1)}$. This means that the presence of only one egoistic player in the group (with $\alpha_i = \beta_i = 0$) suffices to induce all other players to act in a selfish manner, regardless of how inequity averse they are. Put differently, a single egoist rules out any efficiency improvement compared to the *SNE* even if all other $n - 1$ players are highly inequity averse.

Proposition 2 entails a very strong result. It states that the subject with the "weakest preferences" for an equitable outcome dictates the outcome for the whole group. Only if the lowest $\alpha_i$ or the lowest $\beta_i$ are greater than zero do asymmetric equilibria that differ from the *SNE* exist. Of particular interest are equilibria where the smallest $\beta_i$ is greater than zero. In this case the lower bound of the interval given in Proposition 2 is smaller than $x_{SNE}$, that is, there are equilibria "to the left" of the *SNE*. In these equilibria subjects appropriate less than in the *SNE*. Similarly, if the smallest $\alpha_i$ is larger than zero, there exist equilibria in which subjects appropriate more than in the *SNE*. If all players in a group are inequity averse and given the parameters of the Ostrom et al. experiment, the range of possible Nash equilibria is $0 < \hat{x} < 9$ where $\hat{x}_{SNE} = 8$ is *always* an equilibrium independent of $\alpha_i$ and $\beta_i$.[7]

So far we have concentrated on symmetric equilibria. However, there are also asymmetric equilibria. The following proposition provides the details.[8]

*Proposition 3 (Asymmetric Equilibria with Inequity-Averse Subjects)*

*(i) If there are at least k players with* $\dfrac{\beta_i}{\alpha_i} > \dfrac{n-k}{k-1}$ *then there is an equilibrium with less appropriation than in the SNE. In this equilibrium at least k players choose the same appropriation* $\hat{x} < x_{SNE}$; *the other players j choose higher appropriation levels. (ii) If there is no k such that there are at least k players with* $\dfrac{\beta_i}{\alpha_i} > \dfrac{n-k}{k-1}$ *then there is no equilibrium with less appropriation than in the SNE.*

*Corollary 1: If there are* $\dfrac{n}{2}$ *or more selfish players, then there is no equilibrium with less appropriation than in the SNE.*

The intuition of Proposition 3 is straightforward. To get more efficient equilibria than the *SNE* requires that a relatively large fraction of subjects have rather high $\dfrac{\beta_i}{\alpha_i}$ combinations. Notice that because $0 < \beta_i < 1$ and $\alpha_i > \beta_i$, the expression $\dfrac{\beta_i}{\alpha_i}$ is between zero and one. This means that for $k \leq \dfrac{n}{2}$ the only equilibrium is the *SNE*. Only if $k$ is higher than $n/2$ there are $\dfrac{\beta_i}{\alpha_i}$ combinations to ensure a more efficient equilibrium. For example, if the group size is 8, it takes at least 5 nonegoistic players to reach such an equilibrium. In this case the $\dfrac{\beta_i}{\alpha_i}$ combinations of these 5 subjects must be at least $\dfrac{\beta_i}{\alpha_i} > \dfrac{8-5}{5-1} = 3/4$. The more people who are nonselfish, the weaker is the requirement for

$$\frac{\beta_i}{\alpha_i}\left(k=6 \rightarrow \frac{\beta_i}{\alpha_i} > \frac{2}{5}; k=7 \rightarrow \frac{\beta_i}{\alpha_i} > \frac{1}{6}; k=8 \rightarrow \frac{\beta_i}{\alpha_i} > 0\right).$$

Thus, to reach a more efficient outcome, it takes either many subjects with moderate $\dfrac{\beta_i}{\alpha_i}$ combinations or it takes fewer subjects (but still more than $n/2$) with very high $\dfrac{\beta_i}{\alpha_i}$ combinations. Notice that the expression $\dfrac{\beta_i}{\alpha_i}$ rises in $\beta_i$ and decreases in $\alpha_i$. It is therefore more likely to reach more efficient outcomes if subjects have a rather large utility loss from advantageous inequality, and a rather small utility loss from disadvantageous inequality.

To summarize the results of this section: What are the prospects for an out-come that is "better" than that of the *SNE?* If we look at the requirements of Propositions 2 and 3, some skepticism is in place. The requirements are rather tough. In the symmetric case, it takes only one selfish subject to ensure that the only equilibrium is *SNE.* Of course, more efficient equilibria are possible. How-ever, we expect that this is the exception rather than the rule. For example, if we assume that there are about 25 percent purely selfish people (a rather optimistic guess), the chance to have no egoist in a randomly drawn group of 8 subjects is about 10 percent. This means that on average in 1 out of 10 groups, we would possibly expect less appropriation than in the *SNE.*

What about the asymmetric case? On first sight, requirements seem weaker. There are more efficient outcomes even in the presence of a selfish player. How-ever, it takes again more than half of the players who are (i) nonselfish and (ii) who have a rather high utility loss from advantageous inequality compared to their loss in utility that derives from disadvantageous inequality. We have strong doubts that this type of preference is sufficiently frequent. As we have pointed out, we expect that the aversion with respect to disadvantageous inequality is usually much stronger than that arising from advantageous inequality (see also Loewenstein et al., 1989).

Our conclusion, therefore, is that even in the presence of many inequity-averse and reciprocal subjects, the prospects for achieving a more efficient equi-librium than the *SNE* are rather weak in the standard common-pool resource game. This is consistent with much of the reported data, according to which—on aver-age—the *SNE* describes aggregate behavior quite well. In their repeated com-mon-pool resource game, Ostrom et al. (1994) report that average final period appropriation levels in three different groups were 63, 64, and 78 (in case the endowment was 25 tokens) and 60, 63, and 70 (in case the endowment was 10 tokens).[9] These numbers are very close to the standard prediction of 64.

## A Common-Pool Resource Game With Sanctioning Opportunities

In this section we discuss a variant of the standard common-pool resource game. Again, we follow the experimental setup of Ostrom et al., (1992).[10] Their sanctioning institution is built on the standard common-pool resource game dis-cussed earlier. Subjects now first play the standard game, and after each round of play, all subjects receive data on all individual appropriation decisions. Each sub-ject then can decide to sanction any other group member at a certain cost. Techni-cally, any player $i$ can deduct $p_{ij}$ points from player $j$'s payoff at cost $cp_{ij}$ where $c$ is a positive constant smaller than 1. In the reported experiments, different pa-rameter constellations were used where $p_{ij}$ varied from 10 to 80 cents and $cp_{ij}$ varied from 5 to 20 cents.

Because sanctioning is costly in this type of experiments, the standard game theoretic prediction (assuming selfish preferences) is exactly as stated in Proposition 1. The rationale for this prediction is straightforward. Why should a rational and selfish person spend resources to sanction another person in the final stage? Because sanctioning is costly and utility depends only on their own material pay-off, sanctioning is equivalent to throwing away money. Rational subjects are able to perform the necessary backward induction, so everybody knows that nobody will sanction, no matter how egoistical the appropriation behavior on the first stage actually is. Thus appropriation is totally unaffected by the presence of a sanctioning stage.

Contrary to this prediction, Ostrom et al. (1994:176), report the following stylized facts:[11]

- Significantly more sanctioning occurs than according to the standard prediction.
- Sanctioning is inversely related to the cost of sanctioning ($c$).
- Sanctioning is focused primarily on subjects who appropriate the most from the common-pool resource.
- Sanctioning has a modest efficiency-enhancing impact on appropriation behavior (i.e., there is less appropriation than in the *SNE*).
- There is some sanctioning behavior that can be classified as "error, lagged punishment, or 'blind' revenge."
- Taking into account the cost of sanctioning, overall efficiency is similar to the standard common-pool resource without sanctioning opportunities.

This evidence is largely at odds with the homo economicus perspective. Assuming interdependent preferences, however, this evidence can be explained. In particular, our model predicts that defectors will be punished by those players who have sufficiently strong preferences for equity and reciprocity. This punishment serves as a discipline device for the selfish players. As a consequence selfish players have an incentive to act more cooperatively compared to a situation where there is no sanctioning institution. Thus, for a given population (of selfish and inequity-averse subjects), the prospects for more efficient appropriation levels are clearly improved in a common-pool resource environment with sanctioning possibilities. Precise conditions for the existence of equilibria with appropriation levels below the *SNE* are given in the following proposition.

*Proposition 4 (Equilibria with Sanctioning Possibilities)*

*Suppose there is a number $k \leq n$ such that for all players $i \leq k$, the utility parameters $\alpha_i$ and $\beta_i$ satisfy*

$$c < \frac{\alpha_i}{(1+\alpha_i)(n-k)+(k-1)(1-\beta_i)} = \frac{\alpha_i}{(n-1)(1+\alpha_i)-(k-1)(\alpha_i+\beta_i)} \equiv \hat{c} .$$

*We call the players who satisfy this condition "conditionally cooperative enforc-*

*ers" (CCEs). Suppose further that all players $i > k$ obey the condition $\alpha_i = \beta_i = 0$, that is they are selfish. We define $\beta_{min} = min_{i \leq k} \beta_i$ as the smallest $\beta_i$ among the CCEs. Then there is an equilibrium that can be characterized as follows: (i) All*

*players choose* $x \in \left[ \dfrac{a(1 - \beta_{min})}{b(1 + n(1 - \beta_{min}))}, x_{SNE} \right]$. *(ii) If each player does so, there is*

*no sanctioning in the second stage. (iii) If one player chooses a higher appropriation level, then this player is sanctioned equally by all CCEs. The sanctioning equalizes the payoffs of those who sanction and the player who deviates from x. (iv) If more than one player does not play x, an equilibrium of the sanctioning subgame is played.*

Proposition 4 determines the critical condition for an equilibrium in which all players appropriate less than $x_{SNE}$. It states that the cost of sanctioning $c$ must be lower than a certain threshold cost $\hat{c}$ level, which is defined by the preference parameters of the CCEs. The threshold $\hat{c}$, increases in $\alpha_i$, $\beta_i$, and $k$.

The intuition for the positive relation between $\hat{c}$ and $\alpha_i$ is the following: A player with a high $\alpha_i$ experiences a great disutility from disadvantageous inequity. This player is therefore willing to punish a selfish player who appropriates more than $x$ (and therefore earns more than player $i$), even if the sanctioning costs are high.

Why does the critical cost $\hat{c}$ increase in $\beta_i$? Remember that a person with a high $\beta_i$ has a strong aversion against advantageous inequality. Therefore, such a player $i$ will experience a strong disutility from the advantageous inequity toward the other CCEs when he, himself, does *not* sanction a selfish player. Put differently, given that the others spend resources in order to sanction a defector, a person with a high $\beta_i$ will feel solidarity toward these punishers. Thus, although a high $\alpha_i$ leads to punishment because of the inequity toward the deviating person, a high $\beta_i$ leads to punishment in order to reduce the inequity toward those who actually punish.

Finally, why is $\hat{c}$ increasing in $k$? A higher $k$ means there are more subjects who are willing to punish the players who deviate from $x$. Therefore, the desire to be in solidarity with those who punish increases as well, that is, ceteris paribus there will be more punishment.[12]

Notice that when there is a sanctioning opportunity, it is much easier to meet the conditions that sustain a cooperative outcome compared to the standard common-pool resource game. The conditions that $\alpha_i$ and $\beta_i$ have to meet are tougher when sanctioning is ruled out. Put differently, for a given distribution of inequity-averse and selfish players, it may be impossible to reach an equilibrium with a cooperative outcome when sanctioning is impossible, while there are equilibria with a cooperative outcome when sanctioning is possible. Thus, the model does explain why appropriation is more efficient with a sanctioning device than without. It also correctly predicts that those subjects who deviate from the "agreed" appropriation level $x$ will be punished. This is a very important point. Because it is exactly the defectors who get punished, the group can discipline the selfish players. As long as there are enough "norm enforcers," cooperation will be high and stable because the potential deviators face the credible threat of being punished if they behave selfishly. This pattern of punishment behavior can therefore be understood as a norm enforcement device (see Fehr and Gächter, 2000a). We would like to add that the Fehr and Schmidt model as well as the Falk and Fischbacher (1998) model share this feature.[13] Finally, the model does explain why sanctioning activities are inversely related to the cost of sanctioning. This follows immediately from Proposition 4. For a given set of preferences, the equilibria with cooperative outcomes will be the more likely the lower the cost of punishment.

In the experiments analyzed so far, the possibility to sanction clearly improves the prospects of cooperation. It has been argued, however, that the implementation of *explicit incentive devices* such as incentive contracts (which also punish noncompliance) may be counterproductive. In the presence of fairness preferences incentive devices may "crowd out" voluntary cooperation if they are perceived as unfair (see, e.g., Andreoni and Varian, 1999).

In our view, it depends very much on the precise nature of explicit incentives whether they are counterproductive or not. Although reciprocity-based punishments (as observed in the experiment already discussed) and repeated game incentives seem to be compatible with a cooperative atmosphere (Gächter and Falk, 2000), explicit incentive contracts may not always be. If a cooperative atmosphere is important for managing the commons, it is therefore important to balance the potential advantages and disadvantages of explicit incentives.

## The Impact of Communication

In all games discussed so far, subjects interact anonymously and without communication. In reality, however, people often communicate. They discuss problems like "overfishing," make (non)binding agreements on how to behave, and express approval or disapproval through (face-to-face) communication. Unless agreements are binding in a strict sense, however, the standard prediction with respect to behavior remains unchanged. When all people are completely selfish, there is no hope that after the promise not to appropriate excessively, a subject will actually stick to a promise. When it comes down to giving up money just to keep one's promise, standard theory predicts that subjects won't hesitate to pursue their material self-interest. In this sense, the opportunity to communicate is irrelevant for the predicted outcomes just as it was the case with the sanctioning opportunity.

The experimental evidence reported in Ostrom et al. (1994) casts serious doubts at this prediction.[14] They report that subjects "with one and only one opportunity to communicate, obtained an average percentage of net yield above that which was obtained in baseline experiments...without communication (55 percent compared to 21 percent)" (1994:198). Allowing subjects to communicate repeatedly increases efficiency even more (73 percent).[15]

In a meta-study by Sally (1995) on the determinants of cooperative behavior in more than 100 public goods experiments, communication has a significant and positive influence. In one-shot games, cooperation is raised by about 45 percent on average, whereas in repeated games the increase is 40 percent. Communication, however, is an elusive term. In some experiments, subjects really talk to each other, that is they exchange verbal and facial expressions. In other experiments, subjects do not communicate face to face but rather via computer or written notes. In yet other experiments, subjects do not actually talk but simply identify each other, that is, they do not exchange any verbal information. As diverse as the experiments is the discussion on why communication has a positive impact. Kerr and Kaufman-Gilliland (1994), for example, discuss nine different effects communication may have. It is not our aim to address these issues at length. The purpose of this section is to discuss how communication affects decisions when fairness concerns play a role. The two main effects we consider as relevant from this perspective are coordination and the expression of approval and disapproval. Both effects were possible in the common-pool resource communication treatment outlined earlier because subjects could not only exchange information, but also did see and talk to each other.

*Coordination*

Remember that in the standard common-pool resource game with inequity-averse or reciprocal preferences, there may be multiple equilibria. The *SNE* is always one of these equilibria, among others that are more (or less) efficient. To emphasize our point, let us abstract from all details and assume a two-player common-pool resource game. In terms of material payoffs, the game could look like the one expressed in Table 5-1a, that is, the common-pool resource game is similar to a prisoners' dilemma game. Even though it is in their common interest to choose the low appropriation level, both players can individually improve their material payoffs if they choose the high appropriation strategy. This yields the unique equilibrium where both players choose the high appropriation. If both

TABLE 5-1a  A Simple CPR Game without Reciprocal Preferences

|  |  | Player 2 | |
| --- | --- | --- | --- |
|  |  | Low appropriation | High appropriation |
| Player 1 | Low appropriation | 10,10 | 0,15 |
|  | High appropriation | 15,0 | 5,5 |

players have purely selfish preferences, this is the end of the story (and communication has no impact).

In the presence of reciprocal preferences, however, the common-pool resource game is no longer a prisoners' dilemma (see Table 5-1b). The reason is that if both players are sufficiently reciprocally motivated, they don't like to cheat on the other player. If, for example, player 1 chooses the low appropriation strategy, player 2 with reciprocal preferences is better off choosing the low instead of the high appropriation level and vice versa. Even though players forgo some material payoffs, they have a higher utility if they reciprocate the nice behavior of the other player. If player 1 chooses the high appropriation level, however, player 2 has no desire to choose the low level (neither if he is a selfish nor a reciprocal player). Instead, player 2 will in this case also play the high appropriation strategy. As a consequence, there are two (pure) equilibria now, the efficient equilibrium with low appropriations and the inefficient one with high appropriations. Put differently, the prisoners' dilemma game in Table 5-1a with a unique and inefficient equilibrium has turned into a coordination game with one efficient and one inefficient equilibrium. Game theory does not help much in this situation. It simply predicts that some Nash equilibrium will be played, but not which one.[16]

In the presence of multiple equilibria, subjects face a tremendous strategic uncertainty. How shall a person know which strategy the other player will select? It is obvious that communication can have a positive impact in a situation of strategic uncertainty. In fact it has been shown experimentally that communication can help players to coordinate on better equilibria.[17] As an example, take Cooper et al. (1992), who study different coordination games with and without communication. They find that, depending on the precise structure of the coordination game, communication may improve efficiency. This holds even though all announcements are nonbinding. They also show, however, that communication does not always improve coordination. The prospects for improved cooperation depend both on the nature of the game and the nature of the communication process.

In the common-pool resource experiments with communication mentioned earlier, players had intensive opportunities for communication because they could actually talk to each other and were (with some restrictions) allowed to discuss anything they wanted. As reported in Ostrom et al. (1994), subjects usually came

TABLE 5-1b A Simple CPR Game in the Presence of Reciprocal Preferences

|  |  | Player 2 | |
| --- | --- | --- | --- |
|  |  | Low appropriation | High appropriation |
| Player 1 | Low appropriation | 10,10 | 0,9 |
|  | High appropriation | 9,0 | 5,5 |

to the agreement to appropriate a particular amount (e.g., 5 tokens). If this is the case, coordination on "good" equilibria seems possible. Given these extensive communication opportunities and the fact that there is usually a substantial fraction of reciprocal subjects, it seems quite likely that communication raised efficiency because subjects could coordinate their choices on more efficient equilibria.

## Communication as a Sanctioning Device

Social interactions frequently are associated with social approval or disapproval. The anticipation of such social rewards and punishments may have important economic consequences. For example, it may affect the efficiency of team production and the decisions in diverse areas such as tax evasion, the exploitation of the welfare state, criminal activities, union membership, and voting behavior. The behavioral role of social rewards and punishments is stressed in social exchange theory (Blau, 1964). In contrast to pure economic exchanges, social exchanges involve not only the exchange of economic rewards but also the exchange of social rewards. The admiration or the contempt that is sometimes expressed by parents, teachers, professional colleagues, and spectators are prime examples of a social reward. In general social rewards are not based on explicit contractual arrangements but are triggered by spontaneous positive or negative emotions that can be interpreted as approval and disapproval, respectively.

Of course, approval as well as disapproval can be communicated and can have an important impact on individual behavior in a common-pool resource game. People who talk to each other enter a social relationship. Within this relationship, exchange of approval and disapproval is possible. Two assumptions must be met in order to observe more cooperative behavior compared to *SNE*, however. First, there must be subjects who actually care about approval or disapproval and who change their behavior in the expectation of such approval or disapproval. Second, there must be subjects who actually express approval or disapproval. The first condition is obvious. The second condition is important because it is usually not costless to express approval and in particular disapproval. Our point is that *reciprocally motivated subjects* are willing to bear the cost and are willing to reciprocate the cooperative or noncooperative actions by others. Thus, preferences as assumed in our model may explain why communication in combination with the expression of approval and disapproval can have a positive impact on cooperative behavior.

Taken together, we have described two potential channels through which communication may elicit cooperative behavior in the presence of reciprocal preferences. Although the first rests only on the exchange of information, the second is built on the possibility of communication face to face. We would expect, therefore, that communication effects are particularly strong if face-to-face communication is possible (as it is the case in the treatment already discussed).

This is also the conclusion of Rocco and Warglien (1995), who report a study showing that it is the communication face to face that makes the big difference (on this point, see also Frey and Bohnet, 1995; Bohnet and Frey, 1999a; Bohnet and Frey, 1999b; and Ostrom, 1998).

## PUBLIC GOODS: A COMPARISON

So far we have analyzed common-pool resource games. However, many of the arguments apply also to public goods games. In fact, public goods games and common-pool resource games are very similar. Whereas in a common-pool resource game, subjects' decisions impose negative externalities on other subjects, subjects in a public goods game produce positive externalities. In a common-pool resource game, it is nice or kind not to appropriate too much, while in a public goods game it is kind not to contribute too little to the public good. Public goods situations are very important and very frequent in reality.[18] Moreover, there exists a huge experimental literature on public goods games. As we will show, many of the findings reported on common-pool resource problems carry over to those of public goods. In this section we discuss a one-stage public goods game (similar to the standard common-pool resource game) and a two-stage public goods game, where after the first stage, subjects have a sanctioning opportunity (similar to the common-pool resource with sanctioning opportunities).

We start with the following linear public goods game. There are $n \geq 2$ players who decide simultaneously on their contribution levels $g_i \in [0, y], i \in [1, \ldots, n]$, to the public good. Each player has an endowment of $y$. The monetary payoff of player $i$ is given by $x_i(g_1, \ldots, g_n) = y - g_i + a \sum_j g_j$ where $1 / n < a < 1$. Because $a < 1$, a marginal investment into G causes a monetary loss of $(1 - a)$, that is, the dominant strategy of a completely selfish player is to choose $g_i = 0$. However, because $a > 1/n$, the aggregate monetary payoff is maximized if each player chooses $g_i = y$.

Consider now a slightly different public goods game that consists of two stages. At stage 1 the game is identical to the previous game. At stage 2 each player $i$ is informed about the contributions of all other players and can simultaneously impose a *costly punishment* on the other players, just as in the sanctioning common-pool resource game discussed.

What does the standard model predict for the two-stage game? Because punishments are costly, players' dominant strategy at stage 2 is to not punish. Therefore, if selfishness and rationality are common knowledge, each player knows that the second stage is completely irrelevant. As a consequence, players have exactly the same incentives at stage I as they have in the one-stage game without punishments, that is, each player's optimal strategy is to contribute nothing.

To what extent are these predictions of the standard model consistent with the data from public goods experiments? For the one-stage game there are, fortunately, a large number of experimental studies. In a meta-study of 12 experimental studies (with a total of 1,042 subjects participating), Fehr and Schmidt (1999) report that in the final period of public goods games without punishment, the vast majority of subjects play the equilibrium strategy of complete free riding. On average, 73 percent of all subjects choose $g_i = 0$ in the final period. It is also worth mentioning that in addition to those subjects who play exactly the equilibrium strategy, there is often a nonnegligible fraction of subjects who play "close" to the equilibrium.[19] In view of the facts, it seems fair to say that the standard model "approximates" the choices of a big majority of subjects rather well. However, if we turn to the public goods game with punishment, a radically different picture emerges although the standard model predicts the same outcome as in the one-stage game. Figure 5-3 shows the distribution of contributions in the final period of the two-stage game conducted by Fehr and Gächter (2000a). Note that the same subjects generated the distribution in the game without and in the game with punishment. Whereas in the game without punishment, most subjects play close to complete defection, a strikingly large fraction of 82.5 percent cooperates fully in the game with punishment. Fehr and Gächter report that the vast majority of punishments are imposed by cooperators on the defectors and that lower contribution levels are associated with higher received punishments. Thus, defectors do not gain from free riding because they are being punished.

When these results are compared with the evidence from common-pool re-

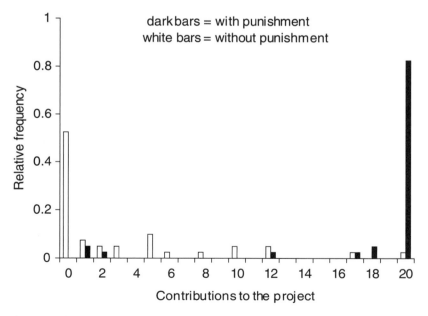

FIGURE 5-3 Contributions to the public good with and without punishment.

source games, a striking similarity arises. In the standard common-pool resource game, average behavior (in final periods) is fairly consistent with the standard prediction. However, if subjects have the opportunity to sanction each other, behavior becomes much more cooperative—even though the standard prediction yields the same outcome. As we have seen in our discussion, our fairness model can explain the evidence in both common-pool resource games. This holds also for the public goods games. The intuition for the one-stage public goods game is straightforward. Only if sufficiently many players have a dislike for an advantageous inequity can they possibly reach some cooperative outcome. As long as only a few players are willing to contribute if others contribute as well, they would suffer too much from the disadvantageous inequality caused by the free riders. Thus, inequity-averse players prefer to defect if they know there are selfish players. To put it differently: The greater the aversion against being the "sucker" the more difficult it is to sustain cooperation in the one-stage game.

Consider now the public goods game with punishment. To what extent is our model capable of accounting for the very high cooperation in this treatment? The crucial point is that free riding generates a material payoff advantage relative to those who cooperate. Because $c < 1$, cooperators can reduce this payoff disadvantage by punishing the free riders. Therefore, if those who cooperate are sufficiently upset by the inequality to their disadvantage, that is, if they have sufficiently high $\alpha$'s, then they are willing to punish the defectors even though this is costly to themselves. Thus, the threat to punish free riders may be credible, which may induce potential defectors to contribute at the first stage of the game.

Notice that according to the present model (and the inequity-aversion approach in general), a person will punish another person if and only if this reduces the inequity between the person and his opponent(s). Therefore, as long as $c < 1$ (as is the case in the common-pool resource problem and the public goods game analyzed previously), the model predicts punishments for sufficiently inequity-averse subjects. If, on the other hand, $c \geq 1$, the Fehr-Schmidt model predicts no punishment at all. This holds regardless of whether we look at public goods games or at the common-pool resource problems. Experimental evidence suggests, however, that many subjects in fact punish others even if punishment does *not* reduce inequity (as is the case if $c \geq 1$). Falk and colleagues (2000b) present several experiments that address this question in more detail. As it turns out, a substantial amount of punishment occurs even in situations where inequity cannot be reduced. For example, in one of their public goods games with punishment, they implemented a punishment cost of $c = 1$. Nevertheless, 46.8 percent of the subjects who cooperated in this game punished defectors. The conclusion from Falk et al. (2000b) is, therefore, that the desire to reduce inequity cannot be the only motivation to punish unkind acts. An alternative interpretation is offered in Falk and Fischbacher (1998), who model punishment as the desire to reduce the unkind players' payoff(s). Their model also correctly predicts punishments in those situations where punishment is costly and cannot reduce inequity.

## DISCUSSION

In the preceding sections, we have demonstrated that with the help of a simple fairness theory, many stylized facts of common-pool resource or public goods experiments can be explained. In fact, the range of experiments that have been successfully analyzed with the help of our fairness theories is even wider. Both the model by Fehr and Schmidt (1999) and that of Falk and Fischbacher (1998) are capable of predicting correctly a wide variety of seemingly contradictory experimental facts. They are, in particular, capable of reconciling the puzzling evidence that in competitive experimental markets with complete contracts, very unfair outcomes that are compatible with the predictions of the pure self-interest model can emerge, while in bilateral bargaining situations or in markets with incomplete contracts, stable deviations from the predictions of the self-interest model, in the direction of more fair and equitable outcomes, are the rule.

The basic behavioral principle that is formalized in the models is that a substantial fraction of the subjects act conditionally on what other subjects do. If others are nice or cooperative, they act cooperatively as well, but if others are hostile, they retaliate. The models also pay attention to the fact that there are large individual differences between subjects. In particular, it is assumed that there are selfish subjects who behave in the way predicted by standard economic theory and reciprocal subjects who exhibit the type of conditional behavior just mentioned. *The interaction of these diverse motivations and the institutional setup is responsible for the observed experimental outcomes.* If there is no institutional rule that externally enforces cooperation or that allows for sanctioning possibilities, the interaction of selfish and conditional subjects frequently leads to noncooperative outcomes. If, on the other hand, subjects dispose of sanctioning possibilities, the reciprocal subjects are able to discipline selfish players. As a consequence, more cooperative outcomes will emerge. This approach goes beyond the standard economic conception, not least because it assigns institutions a much more important role. In the presence of reciprocal and selfish subjects, institutions determine which type of preference is pivotal for the equilibrium outcome. In a sense institutions select the type of player that shapes the final result.

Of course there are several important behavioral factors that we have not addressed or that cannot be explained with the help of the presented theoretical framework. For example, there is a long tradition in social-psychological research that points to the importance of values and altruism as mechanisms to overcome the free-rider incentives inherent to social dilemma situations. Some of the literature associated with concepts of "social motives" or "social value orientation" is discussed in the Kopelman et al. paper (this volume:Chapter 4). Another alternative explanation to the one presented in the current paper comes from the literature on altruism rooted in theories of moral norm activation (Schwartz, 1977) and research on the structure of human values (Schwartz, 1992). This research has been tied to environmental resource management through studies of individual behavior in field situations, mainly measured by attitudinal and self-report indicators in surveys. In Stern et al. (1999), it is shown, for example, that people who accept a movement's values, who believe that things important to those values are threatened and who believe that their actions may help to alleviate the threat, experience a personal norm to support the movement.

The latter explanations as well as the fairness models presented in this paper assume different prosocial motives that mitigate free-rider incentives as suggested by the empirical findings. The theories do not, however, ask about possible evolutionary roots of such prosocial behavior. This important question is addressed in Richerson et al. (this volume:Chapter 12).[20]

Yet another remark is in place. We have emphasized the importance of reciprocity and inequity aversion but have not mentioned the impact of reputation and repeated game effects. Many of the real life common-pool resource or public goods problems are in fact "played" repeatedly. In such repeated interaction, players usually can condition their behavior on past behavior of others. This allows players to build up reputations and to ensure cooperative outcomes, even among selfish players. In the parlance of game theory, this kind of cooperation may be supported as an equilibrium in infinitely repeated games (folk theorems) or in finitely repeated games with incomplete information (see Kreps et al., 1982).[21] Many experiments have demonstrated the efficiency-enhancing effect of repeated versus one-shot interactions. Moreover, it has been shown that reciprocity and repeated game effects interact in a complementary way (Gächter and Falk, in press). In Gächter and Falk's experimental study of a bilateral labor relation, the reciprocal relationship between workers and firms is increased significantly in a repeated interaction compared to one-shot encounters. The driving force behind this "crowding in" of reciprocal behavior is the fact that people who behave selfishly in the one-shot game have an incentive to imitate reciprocity in the repeated game.[22] Thus, in the presence of repeated game incentives, the prospects for cooperative outcomes are expected to be better than according to the one-shot analysis undertaken in this paper.

## NOTES

1 We also refer to the book by Ostrom et al. (1994) that summarizes and discusses the experimental findings. For an overview on experimental results, see also Kopelman et al. (this volume:Chapter 4).

2 The importance of reciprocity has been established in dozens, if not hundreds, of experiments. For rewarding behavior in response to kind acts, see Fehr et al. (1993) or Berg et al. (1995). For punishing behavior in response to hostile acts, see Güth et al. (1982). Recent overviews are provided in Ostrom (1998) and Fehr and Gächter (2000b).

3 For a first attempt to endogenize the choice of refernce agents or standards in a formal model, see Falk and Knell (2000).

4 All propositions are proved in the Appendix to this chapter.

5 The derivation of the social optimum is given in the Appendix to this chapter.

6  Note that Figure 5-2 shows the symmetric case, where the other players' appropriation decisions are equal.

7  Because there is a whole range of equilibria, the question of equilibrium selection arises. This issue is discussed in the later section on coordination.

8  We restrict attention to the cases wherein equilibrium appropriation is less than in *SNE*.

9  We concentrate on the behavior of subjects in the final periods to exclude the possible confound of repeated games effects and to make behvior comparable to our one-shot predictions.

10  A sanctioning institution was first studied by Yamagishi (1986).

11  Moir (1999) studies the impact of monitoring additional to sanctioning. He points out that pure monitoring does not help to overcome excess appropriation. Institutions with a high level of monitoring but a low level of sanctioning may even lead to more apprpriation than institutions without any monitoring. A different design is suggsted by Casari and Plott (1999) in which monitoring and punishment are compacted in a single decision. In their treatment, efficiency is also higher compared to a baseline treatment without monitoring/punishment.

12  Notice that according to Proposition 4, the reason for *c* to increase in *k* is *not* because the costs of punishment can be shared between more punishers.

13  Other models predict a very different pattern of punishment. In Bolton and Ockenfels (2000), for example, punishment is not addressed individually, but directed toward a group average. This could imply, for example, that those who deviate are *not* punished whereas those who cooperate *are* punished. This is at odds with the experimental findings. Moreover, it does not take account of the ptential of reciprocal or equity preferences to establish and enforce social norms. For a detailed discussion of this point, see Falk et al. (2000b).

14  See also Ostrom and Walker (1991).

15  On communication, also see the paper by Kopelman et al. (this volume:Chapter 4).

16  See, however, the literature on equilbrium selection in, for example, Harsanyi and Selten (1988).

17  On coordination game experiments, see Ochs (1995).

18  For an overview on public goods experiments, see Ledyard (1995).

19  The results of the meta-study refer to public goods games where the group size is smaller than 10. There is also an experiment where the group size is substantially higher (40 and 100). In this experiment, contributions in the final period(s) are higher compared to small groups (Isaac et al., 1994).

20  See also Gintis (2000), Sethi and Somananthan (2000), and Huck and Oechssler (1999). See also de Waal (1996), who shows that conditional behavior is observed among chimpanzees. Their food-sharing behavior exhibits some reciprocal pattern: A chimpanzee is ceteris paribus more willing to share food with another champanzee if the latter has shared with the former in the past.

21  Notice that the latter model is built on the assumption that there exist selfish *and* reciprocal (tit-for-tat) types.

22  On the complementary relationship between reputation and reciprocity, also see the paper by Ostrom (1998).

# REFERENCES

Adams, J.S.
    1963  Toward an understanding of inequity. *Journal of Abnormal and Social Psychology* 62:422-436.

Agell, J., and P. Lundborg
    1995  Theories of pay and unemployment: Survey evidence from Swedish manufacturing firms. *Scandinavian Journal Of Economics* 97:295-307.

Andreoni, J., and H. Varian
  1999  Preplay contracting in the prisoner's dilemma. *Proceedings of the National Academy of Sciences* 96:10933-10938.
Berg, J., J. Dickhaut, and K. McCabe
  1995  Trust, reciprocity and social history. *Games and Economic Behavior* 10:122-142.
Bewley, T.
  1998  Why not cut pay? *European Economic Review* 42:459-490.
Blau, P.
  1964  *Exchange and Power in Social Life.* New York: Wiley.
Blount, S.
  1995  When social outcomes aren't fair: The effect of causal attributions on preferences. *Organizational Behavior and Human Decision Processes* 63(2):131-144.
Bohnet, I., and B.S. Frey
  1999a The sound of silence in prisoner's dilemma games. *Journal of Economic Behavior and Organization* 38:43-57.
  1999b Social distance and other rewarding behavior in dictator games: *Comment. American Economic Review* 89:335-339.
Bolton, G.E., and A. Ockenfels
  2000  A theory of equity, reciprocity and competition. *American Economic Review* 90:166-193.
Casari, M., and C. Plott
  1999  Agents Monitoring Each Other in a Common-Pool Resource Environment. Working paper, California Institute of Technology, Pasadena.
Chamess, G., and M. Rabin
  2000  Social Preferences: Some Simple Tests and a New Model. Working paper, University of California Berkeley.
Clark, A.E., and A.J. Oswald
  1996  Satisfaction and comparison income. *Journal of Public Economics* 61:359-381.
Cooper, R., D. DeJong, R. Forsythe, and T. Ross
  1992  Communication in coordination games. *Quarterly Journal of Economics* 107:739-771.
Davis, J.A.
  1959  A formal interpretation of the theory of relative deprivation. *Sociometry* 102:280-296.
Dietz, T., and P.C. Stern
  1995  Toward a theory of choice: Socially embedded preference construction. *Journal of Socio-Economics* 24(2):261-279.
Dufwenberg, M., and G. Kirchsteiger
  1998  A Theory of Sequential Reciprocity. Discussion paper, Center for Economic Research, Tilburg University.
Falk, A., E. Fehr, and U. Fischbacher
  2000a Testing Theories of Fairness: Intentions Matter. Working paper, Institute for Empirical Research, University of Zurich.
  2000b Informal Sanctions. Working paper, Institute for Empirical Research, University of Zurich.
Falk, A., and U. Fischbacher
  1998  A Theory of Reciprocity. Working paper 6, Institute for Empirical Research, University of Zurich.
Fehr, E., and S. Gächter
  2000a Cooperation and punishment in public good experiments—An experimental analysis of norm formation and norm enforcement. *American Economic Review* 90:980-994.
  2000b Fairness and retaliation: The economics of reciprocity. *Journal of Economic Perspectives* 14:159-181.
Fehr, E., G. Kirchsteiger, and A. Riedl
  1993  Does fairness prevent market clearing? An experimental investigation. *Quarterly Journal of Economics* 108:437-460.

Falk, A., and M. Knell
  2000  Choosing the Joneses: On the Endogeneity of Reference Groups? Working paper 53, Institute for Empirical Research, University of Zurich.
Fehr, E., and K. Schmidt
  1999  A theory of fairness, competition, and cooperation. *Quarterly Journal of Economics* 114:817-851.
Festinger, L.
  1954  A theory of social comparison processes. *Human Relations* 7:117-140.
Frey, B.S., and I. Bohnet
  1995  Institutions affect fairness: Experimental investigations. *Journal of Institutional and Theoretical Economics* 151:286-303.
Gächter, S., and A. Falk
  2000  Work motivation, institutions, and performance. In *Advances in Experimental Business Research*, R. Zwick and A. Rapoport, eds. Kluwer Academic Publishers.
  in     Reputation or Reciprocity—Consequences for the Labour Market. Submitted to Scandi-
  press  navian Journal of Economics.
Gintis, H.
  2000  Strong reciprocity and human sociality. *Journal of Theoretical Biology.*
Gordon, S.
  1954  The economic theory of common-property resource: The fishery. *Journal of Political Economy* 62:124-142.
Güth, W., R. Schmittberger, and B. Schwarze
  1982  An experimental analysis of ultimatum bargaining. *Journal of Economic Behavior and Organization* 3(3):367-388.
Hardin, G.
  1968  The tragedy of the commons. *Science* 162:1243-1248.
Harsanyi, J., and R. Selten
  1988  *A General Theory of Equilibrium Selection in Games.* Cambridge, MA: MIT Press.
Homans, G.C.
  1961  *Social Behavior: Its Elementary Forms.* New York: Harcourt, Brace & World.
Huck, S., and J. Oechssler
  1996  The indirect evolutionary approach to explaining fair allocations. *Games and Economic Behavior* 28:13-24.
Isaac, M.R., J.M. Walker, and A.M. Williams
  1994  Group size and the voluntary provision of public goods. *Journal of Public Economics* 54:1-36.
Kerr, N., and C. Kaufmann-Gilliland
  1994  Communication, commitment and coordination in social dilemmas. *Journal of Personality and Social Psychology* 66:513-529.
Kreps, D., P. Milgrom, J. oberts, and R. Wilson
  1982  Rational cooperation in the finitely repeated Prisoners' Dilemma. *Journal of Economic Theory* 27:245-252.
Ledyard, J.
  1995  Public goods: A survey of experimental research. In *Handbook of Experimental Economics*, J. Kagel and A. Roth, eds. Princeton: Princeton University Press.
Levine, D.
  1998  Modeling altruism and spitefulness in experiments. *Review of Economic Dynamics* 1:593-622.
Lind, E.A., and T.R. Tyler
  1988  *The Social Psychology of Procedural Justice.* New York: Plenum Press.

Loewenstein, G.F., L. Thompson, and M.H. Bazerman
    1989   Social utility and decision making in interpersonal contexts. *Journal of Personality and Social Psychology* 57:426-44 1.
Moir, R.
    1999   Spies and Swords: Behavior in Environments with Costly Monitoring and Sanctioning. Working paper, University of New Brunswick.
Ochs, J.
    1995   Coordination problems. In *Handbook of Experimental Economics*, J. Kagel and A. Roth, eds. Princeton: Princeton University Press.
Ostrom E.
    1990   *Governing the Commons: The Evolution of Institutions for Collective Action.* New York: Cambridge University Press.
    1998   A behavioral approach to the rational choice theory of collective action—Presidential address of the American Political Science Association 1997. *American Political Science Review* 92:1-22.
Ostrom, E., R. Gardner, and J. Walker
    1994   *Rules, Games, and Common Pool Resources.* Ann Arbor: University of Michigan Press.
Ostrom, E., and J. Walker
    1991   Communication in a commons: Cooperation without external enforcement. Pp. 287-322 in *Laboratory Research in Political Economy*, T. Palfrey, ed. Ann Arbor: University of Michigan.
Ostrom, E., J. Walker, and R. Gardner
    1992   Covenants with and without a sword: Self-governance is possible. *American Political Science Review* 40:309-317.
Palfrey, T., and J. Prisbrey
    1997   Anomalous behavior in public goods experiments: How much and why? *American Economic Review* 87:829-846.
Parks, C.D.
    1994   The predictive ability of social values in resource dilemmas and public goods games. *Personality and Social Psychology Bulletin* 20(4):431-438.
Pollis, N.P.
    1968   Reference groups re-examined. *British Journal of Sociology* 19:300-307.
Rabin, M.
    1993   Incorporating fairness into game theory and economics. *American Economic Review* 83(5):1281-1302.
Rocco, E., and M. Warglien
    1995   Computer Mediated Communication and the Emergence of Electronic Opportunism. Working paper RCC 13659, Universita degli Studi di Venezia.
Runciman, W.G.
    1966   *Relative Deprivation and Social Justice.* New York: Penguin.
Rutte, C.G., and H.A.M. Wilke
    1985   Preference for decision structures in a social dilemma situation. *European Journal of Social Psychology* 15:367-370.
Sally, D.
    1995   Conversation and Cooperation in Social Dilemmas: A Meta-Analysis of Experiments from 1958 to 1992. *Rationality and Society* 7(1):58-92.
Schwartz, S.H.
    1977   Normative Influences on Altruism. Pp. 221-279 in *Advances in Experimental Social Psychology Volume 10*, L. Berkowitz, ed. New York: Academic Press.

1992  Universals in the content and structure of values: Theoretical advances and empirical tests in 20 countries. Pp. 1-65 in *Advances in Experimental Social Psychology Volume 25*, L. Berkowitz, ed. New York: Academic Press.

Selten, R.
1998  Features of experimentally observed bounded rationality. *European Economic Review* 42:413-436.

Sethi, R., and E. Somananthan
2000  Preference evolution and reciprocity. *Journal of Economic Theory*.

Stem, P.C., T. Dietz, G.A. Guagnano, and L. Kalof
1999  A Value-belief-norm theory of support for social movements: The case of environmentalism. *Human Ecology Review* 6(2):81-97

Van Lange, P.A.M., W. Otten, E.M.N. De Bruin, and J.A. Joireman
1997  Development of prosocial, individualistic, and competitive orientations: Theory and preliminary evidence. *Journal of Personality and Social Psychology*, 73:733-746.

de Waal, F.
1996  *Good Natured. The Origins of Right and Wrong in Humans and Other Animals.* Cambridge, MA: Harvard University Press.

Walker, J., R. Gardner, and E. Ostrom
1990  Rent dissipation in a limited access common-pool resource: experimental evidence. *Journal of Environmental Economics and Management* 19:203-211.

Yamagishi, T.
1986  The provision of a sanctioning system as a public good. *Journal of Personality and Social Psychology* 51:110-116.

# APPENDIX TO CHAPTER 5

*Proof Proposition 1 (Selfish Nash Equilibrium)*

The standard common-pool resource game we look at has the following form:

$$\pi_i = 50 - 5x_i + \left[\frac{x_i}{\sum x_j}\right]\left[23\sum x_j - .25\left(\sum x_j\right)^2\right]$$

Defining e=50, a=18 and b=0.25, we get:

$$\pi_i = 50 + 18x_i - 0.25x_i\Sigma x_j$$

$$\pi_i =: e + ax_i - bx_i\Sigma x_j$$

To find the selfish best reply for $x_i$, we calculate

$$\frac{\partial \pi_i}{\partial x_i} = a - 2bx_i - b\sum_{j\neq i} x_j$$

Setting it equal to zero, we get as best reply $x_i = \max\left(0, \frac{1}{2}\left(\frac{a}{b} - \Sigma_{j\neq 1} x_j\right)\right)$

First suppose that all $x_i^* > 0$. In this case $2bx_i^* = a - b\sum_{j\neq 1} x_j^*$

Summing up the terms for all $i$, we get $2b\sum x_i^* = na - b(n-1)\sum x_i^*$.

Hence $\sum x_i^* = \dfrac{na}{b(n+1)}$.

Entering the sum into $2bx_i^* = a - b\sum x_j^*$ we get $x_i^* = \dfrac{a}{b(n+1)}$. Now consider that there are some players who choose 0. Let $n_0$ be the number of players who choose $x_i$ equal to zero. Then all values above zero must be equal to $\dfrac{a}{b(n-n_0+1)}$. Calculating the best reply for one of the $n_0$ players who originally played 0 now yields a contradiction. QED

**Social optimum:** To find the social optimum, we calculate

$$\frac{\partial\left(\sum \pi_i\right)}{\partial x_i} = an - 2bn\sum_j x_j$$

Hence, in the social optimum, we get $\Sigma_j x_j = \dfrac{a}{2b}$.

*Proof of Proposition 2 (Symmetric Equilibria with Inequity-Averse Subjects)*

First note that if $\sum x_j < \dfrac{a}{b}$ the players who choose the higher appropriation level also have higher payoffs:

$$\pi_j - \pi_i = \left(e + ax_j - bx_j\sum x_k\right) - \left(e + ax_i - bx_i\sum x_k\right) = \left(a - b\sum x_k\right)(xj - x_i)$$

So let us first consider this case.

Suppose all players $j \neq i$ choose $x_j = \hat{x} \leq x_{SNE}$. Because $U_i$ is concave in $x_i$, the best reply is unique. So, to show that $x_i^* = \hat{x}$ is the best reply, it is sufficient to show that it is a *local* optimum. It is clear that $x_i^* \geq \hat{x}$ because otherwise player $i$ could improve his material payoff as well as he could reduce inequity by increasing $x_i$. It remains to check that there is no incentive for $i$ to increase $x_i$ above $\hat{x}$. As the following calculation shows, the derivative from above $\dfrac{\partial U_i}{\partial x_i^+}(x)$ is a linear function in $x$. So, player $i$ has no incentive to increase $x_i^*$ above $\hat{x}$ if this derivative equals at least zero.

$$0 \geq \frac{\partial U_i}{\partial x_i^+} = \frac{\partial}{\partial x_i}\left(\pi_i - \frac{\beta_i}{n-1}\sum_{j,\pi_i > \pi_j}(\pi_i - \pi_j)\right)$$

$$= \frac{\partial}{\partial x_i}(\pi_i - \beta_i(\pi_i - \pi_j))$$

$$= \frac{\partial}{\partial x_i}(\pi_i(1 - \beta_i) + \beta_i\pi_j)$$

$$= \left(a - bx_i - b\sum x_j\right)(1 - \beta_i) - \beta_i bx_i$$

Thus, we get a critical condition for $x_i^*$

$$x_i^* \geq \frac{a(1 - \beta_i)}{b(1 + n(1 - \beta_i))} \tag{5-A}$$

The right-hand side of this inequality is decreasing in $\beta_i$. Thus, the left inequality of the proposition is satisfied, if and only if (5-A) is satisfied for all $i$.

Assume now all other players $j \neq i$ choose $x_j = \hat{x} > x_{SNE}$; $\hat{x} < \frac{a}{nb}$. Now, the critical condition is $\frac{\partial U_i}{\partial x_i^-}(x) > 0$.

$$\frac{\partial U_i}{\partial x_i^-} = \frac{\partial}{\partial x_i}\left(\pi_i - \frac{\alpha_i}{n-1}\sum_{j,\pi_j > \pi_i}(\pi_j - \pi_i)\right)$$

$$= \frac{\partial}{\partial x_i}(\pi_i - \alpha_i(\pi_j - \pi_i))$$

$$= \frac{\partial}{\partial x_i}(\pi_i(1 + \alpha_i) - \alpha_i\pi_j)$$

$$= \left(a - bx_i - b\sum x_j\right)(1 + \alpha_i) + \alpha_i bx_i$$

Thus, we get a critical condition for

$$x_i^* \leq \frac{a(1 + \alpha_i)}{b(1 + n(1 + \alpha_i))} \tag{5-A(1)}$$

The righthand side of this inequality is increasing in $\alpha_i$. Thus, the left inequality of the proposition is satisfied, if and only if 5-A(1) is satisfied for all $i$ and we get the right inequality in the proposition.

It remains to show that there is no equilibrium with appropriation decisions above $\frac{a}{nb}$. We fix $x_j = \hat{x} > \frac{a}{nb}$. The critical condition is $\frac{\partial U_i}{\partial x_i^-}(x) \geq 0$. Because a decrease of the appropriation level now generates inequity in favor of player $i$, we get the following condition:

$$0 \leq \frac{\partial U_i}{\partial x_i^-} = \frac{\partial}{\partial x_i}\left(\pi_i - \frac{\beta_i}{n-1}\sum_{j,\pi_i > \pi_j}(\pi_i - \pi_j)\right)$$

$$= (a - 2bx_i - b(n-1)x^*)(1 - \beta_i) - \beta_i bx_i$$

$$\leq \left(a - 2bx_i - b(n-1)\frac{a}{nb}\right)(1 - \beta_i) - \beta_i bx_i$$

$$= \left(\frac{a}{nb} - 2x_i\right)b(1 - \beta_i) - \beta_i bx_i$$

$$\leq (\hat{x} - 2x_i)b(1 - \beta_i) - \beta_i bx_i$$

Because $\beta_i < 1$, the last term is negative if $x_i$ is close to $\hat{x}$. Hence, there are no equilibria with $x^* > \frac{a}{nb}$. QED

*Proof of Proposition 3 (Asymmetric Equilibria with Inequity-Averse Subjects)*

We first show (ii): Let us assume there is an equilibrium with some $x_i^* < x_{SNE}$. By reordering the players, we can assume that we have $x_1^* \leq x_2^* \leq \ldots \leq x_n^*$. Furthermore, let $k$ be the highest index for which $x_1^* = x_k^*$. Now let's consider $i \leq k$. Because we are in an equilibrium, we have $\frac{\partial U_i}{\partial x_i^+} \leq 0$. Remember that $\pi_j - \pi_i = (a - b\sum x_k)(x_j - x_i)$. So

$$0 \geq \frac{\partial U_i}{\partial x_i^+}(x^*)$$

$$= \frac{\partial}{\partial x_i}\left(\pi_i - \frac{(a - b\Sigma x_k)}{n-1}\left(\beta_i(k-1)(x_i - x_1^*) + \alpha_i\sum_{j>k}(x_j^* - x_i)\right)\right)$$

$$= \frac{\partial \pi_i}{\partial x_i^+} - \frac{(a - b\Sigma x_k)}{n-1}\left(\beta_i(k-1) - \alpha_i(n-k)\right)$$

$$+ \frac{b}{n-1}\left(\beta_i(k-1)(x_i - x_1^*) + \alpha_i\sum_{j>k}(x_j^* - x_i)\right)$$

$$\geq -\frac{(a - b\Sigma x_k)}{n-1}\left(\beta_i(k-1) - \alpha_i(n-k)\right)$$

Hence

$$\beta_i(k-1) - \alpha_i(n-k) \geq 0$$

or

$$\frac{\beta_i}{\alpha_i} \geq \frac{n-k}{k-1}$$

which proves (ii).

Let us now come to the proof of (i). Assume without loss of generality that for $i$ between 1 and k, we have $\frac{\beta_i}{\alpha_i} > \frac{n-k}{k-1}$. This implies $k > \frac{n}{2}$ because $\frac{\beta_i}{\alpha_i} < 1$. We will show that there is an equilibrium with $x_1, = x_2 = ... = x_k < x_{SNE}$. *For* $x \in$ [0, $x_{SNE}$] we define the strategy combination $s(x)$ as follows: We fix $s(x)_i = x$ for $i \leq k$ and choose $s(x)_j$ for $j > k$ as the joint best reply. That means that $s(x)_j$ is a part of a Nash equilibrium in the $(n - k)$-player game induced by the fixes choice of $x$ by the first $k$ players. Because at least half of the players choose $x$, the best reply can never be smaller than $x$ (by increasing the appropriation level below $x$, the material payoff could be increased and the inequity disutility could be decreased as well). If we find $\hat{x}$, such that $\frac{\partial U_i}{\partial x_i^+}(s(\hat{x})) \leq 0$ for $i \leq k$, then $(\hat{x}, ..., \hat{x}, s(\hat{x})_{k+1}, s(\hat{x})_n)$ is the desired equilibrium.

Now

$$\frac{\partial U_i}{\partial x_i^+}(s(\hat{x})) = \frac{\partial \pi_i}{\partial x_i} - \frac{\left(a - b\sum s(\hat{x})_j\right)}{N-1}(\beta_i(k-1) - \alpha_i(N-k)$$

$$+ \frac{b}{N-1}\left(\beta_i(k-1)(x_i - \hat{x}) + \alpha_i\sum_{j>k}(s(\hat{x})_j - x_i)\right)$$

$$= \frac{\partial \pi_i}{\partial x_i} - \frac{\left(a - b\sum s(\hat{x})_j\right)}{N-1}(\beta_i(k-1) - \alpha_i(N-k)$$

$$+ \frac{b}{N-1}\left(\alpha_i\sum_{j>k}(s(\hat{x})_j - x_i)\right)$$

We then get

$$\lim_{\hat{x}\to x_{NE}}\frac{\partial U_i}{\partial x_i^+}(s(\hat{x})) = \frac{\left(a - b\sum s(\hat{x})_j\right)}{N-1}(\beta_i(k-1) - \alpha_i(N-k) < 0$$

Hence, for some $\hat{x}$ near enough to $x_{SNE}$, we get $\frac{\partial U_i}{\partial x_i^+}(s(\hat{x})) < 0$. The strategy com-

bination $s(\hat{x})$ is the desired equilibrium. QED

*Proof of Proposition 4 (Equilibria with Sanctioning Possibilities)*

Proof: We note:

(A) The condition $x \in \left(\frac{a(1-\beta_{\min}))}{b(1+n(1-\beta_{\min})))}, x_{SNE}\right]$ guarantees that $x$ maxi-

mizes the utility for the CCEs if all other players choose $x$.

We call a player a *deviator* who chooses an appropriation $x'$ that results in a higher payoff in the first stage compared to choosing $x$.

(B) If there is a single deviator, then the payoffs for the other players are smaller compared to the situation where there is no deviator.

First, if punishment is executed, the selfish players have no incentive to deviate. Because punishment results in equal payoffs for the CCEs and for the deviator, this payoff is smaller than the payoff in the first stage of the CCE. Hence, a selfish deviator has no incentive to deviate if he risks being punished.

Let us now prove that no CCE has an incentive to change the punishment strategy if a selfish player has not chosen $x$. Let $\pi_p$ be the payoff after punishment for the CCEs and for the deviator. Let $\pi_s$ be the payoff of the selfish players. A CCE player never has an incentive to choose a higher punishment than the equilibrium punishment. This only increases inequity with respect to all players and reduces the material payoff. So let $w$ be a positive number and assume CCE player $i$ chooses a punishment of $p - w$ instead of $p$. We get:

$$U_i = \pi_P + cw - \frac{(n-k-1)\alpha_i}{n-1}(\pi_S(\pi_P + cw)) - \frac{\alpha_i}{n-1}(\pi_p + w - (\pi_p + cw))$$

$$- \frac{(k-1)\beta_i}{n-1}(\pi_p + cw - \pi_p)$$

This is a linear function in $w$. Player $i$ has no incentive to deviate *iff* the derivative with respect to $w$ is negative, so *iff*

$$0 \geq \frac{\partial U_i}{\partial w} = c + c\frac{(n-k-1)\alpha_i}{n-1} - (1-c)\frac{\alpha_i}{n-1} - c\frac{(k-1)\beta_i}{n-1}$$

$$\Leftrightarrow c[(n-1) + (n-k-1)\alpha_i + \alpha_i - (k-1)\beta_i] \leq \alpha_i$$

$$\Leftrightarrow c \leq \frac{\alpha_i}{(n-1)(1+\alpha_i) - (k-1)(\alpha_i + \beta_i)}$$

QED.

# PART II

# PRIVATIZATION AND ITS LIMITATIONS

If it is feasible to establish a market to implement a policy, no policy-maker can afford to do without one. Unless I am very much mistaken, markets *can* be used to implement any anti-pollution policy that you or I can dream up.

(Dales, 1968:100, italics in original)

The two chapters by Tietenberg and Rose challenge an influential body of literature that suggests privatization as a solution for commons dilemmas (Gordon, 1954; Dales, 1968; Hardin, 1968; Crocker, 1966; Montgomery, 1972). In theory, for private goods, markets efficiently determine what, how much, how, and for whom to produce in the current period and over time. Tietenberg and Rose argue that it is difficult to privatize common-pool resources in the real and messy world when property rights are not easily defined and enforced, a prerequisite for efficient market functioning. Tietenberg recommends how and when institutions for privatizing common-pool resources, specifically tradable permits, can be developed. Rose, on the other hand, identifies conditions under which common-pool resources are managed more effectively as common property regimes than by tradable permits.

Chapter 6, by Tietenberg, provides lessons on how and why the optimism about the use of tradable permits in the 1980s changed to a more realistic approach to studying the conditions under which they may bring about a given level of environmental protection at the lowest cost. The chapter examines two aspects of "result efficiency" of this policy instrument: environmental effectiveness and economic effectiveness. However, it also points out the importance of "implementation feasibility." Tradable permits are considered to perform better for common-pool resources with limited negative externalities, a finding echoed by Rose in Chapter 7.

Chapter 7 examines hypotheses regarding the relative performance of common property regimes and tradable environmental allowances, operationalized in terms of their adaptability to (1) changes in resource demand and (2) variability of the resource. The institutional performance is hypothesized to depend on the following factors: (1) size and complexity of the common-pool resource (2) its use (extractive versus additive); and (3) characteristics of resource users and their interactions.

If the problem of common-pool resource overuse lies in ill-defined property rights, then defining property rights would solve the problem. Questions then arise as to what bundle of rights (specifically the right to manage and alienate the common-pool resource) provides the necessary incentives for owners to invest resources to prevent common-pool resource overuse, and who can define property rights and allocate them among individuals. Tradable permits and common property regimes differ across these dimensions.

The level of detail of the right definition and the ability of the regime to vary rates of resource use over time differ significantly between these regimes. Rights can be more detailed and flexible in common property regimes than in tradable permit regimes because they are not traded in the market. In fact, in resources that are complex (exhibit important interactions among various aspects of resource use) and vary over time, Rose points out that common property regimes outperform tradable permits, especially when the users belong to a close-knit, high-trust community.

Tradable permit regimes, on the other hand, develop uniform rules that offer security in market exchange, even allowing for trades among strangers. Therefore, they perform better for large-scale, but noncomplex common-pool resources. However, for complex common-pool resources, Tietenberg points out how rules can be designed to ensure effective working of tradable permits and prevention of resource overuse. He also deals with another criticism of tradable permit regimes: that they sacrifice equity and environmental effectiveness. He suggests that if a society wishes to prevent a concentration of permits in the hands of some resource users, it may limit transferability of the quotas, of course at the cost of lowering economic effectiveness.

Tietenberg examines cases in which a local, state, or national government assigns property rights and allocates them among common-pool resource users. The users are not allocated the complete bundle of rights, but usually only the right to withdraw from the resource (or deposit pollutants into the resource) and the right to sell their allocations to others. Because users do not influence total allocations, the total level of common-pool resource use—and therefore deterioration—depends on governmental decisions. In the case of common property regimes, users usually do not have the right to sell their individual allocations. They can, however, jointly decide the aggregate level of common-pool resource use.

Having said this, it is important to realize that identifying the maximum sustainable use of the resource—a function undertaken by a government in tradable permit regimes and by the user community in common property regimes—is both scientifically difficult (see Wilson, this volume:Chapter 10) and politically sensitive (see McCay, this volume:Chapter 11).

Tietenberg's and Rose's chapters agree on several issues. First, tradable permits perform better for managing simple common-pool resources with few negative externalities. Second, the allocation of rights is a difficult political process that has to be solved in any environmental regime. The allocation process, therefore, deserves special attention in the analysis of "implementation feasibility." Third, both chapters point out the crucial importance of monitoring and enforcement for any institutional arrangement governing common property resources. However, given that tradable permits offer important financial rewards when sold in market, their institutional design must provide for additional monitoring of not only resource use, but also the number of permits and their transfers. This increases the monitoring costs.

In sum, these chapters make a significant contribution to the understanding of under what conditions common-pool resources are better managed through alternative institutional mechanisms. Specifically, they carefully examine the strengths and weaknesses of tradable permit regimes and common property regimes in managing common-pool resources.

## REFERENCES

Crocker, T.D.
  1996  The structuring of atmospheric pollution control systems. In *The Economics of Air Pollution*, H. Wolozin, ed. New York: Norton.
Dales, J.
  1968  *Pollution, Property, and Prices. An Essay in Policy-Making and Economics.* Toronto: University of Toronto Press.
Gordon, H.S.
  1954  The economic theory of a common property resource: The Fishery. *Journal of Political Economy* 62:124-142.
Hardin, G.
  1968  The tragedy of the commons. *Science* 162:1243-1248.
Montgomery, D.W.
  1972  Markets in licenses and efficient pollution control programs. *Journal of Economic Theory* 5:395-418.

# 6

# The Tradable Permits Approach to Protecting the Commons: What Have We Learned?

*Tom Tietenberg*

One of the new institutional approaches for coping with the problem of rationing access to the commons involves the use of tradable permits. Applications of this approach have spread to many different types of resources and many different countries. A recent survey found 9 applications in air pollution control, 75 applications in fisheries, 3 applications in managing water resources, 5 applications in controlling water pollution, and 5 applications in land use control (Organization for Economic Co-operation and Development, 1999:Appendix 1:18-19). And that survey failed to include many current applications.[1]

Tradable permits address the commons problem by rationing access to the resource and privatizing the resulting access rights. The first step involves setting a limit on user access to the resource. For fisheries this would involve the total allowable catch. For water supply it would involve the amount of water that could be extracted. For pollution control it typically specifies the aggregate amount of emissions allowed in the relevant control region. This limit defines the aggregate amount of access to the resource that is authorized. These access rights are then allocated on some basis (to be described) to potential individual users. Depending on the specific system, these rights may be transferable to other users and/or bankable for future use. Users who exceed limits imposed by the rights they hold face penalties up to and including the loss of the right to participate.

These approaches have been controversial.[2] The controversy arises from several sources, but the most important concerns the allocation of the wealth associated with these resources. Although these approaches typically do *not* privatize the resources, as conventional wisdom might suggest, they do privatize at least to some degree access to and use of those resources. Because the access rights can be very valuable when the resource is managed efficiently, the owners of these rights may acquire a substantial amount of wealth. Although the ability to reclaim the previously dissipated wealth for motivating sustainable behavior is an important strength of the system, the ethical issues raised by its distribution among competing claimants are a significant and continuing source of controversy (McCay, 1999).

Another source of controversy involves a broad class of externalities. In general, externalities are effects on the ecosystem or on other parties that are not reflected adequately in the decisions by those holding the access rights. This incomplete internalization of externalities could involve diverse concerns such as adverse effects on species of fish other than those regulated by tradable permits, on the spatial concentration of emissions, or on the consequences of particular upstream water uses on downstream users.

A final source of controversy is ideological. It suggests that because capitalist property rights are the major source of the problem, it is inconceivable that these same rights could be part of the solution.[3]

## OVERVIEW

In this essay I review the experience with three main applications of tradable permit systems: air pollution control, water supply, and fisheries management.[4]

The next section provides a brief summary of the theory behind these programs and both the economic and environmental consequences anticipated by this theory. Some brief points of comparison are made with other competing and/ or complementary formal public policy strategies such as environmental taxes and legal regulation.

The essay proceeds with a description of the common elements these programs share and the design questions posed by the approach. These include the setting of the limit on access, the initial allocation of rights, transferability rules (both among participants and across time) as well as procedures for monitoring and enforcement. It continues by examining how these design questions have been answered by the air pollution, fishery, and water supply applications and how the answers have evolved over time. This evolution has been influenced by changing technology, increased familiarity with the system, and a desire to respond to some of the controversies surrounding the use of these approaches.

The penultimate section examines the hard evidence on the economic and environmental consequences of adopting these approaches. This evidence is juxtaposed with the expectations created by both the economic theory of tradable permits and the theory of choice between co-management and tradable permits by Rose (this volume:Chapter 7).

The final section brings together some tentative lessons that can be drawn from this experience.

# THE BASIC ECONOMIC THEORY

Our inquiry begins by defining what is meant by an optimal allocation of a resource and by extracting the principles that can be used to design economic incentive policies that fulfill the optimality conditions. Optimality theory can help us understand the characteristics of these economic approaches in the most favorable circumstances for their use and assist in the process of designing the instruments for maximum effectiveness.

## The Economic Approach to Optimal Resource Management

What is meant by the optimal allocation of a resource depends on how the "policy target" is defined. Several possible targets have been considered in the literature.[5] Chronologically the first forays into instrument design were based on traditional concepts of economic efficiency. The economically efficient allocation of a resource, defined in partial equilibrium terms, maximizes the net benefits to society, where net benefits are defined as the excess of benefits over costs.[6] Ignoring corner solutions (i.e., when the optimum involves either no use or total use), efficiency is achieved when the marginal benefit of that last unit used is equal to the marginal cost of its provision.

Because the resulting allocation of responsibility is quite sensitive to both spatial and temporal considerations, defining optimality in terms of efficiency imposes a heavy information burden both on modelers and on those charged with the responsibility for implementing the policies. Not only does an efficiency target make it necessary to track the physical relationships underlying the use of the resource, but it also requires monetizing the consequences (both human and nonhuman). Each of these steps is subject to data limitations and uncertainties.

Even when the information burdens associated with the efficiency criterion can be surmounted, it is not universally accepted as an appropriate criterion outside the discipline of economics. Applying this criterion has several somewhat subtle implications, some of which are quite controversial. Take as just one example the class of pollutants having a major impact on human health. The efficiency criterion implies, all other things being equal, targeting more resources toward controlling those emissions that affect larger numbers of people (because the marginal damage caused by a unit of emissions is higher in that setting). This particular allocation of control resources can result in *lower individual* risks for those in high-exposure settings. This contradicts a popular policy premise that suggests that citizens should face *equal individual* risks regardless of where they work or reside.[7]

To respond to both the information and moral concerns with an efficiency approach, the tradable permit approach starts from a sustainability perspective.[8] Whereas efficiency may or may not be consistent with a sustainable allocation, the tradable permits program starts by defining a sustainable target. The sustainable target may or may not be efficient,[9] but it does provide a good opportunity to achieve sustainable outcomes even in cases where efficient allocations may not be compatible with sustainability.[10]

## Value-Maximizing, Sustainable Policy Instruments

One of the insights derived from the empirical literature is that traditional command-and-control regulatory measures, which depended on government agencies to both define the goals and the means of meeting them, were, in many cases, insufficiently protective of the value of the resources.[11] One of the principal theorems of environmental economics demonstrates that under specific conditions, an appropriately defined tradable permit system can maximize the value received from the resource, given the sustainability constraint (Baumol and Oates, 1971,1988).

The logic behind this result is rather simple. In a perfectly competitive market, permits will flow toward their highest valued use. Those that would receive lower value from using the permits (due to higher costs, for example) have an incentive to trade them to someone who would value them more. The trade benefits both parties. The seller reaps more from the sale than she could from using the permit, and the buyer gets more value from the permit than he pays for it.

A rather remarkable corollary (Montgomery, 1972) holds that this theorem is true regardless of how the permits are allocated initially among competing claimants. It is true regardless of whether permits are auctioned off or allocated free of charge. Furthermore, when permits are allocated free of charge, *any* particular initial allocation rule can still support a cost-effective allocation. Again the logic behind this result is rather straightforward. Whatever the initial allocation, the transferability of the permits allows them ultimately to flow to their highest valued uses. Because those uses do not depend on the initial allocation, all initial allocations result in the same outcome and that outcome is cost-effective.

The potential significance of this corollary is huge. It implies that with tradable permits, the resource manager can use the initial allocation to solve other goals (such as political feasibility or ethical concerns) without sacrificing cost-effectiveness. In Alaskan fisheries, for example, some of the quota has been allocated to communities (rather than individuals) to attempt to protect community interests (Ginter, 1995).[12]

## Preconditions

Tradable permits systems may not maximize the value of the resource if the market conditions are not right. Circumstances when the conditions may not be right include the possibility for market power (Hahn, 1984), the presence of high transaction costs (Stavins, 1995), and insufficient monitoring and enforcement.[13] Because tradable permits involve an aggregate limit on access, however, the consequences of market power and/or high transaction costs typically affect costs more than environmental quality. Furthermore, even in the presence of these imperfections, tradable permit programs can be designed to mitigate their adverse consequences.[14]

Without effective enforcement, permit holders who don't get caught may gain more by cheating than by living within the constraints imposed by their allocated permits. In contrast to the two previously mentioned imperfections, this one could lead to the degradation of the resource because the aggregate limit could be breached.

Another important precondition involves the absence of large uninternalized externalities.[15] The presence of uninternalized externalities would imply that maximizing the net benefits of permit holders would not necessarily maximize net benefits for society as a whole, even with a fixed environmental target. For example, fishermen might catch the specified amount of the covered species, but they might use gear that destroys other components of the marine ecosystem. Polluters that reduce a covered pollutant by switching inputs could well increase emissions of another unregulated pollutant. The regulation could serve to protect one environmental resource at the expense of another.

## Comparing Tradable Permits with Environmental Taxes

The mathematics underlying the theorems mentioned also can be used to demonstrate similar theoretical properties for environmental taxes. For every tradable permit system that maximizes the value of the resource, there exists an environmental tax that could achieve the same outcome. In principle, therefore, taxes and tradable permits exhibit a striking symmetry.

In practice, however, this symmetry disappears and striking differences can arise. Once a quantity limit is specified, the government has no responsibility for finding the right price in a tradable permit system; the market defines the price. With a tax system, the government must find the appropriate tax rate—no small task. And with a tax system, the resource rents normally are channeled to the government. With tradable permits, resource users typically retain them. Recent work examining how the presence of preexisting distortions in the tax system affects the efficiency of the chosen instrument suggests that the ability to recycle the revenue (rather than give it to permit holders) can enhance the cost-effectiveness of the system by a large amount. That work, of course, creates a bias toward taxes or auctioned permits and away from "grandfathered" permits (Goulder et al., 1999). How revenues are distributed, however, also affects the attractiveness of alternative approaches to environmental protection from the point of view of the various stakeholders. To the extent that stakeholders can influence policy choice, "grandfathering" may increase the feasibility of implementation (Svendsen, 1999).

Over time the two systems may act quite differently as well if the government decides not to intervene in the market. In a tradable permits system, inflation will merely result in higher permit prices; the limit will remain intact. With taxes the amount of environmental protection will decline over time (as the real value of the tax declines) in the absence of some kind of indexing scheme. Conversely, technical progress that lowers compliance cost will result in more environmental protection under taxes than tradable permits. Finally, the presence of uncertainty about the benefits and costs can lead to a preference of one instrument or the other depending on the nature of the uncertainty (Weitzman, 1974).

## DESIGN CONSIDERATIONS

### Governance Structures

The academic community has emphasized the importance of co-management of environmental resources, with users having a substantial role. This is presumed to increase compliance.[16]

Although tradable permit systems in principle allow a variety of governance systems, the current predominant form in all three applications seems to be a system of shared management, with users playing a smaller role than envisioned by most co-management proposals. For those resource regimes in the United States, it is common for the goals to be set by the government (either at the national or state level) and considerable "top-down" management to be in evidence.

In the case of air pollution, specific quantitative ambient standards are set at the national level, and all programs must live within those limits. In the sulfur allowance program, a national program, the emissions cap also is set at the national level. In the RECLAIM system, the emissions cap was established by the local air quality management district, but the district is subject to the oversight of the national Environmental Protection Agency (EPA) and must show how its choice will enable it to meet the nationally set ambient standards.

Fisheries have a somewhat similar governance arrangement. The Secretary of Commerce and his implementing agency, the National Marine Fisheries Service, use their oversight and approval powers to attempt to assure that locally created approaches meet the various requirements of the Magnuson-Stevens Act, as amended.[17] Unlike the ambient standards, which are quantitatively precise, these objectives are more vaguely specified. That allows the Secretary more discretion, which can be used either to exercise stronger control or to allow more community discretion.[18] Subject to this oversight, regional fisheries councils define both the caps and the rules. Although representatives of access right holders usually are represented on these councils, other groups are represented as well.

Although the use of true co-management in air pollution control is rather rare, some limited forms are beginning to appear in both fisheries and water. Water user associations, for example, play a considerable role in allocating water resources in Chile. Although the *Dirección General de Aguas* has broad authority in water resource management, much of the actual control over river flows is exercised by the *Juntas de vigilancia*, associations made up of all users and users associations on a common section of a river (Hearne, 1998).

The absence of centralized control by California over its groundwater has resulted in the growth of a number of basin authorities controlled by water producers. The transfers of rights that take place among producers of groundwater can be seen as "informal" tradable rights markets.[19] These informal markets appear to be much more likely to involve user-defined rules.

In fisheries, particularly those involving highly sedentary species such as lobsters, substantial local control by users typically is exercised.[20] For example, Maine controls its lobster fishery by means of a zonal system. Fishers within these zones play a considerable role in defining the rules that govern fishing activity within their zone. Though none of the zones currently involve the use of tradable permits, that option is being discussed.

Following the U.S. Congress-imposed moratorium on individual transferable quotas (ITQs), some alternative self-regulation alternatives arose in fisheries. In the Pacific whiting fishery in the Bering Sea, the annual total allowable catch (TAC) of whiting is divided among various sectors, including the catcher-processor vessels, which hold 34 percent of the 1997-2001 TAC (National Research Council, 1999:130). In April 1997, the four companies holding limited entry permits in the catcher-processor sector agreed to allocate the quota among themselves, forming a cooperative for the purpose. To avoid possible antitrust prosecution, a potential barrier to user-based management agreements in the United States, members submitted their proposal to the Department of Justice, which approved it. Though this is not a formal tradable permit, the negotiations over allocations among participants have begun to take on some of the attributes of an informal market.

It should not be surprising that although tradable permit systems potentially allow for a considerable co-management role, only in fisheries and water is there any evidence of an evolution in this direction. The pollution and natural resource cases exhibit an important asymmetry. For air pollution control, the benefits from resource protection fall on the victims of air pollution, not on the polluters who use the resource; from a purely self-interest point of view, resource users (polluters) would be quite happy to degrade the resource if they could get away with it. On the other hand, water users and fishers both can benefit from protection of the resource. Their collective self-interest is compatible with resource protection. This suggests that the incentives for collective action should be, and apparently are, quite different in these two cases.

## The Baseline Issue

In general, tradable permit programs fit into one of two categories: a credit program or a cap-and-trade program. The credit program involves a relative baseline. With a credit program, an individual access baseline is established for each resource user. The user who exceeds legal requirements (say by harvesting fewer fish than allowed or emitting less pollution than allowed) can have the difference certified as a tradable credit.

The cap-and-trade program involves an absolute baseline and trades allowances rather than credits. In this case a total resource access limit is defined and then allocated among users. Air pollution control systems and water have examples of both types. Fisheries tradable permit programs are all of the cap-and-trade variety.

Credit trading, the approach taken in the Emissions Trading Program (the earliest program) in the United States, allows emission reductions above and beyond legal requirements to be certified as tradable credits. The baseline for credits is provided by traditional technology-based standards. Credit trading presumes the preexistence of these standards and it provides a more flexible means of achieving the aggregate goals that the source-based standards were designed to achieve.

Allowance trading, used in the U.S. Acid Rain Program, assigns a pre-specified number of allowances to polluters. Typically the number of issued allowances declines over time and the initial allocations are not necessarily based on traditional technology-based standards; in most cases the aggregate reductions implied by the allowance allocations exceed those achievable by standards based on currently known technologies.

Despite their apparent similarity, the difference between credit- and allowance-based trading systems should not be overlooked. Credit trading depends on the existence of a previously determined set of regulatory standards. Allowance trading does not. Once the aggregate number of allowances is defined, they can, in principle, be allocated among sources in an infinite number of ways. The practical implication is that allowances can be used even in circumstances (1) where a technology-based baseline either has not been, or cannot be, established, or (2) where the reduction is short lived (such as when a standard is met early) rather than permanent.

The other major difference is that cap-and-trade programs generally establish an upper aggregate limit on the resource use, while the credit programs establish only an upper limit for each user. In the absence of some other form of control over additional users, an increase in the number of users can lead to an increase in aggregate use and the eventual degradation of the resource.

## The Legal Nature of the Entitlement

Although the popular literature frequently refers to the tradable permit approach as "privatizing the resource" (Spulber and Sabbaghi, 1993; Anderson, 1995), in most cases it doesn't actually do that. One compelling reason in the United States why tradable permits do not privatize these resources is because that could be found to violate the well-established "public trust doctrine." This common law doctrine suggests that certain resources belong to the public and that the government holds them in trust for the public; they can't be given away.[21]

Economists have argued consistently that tradable permits should be treated as secure property rights to protect the incentive to invest in the resource. Confiscation of rights could undermine the entire process.

The environmental community, on the other hand, has argued just as consistently that the air, water, and fish belong to the people and, as a matter of ethics, they should not become private property (Kelman, 1981). In this view, no end could justify the transfer of a community right into a private one (McCay, 1998).

The practical resolution of this conflict has been to attempt to give "adequate" (as opposed to complete) security to the permit holders, while making it clear that permits are not property rights.[22] For example, according to the title of the U.S. Clean Air Act dealing with the sulfur allowance program: "An allowance under this title is a limited authorization to emit sulfur dioxide. ... Such allowance does not constitute a property right" (104 Stat. 2591).

In practice this means that administrators are expected to recognize the security needed to protect investments by not arbitrarily confiscating rights. They do not, however, give up their ability to change control requirements as the need arises. In particular, they will not be inhibited by the need to pay compensation for withdrawing a portion of the authorization to emit as they would if allowances were accorded full property right status. It is a somewhat uneasy compromise, but it seems to have worked.

## Adaptive Management

One of the initial fears about tradable permit systems is that they would be excessively rigid, particularly in the light of the need to provide adequate security to permit holders. Policy rigidity was seen as possibly preventing the system from responding either to changes in the resource base or to better information. This rigidity could seriously undermine the resilience of biological systems (Holling, 1978).

Existing tradable permit systems have responded to this challenge in different ways depending on the type of resource being covered. In air pollution control, the need for adaptive management typically is less immediate and the allowance typically is defined in terms of tons of emissions. In biological systems, such as fisheries, the rights typically are defined as a share of the TAC. In this way the resource managers can change the TAC in response to changing biological conditions without triggering legal recourse by the right holder.[23] Some fisheries actually have defined two related rights (Young,1999). The first conveys the share of the TAC, while the second conveys the right to catch a specific number of tons of harvest in a particular year. Separating the two rights allows a harvester to sell the right to catch fish in a particular year (perhaps due to an illness or malfunctioning equipment) without giving up the right of future access.[24]

Water has a different kind of adaptive management need. Considerable uncertainty among users is created by the fact that the amount of water can vary significantly from year to year.[25] Because different users have quite different capacities for responding to shortfalls, the system for allocating this water needs to be flexible enough to respond to this variability or the water could be seriously misallocated.

These needs have been met by a combination of technological solutions (principally water storage) and building some flexibility into the rights system. In the American West, the appropriation doctrine that originated in the mining camps created a system of priorities based on the date of first use. The more senior rights then have a higher priority of claim on the available water in any particular year and consequently could be expected to claim the highest price (Howe and Lee, 1983; Livingston, 1998).[26] Other systems, most notably in Australia, use a system of proportionality that resembles the share system in fisheries (Livingston, 1998).

An alternative approach to flexibility with security, the "drop-through mechanism," involves a cascade of fixed-term entitlements, a variation of an approach currently used in the New South Wales fishery (Young, 1999) and proposed for use in controlling climate change (Tietenberg, 1998b). Under this scheme, initial entitlements (call them Series A Entitlements) would be defined for a finite period, but one long enough to encourage investments (say, for the sake of illustration, 30 years; see Figure 6-1). The rights and obligations covered by the Series A entitlements would be known in advance.[27] Periodically (say, for illustration, every 10 years) a comprehensive review would be undertaken that would result in a new set of entitlements (Series B, Series C, and so forth) that also would have a 30-year duration. Emitters holding Series A Entitlements could have the option to switch to the new set of entitlements at any time earlier than the expiration of their Series A Entitlements. Once they switched they would be able to hold Series B Entitlements for their remaining life. This process would continue until it appeared no more reviews were necessary.

## Defining the Aggregate Limits

In all three applications, the limits are defined on the basis of some notion of sustainable use. In air pollution control, the limits are defined to assure that the resulting concentrations fall below the Ambient Air Quality Standards (AAQS).

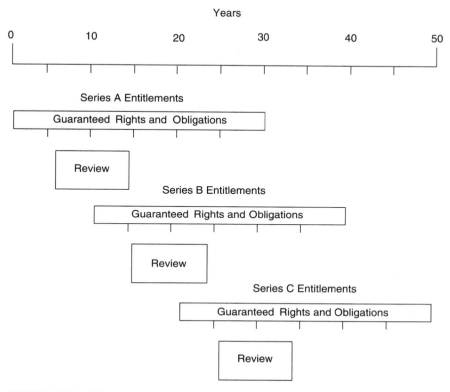

FIGURE 6-1  Building resilience into tradable permit systems.
SOURCE: Based on Figure 7-1 in Young and McCay (1995). Reprinted with permission.

The primary AAQS are defined at levels that protect human health.[28] In water the aggregate limit typically is based on expected water flow (Easter et al., 1998). In formal tradable permit fisheries, the governing body routinely estimates the size of the fish stocks to determine the amount of fish that can be harvested in a given year so that fisheries can be sustained; this amount is termed the "allowable biological catch" (ABC). The catch level that fishermen are allowed to take, the total allowable catch, normally would be equal to or less than the ABC (National Research Council, 1999:3).

## Initial Allocation Method

The initial allocation of entitlements is perhaps the most controversial aspect of a tradable permits system. Four possible methods for allocating initial entitlements are:

- Random access (lotteries)
- First come, first served
- Administrative rules based on eligibility criteria
- Auctions

All four of these have been used in one context or another. Both lotteries and auctions frequently are used in allocating hunting permits for big game. Lotteries are more common in allocating permits among residents while auctions are more common for allocating permits to nonresidents. First come, first served historically was common for water when it was less scarce. The most common method, however, for the applications discussed here is allocating access rights based on historic use.

Two justifications for this approach typically are offered.[29] First, it enhances the likelihood of adoption.[30] Not only does allocating entitlements to historic users cause the least disruption from historic patterns, but it involves a much smaller financial burden on users than an auction[31] (Lyon, 1982; Tietenberg, 1985; Hausker, 1990; Grafton and Devlin, 1996). Second, it allocates permits to those who have made investments in resource extraction. In this sense it serves to recognize and to protect those investments.[32]

In the absence of either a politically popular way to use the revenue or assurances that competitors will face similar financial burdens, distributing the permits free of charge to existing sources could substantially reduce this political opposition. Though an infinite number of possible distribution rules exist, "grandfathered" rules tend to predominate. Grandfathering refers to an approach that bases the initial allocation on historic use. Under grandfathering, existing sources only have to purchase any additional permits they may need over and above the initial allocation (as opposed to purchasing *all* permits in an auction market).

Although politically the easiest path to sell to those subject to regulation, grandfathering has its disadvantages. The presence of preexisting distortions in the tax system implies that recycling the revenue can enhance the cost-effectiveness of the system by a large amount. This implies that from an efficiency or cost-effectiveness perspective, auctioned permits would be preferred to "grandfathered" permits (Goulder et al., 1999).

A second consideration involves the treatment of new firms. Although reserving some permits for new firms is possible, this option is rarely exercised in practice. As a result, under the free distribution scheme new firms typically have to purchase all permits, while existing firms get an initial allocation free. Thus the free distribution system imposes a bias against new users in the sense that their financial burden is greater than that of an otherwise identical existing user. In air pollution control, this "new user" bias has retarded the introduction of new facilities and new technologies by reducing the cost advantage of building new facilities that embody the latest innovations (Maloney and Brady, 1988; Nelson et al., 1993).[33]

Other initial allocation issues involve determining both the eligibility to receive permits and the governance process for deciding the proper allocation.[34] Controversies have arisen, especially in fisheries, about both elements. In fisheries the decision to allocate permits to boat owners has triggered harsh reactions among both crew and processors.

In some fisheries the allocation to boat owners has transformed the remuneration arrangements from a sharing of the risks and revenues from a catch on a predefined share basis to a wage system. Though this transformation can result in higher incomes for crew (Knapp, 1997), the change in status has been difficult to accept for those used to being co-venturers, thereby sharing in both the risk and reward of fishing (McCay et al., 1989; McCay and Creed, 1990).

Processors also have staked their claim for quota (especially in Alaska), albeit unsuccessfully to date (Matulich et al., 1996). The claims are based on the immobility of the processing capital and the fact that allocating quota to boat owners changes the bargaining relationship in ways that could hurt processors (Matulich and Sever, 1999).

Finally, some systems allow agents other than those included in the initial allocation to participate through an "opt-in" procedure. This is a prominent feature of the sulfur allowance program, but it can be plagued by adverse selection problems (Montero, 1999, 2000b).

## Transferability Rules

Although the largest source of controversy about tradable permits seems to attach to the manner in the permits are allocated initially, another significant source of controversy is attached to the rules that govern transferability. According to supporters, transferability not only serves to assure that rights flow to their highest valued use, but it also provides a user-financed form of compensation for those who decide voluntarily to no longer use the resource. Therefore restrictions on transferability only serve to reduce the efficiency of the system. According to critics, allowing the rights to be transferable produces a number of socially unacceptable outcomes, including the concentration of rights, the destruction of community interests, and the degrading of both the environment and traditional relationships among users.

Making the rights transferable allows the opportunity for some groups to accumulate permits. The concentration of permits in the hands of a few either can reduce the efficiency of the tradable permits system (Hahn, 1984; Anderson, 1991; Van Egteren and Weber, 1996) or can be used as leverage to gain economic power in other markets (Misiolek and Elder, 1989; Sartzetakis, 1997). Although it has not played much of a role in air pollution control, it has been a factor in fisheries (Palsson and Helgason, 1995).

Typically the problem in fisheries is *not* that the concentration is so high that it triggers antitrust concerns (Adelaja et al., 1998), but rather that it allows small fishing enterprises to be bought out by larger fishing enterprises. Smaller fishing enterprises are seen as having a special value to society that should be protected.

Protections against "unreasonable" concentration of quota are now common. One typical strategy involves putting a limit on the amount of quota that can be accumulated by any one holder. In New Zealand fisheries, for example, these range from 20 percent to 35 percent, depending on the species (National Research Council, 1999:90-91), while in Iceland the limits are 10 percent for cod and 20 percent for other species (1999:102).

Another strategy involves trying to mitigate the potential anticompetitive effects of hoarding. The U.S. sulfur allowance program does this in two main ways. First, it sets aside a supply of allowances that could be sold at a predetermined (high) price if hoarders refused to sell to new entrants.[35] Second, it introduced a zero-revenue auction that, among its other features, requires permit holders to put approximately 3 percent of allowances up for sale in a public auction once a year.[36]

Another approach involves directly restricting transfers that seem to violate the public interest. In the Alaskan halibut and sablefish ITQ program, for example, several size categories of vessels were defined. The initial allocation was based on the catch record within each vessel class, and transfer of quota between catcher vessel classes was prohibited (National Research Council, 1999:310). Further restrictions required that the owner of the quota had to be on board when the catch was landed. This represented an attempt to prevent the transfer of ownership of the rights to "absentee landlords."

A second concern relates to the potentially adverse economic impacts of permit transfers on some communities.[37] Those holders who transfer permits will not necessarily protect the interests of communities that have depended on their commerce in the past. For example, in fisheries a transfer from one quota holder to another might well cause the fish to be landed in another community. In air pollution control, owners of a factory might shut down its operation in one community and rebuild in another community, taking their permits with them.

One common response to this problem involves allocating quota directly to communities. The 1992 Bering Sea Community Development Quota Program, which was designed to benefit remote villages containing significant native populations in Alaska, allocated 7.5 percent of the walleye pollock quota to these communities (Ginter, 1995). In New Zealand the Treaty of Waitangi (Fisheries Claims) Settlement Act of 1992 effectively transferred ownership of nearly 40 percent of the New Zealand ITQ to the Maori people (Annala, 1996). For these allocations the community retains control over the transfers, and this control gives it the power to protect community interests. In Iceland this kind of control is gained through a provision that if a quota is to be leased or sold to a vessel operating in a different place, the assent of the municipal government and the local fishermen's union must be acquired (National Research Council, 1999: 83).

A final concern with transferability relates to possible external effects of the transfer. Although in theory transfers increase net benefits by allowing permits to flow to their highest valued use, in practice that is not necessarily so if the transfers confer external benefits or costs on third parties.

Such external effects are not rare. In water, for example, transfers from one use to another can affect the quality, quantity, and timing of supply for other downstream users[38] (Livingston, 1998). In air pollution control, transfers can affect the spatial distribution of pollution, and that can trigger environmental justice concerns (Tietenberg, 1995b).[39] In fisheries quota could be transferred to holders with more damaging gear, or a higher propensity for bycatch. In all cases "leakage" provides another possible external effect. Leakage occurs when pressure on the regulated resource is diverted to an unregulated, or lesser regulated, resource, as when fishermen move their boats to another fishery or polluters move their polluting factory to a country with lower environmental standards.

Western U.S. water markets attempt to solve the externality problem by giving any affected party a chance to intervene in the transfer proceeding (Colby, 1995). In the case of a third-party intervention, the transferring parties bear the burden of establishing the absence of damage to third parties. Although this is probably an effective way to internalize the externality, it raises transaction costs significantly and has resulted in many fewer transfers than would have been the case otherwise (Livingston, 1998). Technology is now making an entrance in water markets (the Water Links electronic water exchange in California, for example) to lower transaction costs (Organization for Economic Co-operation and Development, 1999).

One strategy used in U.S. air pollution control policy to resolve the spatial externality problem is regulatory tiering. Regulatory tiering implies applying more than one regulatory regime at a time. Sulfur oxide pollution in the United States is controlled both by the regulations designed to achieve local ambient air quality standards as well as by the sulfur allowance trading program. All transactions have to satisfy both programs. Thus trading is not restricted by spatial considerations (national trades are possible), but the use of acquired allowances is subject to local regulations protecting human health via the ambient standards. The second regulatory tier protects against the harmful spatial clustering of emissions (by disallowing any specific trades that would violate the standards), while the first tier allows unrestricted trading of allowances. Because the reductions in sulfur are so large and most local ambient standards are not likely to be jeopardized by trades, few trades have been affected by this provision. Yet its very existence serves to allay fears that local air quality could be in jeopardy.

## The Temporal Dimension

Standard theory suggests that a fully value-maximizing tradable permit system must have full temporal fungibility, implying that allowances can be both borrowed and banked (Kling and Rubin, 1997; Rubin, 1996). Banking allows a user to store its permits for future use. With borrowing a permit holder can use permits earlier than their stipulated date.

No existing system that I am aware of is fully temporally fungible. Older pollution control programs have had a more limited approach. The Emissions Trading Program allowed banking, but not borrowing. The Lead Phaseout Program originally allowed neither, but part way through the program it allowed banking, at least until the program officially ended and any remaining credits became unusable. The sulfur allowance program has banking, but not borrowing, and RECLAIM has neither (Tietenberg, 1998c).

Why do so few programs have full temporal fungibility? The answers seem to lie more in the realm of politics than economics.

The first concern involves the potential for creating a temporal clustering of emissions. When intertemporal trades are defined on a one-for-one basis, it is possible for emissions to be concentrated in time. Because emissions concentrated in space or time cause more degradation than dispersed emissions (due to a nonlinearity in the dose-response function), regulators have chosen to put a priori restrictions on the temporal use of permits despite the economic penalty that imposes.

A second concern has arisen (particularly in the global warming context) where imposing sanctions for noncompliance is difficult. Some observers have noted that enforcing the cumulative emissions budget envisioned by the Kyoto Protocol on a nation that had borrowed heavily in the earlier years would become increasingly difficult over time (Tietenberg et al., 1998). Given the inherent difficulties in enforcing international commitments under the best of circumstances, opponents of borrowing propose to forestall this difficulty by eliminating any possibility of borrowing. They view the resulting increased compliance cost as a reasonable price to pay for taking the pressure off future enforcement.

## Monitoring and Enforcement

Regardless of how well any tradable permit system is designed, noncompliance can prevent the attainment of its economic, social, and environmental objectives. Noncompliance not only makes it more difficult to reach stated goals, but it sometimes makes it more difficult to know whether the goals are being met.[40]

Although it is true that any management regime raises monitoring and enforcement issues, tradable permit regimes raise some special issues. One of the most desirable aspects of tradable permits, their ability to increase the value of the resource, is a two-edged sword because it also raises incentives for noncompliance. In the absence of an effective enforcement system, higher profitability from cheating could promote illegal activity. Insufficient monitoring and enforcement also could result in failure to keep a tradable permit system within its environmental limit.[41]

Do monitoring and enforcement costs rise under tradable permit programs? The answer depends both on the level of required enforcement activity (greater levels of enforcement effort obviously cost more) and on the degree to which existing enforcement resources are used more or less efficiently. Higher enforcement costs are not, by themselves, particularly troubling because they can be financed from the enhanced profitability promoted by the tradable permit system.[42]

*Monitoring*

In addition to the obvious potential for quota busting that all tradable permit approaches face, fisheries also can face problems with poaching (harvests by ineligible fishermen), unreported highgrading (discarding low-valued fish to make room in the quota for higher valued fish), and bycatch discards (nontargeted species caught and discarded) (National Research Council, 1999:175-180).

Whether these problems are intensified or diminished by the implementation of a tradable permit program depends (in part) on the economic incentives confronting participants. The incentives for highgrading, for example, depend on the magnitude of price differentials for various types and sizes of targeted species. As the price premium for fish of a particular size and type increases, the incentive to use quota for especially valuable fish increases along with the incentive to discard less valuable fish (Anderson, 1994).

Incentives for bycatch can vary considerably as well (Boyce, 1996; Larson et al., 1998). The more leisurely pace of fishing afforded by individual fishing quotas (IFQs) allows fishermen to avoid geographic areas or times when bycatch is more likely.[43] At the same time, the more leisurely pace reduces the opportunity cost of hold space and, consequently, also may provide fishermen with new opportunities to retain a greater proportion of the bycatch as joint products. For example, the halibut fishery encounters significant bycatches of rockfish. Although most rockfish and thornyheads command high exvessel prices, most of this bycatch was discarded during the derby fishery because halibut were even more valuable. A greater portion of this bycatch is now being retained.

On the other hand, implementing an IFQ regime may favor some technologies over others. If the favored technologies typically involve more bycatch, bycatch rates can rise in the absence of enforcement.

Ultimately, therefore, whether highgrading, bycatch, and bycatch discard increase or decrease under an IFQ regime depends on local circumstances, on whether highgrading and bycatch discards are legal (or even required), and on the enforcement response.

Every monitoring system must identify both the information that is needed to monitor the operation of the tradable permit program and the management component that will gather, interpret, and act on this information. Data also should be collected on transfers so that monitoring and analysis of the market can take place. Effective monitoring systems are composed of data, data management, and verification components.

In general, the smooth implementation of a tradable permit program requires two kinds of monitoring data. First, periodic data on the condition of the resource are needed to evaluate the effectiveness of the program over time. These data are used as the basis for adjusting environmental limits as conditions warrant. Second, managers need sufficient data to monitor compliance with the various limitations imposed by the regulatory system.

Monitoring compliance with a tradable permit program requires data on the identity of permit holders, amount of permits owned by each holder, permit, and permit transfers. Where programs have additional restrictions on permit use (such as type of equipment) or on quota transfers (only to "eligible" buyers), the data must be complete enough to contain this information and to identify noncomplying behavior in a timely manner.

One key to a smoothly implemented tradable program is ensuring that all data are input to an integrated computer system that is accessible by eligible users on a real-time basis. Such a system provides up-to-date information on permit use to both users and enforcement agencies. Ideally it also would allow short-notice transfers, such as when a vessel heading for shore has a larger than expected bycatch and needs to acquire additional quota for the bycatch species before landing. Facilitating this kind of flexibility would reduce the enforcement burden considerably by giving permit holders a legal alternative to illegal discarding without jeopardizing the objectives of the program.

The computer system also should provide easy data entry. Card swipe systems, such as those used in the Alaska halibut and sablefish IFQ fisheries, automatically input all the necessary identification data so that only landings (and hence permit use) need to be recorded. It is also possible to have the harvest level recorded directly from the scales (with appropriate adjustments for "ice and slime" or the degree to which the fish are already processed). Entry terminals that are connected to the master computer system should be available at all authorized landing sites.

Technology also has played an important role in the U.S. sulfur allowance system (Kruger et al., 2000). Both the collection and dissemination of the information derived from the continuous emissions monitors is now handled via the Web. Special software has been developed to take individual inputs and to generate information both for the public and for EPA enforcement activities. According to Kruger et al. (2000), the development of this technology has increased administrative efficiency, lowered transaction costs, and provided greater environmental accountability.

Information technology also permits greater accountability by making the information transparent. Evidence suggests that making the information available online to the public may further increase compliance. It also increases the possibilities for public pressure and even legal action from nongovernmental environmental agencies and/or citizens (Tietenberg, 1998a).

To ensure the accuracy of reported data, it is necessary to build a number of safeguards into the program. In fisheries proper control procedures include both onshore and at-sea components. An onshore system of checks normally would include a requirement that sales be made only to registered buyers and that both buyers and quota shareholders co-sign the landing entries. These measures create an audit trail that could be monitored electronically for instances in which a comparison of processed product weight and recorded purchases suggests suspiciously high product recovery rates. The at-sea component would include both onboard observers, where the fishery is profitable enough to bear the cost, and random checks at sea by the appropriate authority (or perhaps by video monitoring). Onboard observers may be particularly important in fisheries where bycatch and highgrading are expected to be problems.

*Enforcement*

A successful enforcement program requires a carefully constructed set of sanctions for noncompliance. Penalties should be commensurate with the danger posed by noncompliance. Penalties that are unrealistically high may be counterproductive if authorities are reluctant to impose them and fishermen are aware of this reluctance. Unrealistically high penalties also are likely to consume excessive enforcement resources as those served with penalties seek redress through the appeals process.

In many cases, predetermined administrative fines can be imposed by the enforcing agency itself for "routine" noncompliance. For example, the Alaskan IFQ programs allow overages of up to 10 percent above the fisherman's remaining IFQ balance to be deducted from the next year's IFQ permit amount. Overages greater than 10 percent are considered a violation and are handled by enforcement personnel. In an ideal system, more serious noncompliance in terms of either the magnitude of the offense or the number of offenses could trigger civil penalties (fines and possible seizure of catch, equipment, and quota). Criminal penalties should be reserved for falsification of official reports and the most serious violations.

Other sanctions are possible. In the sulfur allowance program, for example, those found in noncompliance must not only pay a substantial financial penalty for noncompliance; they must also forfeit a sufficient number of future allowances to compensate for the overage. It is also possible to only allow those in compliance to transfer permits. Any egregious violations can lead to forfeiture of the right to participate in the program at all.

Income levels from fishing generally are bolstered by the implementation of an effective IFQ program. An effective program presumes effective enforcement.

Honest fishermen should be willing to contribute some of their increased rent to ensure the continued existence of an effective IFQ management regime.

## EVALUATION CRITERIA

In assessing the outcomes of these systems I focus on three major categories of effects. The first is implementation feasibility. A proposed policy regime cannot protect the common pool resource if it cannot be implemented or if its main protective mechanisms are so weakened by the implementation process that it is rendered ineffective. What matters is not how a policy regime works in principle, but how it works in practice. The second category seeks to answer the question "How much protection did it offer not only to the common-pool resource, but also other resources that might have been affected either positively or negatively by its implementation?" Finally, what were the economic effects on those who either directly or indirectly use the resource?

### Implementation Feasibility

The record seems to indicate that resorting to a tradable permits approach to controlling resources usually only occurs after other, more familiar, approaches have been tried and failed. In essence the costs of implementing a system like this generally are recognized as large, so incurring such large costs can be justified only when the benefits have risen sufficiently to justify the transition (Libecap, 1990).

Most fisheries that have turned to these policies have done so only after a host of alternative input and output controls have failed to stem the pressure being placed on the resource. A similar story can be told for air pollution control. The offset policy, introduced in the United States for controlling air pollution, owes its birth to an inability of any other policy to reconcile the desire to allow economic growth with the desire to improve the quality of the air.

It is also clear that not every attempt to implement a tradable permit approach is successful. In air pollution control, attempts to establish a tradable permits approach have failed in Poland (Zylicz, 1999), Germany (Scharer, 1999), and the United Kingdom (Sorrell, 1999). Programs in water pollution control generally have not been very successful (Hahn and Hester, 1989).

On the other hand, it does appear that the introduction of new tradable permit programs becomes easier with familiarity. Following the very successful lead phaseout program, in the United States, new supporters appeared and made it possible to pass the sulfur allowance program.[44]

It also seems quite clear that, to date at least, using a grandfathering approach to the initial allocation has been a necessary ingredient in building the political support necessary to implement the approach.[45] Existing users frequently have the power to block implementation, while potential future users do not. This has made it politically expedient to allocate a substantial part of the economic rent that these resources offer to existing users as the price of securing their support. Although this strategy reduces the adjustment costs to existing users, it generally raises them for new users.[46]

The design features of the programs are not stable over time; they evolve with experience. The earliest use of the tradable permit concept, the Emissions Trading Program, overlaid credit trading on an existing regulatory regime and was designed to facilitate implementation of that program. Trading baselines were determined on the basis of previously determined, technology-based standards and created credits could not be used to satisfy all of these standards. For some the requisite technology had to be installed.

More recent programs, such as the Acid Rain and RECLAIM programs, replace, rather than complement, traditional regulation. Allowance allocations for these programs were not based on preexisting technology-based standards. In the case of RECLAIM, the control authority (the South Coast Air Quality Management District) could not have based allowances on predetermined standards even if it had been inclined to do so. Defining a complete set of technologies that offered the necessary environmental improvement (and yet were feasible in both an economic and engineering sense) proved impossible. Traditional regulation was incapable of providing the degree of reduction required by the Clean Air Act.

## Environmental Effects

One common belief about tradable permit programs is that their environmental effects are determined purely by the imposition of the aggregate limit, an act that is considered to lie outside the system. Hence, it is believed, the main purpose of the system is to protect the economic value of the resource, not the resource itself.

That is an oversimplification for several reasons. First, whether it is politically possible to set an aggregate limit may be a function of the policy used to achieve it. Second, both the magnitude of that limit and its evolution over time may be related to the policy. Third, the choice of policy regime may affect the level of monitoring and enforcement and noncompliance can undermine the achievements of the limit. Fourth, the policy may trigger environmental effects that are not covered by the limit.

The demonstration that the traditional regulatory policy was not value maximizing had two mirror-image implications. It implied either that the same environmental goals could be achieved at lower cost or that better environmental quality could be achieved at the same cost. In air pollution control, although the earlier programs were designed to exploit the first implication, later programs attempted to produce better air quality *and* lower cost.[47]

*Setting the Limit*

In air trading programs, the lower costs offered by trading were used in initial negotiations to secure more stringent pollution control targets (acid rain program, ozone-depleting gases, lead phaseout, and RECLAIM) or earlier deadlines (lead phaseout program). The air quality effects from more stringent limits were reinforced by the use of offset ratios for trades in nonattainment areas that were set at a ratio greater than 1.0 (implying a portion of each acquisition would go for better air quality). In addition, environmental groups have been allowed to purchase and retire allowances (acid rain program). Retired allowances represent authorized emissions that are not emitted.

In fisheries the institution of ITQs has sometimes, but not always, resulted in lower (more protective) TACs. In the Netherlands, for example, the plaice quota was cut in half (and prices rose to cushion the income shock) (Davidse, 1999).

*Meeting the Limit*

In theory the flexibility offered by tradable permit programs makes it easier to reach the limit, suggesting the possibility that the limit may be met more often under tradable permits systems than under the systems that preceded them. In most fisheries this expectation seems to have been borne out. In the Alaskan Halibut and Sablefish fisheries, for example, although exceeding the TAC was common before the imposition of an ITQ system, the frequency of excedences dropped significantly after the introduction of the ITQ (National Research Council, 1999).

A recent Organization for Economic Co-operation and Development review (1997:80) concludes:

> The results of individual quota management on resource conservation have been mixed. For the most part, IQs [individual quotas] and ITQs have been effective in limiting catch at or below the TAC determined by management authorities. Catch was maintained at or below the TAC in 24 out of 31 fisheries for which information on this outcome was available. ... In most cases, insufficient monitoring and enforcement allowed catches to exceed TACs.

*Enforcing the Limit*

Sometimes the rent involved in transferable permit programs is used to finance superior enforcement systems. In the sulfur allowance program, for example, the environmental community demanded (and received) a requirement that continuous emission monitoring be installed (and financed) by every covered utility. Coupling this with the rather stringent penalty system has meant 100 percent compliance.

The rents generated by ITQs also have provided the government with a source of revenue to cover the costs of enforcement and administration. In the many of the IQ fisheries in Australia, Canada, Iceland, and New Zealand, industry pays for administration and enforcement with fees levied on quota owners.

Not all uses of tradable permits, however, offer as convincing a solution for the monitoring and enforcement problems. With respect to fisheries, one comprehensive review (Organization for Economic Co-operation and Development, 1997:84) found:

> Higher enforcement costs and or greater enforcement problems occurred in 18 fisheries compared to five that experienced improvements. Enforcement proved particularly difficult in the high value fisheries, in multispecies fisheries, and in transnational fisheries. Support from industry for increased enforcement is common, as quota holders recognize that the illegal fishing by others damages the value of their quota rights and have an incentive to aid authorities with enforcement. ITQ management has led to increased co-operation between fishers and enforcement authorities in several cases, including the New Zealand fisheries in general, and the US wreckfish fishery. ... Underreporting of catch and data degradation was documented for 12 fisheries, but improvements were made in six fisheries.

### Effects on the Resource

In air pollution the programs typically have had a very positive effect on reducing emissions. In both the lead phaseout and ozone-depleting gas programs, the targeted pollutants were eliminated, not merely reduced. Both the acid rain and RECLAIM programs involve substantial reductions in emissions over time (Tietenberg, 1999).

In the fisheries what have been the effects on biomass? The evidence has been mixed. In the Chilean squat lobster fishery, the exploitable biomass has rebounded from a low of about 15,500 tons (prior to ITQs) to a level in 1998 of between 80,000 and 100,000 tons (Bernal and Aliaga, 1999). The herring fishery in Iceland has experienced a similar rebound (Runolfsson, 1999).

On the other hand, one review of 37 ITQ or IQ fisheries found that 24 experienced at least some temporary declines in stocks after instituting the programs. These were largely attributed to a combination of inadequate information on which to set conservative TACs and illegal fishing activity. Interestingly 20 of the 24 fisheries experiencing declines had additional regulations such as closed areas, size/selectivity regulations, trip limits, and vessel restrictions (Organization for Economic Co-operation and Development, 1997:82). These additional regulations apparently were also ineffective in protecting the resource.

### Other Effects

In water one significant problem has been the protection of "instream" uses of water. In the United States, some states only protected private entitlements to water if it was diverted from the stream and consumed. Recent changes in policy and some legal determinations have afforded more protections to these environmental uses of water.

In air pollution control, several effects transcend the normal boundaries of the program. In the climate change program, for example, it is widely recognized (Ekins, 1996) that the control of greenhouse gases will result in substantial reductions of other pollutants as a side effect. Other, more detrimental, effects include the clustering of emissions either in space or time.

In fisheries two main effects have been bycatch and highgrading. Bycatch is a problem in many fisheries, regardless of the means of control. The evidence from fisheries on how the introduction of ITQs affect bycatch and highgrading is apparently mixed. Two reviews (National Research Council, 1999:193; Organization for Economic Co-operation and Development, 1997:83) found that bycatch and highgrading may increase or decrease in ITQ fisheries depending on the fishery.

## Economic Effects

Although the evidence on environmental consequences is mixed (especially for fisheries), it is somewhat clearer for the economic consequences. In the presence of adequate enforcement, tradable permits do appear to increase the value of the commons to which the permits apply. In air pollution control, this takes the effect of considerable savings in meeting the pollution control targets (Hahn and Hester, 1989; Tietenberg, 1990). For water it involves the increase in value brought about by transferring the resources from lower valued to higher valued uses (Easter et al., 1998). In fisheries it not only involves the higher profitability from more appropriately scaled capital investments (resulting from the reduction in overcapitalization), but also from the fact that ITQs frequently make it possible to sell a more valuable product at higher prices (fresh fish rather than frozen fish) (National Research Council, 1999). One review of 22 fisheries found that the introduction of ITQs increased wealth in all 22 (Organization for Economic Co-operation and Development, 1997:83).

In both water and air pollution, the transition was not from an open access resource to tradable permits, but rather from a less flexible control regime to a more flexible one. The transition apparently has been accomplished with few adverse employment consequences, though sufficient data to do a comprehensive evaluation do not exist (Goodstein, 1996).

The employment consequences for fisheries have been more severe. In fisheries with reasonable enforcement, the introduction of ITQs usually has been accompanied by a considerable reduction in the amount of fishing effort. Normally this means not only fewer boats, but also less employment. The evidence also suggests, however, that the workers who remain in the industry work more hours during the year and earn more money (National Research Council, 1999:101).

The introduction of ITQs in fisheries has also had implications for crew, processors, and communities. Traditionally in many fisheries, crew have been co-venturers in the fishing enterprise, sharing in both the risk and reward. In some cases the shift to ITQs has shifted the risk and ultimately shifted the compensation system from a share of profits system to a wage system. Though this has not necessarily lowered incomes, it has changed the culture of fishing (McCay et al., 1989; McCay and Creed, 1990).

Processors can be affected by the introduction of ITQs in a number of ways. First, the processing sector is typically as overcapitalized as the harvesting sector.[48] Because the introduction of ITQs typically extends the fishing season and spreads out the processing needs of the industry, less processing capacity is needed. In addition, the more leisurely pace of harvesting reduces the bargaining power of processors versus fishers. In some areas such as Alaska, a considerable amount of this processing capital may lose value due to its immobility (Matulich et al., 1996; Matulich and Sever, 1999).

Communities can be, and in some cases have been, adversely affected when quota held by local fishers is transferred to fishers who operate out of other communities. Techniques developed to mitigate these effects, however, seem to have been at least moderately successful (National Research Council, 1999:206).

Generally market power has not been a significant issue in most permit markets despite some tendencies toward the concentration of quota. In part this is due to accumulation limits that have been placed on quota holders and the fact that these are typically not markets in which accumulation of quota yields significant monopoly-type powers.[49] In fisheries some concern has been expressed (Palsson, 1998) that the introduction of ITQs will mean the demise of the smaller fishers as they are bought out by larger operations. The evidence does not seem support this concern.[50]

## LESSONS

What can be gleaned from this necessarily brief survey of the theory and implementation experience with tradable permits?

### Evaluation

We begin by identifying the lessons that emerge from our evaluation of the factors affecting the implementation feasibility of transferable permits as well as the environmental and economic effects of their implementation.

• The air pollution programs, on balance, seem to be the most successful in achieving both economic and environmental objectives. In part this seems to be due to the presence of fewer (though certainly not zero) externalities in these programs. Fisheries must cope with potentially severe bycatch problems in multispecies fisheries. Water control authorities must cope with the consequences of trades on downstream users. These small-scale, complex resources with multiple externalities may be better managed by cooperative arrangements.

• The academic community has emphasized the importance of co-management of environmental resources, with users having a substantial role. Although tradable permit systems in principle allow a variety of governance systems, only in fisheries and water is there any evidence of an evolution in this direction. The current predominant form in both air pollution control and fisheries seems to be a system of shared management, with users playing a smaller role than envisioned by most co-management proposals. For those resource regimes located in the United States, it is common for the goals to be set at the national level and considerable "top-down" management to be in evidence. The management of water resources seems closest to user-controlled co-management schemes. In those systems, the rights markets are at the "informal" end of the spectrum.

• Although tradable permit systems in principle allow a variety of governance systems, the only evidence of an evolution toward true co-management has occurred in fisheries and water. The pollution and natural resource cases exhibit an important asymmetry. For air pollution control, the benefits from resource protection fall on the victims of air pollution, not on the polluters who use the resource. From a purely self-interest point of view, resource users (polluters) would be quite happy to pollute the air if they could get away with it. On the other hand, water users and fishers can both benefit from protection of the resource. Their collective self-interest is compatible with resource protection. This suggests that the incentives for collective action should be quite different in these two cases, and this difference could well explain the lower propensity for collective self-governance in the air pollution case.

• A main element of controversy in tradable permits systems involves both the processes for deciding the initial allocation and the initial allocation itself. These problems seem least intense for air pollution and most intense for fisheries. Though a rich set of management and initial allocation options exists, current experience seems not to have been very creative in their use.

• Tradable permit programs are sometimes held to be a relatively rigid approach to resource management. This expectation is created by the belief that once instituted, property rights cannot be changed. In fact, implemented tradable permit programs have exhibited a considerable amount of flexibility. A variety of new design features (such as zero-revenue auctions, bycatch quotas, and drop-through mechanisms) have emerged that are tailored to the characteristics of particular resources. These offer greater flexibility in meeting the needs of particular resource systems. For example, especially flexible adaptive management systems have evolved in programs designed to protect resources that exhibit higher degrees of supply variability (fisheries and water).

• In their most successful applications, tradable permits have been able to simultaneously protect the resources and provide sustainable incomes for users.

Technology advances, such as computerized exchanges, are helping to lower transaction costs, thereby facilitating the capture of more of the rent.

• The two elements that most jeopardize the success of a tradable permits program are inadequate enforcement and uninternalized externalities.

## Unfulfilled Theoretical Expectations

Two important expectations flowing from the economic theory have proved to be an inaccurate characterization of reality:

• The first is the theoretical expectation that transferable permit programs do not effect conservation of the resource because that is handled by the cap. In the theory, setting the cap is considered to be outside the system. Hence, it is believed, the main purpose of the system is to protect the economic value of the resource, not the resource itself. That is an oversimplification for several reasons. First, whether it is politically possible to set an aggregate limit may be a function of the policy used to achieve it. The use of grandfathered permits in the acid rain program, for example, made it possible to establish the limit on sulfur emissions. Second, in both fisheries and air pollution control, the evidence suggests that both the magnitude of the implemented limit and its evolution over time may be related to the policy. The flexibility and lower cost of meeting the limit offered by tradable permits systems can, and has, resulted in the acceptance of more stringent limits. Third, the choice of policy regime may affect the level of monitoring and enforcement, and noncompliance can undermine the achievements of the limit. Experience suggests that depending on the context, transferable permits can either improve or degrade the monitoring and enforcement situation. Fourth, the policy may trigger environmental effects that are not covered by the limit. Activity may be diverted from covered to uncovered resources.

• The second theoretical expectation that falls in the light of implementation experience involves the tradeoff between efficiency and equity in a tradable permits system. Traditional theory suggests that tradable permits offer a costless trade-off between efficiency and equity because, regardless of the initial allocation, the ability to trade assures that permits flow to their highest valued uses. This implies that the initial allocation can be used to pursue equity goals without lowering the value of the resource. In practice, implementation considerations nearly always allocate permits to historic uses, whether or not that is the most equitable allocation. This failure to use the initial allocation to protect equity concerns has caused other means to be introduced to protect equity considerations (such as restrictions of transfers). The additional restrictions generally do lower the value of the resource. In practice, therefore, tradable permits systems have not avoided the trade-off between efficiency and equity so common elsewhere in policy circles.

This evidence seems to suggest that tradable permits are no panacea, but they do have their niche.

## NOTES

1  Two examples of existing programs that did not make the list include the $NO_x$ Budget air pollution control program in the northeastern United States (Farrell et al., 1999) and programs to control conventional air pollutants in several states (Solomon and Gorman, 1998). For a large online bibliography covering these systems, see http://www.colby.edu/personal/t/thtieten/.

2  Consider just three examples. In air pollution control, a legal challenge was brought in Los Angeles during June 1997 by the Los Angeles-based Communities for a Better Environment (Tietenberg, 1995a). In fisheries a challenge was brought against the halibut/sablefish tradable permits system in Alaska (Black, 1997) and Congress imposed a moratorium on the further use of a tradable permits approach in U.S. fisheries (National Research Council, 1999). Though both legal cases ultimately were thrown out, as of this writing the moratorium is still in effect, despite a recommendation by the National Research Council to lift it.

3  One author, for example, compares a tradable permits system to the sale of indulgences in the Middle Ages (Goodin, 1994).

4  For a previous survey that also examines tradable permit systems across resource settings, see Colby (2000).

5  Another characteristic that affects the allocation of control responsibility is the degree to which the pollutant accumulates over time. In the interest of brevity I have not included that case. For an analysis of that case, see Griffin (1987).

6  For a general equilibrium treatment that derives the efficient allocation using a utility framework, see Tietenberg (1973).

7  As an interesting aside, the efficiency approach would tend to minimize health damage for a given level of expenditure, but it would do so by subjecting some individuals to a higher level of individual risk.

8  In this essay, "sustainability perspective" is used to refer to an outcome in which the resource itself is preserved. Sometimes called "environmental sustainability" (Tietenberg, 2000:97), this approach is more restrictive than the conventional notions of weak sustainability and strong sustainability, which maintain the value of the total capital stock and natural capital stock respectively.

9  In U.S. air pollution control, for example, an "acceptable" pollutant concentration level in the ambient air has been established on the basis of human health considerations. For fisheries the total allowable catch is usually defined in terms of the "allowable biological catch." Because neither of these processes involves an explicit calculation of net benefits, they would be efficient only by coincidence.

10  For an excellent formal treatment of the relationship between efficiency and sustainability in both renewable and nonrenewable resource contexts, see Heal (1998).

11  For a detailed explanation of the circumstances leading to the increasing evolution of market-based approaches to pollution control, see Tietenberg et al. (1999).

12  Unfortunately the usefulness of this corollary is limited whenever more than one goal needs to be satisfied by the initial allocation. This is commonly the case, for example, when the resource managers want to use the initial allocation both to build enough support to implement the program and to treat all claimants fairly. The allocations that satisfy each of those two goals may be quite different.

13  Inadequate monitoring and enforcement, of course, plagues all policy instruments, not just tradable permit systems.

14  In the case of market power in fisheries, the maximum number of permits that can be held by any individual or defined group routinely is limited by regulation.(National Research Council, 1999). In the case of transaction costs, it is possible to design administrative systems so as to minimize these costs (Tietenberg, 1998c).

15  Uninternalized externalities plague most other policy instruments as well. This precondition is not meant to differentiate tradable permit systems from other approaches, but rather to point out the conditions under which such systems work more smoothly.

16  Some empirical support for this proposition in implemented programs is beginning to appear. For example, one study of compliance behavior in the United Kingdom fishery (which was not an individual transferable quota, or ITQ, fishery) found that individuals who felt more involved in the management system had a statistically significantly lower probability of a violation (Hatcher et al., 2000).

17  Requirements of the act include the duty to end overfishing, to rebuild overfished stocks, to protect essential fish habitat, to reduce bycatch, and to consider fishing communities (National Research Council, 1999).

18  At least one major analysis of this relationship makes it clear that the Secretary of Commerce and the National Marine Fisheries Service have erred on the side of micromanagement rather than delegating too much authority to the regional councils (National Research Council, 1999:8).

19  Consider the following example from the Raymond Basin in California: "Under the Water Exchange Agreement, each party must offer to the 'exchange pool' its rights to water in excess of its needs for the coming year, at a price no greater than the party's average water production cost. Parties anticipating that their access to water will be inadequate to meet their needs for the coming year submit requests to the exchange pool. The watermaster matches the offers to the requests, with the lowest priced water allocated first, then the next lowest, and so on. The actual allocation does not involve the transfer of water, but rather the right to pump specific quantities of water" (Blomquist, 1992:87-88).

20  McCay (2001) provides examples of other forms of co-management in fisheries. Most of her examples do not involve ITQs, and those that do have limited participation by users.

21  For example, Article XIV of the California Constitution of 1879 denied the ownership of water to individuals and granted them a usufructuary right—the right to the *use* of the water (Blomquist, 1992). The 1981 Water Code in Chile stipulates that water is a national resource for public use, but rights to use water can be granted to individuals (Hearne, 1998).

22  One prominent exception is the New Zealand ITQ system. It grants rights in perpetuity (National Research Council, 1999:97).

23  Compare this case with a case where the rights were defined in tons. If biological conditions indicated the need to lower the TAC significantly, the need to confiscate existing rights might trigger suits seeking compensation against the resource manager.

24  Other systems achieve this result by allowing rights holders to lease the rights to others for a specific period of time.

25  Livingston (1998) reports on an unpublished World Bank survey that found that out of 35 developing countries examined, more than half had rainfall variability of 40 percent.

26  In the western United States, the number of rights expected to be fulfilled in any given year is determined by snowpack measurements and satellite monitoring of streamflows (Livingston, 1998).

27  The scheme is sufficiently flexible that entitlements could rise over time, fall over time, or be constant. The main condition is that the time path be specified for the duration of that particular series.

28  Some programs have additional requirements. In the lead phaseout program, the annual limits declined over time until, in the final year, they went to zero (Nussbaum, 1992). In the RECLAIM program in Los Angeles, the limits decline 8 percent per year (Fromm and Hansjurgens, 1996; Zerlauth and Schubert, 1999).

29  An interesting third possibility emerges from an examination of the air pollution control experience in Chile (Montero, 2000a). Apparently the use of a grandfathered system of allocation, coupled with the high rents from holding those permits, induced a number of previously undiscovered sources to admit their emissions in order to gain entry to the program.

30  For example, assigning rights in this way is considered one factor in how the United States was able to implement a system to control acid rain after many years of failed attempts (Kete, 1992).

31  From the point of view of the user, two components of financial burden are significant: (1) extraction or control costs, and (2) expenditures on permits. Although only the former represent real resource costs to society as a whole (the latter are merely transfers from one group in society to another), to the user both represent a financial burden. The empirical evidence suggests that when a traditional auction market is used to distribute permits (or, equivalently, when all uncontrolled emissions are subject to an emissions tax), the permit expenditures (tax revenue) frequently would be larger in magnitude than the control costs; the sources would spend more on permits (or pay more in taxes) than they would on the control equipment (Tietenberg, 1985).

32  The downside occurs when the investments being rewarded were initiated purely for the purpose of increasing the initial allocation of tradable permits. Not only are these investments inefficient, but rewarding them undermines the ethical basis for an initial allocation based on historic use.

33  The "new source bias" is, of course, not unique to tradable permit systems. It applies to any system of regulation that imposes more stringent requirements on new sources than existing ones.

34  Tradable permits systems are perfectly compatible with the principles of co-management. In this case the community would play a large role in defining the goals and procedures in the system; see National Research Council (1999:135-138).

35  This setaside has not been used because sufficient allowances have been available through normal channels. That doesn't necessarily mean the setaside was not useful, however, because it may have alleviated concerns that otherwise could have blocked the implementation of the program.

36  The revenue is returned to the original permit holders rather than retained by the government, hence the name "zero-revenue auction" (Svendsen and Christensen, 1999).

37  This concern does not arise in all communities because in several fisheries and in air pollution control, the effect of any particular transfer or set of transfers is negligible.

38  These effects may be less pronounced in short river systems. This may be one of the reasons why tradable permit markets in water are so active in Chile (Hearne, 1998).

39  In an unprecedented complaint filed in California during June 1997, the Los Angeles-based Communities for a Better Environment contends that RECLAIM is allowing the continued existence of toxic "hot spots" in low-income communities. Under RECLAIM rules, Los Angeles-area manufacturers can buy and scrap old, high-polluting cars to create emissions-reduction credits. These credits can be used to reduce the required reductions from their own operations. Under RECLAIM most California refineries have installed equipment that eliminates 95 percent of the fumes, but the terminals in question reduced less because the companies scrapped more than 7,400 old cars and received mobile source emission reduction credits, which they credited toward their reduction requirements. The complaint notes that whereas motor vehicle emission reductions are dispersed throughout the region, the offsetting increases at the refineries are concentrated in low-income neighborhoods (Marla Cone, *Los Angeles Times*, as cited in GREENWIRE, 7/23/97:http:/www.eenews.net/greenwire.htm). Though this particular complaint was eventually dismissed by the court, the forces of discontent that gave rise to the suit are far from silenced.

40  In fisheries, for example, stock assessments sometimes depend on the size and composition of the catch. If the composition of the landed harvest is unrepresentative of the actual harvest due to illegal discards, this can bias the stock assessment and the total allowable catch that depends on it. Not only would true mortality rates be much higher than apparent mortality rates, but the age and size distribution of landed catch would be different from the size distribution of the initial harvest (prior to discards). In fisheries this is known as "data fouling."

41  Prior to 1988, the expected positive effects of ITQs did not materialize in the Dutch cutter fisheries due to inadequate enforcement. Fleet capacity increased further, the race for fish continued, and the quotas had to be supplemented by input controls such as a limit on days at sea (National Research Council, 1999:176).

42  Not only has the recovery of monitoring and enforcement costs become standard practice in some fisheries (New Zealand, for example), but funding at least some monitoring and enforcement activity out of rents generated by the fishery already has been included as a provision in the most

recent amendments to the U.S. Magnuson-Stevens Act. The sulfur allowance program mandates continuous emissions monitoring financed by the emitting sources.

43 An IFQ is the right under a limited access system to harvest a specific quantity of fish. ITQs are a form of IFQs in which the rights are transferable.

44 It is frequently suggested that new programs should be of the "cap-and-trade" type because they reduce transaction costs. Although I agree that they reduce transaction costs, it is less clear to me that cap-and-trade programs can always achieve the political will to be implemented without gaining familiarity through the more heavily controlled credit programs. My own reading of the U.S. case suggests that we would not currently have cap-and-trade programs if we had not proceeded first to implement credit programs. These served as a training ground for the various stakeholders before moving to the more flexible programs.

45 One exception is the ITQ program used in Chilean fisheries. Here the permits are allocated by auction (Bernal and Aliaga, 1999).

46 New users have to buy into the system, while existing users retain their traditional entitlement.

46 In an interesting analysis of the cost and emissions savings from implementing an emissions trading system for light-duty vehicles in California, Kling (1994) finds that although the cost savings from implementing an emission trading program (holding emissions constant) would be modest (on the order of 1 percent to 10 percent), the emissions savings possibilities (holding costs constant) would be much larger (ranging from 7 percent to 65 percent).

48 In derby fishing the harvest is landed in a relatively short period of time, creating the need for more peak capacity.

49 In many fisheries, for example, the relevant markets are global, with many different sources of supply. In air pollution the number of participants is typically quite high.

50 An Organization For Economic Co-operation and Development review concludes, "There was very little evidence to support the hypothesis that small scale fishers would be eliminated" (National Research Council, 1999:84).

# REFERENCES

Adelaja, A., J. Menzo, and B. McCay
    1998   Market power, industrial organization and tradeable quotas. *Review of Industrial Organization* 13(5):589-601.

Anderson, L.G.
    1991   A note on market power in ITQ fisheries. *Journal of Environmental Economics and Management* 21(2):291-296.
    1994   An economic analysis of highgrading in ITQ fisheries regulation programs. *Marine Resource Economics* 9:189-207.
    1995   Privatizing open access fisheries: Individual transferable quotas. Pp. 453-474 in *The Handbook of Environmental Economics*, D.W. Bromley, ed. Oxford, Eng.: Blackwell.

Annala, J.H.
    1996   New Zealand's ITQ system: Have the first eight years been a success or a failure? *Reviews in Fish Biology and Fisheries* 6:43-62.

Baumol, W.J., and W.E. Oates
    1971   The use of standards and prices for protection of the environment. *Swedish Journal of Economics* 73:42-54.
    1988   *The Theory of Environmental Policy*. Cambridge, Eng.: Cambridge University Press.

Bernal, P., and B. Aliaga
    1999   ITQ's in Chilean fisheries. In *The Definition and Allocation of Use Rights in European Fisheries: Proceedings of the Second Workshop held in Brest, France, 5-7 May 1999*, A. Hatcher and K. Robinson, eds. Portsmouth, Eng.: Centre for the Economics and Management of Aquatic Resources.

Black, N.D.
   1997    Balancing the advantages of individual transferable quotas against their redistributive ef-
           fects: The case of Alliance Against IFQs v. Brown. *International Law Review* 9(3):727-
           746.
Blomquist, W.
   1992    *Dividing the Waters: Governing Groundwater in Southern California.* San Francisco: ICS
           Press.
Boyce, J.R.
   1996    An economic analysis of the fisheries bycatch problem. *Journal of Environmental Eco-
           nomics and Management* 31(3):314-336.
Colby, B.G.
   1995    Regulation, imperfect markets and transactions costs: The elusive quest for efficiency in
           water allocation. Pp. 475-502 in *The Handbook of Environmental Economics*, D.W.
           Bromley, ed. Oxford, Eng.: Blackwell.
   2000    Cap and trade challenges: A tale of three markets. *Land Economics* 76(4):638-658.
Davidse, W.
   1999    Lessons from twenty years of experience with property rights in the Dutch fishery. Pp.
           153-163 in *The Definition and Allocation of Use Rights in European Fisheries: Proceed-
           ings of the Second Workshop held in Brest, France, 5-7 May 1999*, A. Hatcher and K.
           Robinson, eds. Portsmouth, Eng.: Centre for the Economics and Management of Aquatic
           Resources.
Easter, K.W., A. Dinar, and M.W. Rosegrant
   1998    Water markets: transactions costs and institutional options. *Markets for Water: Potential
           and Performance*, K.W. Easter, A. Dinar, and M.W. Rosegrant, eds. Boston: Kluwer Aca-
           demic Publishers.
Ekins, P.
   1996    The secondary benefits of $CO_2$ abatement: How much emission reduction do they justify?
           *Ecological Economics* 16(1):13-24.
Farrell, A., R. Carter, and R. Raufer
   1999    The NOx Budget: Market-based control of tropospheric ozone in the northeastern United
           States. *Resource and Energy Economics* 21(2):103-124.
Fromm, O., and B. Hansjurgens
   1996    Emission trading in theory and practice: An analysis of RECLAIM in Southern California.
           *Environment and Planning C - Government and Policy* 14(3):367-384.
Ginter, J.J.C.
   1995    The Alaska Community Development Quota Fisheries Management Program. *Ocean and
           Coastal Management* 28(1-3):147-163.
Goodin, R.E.
   1994    Selling environmental indulgences. *Kyklos* 47(4):573-596.
Goodstein, E.
   1996    Jobs and the environment - An overview. *Environmental Management* 20(3):313-321.
Goulder, L.H., I.W.H. Parry, R.C. Williams, III, and D. Burtraw
   1999    The cost-effectiveness of alternative instruments for environmental protection in a second-
           best setting. *Journal of Public Economics* 72(3):329-360.
Grafton, R.Q., and R.A. Devlin
   1996    Paying for pollution - Permits and charges. *Scandinavian Journal of Economics* 98(2):275-
           288.
Griffin, R.C.
   1987    Environmental policy for spatial and persistent pollutants. *Journal of Environmental Eco-
           nomics and Management* 14(1):41-53.

Hahn, R.W.
    1984    Market power and transferable property rights. *Quarterly Journal of Economics* 99(4):753-
            765.
Hahn, R.W., and G.L. Hester
    1989    Marketable permits: Lessons from theory and practice. *Ecology Law Quarterly* 16:361-
            406.
Hausker, K.
    1990    Coping with the cap: How auctions can help the allowance market work. *Public Utilities
            Fortnightly* 125: 28-34.
Heal, G.M.
    1998    *Valuing the Future: Economic Theory and Sustainability.* New York: Columbia Univer-
            sity Press.
Hearne, R.R.
    1998    Institutional and organizational arrangements for water markets in Chile. Pp. 141-157 in
            *Market for Water: Potential and Performance*, K.W. Easter, M.W. Rosegrant, and A.
            Dinar, eds. Boston: Kluwer Academic Publishers.
Holling, C.S.
    1978    *Adaptive Environmental Assessment and Management.* New York: John Wiley & Sons.
Howe, C.W., and D.R. Lee
    1983    Priority pollution rights: Adapting pollution control to a variable environment, *Land Eco-
            nomics* 59(2):141-149.
Kelman, S.
    1981    *What Price Incentives? Economists and the Environment.* Westport, CT: Greenwood Pub-
            lishing Group.
Kete, N.
    1992    The U.S. acid rain control allowance trading system. Pp. 69-93 in *Climate Change: De-
            signing a Tradeable Permit System*, T. Jones, and J. Corfee-Morlot, eds. Paris: Organiza-
            tion for Economic Co-operation and Development.
Kling, C.L.
    1994    Environmental benefits from marketable discharge permits or an ecological vs. economi-
            cal perspective on marketable permits. *Ecological Economics* 11(1):57-64.
Kling, C., and J. Rubin
    1997    Bankable permits for the control of environmental pollution. *Journal of Public Economics*
            64(1):99-113.
Knapp, G.
    1997    Initial effects of the Alaska Halibut IFQ Program: Survey comments of Alaska fishermen.
            *Marine Resource Economics* 12(3):239-248.
Kruger, J.A., B. McLean, and R.A. Chen
    2000    A tale of two revolutions: Administration of the $SO_2$ trading program. In *Emissions Trad-
            ing: Environmental Policy's New Tool*, R.F. Kosobud, ed. New York: Wiley.
Larson, D.M., B.W. House, and J.M. Terry
    1998    Bycatch control in multispecies fisheries: A quasi-rent share approach to the Bering Sea/
            Aleutian Islands midwater trawl pollock fishery. *American Journal of Agricultural Eco-
            nomics* 80(4):778-792.
Libecap, G.D.
    1990    *Contracting for Property Rights.* Cambridge, Eng.: Cambridge University Press.
Livingston, M.L.
    1998    Institutional requisites for efficient water markets. Pp. 19-33 in *Markets of Water: Poten-
            tial and Performance*, K.W. Easter, M.W. Rosengrant, and A. Dinar, eds. Boston: Kluwer
            Academic Publishers.
Lyon, R.M.

   1982   Auctions and alternative procedures for allocating pollution rights. *Land Economics*
         58(1):16-32.
Maloney, M., and G.L. Brady
   1988   Capital turnover and marketable property rights. *The Journal of Law and Economics* 31(1):
         203-226.
Matulich, S.C., R.C. Mittelhammer, and C. Reberte
   1996   Toward a more complete model of individual transferable fishing quotas: Implications of
         incorporating the processing sector. *Journal of Environmental Economics and Manage-
         ment* 31(1):112-128.
Matulich, S.C., and M. Sever
   1999   Reconsidering the initial allocation of ITQs: The search for a Pareto-safe allocation be-
         tween fishing and processing sectors. *Land Economics* 75(2):203-219.
McCay, B.J.
   1998   *Oyster Wars and the Public Trust: Property, Law and Ecology in New Jersey History.*
         Tucson: University of Arizona Press.
   1999   Resistance to Changes in Property Rights Or, Why Not ITQs? Unpublished paper pre-
         sented to Mini-Course, FishRights 99, Fremantle, Australia, November.
   2001   Community-based and cooperative solutions to the "Fishermen's Problem" in the Ameri-
         cas. In *Protecting the Commons: A Framework for Resource Management in the Ameri-
         cas,* J. Burger, R. Norgaard, E. Ostrom, D. Policansky, and B.D. Goldstein, eds. Washing-
         ton, DC: Island Press.
McCay, B.J., and C.F. Creed
   1990   Social structure and debates on fisheries management in the Mid-Atlantic surf clam fish-
         ery. *Ocean & Shoreline Management* 13:199-229.
McCay, B.J., J.B. Gatewood, and C.F. Creed
   1989   Labor and the labor process in a limited entry fishery. *Marine Resource Economics* 6:311-
         330.
Misiolek, W.S., and H.W. Elder
   1989   Exclusionary manipulation of markets for pollution rights. *Journal of Environmental Eco-
         nomics and Management* 16(2):156-66.
Montero, J.P.
   1999   Voluntary compliance with market-based environmental policy: Evidence from the U.S.
         Acid Rain Program, *Journal of Political Economy* 107(5):998-1033.
   2000a  A Market-Based Environmental Policy Experiment in Chile. Working Paper of the Center
         for Energy and Environmental Policy Research, Massachusetts Institute of Technology.
         MIT-CEEPR 2000-005 WP (August).
   2000b  Optimal design of a phase-in emissions trading program. *Journal of Public Economics*
         75(2):273-291.
Montgomery, W.D.
   1972   Markets in licenses and efficient pollution control programs. *Journal of Economic Theory*
         5(3):395-418.
National Research Council
   1999   *Sharing the Fish: Toward a National Policy on Fishing Quotas.* Committee to Review
         Individual Fishing Quotas. Washington, DC: National Academy Press.
Nelson, R., T. Tietenberg, and M.R. Donihue
   1993   Differential environmental regulation: Effects on electric utility capital turnover and emis-
         sions. *Review of Economics and Statistics* 75(2):368-373.
Nussbaum, B.D.
   1992   Phasing down lead in gasoline in the U.S.: Mandates, incentives, trading and banking. Pp.
         21-34 in *Climate Change: Designing a Tradeable Permit System,* T. Jones and J. Corfee-
         Morlot, eds. Paris: Organization for Economic Co-operation and Development.

Organization for Economic Co-operation and Development
  1997    *Towards Sustainable Fisheries: Economic Aspects of the Management of Living Marine Resources.* Paris: Organization for Economic Co-operation and Development.
  1999    *Implementing Domestic Tradable Permits for Environmental Protection.* Paris: Organization for Economic Co-operation and Development.
Palsson, G.
  1998    The virtual aquarium: Commodity fiction and cod fishing. *Ecological Economics* 24(2-3):275-288.
Palsson, G. and A.Helgason
  1995    Figuring fish and measuring men: The individual transferable quota system in Icelandic cod fishery. *Ocean and Coastal Management* 28(1-3):117-146.
Rubin, J. D.
  1996    A model of intertemporal emission trading, banking, and borrowing. *Journal of Environmental Economics and Management* 31(3):269-286.
Runolfsson, B.
  1999    ITSQs in Icelandic fisheries: A rights-based approach to fisheries management. Pp. 164-193 in *The Definition and Allocation of Use Rights in European Fisheries: Proceedings of the Second Workshop held in Brest, France, 5-7 May 1999*, A. Hatcher and K. Robinson, eds. Portsmouth, Eng. Centre for the Economics and Management of Aquatic Resources.
Sartzetakis, E.S.
  1997    Raising rivals' costs: Strategies via emission permits markets. *Review of Industrial Organization* 12(5-6):751-765.
Scharer, B.
  1999    Tradable emission permits in German clean air policy: Considerations on the efficiency of environmental policy instruments. Pp. 141-153 in *Pollution for Sale: Emissions Trading and Joint Implementation*, S. Sorrell and J. Skea, eds. Cheltenham, Eng.: Edward Elgar Publishing Limited.
Solomon, B.D. and H.S. Gorman
  1998    State-level air emissions trading: The Michigan and Illinois Models. *Journal of the Air & Waste Management Association* 48(12):1156-1165.
Sorrell, S.
  1999    Why sulphur trading failed in the UK. Pp. 170-210 in *Pollution for Sale: Emissions Trading and Joint Implementation*, S. Sorrell and J. Skea, eds. Cheltenham, Eng.: Elgar Publishing Limited.
Spulber, N., and A. Sabbaghi
  1993    *Economics of Water Resources: From Regulation to Privatization.* Hingham, MA: Kluwer Academic Publishers.
Stavins, R.N.
  1995    Transaction costs and tradeable permits. *Journal of Environmental Economics and Management* 29(2):133-148.
Svendsen, G.T.
  1999    Interest groups prefer emission trading: A new perspective. *Public Choice* 101(1-2):109-28.
Svendsen, G.T. and J.L. Christensen
  1999    The US SO$_2$ auction: Analysis and generalization. *Energy Economics* 21(5):403-416.
Tietenberg, T.H.
  1973    Specific taxes and the control of pollution: A general equilibrium analysis. *Quarterly Journal of Economics* 87(4):503-522.
  1985    *Emissions Trading: An Exercise in Reforming Pollution Policy.* Washington, DC: Resources for the Future.

1990    Economic instruments for environmental regulation. *Oxford Review of Economic Policy* 6(1):17-33.

1995a   Design lessons from existing air pollution control systems: The United States. Pp. 15-32 in *Property Rights in a Social and Ecological Context: Case Studies and Design Applications*, S. Hanna and M. Munasinghe, eds. Washington DC: The World Bank.

1995b   Tradeable permits for pollution control when emission location matters: What have we learned? *Environmental and Resource Economics* 5(2):95-113.

1998a   Disclosure strategies for pollution control. *Environmental & Resource Economics* 11(3-4):587-602.

1998b   Economic analysis and climate change. *Environment and Development Economics* 3(3): 402-405.

1998c   Tradable permits and the control of air pollution—Lessons from the United States. *Zeitschrift für Angewandte Umweltforschung* 9:11-31.

1999    Lessons from using transferable permits to control air pollution in the United States. Pp. 275-292 in *Handbook of Environmental and Resource Economics*, J.C.J. Van den Bergh, ed. Cheltenham, Eng.: Edward Elgar Publishing Limited.

2000    *Environmental and Natural Resource Economics.* 5th ed. Reading, MA: Addison-Wesley.

Tietenberg, T., M. Grubb, A. Michaelowa, B. Swift, and Z.X. Zhang

1998    *International Rules for Greenhouse Gas Emissions Trading: Defining the Principles, Modalities, Rules and Guidelines for Verification, Reporting and Accountability.* Geneva: United Nations. UNCTAD/GDS/GFSB/Misc.6.

Tietenberg, T.H., K. Button, and P. Nijkamp

1999    Introduction. Pp. xvii-xxvi in *Environmental Instruments and Institutions*, T.H. Tietenberg, K. Button, and P. Nijkamp, eds. Cheltenham:Eng.: Edward Elgar Publishing Limited.

Van Egteren, H., and M. Weber

1996    Marketable permits, market power and cheating. *Journal of Environmental Economics and Management* 30(2):161-173.

Weitzman, M.

1974    Prices vs. quantities. *Review of Economic Studies* 41:447-491.

Young, M.D.

1999    The design of fishing-right systems - The NSW Experience. *Ecological Economics* 31(2):305-316.

Young, M.D., and B.J. McCay

1995    Building equity, stewardship and resilience into market-based property right systems. In *Property Rights and the Environment: Social and Ecological Issues*, S. Hanna and M. Munasinghe, eds. Washington, DC: Beijer International Institute of Ecological Economics and the World Bank.

Zerlauth, A., and U. Schubert

1999    Air quality management systems in urban regions: An analysis of RECLAIM in Los Angeles and its transferability to Vienna. *Cities* 16(4):269-283.

Zylicz, T.

1999    Obstacles to implementing tradable pollution permits: The case of Poland. Pp. 147-165 in *Implementing Domestic Tradable Permits for Environmental Protection.* Paris: Organization for Economic Co-operation and Development.

# 7

# Common Property, Regulatory Property, and Environmental Protection: Comparing Community-Based Management to Tradable Environmental Allowances

*Carol M. Rose*

The days are long gone in which environmentalists have believed that there is a "nature" or "natural world" separable from human activity. Our newspapers are full of stories of the human impact on what might be otherwise fondly thought of as pure "nature," from the ocean depths through the remotest forests to the skies above; the consequences of human agriculture, transportation, manufacture, and resource extraction affect even the most seemingly inaccessible corners of the globe.

Because no part of the world's environment is untouched by human activity, environmental protection must be seen in large measure as a matter of human social organization. But what social organization is possible for dealing with environmental resources? The pessimistic views of Hardin, and his successor Ophuls, have been well-known for decades: On Hardin's analysis, as elaborated by Ophuls, environmental resources are the locus of the "tragedy of the commons," a multiple-person prisoners' dilemma (PD) (Hardin, 1968; Ophuls, 1977; Ullmann-Margalit, 1977). Here it is in the interest of each resource user, taken individually, to exploit the resource *à outrance* while doing nothing to conserve, with the result that otherwise renewable resources instead become wasting assets. In the Hardin/Ophuls view, environmental degradation—overfishing, deforestation, overgrazing, pollution, whatever—is only a bleak set of repetitions of the "tragedy," and only two solutions are possible to stave off the tragic decimation: individual property on the one hand, which internalizes the externalities of common pool exploitation, or "Leviathan" on the other, where governmental directives force individuals to perform in ways that promote the common good (Hardin, 1968; Ophuls, 1977).

The great service of Ostrom and her colleagues has been to contest this unattractive view, and to offer a powerful set of counterexamples of conservationist social institutions. Ostrom and others have pointed out that the problem that Hardin called "the commons" was really a problem of "open access," whereas a common resource that is limited to a particular group of users may suffer no such decimation. Indeed, Hardin's dominating example of the medieval common fields was not tragic at all, but was rather an example of a set of community-based sustainable agricultural practices that lasted for centuries, if not millennia (Cox, 1985; Dahlman, 1980; Ostrom, 1990; Rieser, 1999; Smith, 2000).

As the first chapter of this book notes, there has been considerable variety in the nomenclature that refers to such limited common resources and the community governance processes that manage them,[1] but for purposes of this chapter, I will refer to community-based management regimes for common property resources as "CBMRs." I use this term to convey what I hope is a subtly greater attention to governance institutions and practices, rather than to the common-pool resources that underlie them; obviously, however, the physical and the institutional are intertwined—no doubt giving rise to the difficulties in nomenclature.

Whatever the names and emphases, institutions for managing common resources have become the subject of a growing and rather affectionate literature. This literature includes descriptions and analyses of un-tragic community resource management practices all over the world—Turkish fisheries, Japanese and Swiss grazing communities, ancient and modern Spanish irrigation systems, communal forestry in India and Indonesia, wetlands management by medieval English "fen people," fishing and hunting practices among northern Canadian clan groups, lobster fishing communities in Maine (Berkes, 1995; Bosselman, 1996; Ostrom, 1990).

Obviously, there is a great deal to be said simply for setting the record straight about what the "commons" really mean and have meant over time. But there are larger lessons implicit or explicit in the CBMR literature as well, and they are lessons of a somewhat more political nature. First is the lesson that voluntary social action is possible, and in particular that it is possible as a means to solve resource-related problems. That is to say, contrary to some of the more pessimistic presentations of the dismal science, human beings are not always individual maximizers, getting themselves stuck in the endless repetition of n-person PDs. Instead, quite ordinary people have the psychological, social, and moral wherewithal to arrive at cooperative arrangements on matters of common interest. A second lesson is that bigger is not always better. More particularly, the CBMR literature offers numerous examples in which larger governmental forays into resource management are distinctly inferior to community-based solutions, and in which governmental intervention has badly damaged perfectly workable community systems (Higgs, 1996; Ostrom, 1990; Pinkerton, 1987). In short, the ever-expanding CBMR scholarship argues strongly that nongovernmental, community-based resource management can offer models for efficient and sustainable resource use.

Given the surge of interest in community-based resource regimes, it is curious that their institutional structures do not appear more frequently in legal proposals for the improvement of environmental regulation. This is not because legal scholars are unaware of the literature on community-based common property. Although CBMR scholars for the most part appear to be untouched by legal scholarship, the reverse is not true; legal scholars regularly cite the major studies in a number of contexts, from intellectual property (Merges, 1996) to the burgeoning literature on informal norms (Ellickson, 1991). The legal scholarship on the Internet in particular has drawn analogies to the bottom-up community self-organization that has emerged in much older common resource regimes (Rose, 1998). Nevertheless, aside from a handful of scholars (Bosselman, 1996; Rieser, 1997; Rose, 2000), few in the legal academy have paid much attention to community-based management institutions as potential engines to drive improved *environmental* regulation.

Instead, among legal scholars, the poster children of proposed environmental improvement are a new version of individual entitlements that I will call tradable environmental allowances (TEAs). In TEA regimes, governmental regulators in effect place an upper limit or cap on the total quantity of a given resource that is to be available for use, whether the "use" is extractive or polluting. The regulators then divide the capped total into individual allowances. Henceforth they require all resource users to purchase or trade for whatever allowances they use.

TEAs along this model already have been deployed, to great applause, for the regulation of sulfur dioxide pollution in the United States; they have been used to manage fisheries in Australia, New Zealand, and elsewhere; and they are under much discussion as an element of future international regimes to control greenhouse gases (Rieser, 1997; Stavins, 1998; Tietenberg, 1985; Tietenberg, this volume:Chapter 6).

At least in theory, each TEA regime transforms access to the resource in question into a divisible but finite total quantity, and each individual resource user must pay for every pound of pollution released into the atmosphere or every pound of fish landed; resource use thus becomes in effect a kind of private property that must be acquired through purchase and trade. The property-like characteristics of TEAs are at the heart of their attractiveness. As has been so often argued about more conventional private property, the underlying idea is that if resource users are confronted with the need to purchase TEAs, they will husband resources carefully and will undertake conservation or innovation to substitute for their now expensive resource use (Ackerman and Stewart, 1988; Kriz, 1998; Rose, 1994; Tipton, 1995).

Although TEAs do not entirely vindicate the Hardin/Ophuls view that the choice for governing structures lies *either* with private property *or* with Levia-

than, TEAs—much more than the self-organized CBMRs—do have a Hardin/
Ophuls ring about them. TEAs in effect *combine* Leviathan with private property;
they are state-created private rights, tradable in a market along with other commodities.

Despite the differences between TEA and community-based institutions,
however, these two types of resource management regimes share a basic underlying structure. Neither takes a "hands off" approach to environmental protection.
Quite the contrary, both types of regime contemplate some human use or consumption of renewable resources, whether those resources are wildlife, fish,
grasses, trees, the air mantle, underground aquifers, surface water stocks, or whole
ecosystems. Moreover, although both types of regime contemplate some human
inroads into the resources they regulate, for both regimes the critical issue is to
limit those inroads to moderate "fringe" amounts that are compatible with the
renewal of the underlying cores of the resource stocks. And finally, both types of
regime are fundamentally *property* regimes—individual property in the case of
TEAs, common property in the case of CBMRs; in neither case are resources
open to the world at large, but are rather treated as the domain of their respective
individual or common owners.

Beyond those basics, however, CBMR and TEA regimes often diverge dramatically, so much so that one can see them as alternative ideal types for very
different approaches to property-based environmental management. For example,
in TEA regimes, as in all modern legislative programs, legislators and the public
may discuss explicitly the appropriate permissible "fringe" usage of the resource
in question, that is, the total allowable take or total cap placed on resource use
(Ackerman and Stewart, 1988). In CBMRs, on the other hand, explicit discussions of this sort are less likely to occur, and the total take is more likely to
emerge from established practices—practices that themselves may originate in
efforts to manage interpersonal conflict rather than to engage in explicit management of the larger resource (Bardhan and Dayton-Johnson, this volume:Chapter
3; McCay, this volume:Chapter 11; Seabright, 1993). Even more noticeable are
the very different ways in which individual entitlements are allocated and enforced in the two types of regime. In TEA regimes, regulatory bodies split up the
allowable total take into individual allowances and allocate these allowances;
thereafter the allowance holders may trade among themselves as they wish, subject to monitoring and enforcement by the regulators. In community-based
regimes, on the other hand, the user groups' own practices set individual entitlements. These entitlements generally depend on longstanding residence, reputation, and adherence to community norms—norms that are often very elaborate,
and that are enforced by the community members themselves—and trading is
often quite restricted. I will return shortly to both these subjects, that is, the questions of setting the total take on the one hand, and the structures of entitlements
on the other hand. I will do so because both subjects have some bearing on the
series of comparisons to which I now turn.

In the comparisons that follow, I treat CBMR and TEA regimes as ideal types of property-based environmental management. One caveat: It is important to bear in mind that the community-based common property regimes that are best known are often of long duration, which means that they are apt to be quite traditional, whereas TEA regimes tend to be quite new; this is a factor that can in itself accentuate differences between these regimes. A second caveat runs in the opposite direction: Real life being a more blurred affair than are any "ideal types," one finds in practice that more-or-less community-based regimes sometimes share characteristics with more-or-less tradable allowance regimes. Indeed, as I will illustrate later, some quite promising ideas for modern environmental management try to combine these different approaches. But in setting out CBMR and TEA regimes as more or less "pure" types, I hope to illuminate some of their typical characteristics. More important, I hope to show how these differing typical characteristics map onto different dimensions of current environmental problems, and how they result in quite different strengths and weaknesses as modes of property-based environmental management.

## CBMR AND TEA REGIMES AS MANAGEMENT INSTITUTIONS: VARYING SOLUTIONS UNDER VARYING CONDITIONS

### Resource Size

One important dimension of any environmental issue is simply the *size* of the resource in question. Environmental resources are generally too large for individual ownership. In fact, that is what creates environmental problems: Individual resource uses have spillover or common-pool effects on other persons and resources. Individual landholdings generally lie adjacent to environmental resources such as air, surface water, groundwater, and wildlife. But the landowner who burns trash affects the neighbors' air. Similarly, the landowner who removes trees on his or her own land may diminish nesting bird populations and contribute to an insect explosion throughout the vicinity. Similarly again, the landowner who spills toxins on the ground may pollute an aquifer or a stream that carries the deadly materials miles into the distance. In all these instances, individual uses of individual property spill out into a larger environmental arena. Global environmental issues implicate activities in vastly larger spaces; the most everyday form of combustion anywhere in the world—a motor scooter, a backyard barbecue—may contribute to greenhouse gases that raise global temperatures, lift the levels of the oceans, and contribute to melting tundra (Wiener, 1999).

### CBMRs and Resource Size

The sheer size of many environmental problems may be one reason why community-based institutions have been relatively little noticed as social management regimes for the environmental resources that are typically the candidates for legal regulation. With some limited exceptions, CBMRs tend to encompass activities only on a fairly small scale.

Many CBMRs have been studied in the context of a burgeoning "new institutional economics," a line of scholarship concerning nongovernmental social problem solving of all sorts, and this scholarship suggests some reasons for the generally small scale of community management institutions. An emerging consensus suggests that human beings can overcome PD problems, including the n-person PDs in the form of the "tragedy of the commons"; indeed, this is one of the chief lessons of new common property scholarship. But certain group factors are very helpful in overcoming such problems—especially relatively small numbers in a group, kinship or other intense relationships such as religion, and/or interactions among group members on wide numbers of fronts. Such factors make it possible for group members to monitor one another closely and with relatively low costs, and therewith to form mutually trusting relationships and shared behavioral norms; trust and norms in turn allow people to overcome commons problems, of which environmental problems are one example (Ellickson, 1991; Greif, 1989; Ullmann-Margalit, 1977).

But where environmental problems have large or even global dimensions, as they often do, the small size of many or most community management institutions would appear at first blush to render them irrelevant. The usual range of social interactions around CBMRs seem simply too limited to contain larger environmental damage. Indeed, the coordinated activities involved in such group practices may exacerbate that larger damage. Nineteenth-century whalers, for example, often came from the same towns even though they navigated the globe; at home they enjoyed thick familial and associative relationships. Perhaps not surprisingly, when they were at sea these neighbors generated group customary practices that assisted in the cooperative capture of their large and dangerous prey. But no overarching intergroup social norms ever developed among the far-flung groups of whalers to regulate the total catch of the various types of whales, with the well-known result that a number of the most valuable species were decimated (Ellickson, 1991).

*"Nested" CBMRs*

The small-size pattern does not hold across the board for all community-based management institutions. Ostrom (1990, 1992) gives examples of venerable irrigation networks that have spread over entire watersheds, "nesting" smaller community institutions into larger cooperative entities. For this reason, as Snidal (1995:57) notes, Ostrom regards size as an overrated factor—secondary to institutional structure—in the success or failure of community-based resource governance. Nevertheless as Snidal (1995:59, n.20) also suggests, a number of Ostrom's criteria for successful CBMR institutions do implicitly limit size. Moreover, although communities may enlist (or be enlisted by) wider scale governments for aid in management, in such cases the larger enforcing and/or coordinating entity becomes governmental rather than self-organized (Oye and Maxwell, 1995; Snidal, 1995).

In any event, the primary example of nested CBMRs is irrigation, but irrigation, with its intensive human intervention into natural systems, presents at best an ambiguous case of environmental conservation. Putting that problem to one side, irrigation also may be something of a special case among community management institutions, indeed an exception that suggests why the smaller CBMRs are more generally prevalent. The key probably lies in monitoring. Although the origins of most community-based natural resource management regimes are unknown, if they do emerge from efforts to contain resource competition and dispute, as McCay suggests (this volume:Chapter 11), it is plausible that the activities giving rise to management institutions are generally those in which the members of a community can observe one another's behaviors and their impact on a shared common resource. Resource-related activities involved in irrigating—taking water from ditches, laboring on infrastructure development and upkeep—are especially open to mutual monitoring. Not only can one farmer observe another farmer along the same ditch, but upstream and downstream communities can observe what other communities are doing with respect to water use and infrastructure maintenance (Maass and Anderson, 1978; Ostrom, 1990).

But in the case of many environmental resources—for example, wide-ranging fish or animals, or widely dispersed or invisible pollution—community members are unlikely to observe the impacts of behaviors even within the community, much less the environmental impacts of others on an intercommunity basis. Hence communities may not generate resource-related norms with respect to the entire resource, but at most with respect to some aspect of its use (McCay, this volume:Chapter 11). Perhaps it is for this reason that aside from irrigation, there are few examples of wider scale, nested community management institutions, at least on a self-organizing basis. This is not to say, of course, that larger governmental institutions might not intervene to organize "nested" CBMRs, as Agrawal and Ribot (1999) argue in the case of community forestry in the Kumaon District of India. Insofar as formal governments act as overall managers, CBMRs share an important feature with TEA regimes.

## TEAs and Resource Size

Turning to TEAs and their relation to resource size, TEA regimes are quite different from typical community-based institutions in that they seem particularly attuned to larger resource size. TEA systems are formally structured by governments, and they generally rely on impersonal governmental enforcement rather than social norms. Thus in principle these institutions can encompass environmental problems that coincide with the scope of governments themselves, or even with larger areas subject to intergovernmental agreements.

Quite aside from their governmental origins, there are other important structural reasons why TEAs operate best at larger scales, and rather more poorly at a local level. One of the positive features of TEAs is precisely that they can be traded, so that the allowances tend to flow to those who value them most. But trading works best in large, thick markets. That is why TEAs are feasible for far-ranging gases like sulfur dioxide, where many market participants can participate in trades, but are less easily established for more localized pollutants (Schmalensee et al., 1998).

One intriguing possibility, explored particularly by Rieser in the context of fisheries, again blurs certain aspects of TEAs and CBMRs: She suggests that a TEA regime might allocate at least some quota to communities rather than to individuals (Rieser, 1997). This is an approach that would open the door for community-based institutions under the auspices of TEA regimes, and that could combine the large scale of TEAs with community-based institutions' more nuanced approaches to resource complexity, a subject discussed later in this chapter. If, as Ostrom argues, the key to larger scale community resource management is institutional structure, and if, as Snidal argues and as Berkes (this volume:Chapter 9) describes, community regimes have already relied on larger governments for coordination and enforcement, then TEAs might offer an interesting institutional structure for that coordination, a kind of "nesting" of CBMRs through market-organized institutions.

One final note on resource size: Even when they remain uncoordinated and un-"nested," an enclave community's resource regimes still may be relevant to some important environmental problems, including global ones. Some seemingly large-scale environmental issues are in large part an additive sum of intensely local ones. "Biodiversity loss," for example, is in some measure an umbrella term for a whole series of local losses, from golden-cheeked warblers in Austin, Texas, to radiated tortoises and lemurs in Madagascar (Webster, 1997). In the United States, the most serious threat of species loss is in isolated Hawaii. This pattern is typical; it is precisely the isolated areas that are most likely to have the unique plants and animals that come under siege in a modern economy (Dobson et al., 1997). Moreover, global problems may have at least some localized solutions; greenhouse gases, for example, may be sequestered in local forests, and local forests may be managed through community institutions. Insofar as environmental problems can be subdivided into more local ones, then, even the typically small CBMRs still may be players in the environmental game.

## Resource Complexity

The statement that everything is connected to everything else is a truism in environmentalism. Pull one thread, it is often said, and the entire skein unravels.

If true, this complex interconnectedness would create problems for either TEA regimes or CBMRs, because both contemplate some use of resources. To be environmentally friendly, both TEA regimes and CBMRs must contemplate some constriction on allowable use, that is, constriction to a level that is compatible with renewability of a whole complex network of resources in which the target resource is embedded. Thus the complexity and interactiveness of environmental resources brings us back to an subject mentioned earlier: What is the appropriate level of use, the total "take" or cap on any given environmental resource? I discuss this question briefly before coming back to CBMRs and TEAs.

The fishing industry may have been the first to attempt to answer this question in a disciplined way (Scheiber and Carr, 1997). Toward the end of the 19th century, fishing industry experts hit on the concept of "maximum sustainable yield," an amount that related fishing effort to its effect on the underlying stock; in this analysis, the appropriate limit on total fishing effort was one that could maintain a consistent maximum catch level. Similar ideas soon pervaded forestry practices, as reflected in the U.S. Forest Service's mantra of MUSY (maximum use, sustainable yield). By the 1950s, the great resource economist Gordon refined the model, observing that the appropriate *economic* goal should be not to maximize the yield but rather the "rents," the difference between revenues and the costs of extraction. Gordon's work suggests that instead of the goal of maximum yield, the object of resource management should be "maximum *economic* yield" (MEY), a total take level that is rather more conservationist, and that has become the new conventional wisdom in resource economics (Gordon, 1954; Townsend and Wilson, 1987; Brown, 2000).

But more recent scholarship has cast doubt even on the MEY goal in the environmental context. Once again, fishing gives an example. Although human catch levels clearly influence fish populations, many other things do as well: weather patterns, shifts in water temperature and currents, alterations in food sources and predators, to name just some factors. All these fluctuating elements undermine not only the concept of an ideal climax equilibrium state for any given resource, but also the idea of a smoothly curved relationship between human activity (e.g., fishing or pollution) and resource stock levels (e.g., bountiful fish or clean air). The new "nonequilibrium" thinking suggests that complex and interrelated resources fluctuate in much spikier patterns, and that the best management method may be what is called "adaptive management"—basically, intense use followed by rapid shifts away from the resource at early signs of trouble, allowing the resource to recover (Tarlock, 1994; Townsend and Wilson, 1987).

Now we return to TEAs and CBMRs, beginning with TEAs.

*TEAs and Resource Complexity*

The rhetoric of the total take or cap in TEA regimes often sounds rather close to the resource economists' conventional models. In analyzing "optimal pollution," for example, the goal is often said to be to equate prevention costs and environmental damage at the margin, as illustrated by curves on the conventional charts (e.g., Kaplow and Shavell, 1996). In fact, however, current TEA regimes have set total caps in a manner that departs from economic models and instead generally have taken historic practice as the benchmark. That is, they generally roll back previous use levels by some agreed-on percentage (Heinzerling, 1995; Stavins, 1998; Tipton, 1995).

It is hardly surprising that rollback should be the method of setting the allowable totals for TEA regimes. The introduction of any new environmental regulatory practice generates intense political pressures; this is particularly the case for a regulatory change in which resource users have to pay for something that they previously took "for free" (Libecap, 1989). Rollback is an easy concept to grasp, and it seems to distribute costs with some rough justice. Moreover, rollback can be quite effective to reduce total use; for example, the United States' acid rain control legislation, which instituted TEAs in sulfur dioxide, cut total sulfur dioxide production by quite substantial amounts (Schmalensee et al., 1998). All the same, rollback can hardly be called adaptive management. Although rollback amounts can be rolled back even further in the future, political inertia creates "stickiness" for rapid adaptation once rollback levels are set.

Moreover, another factor also impedes rapid adaption in TEA regimes, bringing us again to a second subject mentioned earlier: the methods of allocating individual entitlements. If TEAs are to bring the usual advantages of property rights—encouraging care and investment by rights holders—then individual allotments must be relatively secure, so that the holders of these rights can rely on them and plan accordingly. Moreover, if TEAs are also to bring the standard benefits of trading and marketability, allowing the entitlements to flow to those who value them most, then these allowances must be relatively simple; simplicity is necessary to allow that these rights to be more or less fungible, and to enable future holders to know what they have. Thus if regulation hedges TEAs with qualifications and conditions, it will undermine both their security and their marketability (Rose, 2000).

This pattern creates something of a flexibility dilemma for TEA regimes. In New Zealand, for example, TEAs for fishing were first set in absolute quantities, but fishery managers quickly realized that if they had to scale back the total allowable catch for the health of the larger fishery, they would face claims for compensation by TEA holders. Noticing this issue, and noticing the reluctance that buy-back programs elicit among politicians, many proponents of TEAs propose that the TEA be set at a given *percentage* of the resource rather than at some fixed *amount*—the solution, incidentally, that New Zealand now has adopted (Clark et al., 1989; Tipton, 1995). But of course this solution entails an unavoidable tradeoff: a percentage-based right, like a short-term right, offers less security and marketability to the holders.

These are not insuperable problems for TEA regimes, and some ingenuity can no doubt help to create a practical balance between flexibility and security, as is the case with other property regimes—even landed property, which is relatively secure but still subject to eminent domain and regulation. But the problems do suggest that TEA regimes may be insufficiently responsive where environmental resources are most densely interactive, complex, and fluctuating; recent commentators, for example, note the difficulties of applying TEAs to the densely interactive resources of wetlands (Salzman and Ruhl, 2000).

Related enforcement problems also derive from the necessarily relatively simple rights structures of TEAs. Because TEAs are designed to be traded, their rights structures must be fairly simple; otherwise they could not be marketed easily. In the air pollution area, TEAs focus on a single pollutant such as sulfur dioxide for the existing regulations for acid rain precursors; perhaps future regulation of greenhouse gases will focus on carbon dioxide. In fisheries too, TEAs also may be defined in some measure of weight for particular species; for example, each TEA corresponds to some number of pounds of quahogs or surf clams. But these relatively simple measures can lead to problems when applied to complex and interactive resources. For example, in fishing, the gross weight of the landed catch may correspond only very inexactly with species conservation. Holders of fishing TEAs know that the larger fish are more profitable than the small ones, and "highgrading" fishermen may actually catch far more fish than their allowances suggest, throwing out the small specimens and keeping the large fish. Just as serious in a complex ecosystem, TEAs in a target species may make fishermen careful about that species, but they may kill with abandon other unmarketable species as "bycatch" (Tipton, 1995; Rieser, 1997; Rose, 2000).

These problems of highgrading and bycatch have been noticed in the literature on fishing TEAs, and although bycatch problems may be less severe under TEA regimes than under some alternative regulations (Tietenberg, this volume:Chapter 6), even strong TEA proponents have suggested that supplemental command-and-control regulation may be required to control these problems (Hsu and Wilen, 1997). As I have noted elsewhere, these problems are examples of a phenomenon that might be classed as "too much property": Creating property rights in one resource may create an imbalance, drawing care and attention to the propertized resources, but crowding out nonpropertized resources (Rose, 1998). Imbalances of this sort are apt to be most serious where resources interact in complex ways. Unless hedged with other regulations or supplementary property regimes, property rights in a single segment of this web could undermine the larger ecosystem. But regulatory hedges complicate the TEA property rights, making them less secure for the holders and less tradable to others.

If TEAs raise questions with respect to their suitability for complex resources, CBMRs, interestingly enough, fare rather better on this dimension.

## CBMRs and Resource Complexity

Even though traditional community resource institutions are far less organized around rational planning, and far more driven by custom and norms, some of their management practices may have certain advantages with respect to adaptability to complex resources. The new, dynamic understanding of environmental resources suggests that intensive use and prompt switching are appropriate adaptive management techniques for complex resource bases (Townsend and Wilson, 1987). Even if not planned to do so, certain traditional resource practices follow this pattern, insofar as hunting, fishing, planting and gathering are undertaken in "pulse" patterns, moving from resource to resource over the course of time (Berkes, 1987; McEvoy, 1986).

This pulse pattern sometimes follows no set of conscious calculations about the whole stock. Indeed, a common traditional belief in hunting and fishing communities is that human activity does *not* affect the stocks of wild animals. Some apparently think it disrespectful to the hunted animals or fish to suggest that they are influenced by human action; instead, the resource stocks are thought to be controlled by the animals themselves, or by God (Berkes, 1987; Brightman, 1987; Carrier, 1987). It would be overly romantic to think that such beliefs constitute a general "respect for nature," or that "respect" for given wildlife resources necessarily entails conservation. Quite the contrary, the idea that the animals control their own numbers may impede any effort to restrain hunting or fishing, and concepts of "respect" may cause opposition to modern resource management techniques such as counting the fish or other wildlife (Berkes, 1987). For this reason, traditional beliefs in some circumstances could contribute to the decimation of particular resources; this may be most likely to happen when traditional practices are confronted with sudden shifts in commercial demand from outsiders, a subject to which I will return shortly. This is not to deny the evidence that some traditional groups have indeed incorporated conservation into their concepts of "respect," perhaps as a result of experiencing and learning from resource depletion shocks (Berkes, 1987; Brightman, 1987). Quite aside from conscious consideration of overall resource stocks, however, traditional hunting, fishing, and gathering practices often rely on diversified resource bases, where pulse patterns of exploitation and relatively low technological methods often leave behind sufficient stocks to regenerate, corresponding in a rough way to more formal adaptive management practices (Berkes, 1987; McEvoy, 1986).

In more settled CBMRs, such as grazing or irrigation regimes, the participants are more likely to be explicit in adjusting their own impacts on underlying resources. This may be because agricultural resources and water levels are more visible than wildlife stocks, and hence the human impact can be monitored and more easily subjected to group discipline. Here, too, some communities' traditional practices respond adaptively to overall resource levels, perhaps more so than in communities dependent on wildlife. For example, Swiss grazing villages limit the right of any resident to "common" more sheep than the resident can feed over the winter; this is a rule that limits individual usage of the common fields and roughly calibrates consumption to the forage available (Netting, 1981). Irrigation communities also carefully adjust individual water appropriation to seasonal water levels (Ostrom, 1990, 1992; Maass and Anderson, 1978).

These adjustments are possible for CBMRs because the individual entitlements in such regimes often are defined in complex ways that incorporate seasonal or resource-related variations—unlike the more fixed TEA entitlements. There is a tradeoff here, as there is in TEAs, but it is made in the opposite direction. Community management practices often show considerable flexibility and responsiveness to dynamic natural change, but at the cost of the security and tradability that promotes investment and innovation; TEAs, on the other hand, promote investment and trade but at the cost of some responsiveness to complex natural change.

## Extraction vs. Pollution

Environmental problems may be grossly divided into two classes: the pollution or "putting-in" issues, and the "taking-out" or extractive issues, such as fishing or hunting or even farming. Curiously enough, in a very rough way, this distinction maps onto TEA regimes and CBMRs. Although there are currently some extractive TEAs in the form of individual fishing quotas, the best known TEAs were created to regulate pollution—that is, the sulfur dioxide TEAs in the U.S. acid rain program (Stavins, 1998). Proposed new applications of TEAs also tend to focus on pollution control, particularly the effort to cut back on global greenhouse gas emissions. In contrast, community-based institutions are generally organized around "taking out" or extractive issues—fishing, hunting, irrigation, agriculture, grazing, and the like.

What are the reasons for this pattern? Any answer is necessarily speculative, but there are some possible reasons for this rough division of labor, some of them harking back to subjects touched on earlier.

First is the factor of regime size, as compared to the size of the common-pool problem it addresses. Pollution problems are typically externalized onto outsiders, in part or in whole. Although community management practices undoubtedly control the ill effects of pollution within the community, participants are unlikely to have much motivation or ability to contain pollution that primarily affects outsiders, except insofar as they are required to do so by interactions with "downstream" communities. Indeed, the very activities that clean up pollution within a community could exacerbate pollution elsewhere, as in pouring wastes into a river or stream. TEAs, on the other hand, are typically organized by larger governmental bodies, and they are aimed precisely at controlling external effects of the use of environmental resources (Esty, 1996). Thus where the environmental issue is pollution, and particularly pollutants that flow far from their source, TEA regimes would seem to be more practicable than CBMRs as property approaches to environmental issues. Some extractive issues may have common-pool effects over large areas (e.g., whale hunting), but many are more localized, as in reef fishing or grazing in particular mountain meadows, and hence they may be managed by the smaller CBMRs.

A second factor is monitoring. Both CBMRs and TEA regimes depend on monitoring; indeed, every property-based regime must have the ability to ascertain whether rights holders stay within their allotments or the entire regime will unravel. But in general, extractive activities are much easier to monitor than are polluting ones. Harvested logs can be observed, the catch from fishing or hunting can be seen, and overuse of grazing fields is noticeable. Pollution, on the other hand, may be entirely invisible. Although some extractive activities may be undertaken surreptitiously (e.g., cheating in taking water from irrigation ditches), CBMRs generally structure rights so that community members can monitor and control one another with respect to this type of overreaching (Ostrom, 1990; Smith, 2000).

Whatever difficulties there may be in monitoring extractive activities, they generally pale by comparison to the problems of monitoring the introduction of pollutants into the air or water or groundwater. Not only does the receiving medium disperse the polluting elements, but insofar as pollutants are invisible and intangible, even polluters themselves may not know what they are doing. Then too, where CBMRs involve relatively small and scientifically unsophisticated communities, as they often do, the participants may lack the technical ability to monitor many forms of pollution. For TEA regimes, monitoring pollutants is also a critical and extremely difficult issue, but larger governments enjoy economies of scale with respect to scientific research (Esty, 1996). Indeed, TEAs have become feasible only as governments have acquired the technical skills to monitor and model pollutants, such as with remote sensing satellites or with sophisticated chemical tags (Rose, 1998; Schmalensee et al., 1998; Tietenberg, this volume:Chapter 6).

A third factor may relate the different feedback effects of "putting in" and "taking out" activities, a point that relates to the new dynamic model of the environment discussed earlier. When certain resources—such as fruits or shellfish—are extracted from a larger ecosystem, deleterious effects may ripple in unexpected feedback loops throughout the entire ecosystem. But here the practical "adaptive management" of CBMRs may be advantageous; their adaptive practices can respond to particular resource shortages by moving on to others before the ill effects of overextraction cause resource crashes, with all the attendant disruptions to the larger ecosystem.

Like extractions, pollutants have ripple effects throughout an ecosystem. For that reason, the *removal* of any given pollutant also can have ripple effects. But in the case of pollution removal, unlike extraction, the ripple effects generally are considered an unalloyed good. For this reason, the simple and single-element focus of TEAs is generally unproblematic with respect to pollution control; the removal of, say, $SO_2$ undoubtedly does have synergistic effects that are not taken into account by TEA holders, but those effects are all positive. But by the same token, the more flexible and multidimensional responses of community management practices may give no particular advantage with respect to pollution removal. Even if the TEA regime reduces pollution in a way that is entirely simple and focused on a single resource, that diminution in pollution is still likely to represent an advance over a more polluted condition. Flexible and multidimensional responses—where CBMRs may have an advantage—are not necessary to create this benefit.

For these various reasons, one might expect CBMRs to be most effective with respect to environmental issues involving "taking out" or resource extraction, whereas TEA regimes are probably at their most effective with "putting in" or pollution problems. No doubt there are exceptions, but in the end, it may not be coincidental that we are more likely to find TEA regimes associated with pollution control, and community-based regimes associated with issues of resource extraction.

## Commerce in Resources

In Western legal regimes, commercially available resources tend to be discussed by reference to a finite and relatively limited number of rights categories. Thus in countries on the European continent, property of rights must be among the "numerus clausus," a defined and closed set of cognizable types of property rights; somewhat similarly, Anglo-American property regimes also provide a number of off-the-rack forms of property, and they sharply discourage efforts to create more complicated forms of property. Recent scholarship suggests that this pattern stems from the fact that in Western legal regimes, a property right—as distinguished from a contractual right—is traded commercially from one person to the next, and then on to the next and the next. Because property is traded to strangers, property rights need to be relatively simple, so that strangers will know what they are getting. (By contrast, contracts can create far more complex forms of rights and duties because these obligations generally affect only the immediate parties, who know the "deal.") Thus for the sake of trades that may take place over many years among complete strangers, Western property rights pare back property rights to a limited number of relatively simple forms (Merrill and Smith, 2000; Rose, 1999).

TEAs, because they are tradable, are subject to these same pressures for simplification. Simplification in TEAs is well known to cause imperfections, however. For example, because of prevailing west-to-east wind patterns, sulfur dioxide TEAs, measured simply in tons, are more damaging if traded toward and exercised in the Midwestern United States than they would be if exercised on the Atlantic coast. This is because a ton of pollutants originating in the Midwest falls in New England, whereas a ton originating on the east coast blows harmlessly out to sea (Revesz, 1996; Salzman and Ruhl, 2000; Stavins, 1998). TEAs could be "vintaged" to take account of locational effects, but if TEAs were hedged with such qualifications, they could split into numerous different markets, creating the usual problems of thin markets (such as holdouts or strategic bargaining) and also creating problems for monitoring ("Did Factory X purchase enough of the right kind of rights for its location?").

Nevertheless, in an imperfect but relatively simple form (e.g., allowances measured simply in tons of emissions), TEAs can be effective devices for dealing with commerce and for incorporating strangers into that commerce. In the larger market made possible by these gross and simple rights definitions, strangers and innovators can purchase and sell TEAs, and officials can monitor and police their use, no matter who the users are. If demand rises and a given environmental resource becomes scarce, the market-based TEA regime responds automatically through a rise in prices. In turn, a price rise may well encourage innovation through conservation or through the introduction of nonpolluting substitutes or more effective and cheaper pollution prevention devices. In these ways, TEA regimes insulate environmental resources from changes in commercial demand.

Once again, it is to be noted that TEAs illustrate the tradeoff between different desirable factors. Like functioning commercial markets in other goods, TEA regimes can accommodate demand shifts through price changes, and they encourage innovation as well as the movement of rights to those who value them most. But the cost of these good things is that TEAs must be relatively simple, and thus they may be adjusted only inexactly to natural environmental conditions.

CBMRs, though sometimes highly adaptive to *natural* change, are much less adaptive to *commercial* changes, and in some ways they may leave environmental resources much more vulnerable to commercial pressure from outsiders. Commerce can open up resources to vastly larger numbers of users outside the community, but unfortunately, community management institutions sometimes seem ill equipped to deal with this phenomenon. A particularly sad example of environmental decimation is now occurring in Madagascar, where endangered radiated tortoises are being hunted out by local gatherers. These animals were once hunted only for occasional feastday meals, but they are now the object of an illegal but seemingly insatiable trade to collectors throughout the world. Local peoples have responded to this leap in commercial demand by hunting as many tortoises as they can today, shrugging off tomorrow's almost certain dearth (Webster, 1997). Because of the suddenness and unexpectedness of this demand—perhaps reminiscent of the European demand for beaver furs several centuries ago in Northern Canada—local peoples seem to have had insufficient time to develop customs or norms that might withstand the onslaught, or that might contain their own contribution to it (Brightman, 1987).

Having said all this, there are certain ways in which the customary practices of community management regimes sometimes do buffer the onslaughts of commerce, precisely because of the impediments that community norms raise to commerce. Whereas TEAs are driven toward relatively simple forms, like most Western property entitlements, entitlements in CBMRs seem to be driven toward complexity. The rights structures in community-based regimes may be fabulously complicated; Papuan fishermen, for example, own overlapping rights to fish in certain places as well as other rights to fish with certain equipment (Carrier, 1987); precontact Maori families owned overlapping rights in objects as small as individual bushes (some had fowling rights, others berrying rights) (Banner, 1999); in medieval Europe as in present-day Swiss villages, villagers owned scattered strips in the fields (Dahlman, 1980; Netting, 1981; Smith, 2000). Even in the more modern irrigation communities of the Philippines, water rights holders also scatter their fields (Ostrom, 1990). Long residence, kinship, extended practice, and the respect of one's fellows are necessary for the full enjoyment of many of these entitlements. Even where an occasional outsider may enter, for example, by buying land or through marriage, he or she is subject to a seasoning process (e.g., Acheson, 1975, 1987; Netting, 1981; Ostrom, 1990).

In short, quite the opposite of TEA regimes, in community-based institutions outsiders find it difficult to enter and insiders cannot sell out easily. What this means, however, is that the participants in the CBMR are stuck with one another because of the very complexity of entitlement structures (Bardhan and Dayton-Johnson, this volume:Chapter 3). Because they are stuck with one another, they are more likely to interact on multiple fronts. In turn, because of those dense interactions, they are more likely to generate the normative structures that help to moderate their own uses of resources (Ostrom, 1990, 2000; Rose, 2000; Ullmann-Margalit, 1977). In that sense, the complexity of CBMRs' entitlement structures is part of a social pattern that may protect environmental resources from depredations not only from insiders, but also from outsiders. Insofar as complex entitlements baffle and thwart outsiders, they also may discourage outsiders from getting their hands on common resources; hence the very anticommercial character of community-based entitlements may protect these resources from commercial shifts.

Historical and contemporary examples suggest, however, that although CBMR practices may impede outside access to the resources that the community considers most central to its well-being, these same practices are not capable of containing unexpected waves of commercial demand for resources not previously considered important or scarce to the community members. The terrible over-hunting of Madagascar tortoises is one example, the decimation of sandalwood in early postcontact Hawaii is another, the historic overtrapping of beaver in the Canadian North is perhaps a third. In all these cases and in others as well, outside commercial demand devastated environmental resources that were nominally in control of a community; indeed, community members were recruited to participate in the decimation (Berkes, this volume:Chapter 9). Perhaps because CBMR regimes are so often governed by norms that emerge over time, a number of these regimes have proved unable to adapt rapidly enough to save some resources from sudden spurts in human demand.

Such failures suggest that between TEAs and CBMRs, TEAs are vastly better prepared to cope with shifts in *human* demand for natural resources. Notice the contrast to resource scarcity coming from *natural* shifts. With respect to the latter, as was discussed earlier, community-based practices may be preferable and may show more of the characteristics of "adaptive management," whether so planned or not. But with respect to commerce, matters are different. TEAs are creatures of a thoroughly commercial understanding of property, and for all its reductionist faults and oversimplifications, this is an understanding that is centrally aimed at accommodating, monitoring, and controlling economic relationships among strangers.

All this suggests that if community-based structures are to be deployed to manage environmental resources that have become commercially valuable in the modern world—such as wildlife in reserve areas—the communities in question may need assistance and possibly restraints from the state in order to shield these communities and their resources from direct contact with that commercial demand.

## Adding It All Up

Putting together all these factors, one is struck by the degree to which TEAs and CBMRs are mirror images, having the opposite strengths and weaknesses. In a table format, and discounting for the extremes of "ideal type" presentation, their respective situational advantages might be laid out as shown in Table 7-1.

Taken together, these contrasting characteristics suggest that, although TEAs have been the flagship for modern property rights schemes in environmental resources, CBMR institutions also have a number of positive features for property-based environmental governance. Indeed, the most positive features of CBMRs

TABLE 7-1 Characteristics and Advantages of Tradable Environmental Allowances (TEA) and Community-Based Management Regimes (CBMR)

|  | TEA Characteristic/Advantage | CBMR Characteristic/Advantage |
|---|---|---|
| Scale | Larger | Smaller (unless "nested" or coordinated) |
| Resource complexity | Simple, single focus | Complex, interactive |
| Practices encouraged | Security of investment, innovation | Adaptation, long-term stability, risk sharing |
| Social structure | Loose, stranger relations | Close knit |
| Adaptation to shifts in natural environment | Less adaptive | More adaptive |
| Adaptation to shifts in human demand | More adaptive | Less adaptive |
| Typical resource application | Pollution (putting in) | Extraction (taking out) |
| Relation to commerce | Accommodates commerce | Vulnerable to commerce |

emerge precisely at the points where TEAs tend to be least effective as environmental protectors, that is, in coping with locally dense, complex natural systems like forests or wetlands (Salzman and Ruhl, 2000).

It is perhaps for such reasons, among others, that modern environmentalists are now experimenting with ways to provide state assistance and control to community-based resource management. One well-known experiment is Zimbabwe's CAMPFIRE (Communal Areas Management Programme for Indigenous Resources), where, under the auspices of state conservation efforts, communities may be treated as wildlife "owners." The expectation, to some degree already fulfilled, is that these communities' members will have an incentive to use their knowledge and skills to save the animals rather than deliver them to poachers, because the community will receive revenues from tourism and sport hunting permits (Anderson and Grewell, 1999). In its general outlines, this program is quite similar to an idea mentioned earlier, allocating fishing TEAs to communities rather than to individuals. It is also similar to the Indian forestry programs mentioned, in which larger government agencies coordinate and "nest" community forestry practices.

A problematic feature of such programs is the degree to which central authorities actually do allow revenues—and hence conservationist incentives—to devolve down to local communities (Agrawal and Ribot, 1999). Such problems illustrate a very important larger point: that success in such mixed regimes depends heavily on the probity and administrative capacities of the larger government. Nevertheless, although rent seeking and frictions undoubtedly occur when state agencies become involved in community-based institutions, wider governmental control over decentralized management has the capacity in principle to take advantage of community institutions' fine-grained resource management practices while helping to overcome their typical weaknesses. Governments can coordinate various communities' efforts and mediate disputes; they can set overall quotas to channel total demand of all the communities; and they can defend community institutions against outsiders. Indeed, even the ancient Spanish CBMRs for irrigation intertwined state officialdom into their community practices, apparently to serve some of these very functions (Glick, 1979; Maass and Anderson, 1978).

Among institutional economists, it is not news that community-based environmental management has some virtues; CBMRs have acquired something of a cheering section among those who study them. There is some reason to be cautious about joining this cheering section unqualifiedly, however. It might be wise to keep in mind a set of critiques that came from past experience, particularly from American legal institutions. In the past, American courts for the most part were implacably hostile to "customary law" and to any efforts to claim a legal place for customary practice. There were some notable exceptions, such as the acceptance of newly-formed customary norms in certain new industries, as among gold miners and whalers. But unlike British courts, American courts refused to accept the claim that longstanding practice, without more, could create legal rights that would govern communities (Rose, 1994).

The reasons were instructive: American courts thought that customary rights were feudal remnants, smacking of the hierarchy of manorial life, at once sclerotic and antidemocratic. They thought that communities should be governed not by the accidents of hoary custom, but rather according to the open constitutional practices of a democratic republic, in which legislation was openly discussed, determined, and changed by elective representatives (Rose, 1994).

An unjaundiced view of modern CBMRs should give rise to some of the same concerns that track through the very inhospitable 19th century American jurisprudence of "customary law." Take, for example, Acheson's attractive, picturesque, and much-cited portrait of the lobster fishermen of Maine's Monhegan Island. The islanders effectively manage the lobster stock as a common property, controlling depredation of nearby lobstering grounds by following customary norms; they allocate fishing rights among themselves and use informal punishment to defend the "perimeter" of their fishing grounds from outsiders (Acheson, 1975, 1987). In a somewhat flintier light, however, these same lobster fishermen appear to be much less attractive: They look xenophobic, hierarchical, thuggish, and thoroughly misogynist. Feminist writers on international human rights echo such concerns in discussing demands for the devolution of governing authority onto fundamentalist religious communities (Shachar, 1998): These communities too seem xenophobic, hierarchical, thuggish, and thoroughly misogynist. This is not to say that all CBMRs should be viewed with suspicion. But some of them should, on democratic grounds, however environmentally friendly they may be.

One of the strongest cases for the recognition and promotion of CBMRs is actually a feature of international human rights: Recognition of community management practices can help to protect traditional peoples who otherwise would be deprived of their longstanding homes and livelihoods altogether. Indeed, much of this deprivation has come through the operation of conventional European-model property regimes, in which traditional community management practices are simply invisible as property (Rose, 1998). As Breckenridge (1992) has pointed out, there are a number of areas in which conservationist concerns overlap with such human rights concerns, and it is in those areas that the recognition of community resource management is most compelling as a basis for the allocation of property rights to the participants.

Even aside from that set of issues, as this article and others have pointed out, there is an environmentalist case to be made for learning from traditional CBMRs. Indeed, there is even a political case to be made for some CBMRs; as Ostrom (1990) stresses, the most long-lived community regimes are likely to have attractive features of member participation, dispute resolution, and intergroup cooperation. But this political case may be strengthened by attempting to devise modern CBMRs in which participation is more egalitarian and potentially more inclusive. Dagan and Heller (2001) argue that we have models of common "liberal property" regimes in cooperatives, condominiums, and even corporations; all these models entail a mix of self-government with the supervision of larger legal institutions. Meanwhile, recent proposals for allocating TEA quotas to communities also incorporate liberalizing reforms for these common property institutions (Rieser, 1997; Rose, 2000).

It may be that the future of CBMRs, with their many environmental strengths, indeed lies in this more liberal direction. What remains to be seen is whether greater liberalization and openness is compatible with the very social practices that give rise to CBMRs' environmental strengths.

## NOTE

1  P. Seabright (1993:114) also has discussed the various designations given to common property institutions and resources.

## REFERENCES

Acheson, M.
  1975  *The Lobster Gangs of Maine.* Hanover, NH.: University Press of New England
  1987  The lobster fiefs revisited: Economic and ecological effects of territoriality in the Maine lobster industry. Pp. 37-68 in *The Question of the Commons: The Culture and Ecology of Communal Resources*, B.J. McCay and J.M. Acheson, eds. Tucson: University of Arizona Press.
Ackerman, B.A., and R.B. Stewart
  1988  Reforming environmental law: The democratic case for market incentives. *Columbia Journal of Environmental Law* 13:171-199.
Agrawal, A., and J. Ribot
  1999  Accountability in decentralization: A framework with South Asian and West African Cases. *The Journal of Developing Areas* 33:473-502.
Anderson, T.L., and J.B. Grewell
  1999  Property rights solutions for the global commons: Bottom up or top down? *Duke Environmental Law and Policy Forum* 10:73-101.
Banner, S.
  1999  Two properties, one land: Law and space in nineteenth-century New Zealand. *Law and Social Inquiry* 24:807-852.
Berkes, F.
  1987  Common-property resource management and Cree Indian fisheries in subarctic Canada. Pp. 66-91 in *The Question of the Commons: The Culture and Ecology of Communal Resources*, B.J. McCay and J.M. Acheson, eds. Tucson: University of Arizona Press.
  1995  Indigenous knowledge and resource management systems: A native Canadian case study from James Bay. Pp. 99-109 in *Property Rights in a Social and Ecological Context: Case Studies and Design Application*, S. Hanna and M. Munasinghe, eds., Washington, DC: World Bank.
Bosselman, F.P.
  1996  Limitations inherent in the title to wetlands at common law. *Stanford Environmental Law Journal* 15:247-337.

Breckenridge, L.
   1992   Protection of biological and cultural diversity: Emerging recognition of local community rights in ecosystems under international environmental law. *Tennessee Law Review* 59:735-785.
Brightman, R.A.
   1987   Conservation and resource depletion: The case of the boreal forest Algonquians. Pp. 121-141 in *The Question of the Commons: The Culture and Ecology of Communal Resources,* B.J. McCay and J.M. Acheson, eds. Tucson: University of Arizona Press.
Brown, G.M.
   2000   Renewable natural resource management and use without markets. *Journal of Economic Literature* 38:875-914.
Carrier, J.G.
   1987   Marine tenure and conservation in Papua New Guinea. Pp. 142-167 in *The Question of the Commons: The Culture and Ecology of Communal Resources,* B.J. McCay and J.M. Acheson, eds. Tucson: University of Arizona Press.
Clark, I.N., P.J. Major, and N. Mollett
   1989   The development and implementation of New Zealand's ITQ Management System. Pp. 117-145 in *Rights Based Fishing,* P.A. Neher, R. Arnason, and N. Mollett, eds. Dordrecht Boston: Kluwer Academic Publishers.
Cox, S.J.B.
   1985   No tragedy of the commons. *Environmental Ethics* 7:49-61.
Dagan, H., and M.A. Heller
   2001   The liberal commons. *Yale Law Journal* 110:549-623.
Dahlman, C.J.
   1980   *The Open Field System and Beyond: A Property Rights Analysis of an Economic Institution.* Cambridge, Eng.: Cambridge University Press.
Dobson, A.P., J.P. Rodriguez, W.M. Roberts, and D.S. Wilcove
   1997   Graphic distribution of endangered species in the United States. *Science* 275:550-553.
Ellickson, R.C.
   1991   *Order Without Law: How Neighbors Settle Disputes.* Cambridge, MA: Harvard University Press.
Esty, D.C.
   1996   Revitalizing environmental federalism. *Michigan Law Review* 95:570-653.
Glick, T.F.
   1979   *Irrigation and Society in Medieval Valencia.* Cambridge, MA: Harvard University Press.
Gordon, H.S.
   1954   The economic theory of a common-property resource: The fishery. *Journal of Political Economy* 62:124-142.
Greif, A.
   1989   Reputation and coalitions in medieval trade: evidence on the Maghribi traders. *Journal of Economic History* 49:857-882.
Hardin, G.
   1968   The tragedy of the commons. *Science* 162:1243-1248.
Heinzerling, L.
   1995   Selling pollution, forcing democracy. *Stanford Environmental Law Journal* 14:300-344.
Higgs, R.
   1996   Legally induced technical regress in the Washington salmon fishery. Pp. 247-277 in *Empirical Studies in Institutional Change,* L.J. Alston, T. Eggerstsson, and D.C. North, eds. Cambridge, Eng.: Cambridge University Press.

Hsu, S.L., and J.E. Wilen
    1997   Ecosystem management and the 1996 Sustainable Fisheries Act. *Ecology Law Quarterly*
           24:799-811.
Kaplow, L., and S. Shavell
    1996   Property rules versus liability rules: An economic analysis. *Harvard Law Review* 109:713-
           790.
Kriz, M.
    1998   After Argentina. *National Journal* 30(49):2848-2853.
Libecap, G.D.
    1989   *Contracting for Property Rights.* Cambridge, Eng.: Cambridge University Press.
Maass, A., and R.L. Anderson
    1978   *...And the Desert Shall Rejoice: Conflict, Growth, and Justice in Arid Environments.* Cam-
           bridge, MA: MIT Press.
McEvoy, A.F.
    1986   *The Fisherman's Problem: Ecology and Law in the California Fisheries 1850-1980.* Cam-
           bridge, Eng.: Cambridge University Press.
Merges, R.P.
    1996   Contracting into liability rules: Intellectual property rights and collective rights organiza-
           tions. *California Law Review* 84:1293-1393.
Merrill, T.W., and H.E. Smith
    2000   Optimal standardization in the law of property: The *numerus clausus* principle. *Yale Law
           Journal* 110:1-70.
Netting, R.M.
    1981   *Balancing on an Alp: Ecological Change and Continuity in a Swiss Mountain Village.*
           Cambridge, Eng.: Cambridge University Press.
Ophuls, W.
    1977   *Ecology and the Politics of Scarcity.* San Francisco: Freeman.
Ostrom, E.
    1990   *Governing the Commons.* Cambridge, Eng: Cambridge University Press
    1992   *Crafting Institutions for Self-Governing Irrigation Systems.* San Francisco: Institute for
           Contemporary Studies Press.
    2000   Collective action and the evolution of social norms. *Journal of Economic Perspectives*
           14:137-158.
Oye, K.A., and J.H. Maxwell
    1995   Self-interest and environmental management. Pp. 191-221 in *Local Commons and Global
           Interdependence: Heterogeneity and Cooperation in Two Domains*, R.O. Keohane, and E.
           Ostrom, eds. London: Sage Publications.
Pinkerton, E.
    1987   Intercepting the state: Dramatic processes in the assertion of local comanagement rights.
           Pp. 344-369 in *The Question of the Commons: The Culture and Ecology of Communal
           Resources*, B.J. McCay and J.M. Acheson, eds., Tucson: University of Arizona Press.
Revesz, R.L.
    1996   Federalism and interstate environmental externalities. *University of Pennsylvania Law
           Review* 144:2341-2416.
Rieser, A.
    1997   Property rights and ecosystem management in U.S. fisheries: Contracting for the com-
           mons? *Ecology Law Quarterly* 24:813-832.
    1999   Prescriptions for the commons: Environmental scholarship and the fishing quotas debate.
           *Harvard Environmental Law Review* 23:393-421.

Rose, C.M.
  1994    *Property and Persuasion: Essays on the History, Theory and Rhetoric of Ownership.* Boulder, CO: Westview Press.
  1998    The several futures of property: Of cyberspace and folk tales, emission trades and ecosystems. *Minnesota Law Review* 83:129-182.
  1999    What government can do for property (and vice versa). Pp. 209-222 in *The Fundamental Interrelationships Between Government and Property*, N. Mercuro and W.J. Samuels., eds. Stamford, CT: JAI Press.
  2000    Expanding the choices for the global commons: Comparing newfangled tradable emission allowance schemes to oldfashioned common property regimes. *Duke Environmental Law and Policy Review* 10:45-72.

Salzman, J., and J.B. Ruhl
  2000    Currencies and the commodification of environmental law. *Stanford Law Review* 53:607-694.

Scheiber, H.N., and C. Carr
  1997    The limited entry concept and the pre-history of the ITQ movement in fisheries management. Pp. 235-260 in *Social Implications of Quota Systems in Fisheries*, G. Palsson and G. Petursdottir, eds. Copenhagen: Nordic Council of Ministers.

Schmalensee, R., P.L. Joskow, A.D. Ellerman, J.P. Montero, and E.M. Bailey
  1998    An interim evaluation of sulfur dioxide emissions trading. *Journal of Economic Perspectives* 12:58-68.

Seabright, P.
  1993    Managing local commons: Theoretical issues in incentive design. *Journal of Economic Perspectives* 7:113-134.

Shachar, A.
  1998    Group identity and women's rights in family law: The perils of multicultural accommodation. *Journal of Political Philosophy* 6:285-306.

Smith, H.E.
  2000    Semicommon property rights and scattering in the open fields. *Journal of Legal Studies* 29:131-169.

Snidal, D.
  1995    The politics of scope: endogenous actors, heterogeneity and institutions. Pp. 47-70 in *Local Commons and Global Interdependence: Heterogeneity and Cooperation in Two Domains*, R.O. Keohane and E. Ostrom, eds. London: Sage Publications.

Stavins, R.N.
  1998    What can we learn from the grand policy experiment? Lessons from $SO_2$ allowance trading. *Journal of Economic Perspectives* 12:69-88.

Tarlock, A.D.
  1994    The nonequilibrium paradigm in ecology and the partial unraveling of environmental law. *Loyola of Los Angeles Law Review* 27:1121-1144.

Tietenberg, T.H.
  1985    *Emissions Trading: An Exercise in Reforming Pollution Policy.* Washington, DC: Resources for the Future.

Tipton, C.A.
  1995    Protecting tomorrow's harvest: Developing a national system of individual transferable quotas to conserve ocean resources. *Virginia Environmental Law Journal* 14:381-421.

Townsend, R., and J.A. Wilson
  1987    An economic view of the tragedy of the commons. Pp. 311-326 in *The Question of the Commons: The Culture and Ecology of Communal Resources*, B.J. McCay and J.M. Acheson, eds. Tucson: University of Arizona Press.

Ullmann-Margalit, E.
   1977   *The Emergence of Norms.* Oxford, Eng.: Oxford University Press.
Webster, D.
   1997   The animal smugglers: The looting and smuggling and fencing and hoarding of impossibly precious, feathered and scaly wild things. *The New York Times Magazine*, 16 February.
Wiener, J.B.
   1999   Global environmental regulation: Instrument choice in legal context. *Yale Law Journal* 108:677-800.

# PART III

# CROSS-SCALE LINKAGES AND DYNAMIC INTERACTIONS

T he dramatic aspects of the commons stand out in the two chapters that make up Part III. Each chapter addresses increasingly crucial issues of cross-scale linkages and dynamic interactions among existing and emergent institutions at the local, national, and international levels. The essence of drama, both as a situation or a series of events involving intense conflict of forces and as stories told through action and dialogue, is embodied in the issues brought to light by Young and Berkes in their respective chapters. Together, these chapters also function as a bridge between earlier chapters in this volume that center on individuals or individual institutions and the following section that focuses on new and emerging issues in research involving common-pool resources and common property management regimes. Young's and Berkes' chapters also foreshadow several theoretical, methodological, and practical challenges presented in the concluding chapter of this volume. In the current era of economic and political globalization, the issues addressed by Young and Berkes are bound to take on increased importance as new institutions emerge to promote and limit diverse globalizing processes, and as these emergent institutions change and interact in new local, national, international, and transnational networks. These two chapters complement each other in important ways. Although Young's paper (Chapter 8) focuses primarily (but not exclusively) on national through global linkages, Berkes' chapter (Chapter 9) is written from the bottom up, that is, it is grounded in local institutions and the ways in which these institutions have been affected by (and affect) higher level institutions at the national and international levels.

Young clearly shows that in order to understand the diverse and multilevel institutions concerned with the management of common-pool resources, these institutions must be situated within their larger biophysical and social contexts and examined in interaction with each other. He maintains that such institutions constantly interact with one another both horizontally (i.e., at the same scale or level of social organization) and vertically (i.e., across scales, from the local through the national and international levels). Young presents a preliminary taxonomy of such interactions in $2 \times 2$ matrix form, which he titles "types of institutional interplay." One dimension of the classification is separated into "functional" and "political" interplay, while the other dimension is divided into "horizontal" and "vertical" interplay. Young's analysis is concerned primarily with vertical, functional interplay among local, national, and international institutions, but as is pointed out in the concluding chapter of this volume, horizontal interactions are becoming increasingly important in the current global context.

Young specifically examines vertical interplay in two distinct areas—terrestrial and marine ecosystems—and the contending positions of national and local institutions and the relationships between international regimes and individual nation states. There is great drama here stemming from the high number of recurring conflicts, clashes, and negotiations between/among actors at various scales or levels of social organization. Compounding the complexity of the interplay between and among these actors is the relative symmetry or asymmetry in the power they hold. Between national and local actors, clashes frequently involve the counterclaims of national governments to what is classified as "public property" and of local actors to what are perceived as common-pool resources. Between international and national organizations, conflicts can emanate from many factors, including a "bad fit" between international agreements and the capacity, compatibility, and competence of national organizations to implement these agreements. Young concludes with the hopeful suggestion that a greater understanding of the dynamics of institutional interplay can be used to design and revamp institutions in order to better ameliorate large-scale, negative environmental impacts.

Berkes' chapter grows out of his decades-long empirical research on local institutions that are concerned with managing common-pool resources. His chapter adds local specificity to the concerns voiced by Young. Berkes delineates a number of cross-scale institutional forms and their effectiveness in being able to link levels of institutions. He begins with a review of the positive and negative effects of higher level institutions on local institutions as a way of demonstrating the importance of examining multiscale institutions and processes. These include impacts due to the centralization of decision making, to shifts in systems of knowledge, to colonization and decolonization, to nationalization of resources, to increased participation in markets, and to the implementation of development policies. Berkes then goes on to delineate a variety of institutional forms that show some promise to facilitate effective cross-scale interactions. Among these are the diverse forms categorized under the umbrella term, "co-management," multistakeholder bodies, development/empowerment/co-management organizations, citizen science, policy communities, and social movement networks. Berkes is careful to point out the importance of expanding analysis beyond the institutional form per se to include an examination of the relevant processes of institutional change. Finally, Berkes proposes to extend the direction of research on the commons through the use of the adaptive management approach (the management equivalent of "learning by doing") and the concept of resilience (i.e., the ability of a system to absorb perturbations). These, he believes, provide a meaningful way to integrate social and natural systems and facilitate movement toward a theory of cross-scale institutional linkages.

# 8

# Institutional Interplay: The Environmental Consequences of Cross-Scale Interactions

*Oran R. Young*

Because individual institutions are highly complex, most analysts focus on specific institutional arrangements, asking questions about the formation, performance, and evolution of these systems on the assumption that a consideration of forces exogenous to individual institutions is not essential for these purposes (Agrawal, this volume:Chapter 2).[1] But as the density of institutions operating in a social space increases, the likelihood of interplay between or among distinct institutions rises. In complex societies, institutional interplay is a common occurrence; the resultant interactions can be expected to loom large as determinants of the performance of individual institutions and of their robustness or durability in the face of various pressures for change. With regard to institutions that address environmental matters—commonly referred to as resource or environmental regimes (Young, 1982a)—this means that interplay is a force to be reckoned with in evaluating whether regimes produce outcomes that are sustainable, much less results that meet various standards of efficiency and equity.

Two sets of analytic distinctions will lend structure to an examination of institutional interplay and help to locate the principal concerns of this chapter within the overall domain of interplay (Young et al., 1999). Institutions interact

This chapter, prepared for the National Academy of Sciences project The Drama of the Commons: Institutions for Managing the Commons, is based in part on a presentation at Session 3.2 on "Institutional Interplay: The Vertical Dimension" at the Open Meeting of the Human Dimensions of Global Environmental Change Research Community, Shonan Village, Japan, June 24-26, 1999. Since then, I have revised the paper several times, expanding its scope substantially, restructuring the logic of the argument it develops, and linking it to the concerns of the international project on the Institutional Dimensions of Global Environmental Change (IDGEC). For more information on IDGEC, visit the project's Web site at http://www.dartmouth.edu/~idgec.

with one another both horizontally or at the same level of social organization (e.g., interactions between trade regimes and environmental regimes operating at the international level) and vertically or across levels of social organization (e.g., interactions between local systems of land tenure and national regulatory systems dealing with matters of land use). The resultant links may generate consequences that are benign, as in cases where regional regimes gain strength from being nested into global regimes, or malign, as in cases where national land use regulations contradict or undermine informal systems of land tenure operating at the local level. Both horizontal and vertical interplay may be more or less symmetrical or reciprocal in nature. Some interactions between distinct institutions are largely unidirectional or asymmetrical. National regulatory regimes that impact local institutions dramatically while being generally insensitive to the impacts of local arrangements exemplify this class of cases. In other cases, interactions are more symmetrical. There are good reasons to believe, for example, that interactions between trade regimes and environmental regimes at the international level, which were once highly asymmetrical, are becoming increasingly symmetrical as environmental regimes gain strength and begin to generate significant consequences for the operation of the global trading system.

Institutions also interact with one another as a result of both functional interdependencies arising from inherent connections and strategic links arising from exercises in political design and management. Functional interdependencies are facts of life. They occur, whether we like it or not, when the substantive problems or activities that two or more institutions address are linked in biogeophysical or socioeconomic terms. The international regimes dealing with the protection of stratospheric ozone and with climate change exhibit inherent links both because chlorofluorocarbons (CFCs), which are the central concern of the ozone regime, are also potent greenhouse gases and because a number of the chemicals that seem attractive as substitutes for CFCs are greenhouse gases as well (Oberthür, 1999). Regimes dealing with the regulation of marine pollution and with the protection of stocks of fish and marine mammals are connected in this inherent sense because the success or failure of efforts to control pollution can be expected to have significant consequences for the well-being of marine ecosystems and the stocks of fish and other organisms they encompass.

Strategic links or interactions involving political design and management, by contrast, arise when actors seek to forge connections between or among institutions intentionally in the interests of pursuing individual or collective goals (Young, 1996). Some exercises in political design are motivated mainly by a desire to enhance institutional effectiveness. Efforts to nest regional regimes (e.g., the various regional seas regimes) into larger or more comprehensive arrangements (e.g., the overall law of the sea), for example, are properly construed as initiatives intended to promote the effectiveness of the smaller scale systems by integrating them into larger systems. Other strategic links reflect conscious efforts to cope with the side effects of arrangements established for other purposes.

Whatever their ultimate results, recent calls for the creation of a World Environ-ment Organization (WEO) owe much to the perception that the operation of the World Trade Organization (WTO) is now producing significant environmental impacts as unintended byproducts of the administration of the global trading sys-tem and that there is a need to create a counterpart to the WTO to level the playing field in interactions among regimes dealing with trade and the environ-ment (Biermann, 2000). In still other cases, strategic links arise as responses to opportunities to improve efficiency by centralizing the supply of services needed to operate two or more distinct institutional arrangements. Funding mechanisms and dispute settlement procedures are obvious cases in point. The Global Envi-ronment Facility (GEF), for example, provides funding both for the climate re-gime and for the regime designed to preserve biological diversity (Sand, 1999). But other services may be subject to such jointness of supply in specific cases.

These distinctions make it possible to locate the central concerns of this chap-ter within the realm of institutional interplay. The emphasis throughout is on vertical interplay or interactions among institutions operating at different levels of social organization. The levels of interest range across the full spectrum from micro-scale or local systems to macro-scale or global systems.[2] For the most part, however, I direct attention to interactions among (1) local institutions and (sub)national institutions and (2) national institutions and international institu-tions. In discussing the consequences of these cross-scale interactions, I start with functional interdependencies. How does the creation of a system of public prop-erty at the national level affect the operation of common-property systems at the local level? How does the character of the national political systems of member states affect the operation of global regimes dealing with issues like climate change or the loss of biological diversity? When functional interdependencies are benign, there is no need to pursue the analysis further. But when these interde-pendencies are malign or favor the interests of some stakeholders over those of others, as they often do, it is natural to move on to a consideration of strategic links. Are there ways to manage cross-scale interactions to minimize conflicts of interest or to maximize efficiency in the pursuit of common goals? In this connec-tion, the chapter seeks to draw lessons from a consideration of functional interde-pendencies that may prove helpful to those concerned with the politics of design and management.

Whereas earlier chapters in this book seek to evaluate existing work, this chapter breaks new ground in the study of institutions governing human/environ-ment relations. Interest in institutional interplay is rising rapidly today. But there is no significant body of literature about such matters to review or sizable collec-tion of data sets to evaluate in addressing this subject. As a result, the account I present in this chapter is necessarily more preliminary and tentative than the analy-ses of earlier chapters. Thus, I proceed by articulating some initial hypotheses about probable consequences of cross-scale interactions and illustrating them with a series of empirical examples. For the most part, these hypotheses rest on utilitarian premises in the sense that they focus on the incentives of key actors as they respond to institutional arrangements or, for that matter, endeavor to manipulate them in ways that further their own interests. The examples show how these processes play out in a range of situations involving human uses of terrestrial and marine resources. But at this stage, they are largely illustrative in nature. My goal is to demonstrate the significance of institutional interplay and to suggest an agenda for future work in this emerging field rather than to arrive at well-tested conclusions about the consequences of institutional interplay in specific settings.

The basic argument of the chapter is easy to state but profound in its implications. The extent to which specific environmental or resource regimes yield outcomes that are sustainable—much less efficient or equitable—is a function not only of the allocation of tasks between or among institutions operating at different levels of social organization but also of cross-scale interactions among distinct institutional arrangements. Understandably, the occurrence of more or less serious conflicts arising from institutional interplay can trigger initiatives on the part of influential actors or interest groups intended to structure the resultant interactions to their own advantage. But such conflicts also can give rise to exercises in institutional design aimed at managing institutional interplay in order to promote the common good or the public interest. In the following sections, I argue that it seldom makes sense to focus exclusively on finding the right level or scale at which to address specific problems arising from human/environment relations. Although small-scale or local arrangements have well-known problems of their own, there are good reasons to be wary of the pitfalls associated with the view that the formation of regimes at higher levels of social organization offers a straightforward means of regulating human activities involving large marine and terrestrial ecosystems. In most cases, the key to success lies in allocating specific tasks to the appropriate level of social organization and then taking steps to ensure that cross-scale interactions produce complementary rather than conflicting actions.

## INTERPLAY BETWEEN (SUB)NATIONAL AND LOCAL RESOURCE REGIMES

Patterns of land use and the sustainability of human/environment relations associated with them are determined, in considerable measure, by the interplay of (sub)national—predominantly modern and formal—structures of public property and local—often informal—systems of land tenure based on common property arrangements.[3] For their part, patterns of sea use and the sustainability of the relevant marine ecosystems are affected greatly by the interplay of (sub)national regulatory systems—legitimized by the creation of exclusive economic zones (EEZs) during the 1970s and 1980s—and subsistence or artisanal practices guiding the actions of local users of marine resources.

For purposes of analysis, this section takes the following preliminary hypotheses as a point of departure. National arrangements afford greater opportunities to take into account the dynamics of large marine and terrestrial ecosystems and to introduce practices involving whole ecosystem management (Sherman, 1992). But regimes organized at the national level also allow for and sometimes promote commodification or, in other words, large-scale, consumptive, market-driven, and frequently unsustainable uses of targeted resources (e.g., timber, fish). These regimes provide arenas in which the interests of powerful, nonresident players often dominate the interests of small-scale local users. Local systems, by contrast, are apt to favor small-scale uses of living resources that evolve over time from the experiences of resident harvesters. Furthermore, local systems are less tied to market systems, and accord higher priority to sustaining local ecosystems over the long term. Because informal/local and modern/national systems commonly coexist—though they seldom enjoy equal standing in relevant political and legal arenas—actual patterns of land use and sea use are affected substantially by cross-scale interactions between these disparate systems operating at different levels of social organization. In the following subsections, I evaluate these hypotheses with reference to terrestrial and marine ecosystems and illustrate the dynamics involved with brief accounts of the use of forest lands in Southeast Asia, grazing lands in the Russian North, and fish stocks in the eastern Bering Sea. But similar forms of interplay involving marine and terrestrial resources occur in many other settings.

## Systems of Land Tenure

The rights of national governments to exercise jurisdiction over all lands and natural resources located within the boundaries of the states in which they operate are widely acknowledged.[4] This is what accords governments the authority to promulgate regulations applying to the activities of both owners of private property and users of common property. But beyond this, governments can and often do assert far-reaching claims to the ownership of land and associated natural resources in the form of public property by virtue of conquest (e.g., Russian ownership of Siberia), the exercise of royal prerogative (e.g., the establishment of crown lands in Sweden), purchase (e.g., the acquisition of Alaska by the United States), inheritance (e.g., Canada's inheritance of crown lands under the British North America Act of 1867), succession (e.g., Indonesia's claims to lands once owned by the Netherlands in the East Indies as an element in the process of decolonization), or some combination of these claims. In most countries, claims to public property are remarkably extensive. Despite the publicity surrounding privatization, the government of the Russian Federation claims most of the land base of Russia as public property. The government of Canada treats the bulk of the country's land base as public property.[5] Even in the United States, widely regarded as a bastion of private property and free enterprise, the federal government alone claims about one-third of the nation's land as public property (Brubaker, 1984).

Yet this is not the whole story regarding systems of land tenure. Although effective control has flowed steadily toward national governments during most of the modern era, many small indigenous or traditional groups residing within states and engaging in distinctive social practices have not relinquished their claims to ownership of large tracts of land and natural resources in the form of common property (Berkes, 1989; Bromley, 1992). Often, these claims overlap or conflict with assertions on the part of national governments to the effect that the areas in question are part of the public domain. Indigenous land claims in British Columbia, for example, cover virtually all the land area of the province. In some cases, national governments have recognized these claims and taken steps to reach settlements with indigenous and traditional claimants. Particularly noteworthy in this connection are the comprehensive claims settlements that the government of Canada has negotiated with northern indigenous peoples over the past several decades and the cooperative arrangements under which the government of Denmark and the Greenland Home Rule handle matters of land use in Greenland. In other cases, the efforts of local communities to assert ownership—or even use— rights have met strong resistance on the part of national governments. The efforts of Sweden's Sami to gain recognition of their rights to use grazing lands constitute a striking case in point (Svensson, 1997). In still other cases, national governments have made little effort so far to take seriously the claims of local communities to rights involving common property. Throughout much of the Russian Federation, where the legacy of collectivization introduced during the period of Soviet rule remains strong, serious land claims on the part of local peoples are just beginning to surface (Fondahl, 1998).

How can these clashes between the claims of national governments to public property and the claims of local communities to common property be resolved? In some cases, such as the settlement of Native land claims in Alaska, the eventual outcome has taken the form of a formal transfer of title to some lands to Native peoples (or organizations acting on their behalf), usually in return for acceptance on the part of these peoples of the extinguishment of residual claims to other areas.[6] As experiences in places like Australia, Canada, Greenland, and Fenno-Scandia make clear, however, the concept of property encompasses a bundle of rights, and the contents of this bundle can be allocated in any of a variety of ways.[7] This has given rise to lively debates about the nature and extent of usufructuary rights in situations where user groups have not been granted full title to land and natural resources. Among the most significant aspects of this debate are issues concerning the rights of national governments to authorize consumptive uses of forests, hydrocarbons, and nonfuel minerals in areas that are important to the conduct of longstanding subsistence or artisanal activities featuring the use of living resources on the part of local peoples.

What difference does the resultant interplay between (sub)national systems of public property and local systems of common property make with regard to overall patterns of land use and to the sustainability of human/environment relations in various areas? The answer to this question emerges from a consideration of differences in the incentives of national policymakers and local stakeholders. For the most part, governments can be expected to look upon public property as a means to promote the national interest through activities inspired by the search for export-led economic growth and the effort to attract foreign direct investment. More often than not, this means treating forests and nonrenewable resources as commodities to be harvested or extracted to meet the demands of world markets. Two other factors reinforce this approach to the use of public property, especially in the developing world and in countries in transition. National governments tend to cater to the interests of politically powerful individuals who have no roots in local areas and who look on concessions covering natural resources located on public property primarily as a means of amassing personal wealth. A particularly virulent form of this phenomenon involves the practice of crony capitalism and the emergence of black markets that many observers of Southeast Asia have described in detail (Dauvergne, 1997a). Environmental groups and some other nongovernmental organizations (NGOs) often endeavor to counter or at least mitigate these forces. But intergovernmental organizations (IGOs), such as the multilateral development banks, whose mandates emphasize the acceleration of economic growth in developing countries, frequently act to reinforce the resultant bias against the preferences of local peoples with regard to patterns of land use (Lipschutz and Conca, 1993). The actions of the World Bank in supporting large-scale irrigation systems, road construction, and nonrenewable resource extraction throughout the developing world offer striking illustrations of this pattern.

It would be a mistake to idealize local peoples as stewards whose social practices do not cause major changes in ecosystems. There is ample evidence to demonstrate that swidden agriculture, the deliberate burning of forest understory, and the harvesting of wildlife all can produce major ecological consequences (Krech, 1999). Unsustainable practices involving the use of natural resources appear to have contributed to the collapse of some small-scale systems in areas as diverse as the Middle East and Central America. But so long as their informal socioeconomic systems remain intact, local peoples do not have strong incentives to harvest timber for export, to extract hydrocarbons or nonfuel minerals to sell on world markets, or to build massive dams to support large-scale irrigation systems and industrial agriculture.[8] Where systems of common property controlled by local users prevail, therefore, we can anticipate that patterns of land use will differ markedly from the patterns likely to arise where systems of public property prevail. In essence, we should expect to find a pronounced tendency toward large-scale exports of products like timber, palm oil, hydrocarbons, and nonfuel minerals in systems where public property arrangements govern the use of land and natural resources. In comparison, local users operating under common property systems are more likely to use land to support subsistence or artisanal lifestyles and to avoid the extractive and developmental patterns characteristic of public property systems. Needless to say, these dynamics will be more complex in those increasingly common situations where the balance between national claims to public property and local claims to common property is contested or in which efforts to resolve such contests have resulted in complex and sometimes confusing allocations of the full bundle of property rights among several distinct groups of claimants. Yet the general trend seems clear.

To see how this reasoning plays out in practice, consider recent developments affecting the forests of Southeast Asia and the grazing lands of northern Russia. As a number of observers have pointed out, the tropical forests of Indonesia, Malaysia, and the Philippines have been harvested in an unsustainable manner over the past several decades (Peluso, 1992; Dauvergne, 1997a). Dauvergne (1997a:2), for example, has shown that "loggers have degraded much of Southeast Asia's old-growth forests, triggering widespread deforestation" and that these activities "irreparably decrease the economic, biological, and environmental value of old-growth forests." Why has this happened? Many commentators have emphasized demand-side considerations, pointing to the role of Japan as a consumer of tropical timber and arguing that Japanese companies often operate close to the margin and have few incentives to promote sustainable uses of Southeast Asian forests. At least as important, however, are supply-side considerations and, more specifically, the rules of the game governing decisions about alternative uses of Southeast Asian forests. A critical link in this story lies in the creation of systems of public property controlled by national governments as part of the process of decolonization and the establishment of independent states in Indonesia, Malaysia, and the Philippines in the aftermath of World War II. In effect, the emergence of public property in these countries constitutes a necessary condition for the pattern of forest degradation that has spread throughout this region. There is nothing in such arrangements that compels national governments to negotiate forest concessions in the quest for export-led growth and to acquiesce in the practices referred to as crony capitalism. But the shifting balance between systems of public property and systems of common property has played a key role in allowing these developments to happen. Local users pursuing long-established lifestyles have no incentives to adopt strategies leading to forest degradation and, in the process, undermining the resource base needed to sustain these lifestyles. Among other things, this explains the views of many activists who see links between campaigns to reform land use practices that cause forest degradation and the struggle to strengthen the rights of indigenous and traditional peoples in countries like Indonesia and Malaysia.[9]

Another illustration involves patterns of land use in northwestern Siberia, where world-class reserves of oil and especially natural gas have been discovered in areas that indigenous peoples, such as the Nenets living on the Yamal Peninsula and the coastal plain of the Pechora River Basin, have long used as communal migration routes and pastures for reindeer (Osherenko, 1995). During the Soviet era, there was little doubt about the choice between hydrocarbon development and the protection of local lifestyles in this region. The national government claimed ownership of the area's land and natural resources as public or state property; oil and gas development was granted priority not only as a means to promote economic development but also as a source of hard currency earnings, and the concerns of the region's indigenous peoples generally were ignored or treated as secondary matters. At the time of its demise, the Soviet Union was the world's largest producer and exporter of natural gas. Yet as Osherenko has shown, recent years have witnessed new developments in patterns of land use in this region (Osherenko, 1995). This is partly a consequence of the collapse of the Soviet Union and the resultant economic decline occurring throughout the Russian Federation. In part, however, it reflects a growing effort on the part of indigenous peoples to reclaim reindeer from the collective and state farms of the Soviet era and to reassert common property rights to the migration routes and grazing lands needed to sustain local economies. From the perspective of these peoples, this pattern of land use is superior to nonrenewable resource development, regardless of world market prices for oil and natural gas.

It is far too soon to make predictions about what the future will bring in this region. The development of gas fields on the Yamal Peninsula, for example, is currently in a state of suspended animation. Rising world market prices, along with a revival of the overall Russian economy, could generate pressure to resume the development of gas fields and transportation corridors in this sensitive area. But it is clear that the shifting balance between national claims to public property and local claims to common property will play a role of considerable importance in determining future patterns of land use in northwestern Siberia.

## Systems of Sea Tenure

The story of sea tenure differs—often quite dramatically—from the account of land tenure set forth in the preceding subsection. Whereas we have no difficulty organizing our thinking around concepts like patterns of land use and systems of land tenure, comparable phrases relating to marine resources—"sea use" and "sea tenure"—have an odd ring to them. Why is this the case? Broadly speaking, it is fair to say that this divergence stems from the fact that there is little history of private property and only limited experience with public property in the ordinary or normal sense of the term when it comes to the management of human uses of marine resources.

Part of the gap between arrangements dealing with land use and their counterparts governing sea use is attributable to the fact that it is often difficult to establish effective exclusion mechanisms applicable to marine resources (Dietz et al., this volume:Chapter 1). This is because marine resources run together in a fluid manner and, in the case of living resources such as fish, often include organisms that move freely from place to place in ways that would frustrate any efforts to establish possessory rights that run with individual or even community owners. Seeking to create private property rights in many fish stocks would be like endeavoring to turn migratory birds into private property in systems of land tenure. Even so, it would be a mistake to exaggerate this argument regarding property rights in marine resources. In cases where the relevant resources are sedentary (e.g., clam and oyster beds), there is a good deal of experience with the creation of property rights, especially in the form of use rights that allow their holders to exclude others from harvesting living resources such as clams, oysters, and even lobsters in designated locations (Acheson, 1987). Even more highly developed are the rights accorded to those who engage in various forms of aquaculture that depend on the existence of secure rights to fish pens and other place-specific marine structures.

As these last observations suggest, moreover, it is important to consider arrangements under which individual elements in the bundle of rights associated with property are relevant, even when there is little prospect of establishing systems based on the full bundles of rights we ordinarily have in mind in thinking about private property and public property. In many situations, for example, use rights to particular fish stocks have been established in forms such as preferences granted to harvesters using particular locations and specific gear types or rights to harvest a specified proportion of the total allowable catch (TAC) for a specific fishery in any given year. The recent emergence of systems of individual transferable quotas (ITQs) in a variety of fisheries is particularly noteworthy in this connection (Iudicello et al., 1999; Tietenberg, this volume:Chapter 6).

In part, the scarcity of systems of private property and public property associated with marine resources arises from limitations on the authority of states to exercise control over marine systems. From the beginnings of the modern states system in the 17th century, states have been treated as territorial units possessing virtually unlimited jurisdiction over terrestrial ecosystems located within their borders but comparatively little jurisdiction over adjacent marine systems (Anand, 1983). Early on, states began to assert some jurisdiction over waters located adjacent to their coasts in the form of a belt known as the territorial sea. For the most part, however, the granting of jurisdiction over the territorial sea was justified largely as an arrangement required for purposes of defense. Under this arrangement, coastal states agreed to allow outsiders to engage in a variety of activities— innocent passage of ships, the laying of submarine cables, overflight by aircraft— taking place within or affecting their territorial seas. Beyond this belt, states considered it impermissible to lay claim to marine systems as public property in the sense of areas actually owned by the state in the same way that the state owns the public domain.

Given this background, it makes sense to look at the 20th century as an era marked by striking expansions of the jurisdiction of coastal states over marine systems in both spatial and functional terms (Juda, 1996). The traditional 3-mile territorial sea has grown to 12 miles, and the establishment of EEZs has granted coastal states jurisdiction over approximately 11 percent of the world ocean and most marine living resources. Justified largely on the basis of arguments regarding conservation or the achievement of sustainable use, the expanded jurisdiction of coastal states over marine systems now extends to the management of a range of activities dealing with the harvesting of both renewable and nonrenewable resources and with the protection of marine systems from various forms of pollution. Even so, it is important to note that the authority of coastal states is still restricted in ways that make it difficult for states to acquire bundles of rights comparable to those applying to terrestrial systems included in the public domain. Coastal states do not have the authority to transfer title to marine systems to private owners in the way that states traditionally have been able to dispose of sizable portions of the public domain. Many governments consider it inappropriate even to collect economic returns from the use of marine resources treated as factors of production, a practice that is routine in situations involving the use of natural resources (e.g., timber, hydrocarbons) located on the public domain. These restrictions have not deterred states from developing regulatory regimes operated by government agencies (or their subunits) and designed to ensure that users of marine resources pay attention to matters of sustainability and environmental quality associated with their activities. Nonetheless, they have produced a situation in which it seems awkward to think in terms of systems of sea tenure.

At the same time, there are substantial parallels between systems of land use and systems of sea use when it comes to the operation of small-scale local arrangements, quite apart from the aggregation of management authority in the hands of the state. In virtually every case, these local arrangements can be thought of as featuring some form of common property (Pinkerton, 1989). Not surprisingly, numerous variations occur, depending on the character of the biogeophysical systems involved, the nature of the harvesting procedures employed, and the content of the cultural norms operative among the members of the group of appropriators. Nonetheless, nearly all these systems have a number of features in common. Although they do not assign full bundles of rights to individual users, they often do grant individuals priority in the use of particular fishing sites or the use of specific gear types. They typically exclude outsiders or, in other words, nonmembers of the community of owners from using the resources in question. They normally feature informal arrangements that evolve on the basis of trial and error and that undergo de facto adjustments over time as a way of adapting to changing conditions in the relevant biogeophysical systems or changing circumstances of the societies within which they operate. Yet the rules in use that comprise these institutional arrangements are ordinarily well understood by members of the relevant user communities, and they are buttressed in most cases by compliance mechanisms that are effective in bringing the behavior of individual appropriators into conformance with the constellations of rights and rules that make up the core of these practices.[10]

How have these small-scale arrangements governing the actions of local users of marine resources performed in practice? As in the case of systems of land tenure, it would be a mistake to idealize indigenous or artisanal systems of sea use. To be sure, anthropologists have succeeded in documenting a sizable number of cases in which these local systems have been sustainable over relatively long periods of time. A particularly intriguing feature of these studies is the exploration of compliance mechanisms (e.g., arrangements featuring taboos) that prove effective from the point of view of guiding the behavior of users toward sustainable practices, even when they are not based on any scientific understanding of the dynamics of the ecosystems in question (Fienup-Riordan, 1990). Nonetheless, there is no basis for assuming that all small-scale systems of sea tenure produce results that are sustainable. Although this is a sensitive and—in some circles—contested matter, there is little doubt that the actual record associated with subsistence and artisanal systems of sea use features a fair number of failures as well as successes, especially in cases involving volatile biogeophysical systems that undergo large-scale nonlinear changes from time to time (Wilson, this volume:Chapter 10).

By the same token, the record compiled by the regulatory regimes created by (sub)national governments to guide uses of marine resources is generally unimpressive. Justified in large part by the need to manage large marine ecosystems on an integrated basis and to bring to bear the insights of science in order to ensure sustainability in the use of marine resources, these regimes have been insufficient to prevent a growing crisis in many of the world's fisheries brought on by an excess of harvesting capacity and an inability—both scientifically and politically—to establish and enforce appropriate quotas or other restrictions governing the consumptive use of living marine resources (McGoodwin, 1991). In fact, national governments have provided regular subsidies to harvesters in a manner that has led to the acquisition of larger and more powerful harvesting capabilities along with heavy debt loads. As this last observation suggests, moreover, the regulatory regimes established by national governments have exhibited a marked tendency to favor the interests of some types of users over others. Thus, large, well-financed, and politically active harvesters generally have profited from the introduction of national systems of sea use in contrast to small-scale subsistence or artisanal harvesters, who have little experience beyond the local level and few of the resources needed to influence national (or even subnational) policies relating to the use of marine resources.

Overall, it is probably fair to say that the result has been a trend toward the commodification of marine resources favoring large commercial operators over small operators, eroding the role of local common property approaches to sea tenure and leading to outcomes that are hard to defend in terms of conservation or even efficiency. Environmental NGOs have become increasingly active in efforts to counter this trend. Recently, moreover, national regulators have begun to experiment with a range of policy instruments (e.g., permits to fish, individual trans-

ferable quotas or ITQs) intended to eliminate or suppress some of the worst features of this commodification (Iudicello et al., 1999). The track record associated
with these efforts is not yet extensive enough to justify firm conclusions. Taken
together, however, it seems fair to conclude that these institutional innovations
show considerable promise at least as responses to the specific problem of overharvesting (National Research Council, 1999a). Yet there is no basis at this stage
for granting high marks to state-based systems of sea tenure with regard to the
production of outcomes that are sustainable over time, much less results that can
be defended on grounds of efficiency or equity.

To see how the interplay between modern/national and more informal/local
systems of sea tenure plays out in practice, consider the situation that has developed in the eastern Bering Sea Region over the past 25 years (National Research
Council, 1996). During the 1970s, the state of Alaska instituted a limited-entry
regime for the inshore fisheries of this area—those fisheries taking place within a
3-mile belt over which the state has jurisdiction—largely in response to declining
harvests of salmon (Young, 1983). Shortly thereafter, the federal government
followed suit by creating a Fishery Conservation Zone (FCZ) together with a set
of regulatory arrangements dealing with the harvesting of all species of fish in an
area extending from the outer boundaries of state jurisdiction to a point 200 nautical miles from the coastline (Young, 1982b). Although it would be incorrect to
argue that these initiatives have had no positive consequences, they have given
rise to a number of unintended side effects largely due to problems of interplay
with other institutional arrangements. The limited-entry system covering inshore
fisheries has disrupted informal arrangements featuring a fluid mix of subsistence
and commercial fishing; placed severe restrictions on the ability of young people
unable to afford the price of a permit to enter the fisheries; and led to a loss of
permits among rural fishers whose financial insecurity causes them to succumb
from time to time to the temptation to sell fishing permits to meet short-term
needs for cash. For its part, the creation of the FCZ in the eastern Bering Sea
precipitated a dramatic rise in the participation of American fishers in this area
and the consequent phasing out of foreign fishers. Because the regime established
to regulate fishing in this area has the status of a national arrangement, the state of
Alaska has been barred from instituting measures to protect local fishers in the
area from competition on the part of large, heavily capitalized fishers based in
Washington and Oregon. The exclusion of foreign fishers from the FCZ caused
them to shift their focus to an area of the central Bering Sea located just outside
the FCZ and known as the doughnut hole.[11] By the early 1990s, the pollock stocks
in this area had collapsed.

During the 1990s, both the U.S. government and the state of Alaska took
some steps to address these unfortunate side effects arising from the institutional
innovations of the 1970s and 1980s. These include the creation of community
development quotas (CDQs) intended to bolster the economies of small, coastal
communities (National Research Council, 1999b) and the negotiation of a six-

nation convention designed to address the problem of overharvesting of pollock in the central Bering Sea (Balton, 2001). Although these are clearly steps in the right direction, it is premature at this stage to conclude that they will solve the problems arising from institutional interplay in the Bering Sea Region. CDQs do not provide a substitute in sociocultural terms for the existence of a strong cadre of local fishers, and the pollock stocks of the doughnut hole have yet to recover sufficiently to activate the management procedures established under the six-nation convention. Accordingly, there is a real danger that the innovations of the 1990s will be assessed in the future as responses that were too little and too late. In any event, it is clear that the growth of coastal state jurisdiction over marine resources and the subsequent emergence of (sub)national systems of sea use have triggered new forms of institutional interplay in this realm whose consequences have been costly not only for many individuals, but also for the welfare of small coastal communities in an area like Alaska.

## INTERPLAY BETWEEN INTERNATIONAL AND NATIONAL ENVIRONMENTAL REGIMES

Turn now to institutional interplay occurring at higher levels of social organization and, more specifically, to the hypothesis that the effectiveness of international environmental regimes—measured in terms of efficiency and equity as well as sustainability—is determined, in considerable measure, by the interplay between rules and decision-making procedures articulated at the international level and the political, economic, and social systems prevailing within individual member states. International regimes normally set forth generic rules applicable to all their members, leaving the implementation of these rules to be handled mainly by public agencies and actors located within individual member states.[12] It follows that the effectiveness of these regimes depends on the performance of national institutions and is likely to vary substantially from one member state to another. Arrangements that perform well when there is a good fit between the provisions of international regimes and the political and economic systems of member states can fail miserably when the interplay between these arrangements is problematic. Following an account of the logic of this hypothesis, this section turns to brief illustrations of this type of interplay in the cases of regimes dealing with tropical timber in Southeast Asia and protected natural areas in the Circumpolar North and of regimes addressing the fisheries of the Barents and Bering Seas. As in the case of interplay between local and (sub)national institutions, similar dynamics occur in many other settings.

### Competence, Compatibility, and Capacity

It is tempting to assume that once countries sign and ratify agreements establishing international regimes, they will proceed to carry out the obligations they assume under these agreements as a matter of course. As numerous studies of national implementation of international obligations have shown, however, there is no basis for making such an assumption (Skjaerseth, 2000; Underdal and Hanf, 2000). Implementation typically varies greatly from one regime to another as well as among individual members of the same regime. Not surprisingly, then, the study of factors influencing implementation at the national level has become an important area of emphasis for regime analysis (Underdal, 1998; Victor et al., 1998; Weiss and Jacobson, 1998). What are the key determinants of whether members succeed in implementing the rules of international agreements within their own jurisdictions and whether they accept the outcomes flowing from decision-making procedures operating under the auspices of international regimes? In some cases, this is essentially a matter of political will. Governments can and do sign agreements they have no intention of implementing. Executive branch officials who sign international agreements in good faith may be unable to persuade legislators to pass implementing legislation and allocate the resources needed to operate these arrangements. Furthermore, changes in the composition of governments can bring to power officials who did not participate in the creation of a regime and have little interest in fulfilling obligations undertaken by their predecessors. At the same time, three sets of factors of a more general nature that bear directly on the matter of institutional interplay have emerged as important considerations in this context. For shorthand purposes, we can label these factors competence, compatibility, and capacity.

Competence is a matter of the political and legal authority needed to implement commitments made at the international level. Competence in this sense is largely a function of the constitutional arrangements prevailing within individual countries. In the United States, for example, international conventions do not become legally binding until they are ratified by a two-thirds majority in the Senate. Even then, the U.S. Constitution does not guarantee that commitments embedded in legally binding conventions will always take precedence over domestic laws (Higgins, 1994). As a result, American negotiators in international forums frequently oppose otherwise attractive institutional arrangements on the grounds that there is little prospect that they can survive the pressures arising from domestic legal and political processes. Small wonder, then, that many other countries find the United States a difficult partner when it comes to the creation and implementation of international regimes. In other cases, the problem arises from the allocation of authority between national and subnational units of government in contrast to the separation of powers among the component parts of national governments. In the Canadian confederation, where authority over many issues resides with the provinces in contrast to the federal government, for example, the government in Ottawa lacks the competence to enter into legally binding commitments at the international level regarding numerous issues, without seeking the explicit consent of the individual provinces.[13]

Compatibility is a matter of the fit or congruence between institutional arrangements set up under the provisions of international agreements and the social practices prevailing within individual member states. Whereas competence is a matter of authority, compatibility concerns standard practices or procedures for handling matters of governance that grow up in political systems over time. Given the decentralized character of international society, there is general agreement on the proposition that member states should be free to implement international commitments within their own jurisdictions in whatever way they choose to do so. But this does not eliminate the problem of institutional compatibility. Consider, by way of illustration, a case in which an international regime calls for the establishment of a system of tradable permits (e.g., permits for exclusive use of bands in the electromagnetic spectrum, permits for extracting minerals from specific sites on the deep seabed, permits for emitting specific quantities of greenhouse gases), while the social practices prevailing within some of the member states are based on the use of command-and-control regulations offering little or no scope for the sorts of incentive mechanisms associated with the creation of tradable permits. To make this concern more concrete, think of the issues now coming into focus with regard to the allocation of slots in the geostationary orbit or bands in the electromagnetic spectrum. For those committed to the proposition that tradable permits are essential to ensure efficiency and, therefore, to secure widespread acceptance of arrangements governing the use of these resources, the advantage of allowing and even promoting the emergence of markets in slots and bands seems beyond doubt. Yet such mechanisms are alien to the political cultures of many countries, and government agencies in these countries are lacking in experience with mechanisms of this sort that would allow them to assimilate such a governance system into familiar and well-understood ways of doing business (Chertow and Esty, 1997; Rose, this volume:Chapter 7; Tietenberg, this volume:Chapter 6).

For its part, capacity is a measure of the availability of the social capital as well as the material resources needed to make good on commitments entered into at the international level (Chayes and Chayes, 1995; Keohane and Levy, 1996). Of course, we are used to paying attention to the problem of capacity in cases where the economic and political systems of developing countries lack the resources needed to shift to alternative technologies (e.g., substitutes for ozone-depleting substances) or to enforce international rules within their jurisdictions (e.g., rules pertaining to trade in endangered species) (Gibson, 1999). But issues of capacity also arise in connection with the actions of advanced industrial countries. In the United States, for instance, international commitments may be treated with benign neglect in cases when no individual agency is willing to take responsibility for their implementation (that is, to become what is known as the lead agency) or when responsible agencies are unable or unwilling to obtain the material resources required to play this role. Consider, in this connection, the contrast between American participation in the regime for Antarctica, where there is no doubt about the role that the National Science Foundation plays as lead agency with regard to matters pertaining to this arrangement, and in the emerging regime for the Arctic, where a dozen or more agencies want a say in what happens but none is able or willing to accept the role of lead agency (Osherenko and Young, 1989).

As this discussion makes clear, international regimes normally operate in social settings featuring substantial institutional heterogeneity among their members. What is more, those responsible for administering international regimes are seldom in a position to resort to what constitutes the normal procedure for handling interplay of this sort between national and subnational governments, a setting in which national governments ordinarily possess the ultimate authority to compel subnational governments to adjust their rules and procedures to ensure that they do not conflict with arrangements designed and implemented at the national level.[14] The result is a mode of operation in which the rules of international regimes are framed in terms that are sufficiently generic to allow officials in individual member states considerable leeway in operationalizing them within their own jurisdictions. Up to a point, this is clearly desirable. National officials are not about to let the managers of international regimes dictate to them, and there is much to be said for allowing individual members to assimilate the rules of international regimes into their own systems in ways they deem appropriate.

The rise of what some observers now call transnational or even global civil society has begun to exert pressure on states to accept common standards in implementing the provisions of environmental regimes within their jurisdictions (Florini, 2000; Princen and Finger, 1994; Wapner, 1997). Nevertheless, the forces described in this subsection accentuate the hypothesis under consideration here to the effect that the consequences of international regimes will be determined in considerable part by the interplay between the regimes themselves and national practices prevailing in individual member states. Among other things, this should lead us to expect considerable variance in the performance of member states when it comes to fulfilling commitments made during processes of regime formation. This variance need not be critical to the overall performance of international regimes. In the case of equipment standards applicable to the construction of oil tankers, for example, the regime can operate effectively so long as a few key member states take the standards seriously (Mitchell, 1994). But in other cases, such as phasing out the production and consumption of ozone-depleting chemicals (French, 1997), it is apparent that it takes conformance on the part of all (or nearly all) to provide effective protection of the relevant natural systems.

## Regimes for Terrestrial Resources

To think concretely about the impact of this form of interplay on patterns of land use, consider the operation of the International Tropical Timber Agreement (ITTA) and the effort to create a Circumpolar Protected Areas Network in the Far North. ITTA, created initially in 1983 and substantially restructured in 1994, is first and foremost a trade agreement in which producers (e.g., Indonesia, Malaysia, and the Philippines) and consumers (e.g., Japan) of tropical timber endeavor to stabilize and regulate the world market in wood products harvested from tropical forests (Humphreys, 1996; Dauvergne, 1997b). What makes this regime interesting from an environmental point of view is the recognition that most harvesting of tropical timber in recent decades has taken the form of highly destructive practices best described as the "mining" of forests and that there is a need to restructure the industry to make it more sustainable.

The centerpiece of the 1994 agreement is a commitment on the part of member states to implement a system of guidelines intended to ensure that both natural and planted tropical forests are managed sustainably and that biological diversity is protected in these forests. To this end, regime members committed themselves to the Year 2000 Objective calling for all tropical timber entering international trade to be produced from tropical forests under sustainable management by the year 2000. Only a few countries succeeded in fulfilling this commitment. Are others likely to be able to meet the standard of the Year 2000 Objective during the near future? Part of the answer depends on the actions of NGOs concerned with this regime (e.g., the Forest Stewardship Council). But the essential key to this issue lies in the interplay between the international regime itself and the national political systems of member countries, such as Indonesia and Japan (Guppy, 1996). At this stage, the prognosis is not particularly encouraging. Given the economic and political turmoil occurring in Southeast Asia in recent years combined with the continuing grip of crony capitalism, the capacity of a country like Indonesia to meet the Year 2000 Objective is limited, and the sanctions associated with nonconformance are likely to prove ineffectual. For its part, the severity of the economic downturn that has plagued Japan in recent years, together with the political influence of the major companies involved in the tropical timber trade, creates a setting that is not conducive to bringing effective pressure to bear on domestic users of tropical timber.

A major goal of the Arctic Environmental Protection Strategy (AEPS)—launched in 1991 but integrated since 1996 into the broader framework of the Arctic Council—is to promote the conservation of flora and fauna in the Circumpolar North (Huntington, 1997). To this end, the AEPS established a Working Group on the Conservation of Arctic Flora and Fauna (CAFF) and provided it with a mandate to take the initiative in devising innovative means to achieve its general goal. Despite the relative weakness of CAFF in terms of formal authority, this initiative has generated a good deal of interest. CAFF has become a forum in which officials from government agencies and representatives of NGOs (e.g., the World Wildlife Fund) interact freely; it has succeeded in capturing and holding the attention of public agencies in a number of member states, and it has emerged as a mechanism for applying universal guidelines relating to biological diversity to the particular circumstances prevailing in the Circumpolar North.[15] One of CAFF's highest priorities has been to promote and oversee the creation of a Circumpolar Protected Areas Network (CPAN) or, in other words, a linked system of parks, preserves, wildlife refuges, and so forth located in all the Arctic countries and organized to provide harmonized management for the entire system (CAFF, 1996). The success of this initiative depends first and foremost on the willingness and the ability of management agencies located within individual member states to collaborate effectively or, in other words, to manage protected natural areas on a coordinated basis. This is where problems begin to arise in connection with this intuitively appealing initiative.

Within some key countries—the United States is a good example—management authority regarding the areas involved resides with a number of distinct agencies (e.g., National Parks Service, Fish and Wildlife Service, Bureau of Land Management) that are not in the habit of cooperating effectively with one another, much less with their counterparts in other countries (Clarke and McCool, 1996). In other countries—the Russian Federation is a prime example—economic and political problems are so severe at this time that little energy and few resources are available for international cooperation. This initiative does not require integrated management across national jurisdictional boundaries; coordinated or harmonized management practices carried out by relevant agencies within each country would suffice. Yet the complexities of institutional interplay between international programs and national practices raise serious questions about the prospects for CPAN.

## Regimes for Marine Resources

Turning now to institutional interplay relating to marine resources in the Barents Sea and the Bering Sea, an even more complex pattern of institutional interplay comes into focus. In effect, the regimes that have emerged in these areas feature interactions between and among three differentiable sets of institutional arrangements: the global rules governing EEZs, the (sub)national regulatory systems that individual coastal states have put in place to govern activities within their individual EEZs, and several regional arrangements created to deal with situations in which the EEZs of individual states either adjoin each other (i.e., the relevant states are adjacent or opposite states) or leave pockets of high seas surrounded by national EEZs. Although the introduction of EEZs was justified in large measure as an institutional innovation required to manage the resources of large marine ecosystems on a sustainable basis, it soon became apparent that this reform created a range of new problems, quite apart from its consequences with regard to the treatment of preexisting problems.

Marine ecosystems do not conform to any legal or political boundaries, however ingenious the effort to delineate them may be. As a result, many states that acquired expanded jurisdiction over the harvesting of living resources in their individual EEZs soon found themselves confronted with a sizable collection of new problems relating to what have become known as straddling stocks (Stokke,

2001). One response to this development, intended mainly to coordinate efforts to manage marine resources located partly within an EEZ and partly in the high seas, is embodied in the Straddling Fish Stocks Agreement, a global arrangement negotiated in the wake of the United Nations Conference on Environment and Development and signed in 1995.[16] Another response, intended to coordinate the efforts of adjacent and opposite states to manage fish stocks common to their individual EEZs and specific areas of the high seas, has taken the form of the creation of a growing collection of regional fisheries regimes.

Two particularly interesting examples of these regional arrangements are the predominantly bilateral Norwegian/Russian regime dealing with the fisheries of the Barents Sea and the somewhat more complex set of arrangements that have emerged in the Bering Sea Region. Not only do these cases exemplify different strategies for dealing with institutional interplay, but they also have produced different outcomes. In the Barents Sea, Norway and Russia capitalized on the creation of EEZs to establish a bilateral regime that has phased out or drastically curtailed participation on the part of fishers from third states and that has put in place a system under which the principal fish stocks of the entire region are managed on an integrated basis (Stokke et al., 1999). This system is not immune to biogeophysical surprises. It has had to adjust to shifting biological conditions (e.g., the location of spring spawning herring), and it has had to cope with severe stresses attributable to the transition from the Soviet Union to the Russian Federation and the subsequent decline in the capacity of Russia to regulate the activities of Russian fishers (Hønneland, 2000; Stokke, 2001). But by and large, this is a case in which the interplay between two sets of national arrangements and an international regime has been managed in such a way as to produce positive results.

The situation that has emerged in the Bering Sea Region, by contrast, illustrates a somewhat less successful response to institutional interplay. Russia and the United States responded to the creation of EEZs by establishing complex but somewhat poorly coordinated national regimes in the western Bering Sea and the eastern Bering Sea, respectively. In addition, the 1990s brought the creation of a regional agreement covering salmon stocks migrating back and forth through the EEZs of the two countries, along with a six-nation agreement dealing with the pollock stocks of the doughnut hole and designed to prevent a recurrence of the collapse of these stocks that occurred in the late 1980s and early 1990s. Unlike the Barents Sea, this is also a region in which NGOs (e.g., Greenpeace, World Wildlife Fund) have become increasingly active. But the results of the complex mosaic arising from these developments are far from reassuring. Both coastal states have experienced problems in controlling harvests of living marine resources within their own EEZs. The pollock stocks of the doughnut hole have not recovered sufficiently to allow for any harvesting under the terms of the international agreement created to manage these stocks. Above all, there are a number of disturbing indications that anthropogenic forces have played a role in triggering severe stresses affecting the Bering Sea ecosystem as a whole (National Marine Fisheries Service, 1997; National Research Council, 1996; World Wildlife Fund and The Nature Conservancy of Alaska, 1999). These include startling declines in populations of several unharvested species, such as sea lions, northern fur seals, sea otters, and red-legged kittiwakes, as well as some harvested species, such as spectacled eiders and several species of geese. No doubt, it would be wrong to point to problems of institutional interplay as the sole cause of these troubling developments. But it is hard to avoid the conclusion that difficulties plaguing efforts to manage the interplay of institutional arrangements across levels of social organization constitute a significant feature of this story.

## IMPLICATIONS AND TAKE-HOME MESSAGES

The principal conclusion to be drawn from the analysis set forth in the preceding sections is that cross-scale interactions among resource regimes generate an inescapable tension. Higher level arrangements offer opportunities to consider functional interdependencies in large marine and terrestrial ecosystems and to devise regimes based on the precepts of ecosystems management. Yet substantial costs often are associated with the creation of higher level arrangements that take forms such as an inability to come to terms with local variations in biogeophysical conditions and a lack of sensitivity to both the knowledge and the rights and interests of local stakeholders.

Those operating at higher—national or international—levels typically are compelled to devise and promulgate structures of rights and regulatory rules in terms that are broadly encompassing and generic. This may cause few problems in dealing with marine and terrestrial ecosystems that are homogeneous. But problems mount rapidly where there are local variations both in pertinent biogeophysical conditions (e.g., the population dynamics of fish stocks) and in patterns of human uses of natural resources (e.g., hunting and herding practices). In the absence of effective procedures for cross-scale coordination, the result is apt to be a proliferation of formal rights and rules that are poorly suited to local circumstances or the emergence of systems so encrusted with local exceptions and informal interpretations that they become unworkable.

Similar observations are in order regarding the rights and interests of various groups of stakeholders. Moving to higher levels of social organization can open up opportunities for increased efficiency in the use of resources and for more comprehensive approaches to equity. But the costs associated with such developments are apt to be substantial. National regimes increase the influence of economically and politically powerful actors (including nongovernmental organizations) who do not reside within the ecosystems they exploit, who can move their operations with relative ease to new areas once the resources of one area are exhausted, and who favor the exploitation of resources that are tradable in (often international) markets. For their part, international regimes often cater to the interests of multinational corporations that have operations located in many places and that have no long-term commitment to the ecological welfare of particular areas and the social welfare of those who reside permanently in these areas. Under the circumstances, it is easy to see that shifts to higher levels of social organization, justified in order to manage large marine and terrestrial ecosystems in a holistic manner, can and often do lead to changes in patterns of land use and sea use that raise profound questions—not only in terms of sustainability but also in terms of normative concerns, including equity and efficiency.

Problems of this sort often trigger exercises in political design, and the vigor of the current debate about what has become known as the subsidiarity principle is testimony to the importance attached to finding effective ways to cope with the environmental consequences of cross-scale interactions. But the subsidiarity principle, which calls for management authority to be vested in the lowest level of social organization capable of solving pertinent problems, does not offer much help in coming to terms with the problems of vertical interplay addressed in this chapter. National and even international arrangements are needed to manage human activities relating to large marine and terrestrial ecosystems. Yet the dangers inherent in moving from local to national and from national to international regimes are severe. What is needed, in situations of this type, is a conscious effort to design and manage institutional arrangements that recognize different types of knowledge and protect the rights and interests of local stakeholders, even while they introduce mechanisms at higher levels of social organization required to cope with the dynamics of ecosystems that are regional and even global in scope.

This is not a task to be handled through efforts to determine the proper level of social organization at which to vest management authority. A more promising response to this tension involves the establishment of arrangements that many analysts have explored in recent years under the rubric of co-management (Berkes, this volume:Chapter 9; Osherenko, 1988). In the typical case, co-management involves the creation of environmental or resource regimes featuring partnerships between local users of natural resources and agencies of (sub)national governments possessing the formal authority to make decisions about human activities involving marine and terrestrial ecosystems as well as the material resources needed to administer management systems. This intrinsically appealing approach eventually may lead to a range of social practices that are of lasting significance in dealing with specific problems of vertical interplay. But it would be premature to jump to any such conclusion at this stage. Co-management is in danger of becoming a catch-all or residual category containing a ragtag collection of tenuously related approaches to resource management. Even in dealing with the interplay between local and national arrangements, experience on the ground with co-management is limited, and we are far from the formulation of well-tested propositions about the determinants of success and failure in the creation and operation of co-management regimes. It is anything but clear whether experience with co-management in dealing with local/national interactions can be scaled up to offer an effective method of organizing the interplay between (sub)national and international regimes. These observations are not meant to belittle the significance of co-management as a strategy featuring exercises in political design intended to manage problems arising from functional interdependencies; many analysts currently are engaged in interesting studies of co-management. Nonetheless, there is much to be done before we can assert that substantial progress is being made in structuring institutions in such a way as to eliminate or at least alleviate the tensions arising from cross-scale interactions.

We must bear in mind as well that the creation of institutions at every level of social organization is a political process centering on what can be described as institutional bargaining (Young, 1994). Whatever their consequences in terms of considerations like sustainability or efficiency, environmental or resource regimes always have significant consequences for the interests of those—nonstate actors as well as states—subject to their rules and decision-making procedures. It should come as no surprise, therefore, that individual actors often work hard to advance their own causes in processes of regime formation and that outcomes are likely to reflect the political influence of major participants or coalitions of participants in these processes.[17] This is not to suggest that efforts to design institutions that will advance social goals like sustainability or efficiency are bound to become exercises in futility. In fact, institutional bargaining has several features that make it more open to design considerations than conventional distributive bargaining (Young, 1994). There is reason to believe that we can gradually develop a repertoire of best practices in this field through comparative studies of resource regimes and even the conduct of social experiments. Yet there is no escaping the fact that regime formation is better understood as a political process in which bargaining strength plays a central role than as an exercise in social engineering in which apolitical design principles predominate.

## CONCLUSION

The argument of the substantive sections of this chapter is intended to initiate a study of the roles that cross-scale interactions among distinct institutions play in the overall picture of human/environment relations. The cases of land use and sea use are particularly interesting in this connection because patterns of land and sea use are directly and intimately linked to large-scale environmental changes, such as the loss of biological diversity and climate change. But similar issues of institutional interplay arise in conjunction with other concerns, including human uses of atmospheric and hydrological systems. There is no assumption here that institutions in general or the interplay among distinct institutions in particular can account for all the variance in human impacts on atmospheric, hydrological, marine, or terrestrial systems. On the contrary, institutional drivers interact with other forces in complex ways; one of the main challenges facing those interested in the human dimensions of environmental change is to sort out the relative significance or weight of institutional drivers and other driving forces.

Yet an emphasis on the role of institutions in this connection has great appeal, so long as care is taken to avoid the assumption that institutional arrangements operate in a vacuum, producing results without regard to the character of the broader biogeophysical and socioeconomic settings in which they operate. The content of prevailing institutions is subject to intentional reform, a fact that opens up the opportunity to engage in design efforts in the interests of minimizing the negative consequences of existing institutions and supplementing or even replacing these arrangements in order to mitigate or adapt to environmental changes. The message of this chapter regarding efforts to design arrangements to minimize problems arising from institutional interplay is one of great caution but certainly not of pessimism. Even if we succeed in identifying institutional forces giving rise to environment problems, there is no guarantee that we can take the steps—including exercises in political design and management—needed to alter the operation of prevailing arrangements in a well-planned fashion. Nonetheless, the prospect that (re)designing institutions can play a role in controlling or managing environmental changes provides a compelling reason to invest time and energy in enhancing our understanding of the dynamics of institutional interplay.

## NOTES

1  Like other participants in this project, I use the term institution to mean an assemblage of rules, decision-making procedures, and programs that gives rise to a social practice, assigns roles to participants in this practice, and governs interactions among occupants of these roles.

2  In this account, I use the concept of level of social organization as a means of describing scale delimited in spatial terms. Where interplay occurs between institutions operating at the local level and institutions operating at the national level, for example, I describe the resultant interplay as an instance of cross-scale interactions.

3  In the discussion to follow, public property refers to land/sea and associated natural resources owned by the state; private property refers to land/sea and natural resources belonging to individual members of society; and common property refers to land/sea and natural resources owned jointly by the members of an identifiable community. In analyzing the consequences of interactions among different structures of property rights, I consider various types of natural resources and environmental services, including but not limited to what are commonly known as common-pool resources.

4  Both Principle 21 of the 1972 Stockholm Declaration and Principle 2 of the 1992 Rio Declaration, for example, declare that "States have…the sovereign right to exploit their own resources…"

5  Recent settlements of comprehensive claims with aboriginal peoples in the Canadian North have reduced the scope of public property somewhat and, at the same time, introduced some interesting arrangements featuring more complex systems of land tenure. Even so, public land remains the norm in Canada.

6  In the Alaska Native Claims Settlement Act of 1971 (P.L. 92-203), for example, the U.S. government awarded title to nearly 44 million acres of land to Native corporations but, at the same time, declared that "All aboriginal titles, if any, and claims of aboriginal title in Alaska . . . including any aboriginal hunting or fishing rights that may exist, are hereby extinguished" (Sec. 4b).

7  For an early, but still helpful treatment of property systems as social institutions, see Hallowell (1943).

8  In cases where preexisting subsistence practices have given way to mixed economies, local peoples may experience a growing need to exploit natural resources to generate a flow of cash.

9  For evidence of similar interactions occurring in other parts of the world, see Gibson et al. (2000).

10  For an extended account of the role of rules in use and the relationship between such rules and formal rules see, Ostrom (1990).

11  The doughnut hole constitutes a pocket of high seas wholly surrounded by the EEZs of Russia and the United States.

12  Some recent arrangements (e.g., the ozone and climate regimes) distinguish among classes of members on the basis of what has become known as the principle of common but differentiated responsibility.

13  A concrete case in point involves the harvesting of whales. Under the Canadian Constitution, the formal authority to set harvest quotas for any harvests of whales resides with the provinces.

14  In some countries—the United States is a good example—there is a long history of tension regarding the allocation of authority between the national government and various subnational governments.

15  Updates on the work of CAFF appear regularly in the *Arctic Bulletin,* published four times a year under the auspices of the World Wildlife Fund Arctic Program.

16  The Straddling Fish Stocks Agreement also deals with other issues, such as the management of consumptive uses of highly migratory species (e.g., tunas).

17  Note, however, that it would be wrong to equate bargaining strength with structural power or power in the purely material sense (Young, 1994).

# REFERENCES

Acheson, J.M.
    1987  The lobster fiefs revisited: Economic and ecological effects of territoriality in Maine lobster fishing. Pp. 37-65 in *The Question of the Commons: The Culture and Ecology of Communal Resources*, B.J. McCay and J.M. Acheson, eds. Tucson: University of Arizona Press.
Anand, R.P.
    1983  *Origin and Development of the Law of the Sea: History of International Law Revisited.* The Hague: Martinus Nijhoff.
Balton, D.A.
    2001  The Bering Sea Doughnut Hole Convention: Regional solution, global implications. Pp. 143-177 in *Governing High Seas Fisheries: The Interplay of Global and Regional Regimes*, O.S. Stokke, ed., London: Oxford University Press.
Berkes, F., ed.
    1989  *Common Property Resources: Ecology and Community-Based Sustainable Development.* London: Belhaven.
Biermann, F.
    2000  The case for a world environment organization. *Environment* 42(9):22-31.
Bromley, D.W., ed.
    1992  *Making the Commons Work: Theory, Practice, and Policy.* San Francisco: ICS Press.
Brubaker, S., ed.
    1994  *Rethinking the Federal Lands.* Washington, DC: Resources for the Future.
Chayes, A., and A.H. Chayes
    1995  *The New Sovereignty: Compliance with International Regulatory Agreements.* Cambridge, MA: Harvard University Press.

Chertow, M.R., and D.C. Esty
   1997   *Thinking Ecologically: The Next Generation of Environmental Policy.* New Haven, CT:
          Yale University Press.
Clarke, J.N., and D.C. McCool
   1996   *Staking Out the Terrain: Power and Performance among Natural Resource Agencies.*
          Albany: State University of New York Press.
Dauvergne, P.
   1997a  *Shadows in the Forest: Japan and the Politics of Timber in Southeast Asia.* Cambridge,
          MA: MIT Press.
   1997b  A model of sustainable trade in tropical timber. *International Environmental Affairs* 9:3-
          21.
Fienup-Riordan, A.
   1990   *Eskimo Essays: Yu'pik Lives and How We See Them.* New Brunswick, NJ: Rutgers Uni-
          versity Press.
Florini, A.M., ed.
   2000   *The Third Force: The Rise of Transnational Civil Society.* Tokyo and Washington, DC:
          Japan Center for International Exchange and Carnegie Endowment for International Peace.
Fondahl, G.
   1998   *Gaining Ground: Evenkis, Land, and Reform in Southeastern Siberia.* Boston: Allyn and
          Bacon.
French, H.F.
   1997   Learning from the ozone experience. Pp. 151-171 in *State of the World 1997*, L.R. Brown,
          C. Flavin, and H. French, eds. New York: W.W. Norton.
Gibson, C.C.
   1994   *Politicians and Poachers: The Political Economy of Wildlife Policy in Africa.* Cambridge,
          Eng.: Cambridge University Press.
Gibson, C.C., M.A. McKean, and E. Ostrom, eds.
   2000   *People and Forests: Communities, Institutions, and Governance.* Cambridge, MA: MIT
          Press.
Guppy, N.
   1996   International governance and regimes dealing with land resources from the perspective of
          the North. Pp. 136-162 in *Global Environmental Change and International Governance*,
          O.R. Young, G.J. Demko, and K. Ramakrishna, eds. Hanover, NH: University Press of
          New England.
Hallowell, A.I.
   1943   The nature and function of property as a social institution. *Journal of Legal and Political
          Sociology* 1:115-138.
Hesenclever, A, P. Meyer, and V. Rittberger
   1997   *Theories of International Regimes.* Cambridge, Eng.: Cambridge University Press.
Higgins, R.
   1994   *Problems and Process: International Law and How We Use It.* Oxford, Eng.: Clarendon
          Press.
Hønneland, G.
   2000   *Coercive and Discursive Compliance Mechanisms in the Management of Natural Re-
          sources: A Case Study from the Barents Sea Fisheries.* Dordrecht, Neth.: Kluwer Aca-
          demic Publishers.
Humphreys, D.
   1996   Hegemonic ideology and the International Tropical Timber Organization. Pp. 215-233 in
          *The Environment and International Relations*, J. Volger and M. Imber, eds. London:
          Routledge.

Huntington, H.P.
1997 The Arctic Environmental Protection Strategy and the Arctic Council: A Review of United States Participation and Suggestions for Future Involvement. Report prepared for the Marine Mammals Commission.

Iudicello, S., M. Weber, and R. Wieland
1999 *Fish, Markets, and Fishermen: The Economics of Overfishing.* Washington, DC: Island Press.

Juda, L.
1996 *International Law and Ocean Use Management: The Evolution of Ocean Governance.* London: Routledge.

Keohane, R.O., and M.A. Levy, eds.
1996 *Institutions for Environmental Aid.* Cambridge, MA: MIT Press.

Krech, S.
1999 *The Ecological Indian: Myth and History.* New York: W.W. Norton.

Levy, M.A., O.R. Young, and M. Zürn
1995 The study of international regimes. *European Journal of International Relations* 1(September):267-330.

Lipschutz, R.D., and K. Conca, eds.
1993 *The State and Social Power in Global Environmental Politics.* New York: Columbia University Press.

McGoodwin, J.
1990 *Crisis in the World's Fisheries: People, Problems, and Policies.* Stanford, CA: Stanford University Press.

Mitchell, R.B.
1994 *Intentional Oil Pollution at Sea: Environmental Policy and Treaty Compliance.* Cambridge, MA: MIT Press.

National Marine Fisheries Service
1997 Bering Sea Ecosystem – A Call to Action. Draft white paper dated September 21.

National Research Council
1996 *The Bering Sea Ecosystem.* Washington, DC: National Academy Press.
1999a *Sharing the Fish. Toward a National Policy on Individual Fishing Quotas.* Washington, DC: National Academy Press.
1999b *The Community Development Quota Program in Alaska and Lessons for the Western Pacific.* Washington, DC: National Academy Press.

Oberthür, S.
1999 Linkages Between the Montreal and Kyoto Protocols. Unpublished paper prepared for United Nations University conference on Synergies and Coordination Between Multilateral Environmental Agreements.

Osherenko, G.
1988 Can comanagement save Arctic wildlife? *Environment* 20(July-August):6-13, 29-34.
1995 Property rights and transformation in Russia: Institutional change in the Far North. *Europe-Asia Studies* 47:1077.

Osherenko, G., and O.R. Young
1989 *The Age of the Arctic: Hot Conflicts and Cold Realities.* Cambridge, Eng.: Cambridge University Press.

Ostrom, E.
1990 *Governing the Commons: The Evolution of Institutions for Collective Action.* Cambridge, Eng.: Cambridge University Press.

Peluso, N.L.
1992 *Rich Forests, Poor People: Resource Control and Resistance in Java.* Berkeley: University of California Press.

Pinkerton, E., ed.
   1989   *Co-operative Management of Local Fisheries.* Vancouver: University of British Columbia
          Press.
Princen, T., and M. Finger
   1994   *Environmental NGOs in World Politics: Linking the Local and the Global.* London:
          Routledge.
Rittberger, V., ed.
   1993   *Regime Theory and International Relations.* Oxford, Eng.: Clarendon Press.
Sand, P.H.
   1999   Carrots without sticks? New financial mechanisms for global environmental agreements.
          *Max Planck Yearbook of United Nations Law* 3:363-388.
Sherman, K.
   1992   Large marine ecosystems, Pp. 653-673 in *Encyclopedia of Earth System Science.* Vol. 2.
          New York: Academic Press.
Skjaerseth, J.B.
   2000   *North Sea Cooperation: Linking International and Domestic Pollution Control.* Manches-
          ter, Eng.: Manchester University Press.
Stokke, O.S., ed.
   2001   *Governing High Seas Fisheries: The Interplay of Global and Regional Regimes.* London:
          Oxford University Press.
Stokke, O.S., L.G. Anderson, and N. Mirovitskaya
   1999   The Barents Sea Fisheries. Pp. 91-154 in *The Effectiveness of International Environmental
          Regimes: Causal Connections and Behavioral Pathways*, O.R. Young, ed. Cambridge,
          MA: MIT Press.
Svensson, T.G.
   1995   *The Sami and Their Land.* Oslo: Novus forlag.
Underdal, A. ed.
   1995   *The Politics of International Environmental Management.* Dordrecht, Neth.: Kluwer Aca-
          demic Publishers.
Underdal, A., and K. Hanf, eds.
   1998   *International Environmental Agreements and Domestic Politics: The Case of Acid Rain.*
          Aldershot, Eng.: Ashgate.
Victor, D.G., K. Raustiala, and E.B. Skolnikoff, eds.
   1998   *The Implementation and Effectiveness of International Environmental Commitments:
          Theory and Practice.* Cambridge, MA: MIT Press.
Wapner, P.
   1997   Governance in global civil society. In *Global Governance: Drawing Insights from the
          Environmental Experience*, O.R. Young, ed. Cambridge, MA: MIT Press.
Weiss, E.B., and H.K. Jacobson, eds.
   1998   *Engaging Countries: Strengthening Compliance with International Environmental Ac-
          cords.* Cambridge, MA: MIT Press.
Working Group on the Conservation of Arctic Flora and Fauna (CAFF)
   1994   Circumpolar Protected Areas Network (CPAN) – Strategy and Action Plan. *CAFF Habitat
          Conservation Reports No. 6.*
World Wildlife Fund and The Nature Conservancy of Alaska
   1999   *Ecoregion-Based Conservation in the Bering Sea.* Anchorage: World Wildlife Fund and
          The Nature Conservancy.
Young, O.R.
   1982a  *Resource Regimes: Natural Resources and Social Institutions.* Berkeley: University of
          California Press.

1982b  The political economy of fish: The Fishery Conservation and Management Act of 1976. *Ocean Development and International Law* 10:199-273.

1983   Fishing by permit: Restricted common property in practice. *Ocean Development and International Law* 13:121-170.

1994   *International Governance: Protecting the Environment in a Stateless Society.* Ithaca, NY: Cornell University Press.

1996   Institutional linkages in international society. *Global Governance.* 2:1-24.

Young, O.R., A. Agrawal, L.A. King, P.H. Sand, A. Underdal, and M. Wasson.

1999   *Institutional Dimensions of Global Environmental Change (IDGEC) Science Plan. IHDP Report No. 9.* Bonn: International Human Dimensions Programme.

# 9

# Cross-Scale Institutional Linkages: Perspectives from the Bottom Up

*Fikret Berkes*

The balance of evidence from the commons literature of the past few decades is that neither purely local-level management nor purely higher level management works well by itself. Rather, there is a need to design and support management institutions at more than one level, with attention to interactions across scale from the local level up. Here we use *cross-scale interactions* to refer to linking institutions both *horizontally* (across space) and *vertically* (across levels of organization). Cross-scale institutional linkages mean something more than management at several scales, isolated from one another. Issues need to be considered simultaneously at several scales when there is coupling or interaction between scales. Indeed, many cases of resource and environmental management are *cross-scale* in both space and time.

For example, many inshore tropical fisheries in island nations of the world, such as in the Caribbean, southeast Asia, and Oceania, are carried out by small-scale fishing units that do not range more than a day from a home port (Berkes et al., 2001). Fishers follow community norms, and if there is any regulation in these fisheries, it is community-based. However, many of the stocks they fish range into areas harvested by other groups around the island, and should ideally be managed over a larger area covering the whole island or several islands. Of course, some of the stocks, such as tunas, also travel across the national boundaries of these island states. The Caribbean flying fish stock, for example, ranges through at least six island nations, and requires bilateral and multilateral agreements for its management (Berkes et al., 2001).

Clearly, such fisheries cannot be managed at a single scale but rather must be managed at multiple scales. As there is coupling between scales, management institutions need to be linked both horizontally across geographic space and vertically across levels of organization. Furthermore, in a globalized world, the need for cross-scale institutions and vertical linkages becomes even greater. Globalization intensifies coupling and renders local institutions increasingly vulnerable. Local rules with emphasis on "how" people should fish rather than "how much" (Wilson et al., 1994) break down in most modern commercial fisheries subject to national and international market pressures, requiring other measures such as quota management to cap the quantity of harvest (Hilborn and Walters, 1992) and the crafting of new and different kinds of institutions.

The focus on institutions emerges from the commons literature documenting a rich diversity of ways in which rules can be made to avert the commons dilemma. Much of this literature refers to local-level commons institutions, and the bulk of the scholarship is concerned with community-based management. There are commons issues at the global level as well, and at various levels from the local to the global, with a growing literature base. However, the *links* between the various scales of commons management have not received much attention. Yet these links and the cross-scale institutions that provide them are important in their own right.

Given the significance of cross-scale institutional linkages and their dynamics, surprisingly little research has been carried out in this area. There is a large literature on common property institutions, and a growing base of mostly empirical literature on co-management, or the sharing of management power and responsibility between the government and local-level institutions, but relatively little on cross-scale institutions per se. Ostrom (1990) has proposed a set of seven design principles, plus an eighth for nested systems, that appears to characterize robust common property institutions. These principles have been widely used to guide research, despite perceived shortcomings (e.g., Steins et al., 2000). Agrawal (this volume:Chapter 2) argues that the number of factors that may be critical for commons management may more likely be on the order of 35, and that existing theory is short in specifying what makes for sustainable commons management. Young (this volume:Chapter 8) has drawn attention to the importance of partnerships between or among different levels of agencies, and the potential of such arrangements in dealing with problems of vertical linkages in institutional interplay. He points out, however, that we are far from the formulation of well-tested propositions about the determinants of success and failure in these cross-scale management regimes.

The subject of this chapter is the investigation of cross-scale institutional linkages, including co-management arrangements, and the exploration of new research directions. Within this larger goal, the objectives are (1) to identify promising institutional forms for linking across levels of institutions, and (2) to investigate the dynamics of cross-scale institutions in reference to adaptive management and resilience.

The chapter begins with a review and synthesis of the impacts of higher level institutions (national and international) on local-level institutions, as a way of introducing the importance of vertical and horizontal linkages. It summarizes a variety of ways in which larger scale institutions can interfere with or support smaller scale ones. This part deals with some of the same issues as the chapter by Young, except that Young approaches the problems by linking the national level to the global, whereas this chapter takes a perspective from the bottom up. The second section of the chapter proceeds to identify some institutional forms that facilitate cross-scale resource and environmental management, noting that there is as yet no accepted typology of these emerging cross-scale linking institutions. Some of these institutions are captured by the catch-all term, co-management. However, the chapter argues, this term hides complexity and is inadequate to encompass the full range of cross-linking institutions. As well, there is a need to move beyond the static analysis inherent in looking merely at institutional *forms*; we need to investigate processes of institutional change.

Hence, the third section focuses on the dynamics of cross-scale institutional linkages and the issue of scale. It develops the argument that the adaptive management approach may be useful in building a theory of cross-scale institutional linkages. A key concept is resilience, used here to refer to the ability of a system to absorb perturbations and to build capacity for self-organization, learning, and adaptation. Resilience thinking is a useful tool to link social systems and natural systems (Berkes and Folke, 1998). It helps to investigate scale issues not only from the institutional point of view per se, but also in regard to the fit between institutional scales and the ecosystem that generates resources at multiple scales (Folke et al., 1998).

Given that cross-scale institutional linkages have not been explored extensively, this chapter offers not a definitive review, but some concepts and hypotheses that may serve as a starting point for further research and theory development. The research agenda that comes out of this chapter is at an early rather than a mature stage.

The scope of the chapter is local to national, focusing on the link between local institutions and higher level government entities. Various cross-scale management issues involving different levels of government, for example, between federal- or state-level agencies or between the European Union and its member states, are outside the scope of this chapter. Also beyond the scope is the growing literature in political science and public administration on the relationships among national, state, and local levels of government.

## EFFECTS OF HIGHER LEVEL INSTITUTIONS ON LOCAL INSTITUTIONS

The commons literature is full of examples of the impacts of the state on local institutions. Some of the mechanisms or processes by which higher level institutions impact local institutions include centralization of decision making; shifts in systems of knowledge; colonization; nationalization of resources; increased participation in national and international markets; and national-level development projects. Table 9-1 provides examples of each of these six classes of impacts; here we expand on the first two.

Excessive centralization of resource management is not confined to countries with centrally planned economies such as the former Soviet Union. It is found in nearly all governments in which resource management functions have been taken over by a managerial elite. However, such centralization has not occurred uniformly across resource types and geographic areas. For example, in the adjacent Canadian provinces of Ontario and Quebec, the development of provincial resource management agencies took different paths. In Ontario, the provin-

TABLE 9-1 Effects of Higher Level Institutions on Local Institutions

| Class of Impacts | Examples |
| --- | --- |
| Centralization of decision making | The former Soviet Union centralized decision making for rational resource management and for setting production targets, sweeping away, in the process, local management systems and institutions such as the *artels* of the Ural Cossacks for managing fisheries of the Caspian Sea region (Kropotkin, 1914). |
| Shifts in systems of knowledge | From the 1950s onward, caribou management in the Canadian Arctic came to be based primarily on quantitative population models. Science replaced aboriginal management systems based on accumulated local observations and ethical rules (Berkes, 1999). |
| Colonization | To create revenues from timber extraction, the colonial regime in India dismantled local institutions for forest and grazing land management, moving the locus of control to the center (Gadgil and Guha, 1992). |
| Nationalization of resources | The Government of Nepal nationalized forests in 1957 (to curb deforestation), but the result was the creation of de facto open access because the government measure disempowered local institutions that had functioned in forest resource sharing (Messerschmidt, 1993). |
| Increased participation in markets | To take advantage of the demand for prawns in the international market, the government subsidized trawlers in the 1960s and the 1970s in Kerala, India, in an area previously dominated by small-scale, nonmechanized boats, touching off a social crisis and a resource crisis (Kurien, 1992). |
| Development policies | On lands occupied and used by Barabig pastoralists in Tanzania, state policies for the development of wheat agriculture, supported by international development agencies, resulted in the destruction of local institutions for sustainable land use (Lane, 1992). |

cial agency was already well organized by the end of the 1940s. Through the strong presence of this agency, wildlife management was centralized early on, even in northern Ontario, which is occupied predominantly by aboriginal groups. By contrast, in Quebec, the government management agency was only weakly present in the north, even as late as the 1970s. Perhaps as a result, local institutions for wildlife management were strongly present in the Cree areas of northern Quebec as late as the 1980s and effectively managed wildlife (Drolet et al., 1987). By contrast, in the Cree areas of northern Ontario, such institutions were almost nonexistent, presumably because they had been swept away by centralization (Berkes et al., 1991).

The replacement of local institutions by centralized ones often involves a change in the way knowledge is used for management. Local institutions tend to use their own folk knowledge, often referred to as local knowledge, indigenous knowledge, or traditional ecological knowledge, whereas centralized management agencies tend to use internationally accepted scientific practice and often assume away local knowledge and practice (Berkes, 1999; Williams and Baines, 1993). The shift of knowledge systems is one of the major impacts of government-level institutions on local institutions because it is often accompanied by a change in control over a resource. The differences between the two systems of knowledge can be substantial in the way resources are viewed.

A case in point is caribou management in the Canadian North (Berkes, 1999). A number of studies indicate that aboriginal hunters from the Arctic and the Subarctic monitor caribou distributions, migration patterns and their change, predator presence, individual behavior, sex and age composition of the herd, and fat deposits in animals. The Western science of caribou management also monitors much the same things, but there is a fundamental difference: decision making in scientific management is based primarily on population models. The aboriginal system, by contrast, is based on local observations and ethics, assumes that caribou are not predictable or controllable, and does not try to use harvest or population size estimates. Rather, it pays relatively high attention to fat content (an excellent integrative indicator of caribou health) and uses a *qualitative* mental model that provides hunters with an indication of *trends* over time.

This qualitative model reveals the *direction* (increasing or decreasing) in which the population is headed, without requiring the estimation of the population size itself (Berkes, 1999). This locally developed, aboriginal approach to management has potential to result in good resource management, but it is different from scientific management. Centralization of management leads to a shift in the knowledge system used. Government management of resources, based on universal science rather than on locally developed knowledge, undermines the knowledge systems, as well as the institutions, of northern aboriginal groups. Hence, the centralization of resource management and the assertion of "government's science" becomes a political tool for the control of the local indigenous populations (Freeman, 1989).

In the list of impacts of higher level institutions on local-level ones in Table 9-1, many of the examples seem to show negative impacts. However, the designation of impacts as "negative" or "positive" is a value judgment. For example, impacts of modernization and economic development on local institutions may be seen as negative by some and positive by others. Increased participation in markets may result in the shift of control over a resource from the local institutions to the outside. But there are also counterexamples in which commercialization of a subsistence resource has resulted in the strengthening of local-level institutions.

An example is the evolution of the family-controlled beaver trapping territory system in eastern James Bay, Canada, with the advent of the fur trade after the 18th century. The ethnohistorical evidence is not conclusive, but Berkes (1989a) speculated that as the beaver resource became more valuable and scarce, tighter controls became necessary, shifting a loosely controlled communal system of use into a family-controlled system with a senior trapper ("beaver boss") in charge. A model was proposed in which the local institutional strength was governed by two driving forces, the intensification of resource use (as a result of trade or other factors) and the incursion of outsiders (commercial hunters as stimulated by high fur prices) into the system. The model was consistent with the historical record of three cycles of exploitation—each of which involved creation of open access, resource decline, reassertion of local controls, and resource recovery (Berkes, 1989a).

In general, historical factors are often important in determining whether the impacts of higher level institutions on the local ones are positive or negative. A distinction should be made between *processes* and their *outcomes*. As pointed out by S. Stonich and P.C. Stern (personal communication, August 2000), a process such as decolonization might have either positive or negative impacts on local institutions, depending on how it is carried out. If it results in the centralization of power, for example, it would seem likely to have negative effects on local institutions. The same can be said about the process of commercialization of a subsistence resource. The speed of change may be one important factor; a local institution is more likely to adapt to a perturbation over a period of decades than a period of months. Ecological considerations, such as the level of exploitation of a resource as compared to its natural rate of replenishment, are also important. The locus of control of the resource is yet another factor. However, we do not have in hand well-tested propositions about the determinants of the outcome.

Higher level institutions can also impact local-level ones through deliberate interventions. The commons literature includes many examples of how certain forms of state involvement may strengthen or rejuvenate local-level institutions. These include state recognition of local institutions; development of enabling legislation; cultural revitalization; capacity building; and local institution building (see Table 9-2). Here we expand on the first and touch on the second of these five items.

TABLE 9-2 Strengthening Local-Level Institutions for Cross-Scale Interaction

| Classes of Activities | Examples |
|---|---|
| State legitimization of local institutions | If resource users have the right to devise their own institutions without being challenged by external authorities, they can enforce the rules themselves. This is the principle of "minimal recognition of rights to organize" (Ostrom, 1990). |
| Enabling legislation | Legislation that makes it possible, or creates the legal preconditions, in this case, for the effective functioning of local-level institutions. Enabling legislation may be used to provide legitimacy for locally devised rules, or it may in other ways empower local institutions (Peters, 1986). |
| Cultural and political revitalization | Resistance to dominant culture and political force; sometimes used to refer to broader social and political action in which the dominant group is overthrown, not only in form but also in ideology (Smith, 1999). Revitalization movements may be about empowerment and cultural rediscovery, as well as revival of local institutions. |
| Capacity building | The sum of efforts needed to nurture, enhance, and utilize the skills and capabilities of people and institutions at all levels—nationally, regionally, and internationally. It does not seek to resolve specific problems but rather seeks to develop the capacity within communities, governments, and other organizations to resolve their own problems (National Round Table on the Environment and the Economy, 1998). |
| Institution building | Institutions can be crafted (Ostrom, 1992). Local institutions for the commons also may arise spontaneously, but this often takes time. Local institutions may be helped along by creating a favorable environment that speeds up their development. Some NGOs specialize in such institution building. |

Legitimization or recognition of local-level institutions is a well-known theme in the commons literature. Among the design principles illustrated by long-enduring common property institutions analyzed by Ostrom (1990:90) is "the right of appropriators to devise their own institutions" without being challenged by external authorities. This, as Ostrom puts it, is the "minimal recognition of rights to organize." If government recognizes locally developed rules, community institutions are in a better position to enforce these rules themselves. In some cases, the state may go further and legally recognize local rules. However, this has some disadvantages as well as advantages. The inherent risk in codifying local rules, such as those of marine tenure systems, is that writing them down may "freeze" them in space and time, thereby reducing their flexibility (Baines, 1989).

Some of the aboriginal land claims settlements in Canada, New Zealand, and Australia provide examples of state recognition of local institutions. For example, Canada's James Bay and Northern Quebec Agreement of 1975 explicitly and legally recognizes the hunter-trapper organizations of the Cree and their jurisdiction over certain kinds of resources, mainly fish and wildlife, and their management (Berkes et al., 1991). Government legislation that provides for state recognition of local institutions may be considered enabling legislation. The importance of enabling legislation was recognized early on by commons specialists, as reflected in the consensus of participants in the closing comments of the 1985 Conference on Common Property Resource Management (Peters, 1986).

Additional mechanisms to strengthen local-level institutions are provided by revitalization movements, capacity building, and local institution building. There is no widely accepted classification of these interventions and changes, and the classes can no doubt be subdivided further. However, detailed typologies necessarily will be fuzzy and of limited value. Perhaps more important, the consideration of mechanisms in support of local institutions highlights the dynamic nature of institutions. Ostrom's idea that there is a bank or a "capital" of institutions from which institutions actually can be *crafted* (Ostrom, 1992) serves to highlight the dynamics of institutions.

Even though the literature is rich in cases, we lack theory and guiding principles, perhaps through the identification of driving forces, to build or strengthen local institutions. Promising lines of inquiry will perhaps emerge out of commons dilemma experiments (Kopelman et al., this volume:Chapter 4) and common-pool resource games (Falk et al., this volume:Chapter 5), as well as out of carefully constructed multivariate research on commons management regimes (Bardhan and Dayton-Johnson, this volume:Chapter 3). However, these approaches are unlikely to be sufficient by themselves because the historical and cultural context of cases is so important.

Critics have pointed out that some of the commons literature tends to concentrate on local-level institutions to the exclusion of the outside world that impacts them and shapes them (e.g., Steins et al., 2000). There is not much debate there; impacts of higher level institutions are clearly pervasive. Commons management cannot be done only at the local or the national level; it is cross-scale, with the larger scale institutions interfering with or supporting smaller scale ones through a diversity of mechanisms. We turn next to consider some institutional forms that facilitate interactions across scales of organization, and examine how institutions at various levels can be vertically linked, how they come into existence, and, in some cases, how they change.

## PROMISING INSTITUTIONS FOR CROSS-SCALE LINKAGES

In recent years a literature has developed on forms of institutions with potential for cross-scale linkages. One of these forms is co-management (Jentoft, 1989; Pinkerton, 1989). Others include multistakeholder bodies; institutions oriented for development, empowerment, and co-management; the emerging class of institutions for "citizen science"; policy communities; and social movement networks. Much of this literature has not yet been connected to the commons research community, and the same can be said about the literature on public participation (e.g., Renn et al., 1995; Dietz and Stern, 1998). Table 9-3 lists some characteristics of each type. A seventh and somewhat different set concerns research and management approaches that enable cross-scale linkages. We discuss each in turn.

### Co-Management Arrangements Between Communities and Governments

The simplest kind of cross-scale institutional linkage is the one that connects local-level management with government-level management in partnerships. Literature contains examples of co-management arrangements in a diversity of regions with a number of resource sectors. Many co-management initiatives are in progress in the areas of fisheries, wildlife, protected areas, forests, and other resources in various parts of the world, from Joint Forest Management in India (Poffenberger and McGean, 1996) to the implementation of aboriginal resource rights in the United States, Canada, New Zealand, and Australia.

Often there are legal reasons for instituting co-management arrangements, as in aboriginal land and resource claims (Singleton, 1998). But another reason for the growing interest is that effective resource management often requires partnerships to combine the strengths of government-level and local-level resource management and to mitigate the weaknesses of each (Pomeroy and Berkes, 1997). In some cases, as in the Philippines coastal fisheries, the development of co-management is related to the government's problems with enforcement (Pomeroy, 1995). Conflict resolution is another primary reason for co-management arrangements, as documented in a Costa Rican coastal national park (Weitzner and Fonseca Borras, 1999). This is consistent with McCay's observation (this volume:Chapter 11) that commons institutions often serve the purpose of conflict resolution.

Figure 9-1 shows the linkages in two co-management arrangements. The first (Figure 9-1a) is the Beverly-Qamanirjuaq Caribou Co-Management Board in northern Canada. Although this is not a co-management arrangement under land claims and not legally binding, it is a longstanding body (since 1982), and it is considered successful in resolving disputes and in enabling effective local input into what used to be a centrally managed resource (Kendrick, 2000). The second example (Figure 9-1b) is a formally legislated aboriginal land claims

TABLE 9-3 Characteristics of Some Institutional Forms That Enhance Cross-Scale Interplay

| Institutional Form | Vertical Linkages | Power Sharing | Area of Emphasis | Examples |
|---|---|---|---|---|
| Co-management | Local-level users with the government level | Formal power sharing in partnership | A mechanism to enable local-level users to participate in management | Aboriginal land claims agreements |
| Multistakeholder bodies | Multiple user groups and interests with the government level | Often advisory | Often a tool for public participation | Model Forest stakeholder groups; see Table 9-4 |
| Development empowerment co-management organizations | Often a three-way relationship with users, NGOs, and government agencies | Rarely formal power sharing | Social development, empowerment | Bangladesh fisheries; see Figure 9-2 |
| Citizen science | Local activist groups with government agencies | Information and policy partnerships but rarely formal power-sharing | Citizen activism for environmental management | Watershed associations in Minnesota |
| Policy communities | The local level with the regional and international | No formal power sharing | Solving regional problems, with local input | Epistemic communities in the Mediterranean Action Plan |
| Social movement World networks | Emphasis on horizontal linkages; some vertical linkages | No formal power sharing | North-South linkages to address impacts of higher level institutions | The Third Network and the World Trade Organization agreement on trade-related IPRs |

settlement, the James Bay and Northern Quebec Agreement (Berkes, 1989b). In the second example, co-management applies not only to one species, as in the caribou case, but to an area with all the resources in it. As Figure 9-1 illustrates, in both cases, the co-management arrangement provides vertical linkages, not only between the local level and the government, but also with regional and provincial governments as appropriate.

(a)   Beverly-Qamanirjuaq Caribou Co-Management Board

Federal government
(Canadian Wildlife Service)

Provincial and territorial governments
(Northwest Territories, Nunavut, Manitoba, Saskatchewan)

Local governments
(Dene and Inuit caribou user communities)

(b)   James Bay and Northern Quebec Agreement

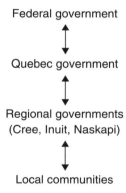

Federal government

Quebec government

Regional governments
(Cree, Inuit, Naskapi)

Local communities

FIGURE 9-1 Cross-scale linkages in co-management arrangements: (a) Beverly-Qamanirjuaq Caribou Co-Management Board, and (b) James Bay and Northern Quebec Agreement, Canada.

Figure 9-1 shows the outline of vertical arrangements but hides the details of the actual interactions involved in the cases, from the signing of the agreement to its implementation. A co-management *agreement* goes only part of the way to produce a viable arrangement. Simply put, there is little incentive for government agencies to share the power they hold (Lele, 2000). There are good reasons to be skeptical of all claims of successful co-management—at least, of easy successes.

Detailed studies, such as Singleton's (1998) on the Pacific Northwest salmon co-management arrangements following the Boldt decision of 1974, indicate that building trust between the parties may require a long time, on the order of a decade. Longer term studies, such as those by Singleton (1998) and Kendrick (2000), characterize co-management, not as an end point, but as a process of mutual social learning in which each side learns from and adjusts to the other over a period of time.

### Multistakeholder Bodies

A second and related form of cross-scale linkage is multistakeholder bodies. Characteristically, multistakeholder bodies link multiple user groups and interests, local and regional, with the government, and provide a forum for conflict resolution and negotiation among users. Table 9-4 provides a number of examples of stakeholder bodies. Some are established formally, as in the case of the Barbados, Norway, and U.S. examples. Some authors see stakeholder bodies, as compared to co-management arrangements with specific groups, as diffusing the powers to be shared. According to Murphree (1994), stakeholder groups "can easily transform interest into a conceptual collective by a vast and amorphous circle of stakeholders."

Many stakeholder bodies are ineffective for these reasons: They are too easy to set up; they can turn into "talkshops"; and they can be used by governments as a forum to sound out ideas or as a mechanism to defuse an imminent conflict, without conceding any real shared management power to the parties. There are other cases, however, in which multistakeholder bodies have made a significant impact on the way management is carried out, as in U.S. Regional Fishery Management Councils (McCay and Jentoft, 1996). There are yet other cases in which multistakeholder groups have legally defined powers of management, as in the Lofoten cod fishery in Norway (Jentoft, 1989). Multistakeholder bodies are not always easy to distinguish from co-management. For example, the Lofoten regime is usually described as co-management, even though management powers on the users' side is vested in a number of competing gear-groups and not in an institution that represents the fishers per se (Jentoft, 2000).

### Development, Empowerment, Co-Management Arrangements

This form of linkages seems distinct from the first two sets in terms of the emphasis on community development and empowerment, with co-management as an incidental outcome. These arrangements often involve nongovernmental organizations (NGOs) or other capacity-building bodies. Often there are horizontal as well as vertical cross-scale linkages. Figure 9-2 illustrates four different arrangements of communities, government agencies, and NGOs in a pilot project designed to empower fishing communities in Bangladesh to take over their own

TABLE 9-4 Examples of Multistakeholder Bodies

| Examples | Description |
| --- | --- |
| Committee on Resources and the Environment (CORE), British Columbia, Canada | CORE established several roundtables in the mid-1990s to act as advisory bodies to the environment minister in the planning for a diversity of forest uses, reflecting "full range of public values." Each roundtable had representation from some 20 user groups. |
| Manitoba Model Forest, Canada | One of 10 model forests across Canada (and similar to others in an international network), set up as a demonstration project for the sustainable use of a forested ecosystem; includes a multistakeholder group consisting of the various users and communities who live in the area. |
| Lofoten Cod Fishery, Norway | A co-management arrangement of long standing (Lofoten Act, 1895) in which the Norwegian government has devolved the fishery to the users (Jentoft, 1989). District committees of fishermen make yearly regulations and deal with user-group conflicts. Organized on gear-group representation and predominantly union based (Jentoft, 2000). |
| Barbados Fisheries Advisory Committee | A seven-member body set up by the Fisheries Act to advise the minister; it includes the various sectors of the fishing industry—fishermen, fish processors, boat owners, and fish vendors (McConney and Mahon, 1998). |
| U.S. Regional Fishery Management Councils | One of several regional bodies consisting of government officials and members of the public who reflect various fishery and coastal environmental interests. Charged with developing management plans for fisheries of the EEZ (McCay and Jentoft, 1996). |
| Great Barrier Reef Management Authority, Australia | The Great Barrier Reef Marine Park Act of 1975 established an authority that has the responsibility to seek out regional stakeholders to discuss management plans. Bodies representing the various uses of the reef, with priority going to those most dependent on the park's resources, assist with ecosystem-based management of the larger reef area (Kelleher, 1996). |

fishing licenses from the government, rather than working for license-holding middlemen (Ahmed et al., 1997).

In more than a decade of institutional experimentation with pilot projects in Bangladesh, four strategies could be recognized. In the government agency-led strategy, development assistance was channeled directly through the government body, Bangladesh Department of Fisheries (Figure 9-2a). However, it soon became apparent that long-term development work in the communities did not fit with the 3-year rotation of civil servants. Hence, the strategy changed after a few years in favor of an NGO-led approach. In some of the communities in the pilot project phase, the NGO played a go-between role (Figure 9-2b). In others, the

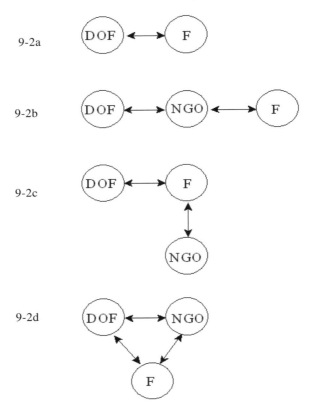

FIGURE 9-2 Different strategies in development, empowerment, and co-management arrangements from a project in Bangladesh (adapted from Ahmed et al., 1997).
NOTE: Bangladesh Department of Fisheries, DOF; fishing community, F; one of four Bangladesh nongovernmental organizations—Bangladesh Rural Advancement Committee (BRAC), Proshika, Caritas, Grameen—NGO.

community group interacted directly with the government; the NGO provided support for the community, with the potential of phasing itself out when the community became self-sufficient to conduct its own affairs (Figure 9-2c). In the government agency/NGO strategy, one field officer each from the government and the NGO worked jointly with the community, to give a three-way relationship for development and resource management until the NGO was ready to phase itself out (Figure 9-2d). These various strategies resulted in a rich variety of cross-scale linkages, including the vertical linking of the local level to the government level.

Is the Bangladesh case perhaps unique? Another long-term development case, this one from St. Lucia, West Indies, and involving a regional NGO that specializes in coastal resources and rural development, shows a great deal of similarity to the Bangladesh case despite differences in geographic area and in the nature of resource and development issues (Renard, 1994; Smith and Berkes, 1993). Of particular interest is the potential for the transmission of the development-empowerment experience horizontally from one group to another; from one area to another; from one country to another (in the Bangladesh/Grameen case); and from one resource sector (e.g., fisheries, forests, protected areas) to another (in the St.Lucia case).

## Citizen Science

A fourth class for cross-scale linkages is emerging institutions for what might be termed "citizen science." Examples include environmental stewardship groups in Canada (Lerner, 1993), regional associations for watersheds and lake water quality in Sweden (Olsson and Folke, 2001), watershed associations in Minnesota (Light, 1999), and People's Biodiversity Registers in India (Gadgil et al., 2000). As a class, citizen science is characterized by citizen activism for environmental management and by the involvement of environmental NGOs, and hence it differs in its primary focus from development-empowerment organizations.

Many citizen science cases come from industrially developed countries that have strong traditions of civil society and well-developed environmental movements. As a class, they tend to use a mix of scientific knowledge and local observations, as in the Swedish case. Although many of the citizen science examples come from Western societies, there are some notable exceptions.

In India, "people's science movements" have a history that goes back to the 1960s in the southern state of Kerala. In the 1970s, they took the form of alternative resource and environmental assessments with inputs from university scientists. Out of this emerged in the 1980s an activity called the village-level resource mapping program. People's Biodiversity Registers (PBR) is a program that developed in the mid-1990s in a number of states in India, involving hundreds of communities. It aims to document rural and forest-dwelling people's understanding of living organisms and their ecological setting, ongoing ecological change, their own development aspirations, and how they would like to see resources managed. The PBR program, using Participatory Rural Appraisal (PRA)-type methodologies and linking the local to the regional (and potentially to the national and the international), is probably the largest people's science movement (Gadgil et al., 2000).

## Policy Communities and Social Movement Networks

A number of institutions provide cross-scale linkages by connecting local issues with regional and international agencies. A relatively well-known set of institutions of this kind is what Haas (1992) has termed epistemic communities.

An example is the group of scientists, government experts, and NGO representatives who enabled the Mediterranean Action Plan. Members of such communities share principled beliefs, notions of validity, and policy goals that cut across political boundaries. Haas points out that the Mediterranean Action Plan brought together countries that are often in conflict, indicating that epistemic communities were significant in overriding such differences. More broadly, all policy issues bring together a "community" of players, from governments and other arenas. Thus, some scholars consider epistemic communities to be a subset of policy communities (Coleman and Perl, 1999). Others consider epistemic communities as unique, willful groups of individuals driven by their internalized beliefs about causation.

Auer (2000) pointed out that scholars in international relations have been investigating the environmental policy competencies of NGOs and intergovernmental organizations. Increasingly, nonstate actors, especially NGOs, are seen to be undertaking functions that states are either unwilling or unable to do. In addition to facilitating environmental negotiations between states, as in Haas' epistemic communities, NGOs can perform key information gathering, dissemination, advocacy, and appraisal functions, thus facilitating cross-scale linkages.

The international *Institutional Dimensions of Global Environmental Change* project science plan discusses institutions for linking the local and the regional in two areas of the world, Southeast Asia and the Arctic (Young, 1999). The arctic region includes cross-scale institutions such as the Arctic Council and the Inuit Circumpolar Conference (ICC), which connect the Inuit people of several countries, thus providing horizontal as well as vertical linkages.

Cross-scale institutions like the ICC may be characterized more properly as social movement networks, rather than as policy communities. Such networks create links between local institutions in the South (developing countries) and supportive groups in the North (industrialized countries). For example, the Third World Network (2001) consists of citizen groups in the developing world and supportive groups in the North involved in environment/development issues in which international institutions such as the World Trade Organization have local impacts. The Third World Network addresses issues such as the protection of intellectual property rights of farmers and other biodiversity users against the patenting of life forms.

## Collaborative Research and Management
## That Enable Cross-Scale Linkages

Research and researchers may have an impact on the institutions they study, especially if the approaches used tend to have a stimulating effect on cross-scale linkages. It may be useful to consider these collaborative research and management approaches (Blumenthal and Jannink, 2000) as a separate set because the emphasis is on a technique, rather than on a structure or an outcome, as in those in

Table 9-3. Table 9-5 lists four such approaches. Each has potential to provide linkages between the local level and the regional. Ecosystem-based management and adaptive management pay explicit attention to ecological-scale issues. Bioregionalism, which is a body of practice and not a collaborative methodology per se, is a special case of ecosystem-based management. It is of special interest

TABLE 9-5  Research and Management Approaches That Enable Cross-Scale Linkages

| Approach | Description |
|---|---|
| Ecosystem-based management or ecosystem management | Has come to include human uses of resources. The U.S. Forest Service adopted ecosystem management in 1992 as its official policy for managing national forests, and some other agencies followed suit. The policy came about mainly in response to increased spatial scales of management that require interagency and local landowner cooperation (Grumbine, 1994). However, how well ecosystem management may serve as an institution of cross-scale linkages is an open question. |
| Adaptive management | The scientific version of learning by doing. It uses the tools of systems modeling and iterative hypothesis testing, "adapting" management prescriptions by treating policies as hypotheses. Adaptive management typically focuses at the level of a local ecosystem. However, because different ecological interactions and resource use patterns occur at different scales, adaptive management, at least in the more recent applications, takes an explicitly cross-scale approach (Walters, 1986; Holling et al., 1998). |
| Participatory Rural Appraisal (PRA) | Derives from Rapid Rural Appraisal (RRA) and Agroecosystem Analysis (AEA), both in the area of development, and first appeared in the late 1970s (Chambers, 1994). All three methodologies help link up the scale of individual farms and villages with the regional scale of development. PRA is distinguished by its insistence on a grassroots, "farmer-first" approach, empowering decision making and application at the local level. |
| Participatory Action Research (PAR) | Similar to "action anthropology," shares with PRA the emphasis on the empowerment of users at the local scale. PAR places research and researchers at the service of the community; researchers help the community to carry out its own research agenda, in accordance to its own priorities and values (Chambers, 1994). |

because of its explicit emphasis on matching the scale of livelihood systems to that of the ecosystem in which the group lives.

Participatory Rural Appraisal (PRA) and Participatory Action Research (PAR) both focus on research that empowers local communities (Chambers, 1994). These two approaches have a great deal in common with development-empowerment organizations. In fact, many of the NGOs that operate in the development area use PRA techniques routinely in sharing information across scale. By contrast, under the rules of PAR, researchers are at the service of the community, no more, no less. There is no information-sharing mandate, nor is there a mandate for cross-scale interplay, except at the initiative of the communities themselves.

The eclectic list of institutional forms for cross-scale interaction covered in this section is no doubt incomplete. For example, where do "encompassing organizations" (McCay, this volume:Chapter 11) fit in? Different typologies may be constructed by others, perhaps related to different disciplinary perspectives in planning, sociology, anthropology, political science, development, and other areas. The main point here is that there is, in fact, a diversity of cross-scale institutional forms in existence. The task is not so much to refine this list, but to increase the size of the commons practitioner's tool kit by showing that "co-management" can be unpacked into a range of types of linkages and institutions. In this regard, the chapter parallels Tietenberg's (this volume:Chapter 6) effort to expand the notion of tradable permits into a range of tradable rights and institutions.

What is exciting about these developments is that nearly all of these cross-scale institutions are new. In the 1980s, there was a great deal of concern in commons circles about the demise of many commons institutions. Was it a matter of time before all local-level commons institutions were swept away by government management and inexorable open access a la Hardin? What we have found in the past two decades is that institutions are emerging at least as fast as others are disappearing, and that these include cross-scale institutions as well as local-level ones. However, we know precious little about this dynamic. Is diversity the source of creation? Is it institutional capital? There is a need for studies that focus on institutional aspects of cross-scale management. More systematic information is needed on co-management and other cross-scale institutions, their reasons for success and failure, institution building, capacity building, and the design of supportive policies.

## DYNAMICS AND SCALE IN CROSS-SCALE INTERACTIONS

What promising lines of inquiry are there for new research directions? In particular, what are some of the promising venues regarding scale and dynamics in researching cross-scale institutional linkages? As a way of introducing the importance of cross-scale institutional linkages, this chapter has reviewed the conditions under which the involvement of the state may facilitate or impede local management. It has also explored several institutional forms with the potential to improve cross-scale linkages, noting the fluid and diverse nature of these forms. In fact, a rich diversity of institutional forms exists for linking local-level or community-based institutions with those at the regional, national, or international levels. As shown by the Bangladesh fisheries example, these institutional forms are highly dynamic, changing from area to area and year to year.

In addition to management regimes involving these institutional forms, cross-scale linkages also may be enhanced through the use of certain research and management approaches. Of the four such approaches considered in Table 9-5, adaptive management is of particular interest because of its explicit attention to scale and dynamics and because of its potential as a tool for linking social systems and natural systems. This section develops the contention that the adaptive management approach, with a consideration of resilience, is useful for both the theory and practice of cross-scale linkages.

## Adaptive Management

Adaptive management was designed to integrate *uncertainty* into the decision-making process, and to ensure that policy makers and managers could *learn* from their successes as well as failures. As a resource management approach and planning tool, it was initially more technocratic than participatory (Holling, 1978). According to Lee (1999), it still "appears to be a 'top down' tool useful primarily when there is a unitary ruling interest able to choose hypotheses and test them." But because it emphasizes learning by doing, feedback relations, and adaptive processes, it has become a particularly promising approach to study the dynamics of systems, both natural and social. Initially concerned with the dynamics of ecosystems, adaptive management also has been applied to the study of the dynamics of linked social and natural systems (Berkes and Folke, 1998).

As used by Hilborn and Walters (1992), adaptive management requires the following six components: (1) identification of alternative hypotheses; (2) assessment of whether further steps are needed to estimate the expected value of additional information; (3) development of models for future learning about hypotheses; (4) identification of policy options; (5) development of performance criteria for comparing options; and (6) formal comparison of options. Together, these steps provide the tools to deal with uncertainty and lay a foundation for learning. We deal with each in turn.

Steps (1), (2), and (4) explicitly require the manager to integrate uncertainty into the management strategy. This is a distinct break from the notion that science can deliver the information needed for resource management, simply and unambiguously. Adaptive management assumes inherent uncertainty in ecosystems and recognizes the limits of knowledge. There are scientific uncertainties that are too expensive or time consuming to resolve, as well as others that are not resolvable due to the inherent uncertainty and unpredictability of nature (Wilson, this volume:Chapter 10).

The rationale for the consideration of uncertainty comes mainly from the recognition that natural systems and social systems are seldom linear and predictable, and from systems theory that emphasizes connectedness, context, and feedback. Processes in ecology, economics, and many other areas are dominated by nonlinear phenomena and an essential quality of uncertainty. These observations have led to the notion of *complexity*, developed through the work of many people and groups, notably the Santa Fe Institute (2001). In complex systems, small changes can magnify quickly and flip a system into one of many alternative paths. Such systems organize around one of several possible equilibrium states or attractors. When conditions change, the system's feedback loops tend to maintain its current state—up to a point. At a certain level of change in conditions (threshold), the system can change rapidly and catastrophically. Just when such a flip may occur and the state into which the system will change are rarely predictable (Holling, 1986).

Turning to the issue of learning, steps (3), (5), and (6) of adaptive management require that managers learn from the outcome of the decisions made. Adaptive management emphasizes learning by doing, and this is accomplished by treating policies as hypotheses and management as experiments from which managers can learn. Organizations and institutions can learn as individuals do, and hence adaptive management is based on social learning. Lee (1993) details such social learning based on the extensive experience with the Columbia River basin, a region full of cross-scale institutions. By emphasizing the interaction between management institutions and the biophysical system, Lee (1993) argues that one cannot expect to manage the environment unless one understands the effects of this interaction.

The goal of adaptive management is different from conventional management. In adaptive management, the goal is not to produce the highest biological or economic yield, but to understand the system and to learn more about uncertainties by probing the system. Feedback from management outcomes provides for corrections to avoid thresholds that may threaten the ecosystem and the social and economic system based on it. Thus, adaptive management depends on feedbacks from the environment in shaping policy, followed by further systematic experimentation to shape subsequent policy, and so on; the process is iterative (Holling, 1986; Holling et al., 1998).

Adaptive management is an understudied area in commons research, except perhaps in fisheries. Lee's (1993) work shows how the study of institutions and participatory processes can be combined with research on adaptive management. Many interdisciplinary scholars are looking for adaptive management-style alternatives to conventional scientific approaches in dealing with problems of complex systems. For example, in the area of sustainability, Kates et al. (2001) argue that "sustainability science must differ fundamentally from most science as we know it." The common sequential analytical phases of scientific inquiry, such as conceptualizing the problem, collecting data, developing theories, and applying the results, need to become, in the emerging sustainability science, parallel functions of social learning, adaptive management, and policy as experiment.

In particular, this new kind of science recognizes the need to act before scientific uncertainties can be resolved (Dietz and Stern, 1998). This is not only because it is difficult to get experts to agree on something but also because some uncertainties are not resolvable by science. Hence, as McCay (this volume:Chapter 11) suggests, it becomes important for commons management to design institutions and processes that bring scientists and resource users to work together. For example, the participation of fishers in decision making not only increases the likelihood that they "buy into" management decisions, but it also makes sure that the parties share the risk in decision making in an uncertain world, a much humbler role for the manager (Berkes et al., 2001).

## Resilience

Partnerships of managers and users do not resolve scientific uncertainties, but they help place those uncertainties in an institutional context that encourages building trust among parties, learning by doing, and developing the capacity to respond—in short, building resilient institutions. Resilience is a central idea in the application of adaptive management. It has three defining characteristics. Resilience is a measure of (1) the amount of change the system can undergo and still retain the same controls on function and structure; (2) the degree to which the system is capable of self-organization; and (3) the ability to build and increase the capacity for learning and adaptation (Resilience Alliance, 2001).

Resilience is an emergent property in complex systems terminology, that is, a property that cannot be predicted or understood simply by examining the system's parts. Resilience is a crucially important property of a system because the loss of resilience moves a system closer to a threshold, threatening to flip it from one equilibrium state to another. Just when the system will reach the threshold is difficult to predict; such changes constitute surprises or events which, even in hindsight, could not have been predicted (Holling, 1986). Conversely, increased resilience moves a system away from thresholds. Highly resilient systems can absorb stresses and perturbations without undergoing a flip; they are capable of self-organization and have the ability to build and increase capacity for learning and adaptation.

The idea of resilience has been applied mostly to ecosystem dynamics to study renewal cycles, equilibrium shifts, and adaptive processes in general. Use of the resilience idea is based on the assumption that cyclic change is an essential characteristic of all social and ecological systems. For example, resource crises (such as a forest fire) are important for the renewal of ecosystems. Such renewal occurs through an "adaptive cycle" often consisting of exploitation-conservation-release-reorganization phases. Adaptive cycles are driven by naturally occurring crises. If renewal is delayed or impeded, a larger and more damaging crisis eventually occurs, endangering the structure and function of the system and its ability for self-organization. For example, strict fire controls in forests and parks prevent renewal. They can result in the accumulation of fuel loads (leaf litter) on the forest floor, making the forest susceptible to "fires of an extent and cost never experienced before" (Holling, 1986:300), such as the fire that swept nearly half of Yellowstone National Park in 1988.

The resilience idea has been applied to linked social-ecological systems. In a study of environmental management in several large ecosystems, Gunderson et al. (1995) found a close coupling between ecosystem crises and crises in the governmental agencies in charge of managing them. In several of the cases, environmental crises led to institutional crises, as in Chesapeake Bay and Florida Everglades, and solutions were accompanied by institutional learning and renewal. How far can the link between ecosystems and institutions be pursued? Resource crises do not *always* lead to institutional crises and renewal, as in the case of the Newfoundland cod collapse (Finlayson and McCay, 1998). However, there is considerable evidence to support the idea that crises do play a useful role in some cases by triggering renewal and reorganization in both ecosystems and institutions, thus building resilience (Gunderson et al., 1995).

Such considerations can lead to new empirical and theoretical work on linkages between social and ecological systems, and on the question of what produces adaptive capacity in institutions. Levin et al. (1998) and Levin (1999) have emphasized two clusters of features that make for a resilient system. One is the presence of effective and tight feedback mechanisms or a coupling of stimulus and response in space and time. For example, it is relatively easy to get a neighborhood association to act on a problem. But as problems become broader in scale (e.g., regional air pollution), the feedback loops become looser and the motivation to act becomes less.

Creating appropriate incentive structures can be done by tightening cost and benefit feedback loops, for example, by assigning property rights. In some cases when the market can work properly and social costs are taken into account, privatization is an effective measure (Levin et al., 1998). In other cases, the transfer of communal property rights to local groups can be effective. For example, under the Joint Forest Management program, local controls and profit-sharing arrangements between government and villagers restored the productivity of previously degraded forest areas in West Bengal, India (Poffenberger and McGean, 1996). Similarly, the transfer of property rights to local groups has fostered wildlife conservation in parts of Africa (Murphree, 1994).

A second feature of a resilient system is the maintenance of heterogeneity, and the availability of a diversity of options for selection to act on as conditions change. The resilience of any complex adaptive system is embodied in the diversity of its components and their capacity for adaptive change. Heterogeneity helps maintain redundancy of function. Such redundancy would not be important if systems had one equilibrium state and conditions were relatively static. But often they are not. "Redundancy and heterogeneity are hand and glove; much redundancy is reflected, for example, in the heterogeneity within functional groups of species performing similar ecological roles" (Levin, 1999:202). What do these observations mean for social and institutional resilience?

The diversity of options idea is similar to Ostrom's (1992) institutional capital. It is consistent also with Adger's (2000) analysis of the resilience of institutions, and his emphasis on social capital, inclusivity of the institution, and the degree of development of trust relations among the parties. The heterogeneity/ redundancy idea brings an insight to the interpretation, for example, of the diversity of reef and lagoon tenure systems and other common property institutions observed in Oceania (Baines, 1989; Williams and Baines, 1993), and the folly of replacing such a diversity with a simple scientific resource management measure such as fisheries quotas (Wilson et al., 1994). Regarding cross-scale institutions, the insight from resilience and diversity is that it makes sense to continue to develop different kinds of co-management arrangements and other institutional forms. There is no such thing as an optimum arrangement that can be replicated everywhere.

Resilience thinking helps commons researchers to look beyond institutional forms, and ask instead questions regarding the adaptive capacity of social groups and their institutions to deal with stresses as a result of social, political, and environmental change. One way to approach this question is to look for informative case studies of change in social-ecological systems and to investigate how societies deal with change. From these cases, one can hope to gain insights regarding capacity building to adapt to change and, in turn, to shape change. These are, in fact, the objectives of a team project in progress (Folke and Berkes, 1998).

The resilience approach provides a promising entry point to move from static analysis of cross-scale linkages to the study of institutional dynamics. In highlighting change, it forces a reversal of the conventional equilibrium-centered thinking. As van der Leeuw (2000) puts it, rather than assuming stability and explaining change, one needs to assume change and explain stability. Adaptive management and resilience have been used to study the interactions of regional, national, and state-level agencies (Gunderson et al., 1995) and cross-scale interactions involving citizen participation in regional environmental management (Lee, 1993).

## CONCLUSION

The chapter began with a review and synthesis of the impacts of higher level institutions on local-level institutions, as a way of introducing the importance of vertical and horizontal linkages and detailing the variety of ways in which larger scale institutions can interfere with or, alternatively, support smaller scale ones. The second section pointed to some promising and emerging institutional forms for cross-scale linkages, concluding that co-management, as a catch-all term, is not adequate to encompass the full range of cross-linking institutions. The section emphasized that the consideration of institutional *forms* readily leads to the question of institutional *dynamics*. The third section raised the question of dynamics of institutions as a major subject area for future research, and made a case for the adaptive management approach, with a consideration of resilience, for building a theory of cross-scale institutional linkages.

Much of commons research brings together social sciences and natural sciences, and uses research methods and approaches from a variety of disciplines. But there is a need for tools to enable commons researchers to deal with people and environment as an integrated system. In particular, there is a need to study how institutions may respond to environmental feedbacks. The emphasis of adaptive management on feedback learning is important in this regard. As a key concept of adaptive management, resilience provides a window for the study of change, emphasizing learning, self-organization, and adaptive capacity. More work is needed on how societies and institutions develop ecological knowledge to deal with environmental change and, in turn, how they can act to shape change.

Emphasis on adaptive change and resilience is useful to deal with the dynamics of institutional change in relation to the dynamics of ecosystems and the goods and services they provide. Ecosystems generate natural resources and services (e.g., clean air and water) at multiple scales. But jurisdictional boundaries rarely coincide with ecosystem boundaries. Needed are cross-scale institutions that are in tune with the scales at which ecosystems function. The fact that there is often a mismatch in scale between institutions and ecosystems is considered part of the reason for resource mismanagement (Folke et al., 1998). Thus, a major task is to design cross-scale institutional linkages in a way that facilitates self-organization in cycles of change, enhances learning, and increases adaptive capacity.

Cases in the book, *Linking Social and Ecological Systems*, show that local-level institutions learn and develop the capability to respond to environmental feedbacks faster than do centralized agencies (Berkes and Folke, 1998). Thus, if management is too centralized, valuable information from the resource, in the form of feedbacks, may be delayed or lost because of the mismatch in scale. However, if management is too decentralized, then the feedback between the user groups of different resources, or between adjacent areas, may be lost. One way to tighten the feedback loops is to assign property rights to resources, thus creating incentives for sustainable resource use. The assigning of property rights may be a necessary condition but perhaps not a sufficient condition for sustainability.

Resource management systems cannot readily be scaled up or scaled down. As Young (1995) put it, "macro-scale systems are not merely small-scale systems writ large. Nor are micro-scale systems mere microcosms of large-scale systems." Because of the interactions between scales (e.g., the island nations fishery example), the appropriate level at which a commons issue should be addressed is never very clear. Instead of looking for the one "correct" scale for analytical purposes, it may be useful to start with the assumption that a given resource management system is multiscale, and that it should be managed at different scales simultaneously.

Such approaches are important in dealing with larger scale commons issues as well. In the area of global change, for example, researchers have started to address the question of match between multiscale institutions and ecosystems (Folke et al., 1998). These studies open up new areas of commons investigations by suggesting that the persistence of resource and environmental degradation may in part be related to cross-scale institutional pathologies, mismatches in scale, and lack of attention to cross-scale linkages.

# REFERENCES

Adger, W.N.
    2000    Social and ecological resilience: Are they related? *Progress in Human Geography* 24:347-364.
Ahmed, M., A.D. Capistrano, and M. Hossain
    1997    Experience of partnership models for the co-management of Bangladesh fisheries. *Fisheries Management and Ecology* 4:233-248.
Auer, M.
    2000    Who participates in global environmental governance? Partial answers from international relations theory. *Policy Sciences* 33:155-180.
Baines, G.B.K.
    1989    Traditional resource management in the Melanesian South Pacific: A development dilemma. Pp. 273-295 in *Common Property Resources*, F. Berkes, ed. London: Belhaven.
Berkes, F.
    1989a   Cooperation from the perspective of human ecology. Pp. 70-88 in *Common Property Resources*, F. Berkes, ed. London: Belhaven.
    1989b   Co-management and the James Bay Agreement. Pp. 189-208 in *Co-operative Management of Local Fisheries*, E. Pinkerton, ed. Vancouver: University of British Columbia Press.
    1999    *Sacred Ecology: Traditional Ecological Knowledge and Resource Management.* Philadelphia and London: Taylor & Francis.
Berkes, F., and C. Folke, eds.
    1998    *Linking Social and Ecological Systems. Management Practices and Social Mechanisms for Building Resilience.* Cambridge, Eng.: Cambridge University Press.
Berkes, F., P.J. George, and R.J. Preston
    1991    Co-management. *Alternatives* 18(2):12-18.
Berkes, F., R. Mahon, P. McConney, R.C. Pollnac, and R.S. Pomeroy
    2001    *Managing Small-Scale Fisheries: Alternative Directions and Methods.* Ottawa: International Development Research Centre.
Blumenthal, D., and J.L. Jannink
    2000    A Classification of Collaborative Management Methods. *Conservation Ecology* 4(2):13. Available: http://www.consecol.org/vol4/iss2/art13. [Accessed October 2001].
Chambers, R.
    1994    The origins and practice of participatory rural appraisal. *World Development* 22:953-969.

Coleman, W.D., and A. Perl
  1999  Internationalized policy environments and policy network analysis. *Political Studies* 47:691-709.
Dietz, T., and P.C. Stern
  1998  Science, values, and biodiversity. *BioScience* 48:441-444.
Drolet, C.A., A. Reed, M. Breton, and F. Berkes
  1987  Sharing wildlife management responsibilities with native groups: Case histories in Northern Quebec. Pp. 389-398 in *Transactions of the 52nd North American Wildlife and Natural Resources Conference.*
Finlayson, A.C., and B.J. McCay
  1998  Crossing the threshold of ecosystem resilience: The commercial extinction of northern cod. Pp. 311-337 in *Linking Social and Ecological Systems*, F. Berkes and C. Folke, eds. Cambridge, Eng.: Cambridge University Press.
Folke, C., and F. Berkes
  1998  *Understanding Dynamics of Ecosystem-Institution Linkages for Building Resilience.* Beijer Discussion Paper 112. Stockholm: Beijer International Institute of Ecological Economics.
Folke, C., L. Pritchard, Jr., F. Berkes, J. Colding, and U. Svedin
  1998  *The Problem of Fit Between Ecosystems and Institutions.* IHDP Working Paper 2. Bonn: International Human Dimensions Programme on Global Environmental Change.
Freeman, M.M.R.
  1989  Gaffs and graphs: A cautionary tale in the common property resource debate. Pp. 92-109 in *Common Property Resources*, F. Berkes, ed. London: Belhaven.
Gadgil, M., and R. Guha
  1992  *This Fissured Land. An Ecological History of India.* Delhi: Oxford University Press.
Gadgil, M., P.R. Seshagiri Rao, G. Utkarsh, P. Pramod, and A. Chhatre
  2000  New meanings for old knowledge: The People's Biodiversity Registers programme. *Ecological Applications* 10:1251-1262.
Grumbine, R.E.
  1994  What is ecosystem management? *Conservation Biology* 8:27-38.
Gunderson, L.H., C.S. Holling, and S.S. Light, eds.
  1995  *Barriers and Bridges to the Renewal of Ecosystems and Institutions.* New York: Columbia University Press.
Haas, P.M.
  1992  Introduction: Epistemic communities and international policy coordination. *International Organization* 46:1-35.
Hilborn, R., and C. Walters
  1992  *Quantitative Fisheries Stock Assessment: Choice, Dynamics and Uncertainty.* New York: Chapman and Hall.
Holling, C.S., ed.
  1978  *Adaptive Environmental Assessment and Management.* New York: Wiley.
Holling, C.S.
  1986  The resilience of terrestrial ecosystems: Local surprise and global change. Pp. 292-317 in *Sustainable Development of the Biosphere*, W.C. Clark and R.E. Munn, eds. Cambridge, Eng.: Cambridge University Press.
Holling, C.S., F. Berkes, and C. Folke
  1998  Science, sustainability and resource management. Pp. 342-362 in *Linking Social and Ecological Systems*, F. Berkes and C. Folke, eds. Cambridge, Eng.: Cambridge University Press.
Jentoft, S.
  1989  Fisheries co-management. *Marine Policy* 13:137-154.
  2000  The community: A missing link in fisheries management. *Marine Policy* 24:53-59.

Kates, R.W., W.C. Clark, and R. Corell, J.M. Hall, C.C. Jaeger, I. Lowe, J.J. McCarthy, H.J. Schellnhuber, B. Bolin, N.M. Dickson, S. Faucheux, G.C. Gallopin, A. Gruebler, B. Huntley, J. Jäger, N.S. Jodha, R.E. Kasperson, A. Mabogunje, P. Matson, H. Mooney, B. Moore III, T. O'Riordan, and U. Svedin
    2001    Sustainability Science. Statement of the Friibergh Workshop on Sustainability Science. *Science* 292: 641-642.

Kelleher, G.
    1996    Public participation on "The reef." *World Conservation* 2:19.

Kendrick, A.
    2000    Community perceptions of the Beverly-Qamanirjuaq Caribou Management Board. *Canadian Journal of Native Studies* 20:1-33.

Kropotkin, P.
    1914    *Mutual Aid: A Factor in Evolution.* Boston: Extending Horizons.

Kurien, J.
    1992    Ruining the commons and responses of the commoners. Pp. 221-258 in *Grassroots Environmental Action*, D. Ghai and J.M. Vivian, eds. London and New York: Routledge.

Lane, C.
    1992    The Barabig pastoralists of Tanzania: Sustainable land in jeopardy. Pp. 81-105 in *Grassroots Environmental Action*, D. Ghai and J.M. Vivian, eds. London and New York: Routledge.

Lee, K.N.
    1993    *Compass and Gyroscope: Integrating Science and Politics for the Environment.* Washington, DC: Island Press.
    1999    Appraising Adaptive Management. *Conservation Ecology* 3(2):3. Available: http://www.consecol.org/vol3/iss2/art3. [Accessed October 2001].

Lele, S.
    2000    Godsend, sleight of hand, or just muddling through: Joint water and forest management in India. Overseas Development Institute. *Natural Resource Perspectives* 53:1-6.

Lerner, S., ed.
    1993    *Environmental Stewardship: Studies in Active Earthkeeping.* Geography Publication Series No. 39. Waterloo, Ont.: University of Waterloo.

Levin, S.A.
    1999    *Fragile Dominion: Complexity and the Commons.* Reading, MA: Perseus Books.

Levin, S.A., and S. Barrett
    1998    Resilience in natural and socioeconomic systems. *Environment and Development Economics* 3:225-236.

Light, S., compiler
    1999    *Citizens, Science, Watershed Partnerships, and Sustainability in Minnesota: The Citizens' Science Project.* Minneapolis: Minnesota Department of Natural Resources.

McCay, B.J., and S. Jentoft
    1996    From the bottom up: Participatory issues in fisheries management. *Society and Natural Resources* 9:237-250.

McConney, P., and R. Mahon
    1998    Introducing fishery management planning to Barbados. *Ocean & Coastal Management* 39:189-195.

Messerschmidt, D.A.
    1993    *Common Forest Resource Management: Annotated Bibliography of Asia, Africa and Latin America.* Rome: Food and Agriculture Organization Community Forestry Note 11.

Murphree, M.
    1994    The role of institutions in community-based conservation. Pp. 403-447 in *Natural Connections: Perspectives in Community-based Conservation*, D. Western and R.M. Wright, eds. Washington, DC: Island Press.

National Round Table on the Environment and the Economy
    1998    *Sustainable Strategies for Oceans: A Co-Management Guide.* Ottawa: National Round
            Table on the Environment and the Economy.
Olsson, P., and C. Folke
    2001    Local ecological knowledge and institutions for ecosystem management of crayfish popu-
            lations: A case study of Lake Racken watershed, Sweden. *Ecosystems.*
Ostrom, E.
    1990    *Governing the Commons. The Evolution of Institutions for Collective Action.* Cambridge,
            Eng.: Cambridge University Press.
    1992    *Crafting Institutions for Self-Governing Irrigation Systems.* San Francisco: ICS Press.
Peters, P.
    1986    Concluding statement. Pp. 615-621 in National Research Council, *Proceedings of the
            Conference on Common Property Resource Management.* Washington, DC: National
            Academy Press.
Pinkerton, E., ed.
    1989    *Co-operative Management of Local Fisheries.* Vancouver: University of British Columbia
            Press.
Poffenberger, M., and B. McGean, eds.
    1996    *Village Voices, Forest Choices: Joint Forest Management in India.* Delhi: Oxford Univer-
            sity Press.
Pomeroy, R.S.
    1995    Community based and co-management institutions for sustainable coastal fisheries man-
            agement in southeast Asia. *Ocean & Coastal Management* 27:143-162.
Pomeroy, R.S., and F. Berkes
    1997    Two to tango: The role of government in fisheries co-management. *Marine Policy* 21:465-
            480.
Renard, Y.
    1994    *Community Participation in St. Lucia.* Washington, DC: Panos Institute, and Vieux Fort,
            St. Lucia: Caribbean Natural Resources Institute.
Renn, O., T. Webler, and P. Wiedemann
    1995    *Fairness and Competence in Citizen Participation: Evaluating Models for Environmental
            Discourse.* Dordrecht, Neth.: Kluwer.
Resilience Alliance
    2001    The Resilience Alliance: A Consortium linking ecology, economics, and social insights
            for sustainability. Available: http://www.resalliance.org/programdescription. [Accessed
            September, 2001].
Santa Fe Institute
    2001    Homepage. Available: http://www.santafe.edu [Accessed September, 2001].
Singleton, S.
    1998    *Constructing Cooperation: The Evolution of Institutions of Comanagement.* Ann Arbor:
            University of Michigan Press.
Smith, A.H., and F. Berkes
    1993    Community-based use of mangrove resources in St. Lucia. *International Journal of Envi-
            ronmental Studies* 43:123-131.
Smith, L.T.
    1999    *Decolonizing Methodologies. Research and Indigenous Peoples.* London: Zed Books.
Steins, N.A., V.M. Edwards, and N. Roling
    2000    Re-designed principles for CPR theory. *Common Property Resource Digest* 53:1-5.
Third World Network
    2001    Homepage. Available: http://www.twnside.org.sg/access.htm [Accessed September,
            2001].

van der Leeuw, S.E.
   2000   Land degradation as a socionatural process. Pp. 190-210 in *The Way the Wind Blows: Climate, History and Human Action*, R.J. McIntosh, J.A. Tainter, and S.K. McIntosh, eds. New York: Columbia University Press.
Walters, J.C.
   1986   *Adaptive Management of Renewable Resources.* New York: McGraw-Hill.
Weitzner, V., and M. Fonseca Borras
   1999   Cahuita, Limon, Costa Rica: From conflict to collaboration. Pp. 129-150 in *Cultivating Peace: Conflict and Collaboration in Natural Resource Management*, D. Buckles, ed. Ottawa: International Development Research Centre.
Williams, N.M., and G. Baines, eds.
   1993   *Traditional Ecological Knowledge: Wisdom for Sustainable Development.* Canberra: Centre for Resource and Environmental Studies, Australian National University.
Wilson, J.A., J.M. Acheson, M. Metcalfe, and P. Kleban
   1994   Chaos, complexity and communal management of fisheries. *Marine Policy* 18:291-305.
Young, O.
   1995   The problem of scale in human/environment relationships. Pp. 27-45 in *Local Commons and Global Interdependence*, R.O. Keohane and E. Ostrom, eds. London: Sage.
Young, O., ed.
   1999   *Institutional Dimensions of Global Environmental Change. Science Plan.* Bonn: International Human Dimensions Programme on Global Environmental Change.

# PART IV

# EMERGING ISSUES

The themes of change and emergence organize the three chapters in Part IV. First, change and emergence are central themes in each of the chapters. Second, the chapters reflect change in the character of theory of the commons and the emergence of new questions and new theoretical approaches. Thus they are not a dénouement in the sense of a resolution for the drama of the commons. But consider that "denouement" is derived from the French "to untangle." These chapters offer innovative perspectives that deconstruct some of the knottiest questions regarding the drama of the commons—How can we deal with complexity and uncertainty? How do the institutions that we use to manage commons emerge? What are the processes shaping cultures that manage or mismanage commons? These chapters don't answer these questions. They offer new ways of thinking about these issues, allowing us to see the threads that compose the knot and thus take the first steps toward unraveling it.

Wilson (Chapter 10) begins with a classic tale of mismanagement—the collapse of oceanic fisheries. This tragedy of the commons is so familiar that it might seem it doesn't bear retelling. But Wilson offers a fresh lens. He focuses on the scientific uncertainty that is inevitable when dealing with complex ecosystems and asks how that uncertainty interacts with the institutions for commons management to produce dire ends. Wilson's contribution is to suggest ways to think beyond the mismatch between scientific understanding and institutional dynamics. He suggests that we can learn to manage in the face of great uncertainty by developing institutions that are scaled to the systems they are intended to manage. Complex adaptive systems such as fisheries do not behave randomly even if they cannot be adequately modeled. Rather, drawing on notions from chaos and complexity theory, Wilson suggests that such systems shift from one state to another and that within each state, the behavior of the system is relatively predictable. The trick is to predict the state shifts. He argues that this can be accomplished by matching the span of control of management and monitoring institutions to the scope of the system. If the complex whole is composed of less complex parts, then the behavior of the system might be understood by developing institutions focused on the understandable parts rather than the impenetrable totality. As Wilson notes, we can't know in advance that any particular institutional arrangement will work. Indeed, presuming we know what will work is part of the hubris that led to the tragedy of the commons in marine fisheries. But he offers some design hypotheses that should prove useful in guiding both institutions and researchers.

McCay (Chapter 11) examines why commons institutions might emerge in some circumstances and not in others. She outlines the many conditions that must be in place for an institution to emerge and engage commons users. This provides a rich set of hypotheses for future research. In addition, she notes that in many cases, commons management institutions may emerge for reasons that are quite distinct from any desire to manage the commons in a sustainable way. She offers the provocative idea that groups may learn to manage commons more to minimize conflict than to conserve a resource. In the last half of the chapter, she examines theoretical stances in human ecology. Students of the commons should not take this as an academic exercise. McCay notes that researchers who go to the field with strong preconceptions about what they are studying and what might explain it may miss what is really happening. She acknowledges that theory and problem selection are essential parts of the research enterprise. But epistemological naïveté can lead to research designs that yield far less than could be obtained by deeper theoretical thinking at the outset. Ideas drawn from the economics of flexibility and event ecology provide the base for a more sophisticated approach to conceptualizing research problems around the commons.

Richerson, Boyd, and Paciotti (Chapter 12) draw on recent work in the Darwinian theory of cultural evolution to suggest why human groups may be able to manage commons and what the limits to such efforts may be. The approach they advocate, contra sociobiology, suggests that in organisms with culture, altruism may be quite common. Rather than the tragedy of the commons, we would expect a comedy of the commons in which people cooperate. But concern for the common good may not extend to all others—there is reason to believe that altruism may extend only to individuals perceived as members of the same social group. Culture determines who is "in" and who is "out." So the problem in designing institutions to manage commons is a problem of creating a shared definition of the "in" group and eliciting solidarity toward it. Like Wilson and McCay, the arguments in this chapter draw on some rather deep currents in contemporary theory. The ideas that emerge are not esoteric but provide guidance for both empirical research and institutional design.

These three chapters reflect the growing maturity of research on the commons. From the start, commons research was addressing core issues in social theory by examining the balance between altruism and self-interest and the alternative ways in which people might be rational. The rich theoretical and empirical work that followed allows even more complex questions about emergence, transformation, and dynamics to be raised. This entrains some theoretical approaches at the cutting edge of the social sciences: complexity theory, event ecology, and cultural evolutionary theory. Although these chapters are only the opening lines, they hold promise of a very exciting next act in the drama of the commons.

# 10

# Scientific Uncertainty, Complex Systems, and the Design of Common-Pool Institutions

*James Wilson*

This paper addresses the question of how we cope with scientific uncertainty in exploited, complex natural systems such as marine fisheries. Ocean ecosystems are complex and have been very difficult to manage, as evidenced by the collapses of many large-scale fisheries (Boreman et al. 1999; Ludwig et al., 1993; National Research Council, 1999). A large part of the problem arises from scientific uncertainty and our understanding of the nature of that uncertainty. The difficulty of the scientific problem in a complex, quickly changing, and highly adaptive environment such as the ocean should not be underestimated. It has created pervasive uncertainty that has been magnified by the strategic behavior of the various human interests who play in the game of fisheries management.

This paper argues that scientific uncertainty in complex systems creates a more difficult conservation problem than necessary because (1) we have built into our governing institutions a very particular and inappropriate scientific conception of the ocean that assumes much more control over natural processes than we might hope to have (i.e., we assume we are dealing with an analog of simple physical systems), and (2) the individual incentives that result from this fiction, even in the best of circumstances, are not aligned with social goals of sustain-

I would like to thank the many people who have commented on various drafts of this chapter. Spencer Apollonio, Jefferson White, Gisli Palsson, Teresa Johnson, Deirdre Gilbert, Yong Chen, Robin Alden, Ted Ames, Elinor Ostrom, William Brennan, Jennifer Brewer, and Carolyn Skinder have all made helpful comments and often have caused me to rethink and rework many of the ideas in the chapter.

ability. As a result, I believe we have slowed significantly the process of learning about the ocean, defined scientific uncertainty and precautionary acts in a way that may turn out to be highly risky, and created dysfunctional management institutions. This chapter suggests we are more likely to find ways to align individual incentives with ecosystem sustainability if we begin to view these systems as complex adaptive systems. This perspective alters especially our sense of the extent and kind of control we might exercise in these systems and, as a result, has strong implications for the kinds of individual rights and collective governance structures that might work.

## AN EXAMPLE FROM THE NEW ENGLAND FISHERIES

When ocean fisheries management began after World War II, practical scientific and political concerns dictated a large-scale, single-species approach to management. International fisheries management institutions were given very large geographical jurisdictions, few resources, and little real governance authority. Yet they were asked to develop regimes for the conservation of ocean resources. The scientific problem these institutions and the scientists working for them confronted was extraordinarily difficult, especially given the problems and costs of observation and the relatively undeveloped state of ecological theory at that time.

Consider how one might have started, at that time, to conceptualize a complex system that can be perceived only in the most indirect, costly, and occasional way. The fisheries scientists of that time chose a reductionist approach that emphasized sophisticated mathematical modeling of individual populations. It was consistent with scientific understanding of natural systems, with their (hoped-for) ability to measure and quantify, and with the authority given to the agencies for which they were working.[1] In particular, the conception was to concentrate on area- and species-specific populations (stocks) located within broadly identified fishing areas or ecosystems. The International Commission for the Northwest Atlantic Fisheries (ICNAF), for example, broke its enormous jurisdiction into numerous smaller, but still very large, statistical areas that were thought to correspond with major ecological or fishing areas, such as, Georges Bank, the Gulf of Maine, the Grand Banks of Newfoundland, the Scotia Shelf, and so on. Its scientific efforts concentrated almost exclusively on the commercial species of interest to the parties of ICNAF (Halliday and Pinhorn, 1990).

From both a scientific and institutional perspective, it is difficult to argue that these early approaches were "wrong," given the constraints and the complexity of ocean ecosystems. Nevertheless, a scientific pattern was established—a kind of intellectual path dependency that persists today.[2]

With the advent of extended national fisheries jurisdiction in 1977, both the United States and Canada adopted with almost no changes the single-species scientific perspective and scale of application that had developed under ICNAF.[3] In both countries, initial fisheries management plans were simply a continuation of a course that had been set by ICNAF. Even today the United States and Canada use the same statistical areas and definitions that were defined in the early 1950s. Except for refinements in statistical procedures, longer data series, the attention to some new species, and much more complete recording of fishing mortality, essentially the same methodology—certainly the same fundamental theory—is still used to assess the status of each stock and reach recommendations about acceptable levels of catch.

The most significant inheritance from the international era, however, was and is the scientific approach that simplifies the reality of complex ocean systems by treating each individual species as if it were an independent or isolated entity. The core of single-species theory is the belief that the future size of individual stocks is strongly related to spawning stock biomass, which, in turn, is strongly determined by how much fishing occurs. The relationship between fishing and spawning stock size is clear and easy to measure. But the theorized relationship between the spawning stock and recruitment is generally unknown and only claimed to exist for a few stocks, and then only at very low population sizes (Hall, 1988; Myers et al., 1995).[4] In spite of the absence of confirming evidence, fisheries scientists are firmly convinced that the sustainability of each population depends on the maintenance of an adequate spawning stock biomass.

Consequently, in the day-to-day management of fisheries, there is no attempt to predict recruitment. It is simply hoped, or assumed, that recruitment will proceed at a rate that is close to the average for some recent time period—one or two decades. Fisheries scientists advise managers about desirable catch rates, or amounts, in terms of what they estimate will produce the best yield from the year classes already in the water while maintaining a reasonable level of spawning stock biomass. There is an implicit but strong assumption that ecological interactions are minimal and not disturbed in any fundamental way by simultaneously fishing all or many species at moderate or even high rates. In addition, there are very difficult measurement and estimation problems. Errors of measurement on the order of 30 to 50 percent are common (Hilborn and Walters, 1992; Walters, 1998). As William Fox, science director of the National Marine Fisheries Service (NMFS), puts it, "there's a bit of experience involved, not something that can be repeated by another scientist. It's not really science; it's like an artist doing it—so a large part of your scientific advice comes from art" (Appell, 2001). Most fisheries scientists are reasonably well aware of the shortcomings of the theory and uncertainties regarding measurements and estimates of population size.

## THE RESPONSE TO UNCERTAINTY

When these uncertainties became apparent in the early years of extended jurisdiction, they were met by a few interested parties in the fishing industry with honest expressions of skepticism and, more commonly, with gaming strategies that reflected the interests and circumstances of various individuals and groups. The nonstrategic industry response came in the form of a rather inarticulate skepticism about the underlying theory concerning the relationship between the spawning stock and subsequent recruitment and about how best to conserve, or sustain, the resource (Smith, 1990). I do not believe this argument ever was recognized by government scientists simply because it was not contained within a formally stated doctrine (or maybe it was that "paradigmatic" talking past one another, or incomprehension, that Kuhn, 1962, discusses). Nevertheless, this argument was inextricably bound up with the industry's highly critical and strategic response to scientists' uncertainty about estimates of (changes in) stock sizes. These estimates are especially important to industry because they are the basis for short-term policy setting regarding allowable catches and other rules restraining fishing.

Furthermore, because the New England industry at that time was essentially an open-access industry, it had the usual tendency toward a strongly myopic perspective. Industry arguments tended to be supported by a large amount of anecdotal evidence. Almost without exception this evidence was marshaled to show economic hardship and to argue against biological estimates of scarcity and, of course, the need for reduced fishing efforts. Given the patchy nature of the resource and fishermen's finely honed skills at locating those patches, statements about localized abundance did not impress NMFS scientists, who were doing their best to carry out surveys based on stratified random sampling of the resource. Economic hardship arguments were simply interpreted as exaggerated claims that reflected the expected zero-profit state of the industry given open access.

However, members of the management council,[5] who were nearly all nonscientists, were influenced by both the biological and economic hardship arguments. They shared the values of those users or, at least, gave them credence and, as a result, did tend to discount or modify scientific advice in the direction of higher harvests or fewer restrictions. The results of council deliberations were almost always less restrictive, or at least different, regulations than those recommended by NMFS scientists. From the perspective of NMFS scientists, it was as if the council, when given a confidence limit around a recommended catch level, would always choose the higher end of that range rather than the average or an even more conservative level. According to those scientists, the council lacked the political will to act in a way that would conserve the stocks (Rosenberg et al., 1993).

NMFS and the environmental community became very frustrated at the council's unwillingness to act (or, at least, to act in the way they wanted).[6] They viewed the council's response to this uncertainty as a sure way to gradually, if not quickly, erode the stocks. NMFS officials, in either explicit or tacit agreement (with one another), appear to have decided that the relatively democratic processes of the council could not be relied on to achieve the greater good of conservation. Especially problematic was the council's perceived tendency to sacrifice biological restraint in order to solve politically important economic problems.

NMFS mounted a campaign to require the use of only quantitative data in council decision making, began to provide only point estimates of stock size and changes, and did its best to separate biological decisions from what were called allocative decisions (e.g. NOAA, 1986 [also known as the Calio report]; 1989 [602 guidelines and overfishing definitions]; Sustainable Fisheries Act [Public Law No. 104-297, 110 Stat. 355, 1996]. At the same time, the regulatory process increasingly became the object of court complaints in which NMFS was forced to defend its decisions (really its decisions to accept the advice of the councils). These challenges frequently questioned NMFS science (that is, estimates of changes in population size, not the basic theory) and were most easily met in court by thorough quantification of the basis for the decision. As a result, a strong bias seemed to enter into the choice of regulatory tools. Rules that were easily quantified were strongly preferred. Rules that were more difficult to quantify or that could not be analyzed easily within the context of the standard set of management models were not. For example, industry often proposed spawning area closures. Just as often NMFS opposed these suggestions with statements that no benefit could be shown or that "it doesn't matter when you kill the fish."

In short, every effort was made to insulate the regulatory process from the problems posed by scientific uncertainty. The preferred approach of NMFS and a number of environmental groups was to give experts (i.e., NMFS) control over biological objectives and the councils control over who got what—the allocation problem (NOAA, 1986). They hoped that through this approach, biological objectives would not be sacrificed even though it would leave the public (i.e., the councils) to engage in a dogfight over who got what.

This response to the political problems raised by scientific uncertainty is not uncommon; one has to assume that this policy approach was adopted in a good-faith attempt to promote the conservation and sustainability of our fisheries. After all, even if it was realized that current theory was inadequate, it was still the only theory—the only guidance—available, and given the perceived threats to the stocks and a perceived need to act, avoidance of a discussion of scientific uncertainty might have seemed justified.

However, given the inability to verify the core relationship in the theory, this kind of approach to the uncertainty problem carries unusual risks. Precautionary management steps taken on the assumption that the single-species "spawning stock/recruitment" line of causation is the operative long-term determinant of sustainability may turn out to be highly risky if other ecological factors (e.g., habitat, spatial distributions of local stocks, population behavior, trophic hierarchy, and so on, which tend to be ignored in the single-species scientific agenda) are determinative of species abundance. Under these circumstances, the usual prescription of single-species management—to fish moderately—still could lead to overfishing through the piece-by-piece loss of local stock spawning groups (Ames, 1998; Hutchings, 1996; Rose et al. 2000; Stephenson, 1998; Wroblewski, 1998; Wilson et al., 1999), through the destruction of essential habitat (Watling and Norse, 1996), through a gradual reduction in average trophic level (Pauly et al., 1998), and/or through the reduction or destruction of other ecological factors important to sustainability. In short, restraints appropriate to a single-species approach might simply perpetuate the problem. Taking uncertainty out of the public discussion may deprive us of the only defense we have against the even greater and more catastrophic uncertainty arising from an incomplete or incorrect understanding of the system. Removing uncertainty from the public discussion can be expected to retard our ability to learn, risk the credibility of science and the governance process on unproven theory, and most of all, diminish our long-term ability to conserve the resource (Rosa, 1998a).

The New England experience has been repeated in one form or another all around the globe. It is a problem that afflicts the advisory processes of the New England Council, but it has been just as difficult for the consultative processes of Canada and other countries (e.g., Finlayson, 1994). The problem this history raises is whether a democratic process or any collective process that gives serious weight to user input is capable of dealing with environmental uncertainty in a way that conserves resources. Or is it the case that the strategic response to uncertainty of the various individuals and groups and the resulting difficulty of building trust effectively forecloses successful negotiation of agreements concerning mutual restraint?

The argument of this chapter is that we can probably deal with uncertainty in an open democratic fashion, but that we have to be clear about the kind of uncertainty we face and the design of the institutions we build for dealing with that uncertainty. We can create institutions nicely tailored to a particular scientific theory and preconception of the nature of the uncertainty (we believe) we face, or we can design institutions on an alternative basis, one that assumes as little as possible about the nature of causal relationships and emphasizes the role of collective learning and institutional evolution. The appropriateness of one or the other approach would appear to depend on the state of our scientific knowledge or, alternatively, our ability to test and validate. The next sections of the chapter turn to a brief discussion of the view of uncertainty in a normal, reductionist scientific environment and how one's view of uncertainty changes in the context of a complex adaptive system.

## CONVENTIONAL VIEW OF UNCERTAINTY

As Pahl-Wostl (1995:196) writes, "Judged from a traditional point of view, uncertainty and the lack of predictive capabilities equal ignorance. Such thinking still pervades most scientific practice. It determines how knowledge is valued, what type of knowledge is required for decision making. It has shaped both scientific and political institutions. Such a view is inadequate to deal with the complexity of the environmental problems facing us today." Generally we think of three types of uncertainty in the study of natural systems (Walters, 1986:162). There is the uncertainty that arises from exogenous disturbances—noise. There is uncertainty about the values of system parameters, and, finally, there is uncertainty about system structure—sometimes called model uncertainty. A quantitative measure of the first two kinds of uncertainty, according to the American Heritage Dictionary, is simply "the estimated amount or percentage by which an observed or calculated value may differ from the true value." Implicit in this definition is the assumption that we know or believe we know the basic cause-and-effect relationships—the system structure—in the fishery or whatever we are studying.

In these circumstances, what stands for good science is the ability to detect relationships in what might otherwise appear to be noise and/or to narrow the uncertainty about our knowledge of the value of the parameters of the system. Normally, the smallest confidence interval around parameter estimates is generally believed to be the best science. It is through a continuing scientific process that we reduce or resolve parametric uncertainty. The instance of model uncertainty is also best addressed through a scientific process, but in this case one that consists of the discovery of causal relationships. Once that discovery occurs, the problem of uncertainty melds almost indistinguishably into the statistical process associated with parametric uncertainty.[7]

From the social point of view, uncertainty is not a desirable state of affairs but it is not especially problematic when science is in a position to learn rapidly. Repeated, consistently good predictions tend to validate the theory and to create trust and a willingness to invest in still more precise knowledge. Eventually, issues that previously might have been subject to strategic, self-interested argument (e.g., whether my steel or yours is better for use in a bridge) instead can be referred to experts for a disinterested (or public interested) decision. Normal peer review for quality control is generally a sufficient safeguard. In these circumstances, relatively insular expert-driven institutions operating under an umbrella of legislative objectives and standards are efficient and consistent with public interest. These are the kind of arrangements we generally make for building, bridge, auto, and pharmaceutical safety, among other things.[8]

The history of technological advance over the past 200 years illustrates the power of this method. But unlike civil engineering and the many other fields that have flourished using a reductionist approach, the sciences dealing with complex natural and human systems such as marine fisheries have not been able to develop a track record that generates broad social trust. Walters was (at least in 1986:162-163) very pessimistic about our ability to deal with these kinds of systems: "I doubt that there can, in principle, be any consensus about how to plan for the inevitable structural uncertainties that haunt us, any more than we can expect all human beings to agree on matters of risk taking in general."

## UNCERTAINTY IN COMPLEX ADAPTIVE SYSTEMS

The growth of understanding of complex adaptive systems in the past two decades suggests we may be dealing with ecological and human systems whose structure and dynamic behavior bear little resemblance to the equilibrium, single-species environment characterized by conventional resource theory. If we conceptualize fishery systems from the complex systems perspective, we are likely to approach the uncertainty (and the institutional design) problem in a way very different from the conventional.

In a Newtonian world, the stability of cause-and-effect relationships makes it possible to pursue reductionist science. This stability makes the observation and measurement of system relationships reliable and, more importantly, allows us to accumulate useful knowledge and to intervene in the system with predictable outcomes at whatever scale we find appropriate to our needs. As mentioned earlier, there is no doubt that many parts of our world fit this paradigm well. What is problematical about complex systems in this regard are their pervasive nonlinear, causal relationships (Holling, 1987). At any time a large number of factors may influence the outcome of a particular event, each one to a greater or lesser extent; at another time, the strength of those same causative factors on the same event may be very different. The result is a decline in predictability and/or often a shift in the scale or dimension of predictability (e.g., Levin, 1992; Costanza and Maxwell, 1994; Pahl-Wostl, 1995; Ulanowicz, 1997).

This happens simply because the relative intensity of causal relations in the system changes from time to time. Extreme examples are the regime shifts such as have occurred in response to fishing and/or environmental changes in many places around the world (e.g., Dickie and Valdivia, 1981 [Peru]; Boreman et al., 1999 [Grand Banks and Georges Bank]). Under these circumstances similar species may be present, but in such radically altered proportions that predictions based on extrapolations of past relationships would be far off the mark. Certainly, if one were in a position to compare the entirety of the two systems (before and after the shift) as if they were stable systems, one probably would find strong dissimilarities in the intensity and relative importance of the interactions among components.

Examples less extreme than regime shifts take place as the normal course of events in complex systems. Components in the system are continually adapting and evolving (not simply changing magnitude) in response to developments within the system itself (e.g., fishermen's response to a change in regulations, changes in the species distribution, or the driving forces in an economic system). Not only are we faced with ignorance about the strength of any particular causative relationship because of the pervasive nonlinearities of the system, but we can no longer be sure that a particular causative agent still enters the equation. These characteristics of complex adaptive systems clearly limit our ability to extrapolate on the basis of past system states and, consequently, the feasibility of prediction as usually defined from a reductionist scientific perspective (Pahl-Wostl, 1995). Recognition of the instability of the parameters of complex adaptive systems expands our understanding of the possible scope of our ignorance (Ulanowicz, 1997).

Nevertheless, there is perceptible order in these systems. This order can be understood and that understanding allows for the formation of a vision (a fuzzy prediction) of the future. Over time the order is exhibited in what many authors refer to as dynamic, or characteristic, patterns (Pahl-Wostl, 1995; Levin, 1999). I would describe this order as recurring similar patterns, never quite the same, sometimes startlingly novel because of the changing and adapting elements of the system, but also usually distinguishable from patterns in other systems (Holling, 1987).

Recognition of the patterns of change in a particular complex system can lead to an understanding of that system. That is, we can view patterns as historical events and understand the mechanisms that led to a particular outcome. But this understanding may provide us with the ability to predict in only the most qualitative ways—especially when we get beyond the immediate (inertial) term.

This characteristic of complex systems raises fundamental and difficult questions: How can we cope with or successfully intervene in ways that sustain the resources of these systems over the long run if we cannot predict the long-term consequences of our own actions? More importantly, how can we hope to make collective decisions in these circumstances? Won't honesty about our lack of knowledge lead to a situation in which groups or individuals can honestly question and oppose restraint because it is costly in the short run and with unproven benefits in the long run? In short, if we are in a world of complex systems, does the absence of predictability mean that we have no rational basis for making conservation decisions?

## LEARNING IN COMPLEX ADAPTIVE SYSTEMS

In complex ocean systems, learning the appropriate kind and extent of restraint required for sustainability is definitely a more difficult problem than one might be led to believe from a single-species theoretical perspective. Conventional resource management theory and practice is founded on the presumption that it is possible to simplify and predict fisheries systems at the scale of individual stocks using the same methods that have been applied so successfully to physical systems. If managers could predict in this way, even with wide confidence limits, they would be in a position to manipulate outcomes in the system. They would be able to create meaningful property rights and enter into implicit, or explicit, contracts with fishers (e.g., "If you harvest only $x$ amount today, then in the following year[s] there will be $y$ amount [plus or minus] available to harvest"). These contracts would tend to be enforceable because individual incentives would be aligned with social goals and, as a result, would tend to lead to sustainable resources (Scott, 1992). Unfortunately, this kind of straightforward quid pro quo, top-down, contractual methodology is likely to be effective only when we can quickly learn, predict, and control outcomes.

The lack of predictive ability in complex systems clearly impairs this kind of straightforward contractual methodology; nevertheless, because these systems can be understood in some sense, the basic economic idea of a valuable return to restraint remains viable. The key to understanding the appropriate kinds of restraint lies in the recognition of patterns.

Imagine a world of many possible system states that change from one to another in recognizable, but generally novel, patterns and each with different causative relationships. The system's propensity for one or another state, then, depends on the probability that a particular set of causative relationships with a particular set of values will appear at any point in time (Ulanowicz, 1997). In circumstances that are close in time or space, one might expect similarity of system states simply because of inertia. As time accumulates (or separating distance becomes greater), there is more scope for change in the circumstances of the system and less predictability. This does not mean the system in a particular place continues to diverge forever from its earlier state; it simply means that the set of possible system states changes.

Part of the reason for recurring patterns may be found in the differing response times (i.e., fast and slow) of the variables in the system (e.g. Simon, 1969; Allen and Starr, 1982; O'Neill et al., 1986; Holling, 1987).[9] For example, the highly fecund fish of the ocean can change their numbers dramatically over the course of a single spawning cycle. Other organisms in the system—sponges, corals—may exhibit changes of similar magnitude, but only over a much longer period of time. Generally aspects of the system that are slow to grow or develop or evolve—population age structures that include older animals, physical structures such as corals, tube worm colonies, learned and genetic behavioral aspects of populations such as migration routes and spawning sites—can be expected to constrain the faster elements in the system.[10] Put differently, the timing and flows of energy among the population components of the system are constrained by the attributes or structure of the slow or relatively constant components of the system.

If the values of these slow, longer term variables change, the set of possible system configurations changes as well; if the longer term variables remain relatively constant or nearly so, the short term is characterized by recurring configurations derived from a limited set of system states. Thus, one would expect a system in which long-term variables such as habitat and abiotic factors remained unchanged, to generate an always-changing set of similar system states (Pahl-Wostl, 1995). (Seasonal patterns, for example, are an obvious and easy pattern to discern.) It follows that destruction or erosion of long-term constraining variables, such as, habitat, trophic structure, and behavioral factors such as a learned migration pattern, would be expected to change the set of possible system states so that it includes states unlike those experienced previously and, consequently, reduces the ability to perceive patterns and learn.[11]

In his book, Emergence, Holland (1998) describes the learning process a computer[12] (and presumably humans) must go through to learn the game of checkers. He describes checkers as a very simple example of a complex adaptive system. Checkers has a limited number of pieces subject to a very few rules of movement, and its slow variables (the rules of the game, the size of the board, the kinds of pieces) are comfortably constant. Yet checkers is very difficult to predict and yields an immense number of possible board states. After only the first few moves of a game, it is unlikely that even an experienced player will encounter board configurations identical to those he's seen before. The state of the "system"—the configuration of the board—is nearly always novel, but patterns of configurations more or less similar to those experienced previously are likely. The train of causation in the system is not stable, varying with each configuration of the board. Feedback about one's interventions in the system is rarely clear. A "good" move can only be interpreted as such after the game has ended; it is entirely possible that a "double jump" might have led to the loss of a game or that a "poor" move might have set up a winning sequence. Looking ahead to try to predict the outcome of one among a set of alternative moves is an exercise that can yield only an ambiguous answer. So how do we learn to play checkers? Or in our case, how do we learn about the impact of human actions in the ecosystem?

As mentioned earlier, the fundamental basis for learning and prediction in this kind of environment is the recognition of patterns. Because of the multiplicity and novelty of board configurations, and especially because of the adaptive behavior of one's opponent, outcomes from any given decision cannot be expected to be the mean of outcomes of past similar situations. The adaptive behavior of the player's opponent introduces a strong tendency for surprise and unintended results, especially for a player with a naïve statistical strategy.

Holland (1998) describes a number of measures that help the player assess and evaluate the current configuration of the board (for example, simple measures such as "pieces ahead," "kings ahead," and "net penetration beyond center line"). The same set of measures can be used to assess the likely outcome of alternative moves the player faces. In other words, the player can think through the possible board configurations—two, three, or more moves ahead—that might arise from each alternative move. Conservative and generally more successful assessments assume the other player knows at least as much about the game as the player making the assessment. A kind of worst case precautionary principle applies. Some alternatives lead to clearly undesirable outcomes, others to outcomes that might be tolerable, and still others to outcomes that might improve the player's position in the game. These assessments constitute a set of alternative visions of the future and are the basis of the player's choice of moves. Decisions are in no way perfect and are especially dependent on the player's experience, but their imperfection is far less debilitating than, say, those facing a player who is using a statistical approach (and who is aware of his opponent's guile).

## From Checkers to Ecosystems

Holland's checkers game is very interesting and illuminating in its description of how one learns and especially how one develops a vision of the future when causal relations are not stable. But the premise that learning can take place in this way appears to be based on circumstances that—in the case of checkers—are relatively tractable. In particular, learning checkers appears to be eased by the existence of a limited number of system states or board configurations, the ability to construct relatively clear criteria for assessment of possible futures, and the player's ability to quickly and with low cost acquire experience with the system (including opponents).

Checkers, unlike an ecosystem, does not contain variables of differing time steps, which means that even though the number of board configurations can be very large, that set does not change. If the size of the board could change, or the rules governing the movement of pieces could mutate, checkers might become an extraordinarily difficult game to play well. One could learn to play well under one set of circumstances, and then a mutation of the rules governing movement of pieces or board size might erase or invalidate much of what had been learned to that point. Both the number of possible system states and the number of observations required to recognize patterns typical of the game (in both its new and old states together) would increase greatly. And learning would slow down.

In an ocean ecosystem, if one considers all possible population levels and parameter states, the likelihood of ever observing identical configurations of the system would appear to be rare. On the other hand, the possibility of observing similar, recognizable configurations if the long time step variables in the system are stable (e.g., climate, habitat, particular behavioral patterns) seems much higher. That is, if habitat and other relatively stable, long time step variables of the system remain in place over time, one might expect the system to have a strong propensity to settle into a set of configurations or patterns similar to those that have been observed in the recent past. For example, even though population numbers may be highly variable, the identity of the Gulf of Maine ecosystem is apparent to a fisherman or scientist who has worked there his whole life. Like the checkers player, fishermen learn to recognize system patterns and have some sort of vision of the future, including a hard-to-prove sense of what effects humans have on the system.

Equally difficult compared with checkers is the establishment of social goals. In an ecosystem neither the ultimate nor the proximate goal are clear. Both depend on the structure of rights and the process of governance. A typical open-access regime contains private rights that nearly always generate individual incentives for short-run, profit-maximizing objectives that have little to do with conservation. Other rights regimes are capable of setting more rational long-run goals but face formidable problems about how to achieve those goals. The simple assumption that resource property rights (of almost any sort) will lead to a collective interest in conservation is not obvious in a complex system, as is argued later in this chapter. In other words, unlike checkers, the goal of the game emerges from the rights structure, or rules, used to play the game. This makes the process of deciding what kinds of restraint are appropriate even more difficult.

Whatever management or governance regime is established, it is likely to arrive at a very imprecise vision of the future and of the ways human activity shapes the system. This limited vision of the future is not scientific in the usual use of the word, but it is far more valuable than a sense that the future is totally unpredictable and not subject to influence. For the individual fisherman, its special value lies in the fact that it limits the set of system states that might reasonably be expected to occur in the future (Palsson, 2000). Consequently, his current actions are not immobilized by a sense that the number of outcomes is huge and every outcome equally possible. For example, if one expects certain seasonal patterns, even though they may be strong or weak, late or early, and if one expects certain species to be present even though their abundance may be great or little, the limited set of possible futures represented by these expectations makes preparation for the eventualities of the future possible (Acheson, 1988; Wilson, 1990). If it were not possible to narrow the set of conceivable futures, current action would lack any rational basis unless it were totally myopic and reactive. This limited individual vision of the future is important because it leads to a sense of what kinds of collective restraint are required.

In short, a limited set of familiar system patterns permits the formation of individual visions of the future. This vision is the basis for forward-looking adaptive behavior (Palsson, 2000). It is the rational foundation for investment in both physical and human capital and, importantly, is the basis for restraint with regard to current harvests or harvest activity. These individual visions of the future are the ultimate basis for the construction of a social objective. Consequently, from this perspective, maintenance of familiar system patterns (i.e., of the conditions necessary for "normal" system configurations) becomes the principal objective of management. Maintaining the "old" structures in the system (subject to Holling's caveat) becomes the principal means to achieve that objective.

This is a very different view of the basis for restraint than that contained in conventional resource theory. Theories based on a presumption of full (or stochastic) knowledge of causal relationships almost invariably emphasize quantitative prescriptions involving the fast variables in the system (e.g., quotas for the amount of fish caught, number of boats allowed to harvest, and so on) and ignore or assume constant the slow variables in the system. On the other hand, an approach that emphasizes "familiar patterns" suggests a focus on policies designed to maintain those aspects of particular populations and other system components that are long term in nature (e.g., the age structure of populations, learned behavior for migrations, and spawning sites that might be destroyed by loss of local components of metapopulations, habitat, and so on).

The argument is that preservation of the long time step variables—the factors that determine the short-term configurations of the system—is where the emphasis on restraints should be placed because that is where feedback and predictability, such as they are, are available to us. This implies relatively constant rules changed only infrequently. It does not suggest a feverish chasing after the fast variables in the system in an attempt to fine tune. The other side of this same coin is the fisherman's sense that if current conditions in the system were different—that is, if the structure of long-term variables was different—the expected set of system states also would be unfamiliar and larger and would make learning about the system and economical adaptation to future system states much more difficult. This would reduce the rational basis for restraint and make it much more difficult to achieve a scientific understanding of the system.

In summary, this perspective from complex systems theory leaves us with a sense that we have a very modest, very short-term capability for prediction at the species level, and an even more modest ability to control outcomes at that level in complex systems. We clearly influence the system, but the specificity of outcomes (especially in terms of short time step variables such as recruitment changes in population size) resulting from our actions is likely to escape us. Nevertheless, we can develop imperfect visions of the future of the system, visions that put boundaries on the probable configurations of the system. There may be certain configurations of some elements of the system—the long time step variables—which, if we take steps to protect, we can expect to lead to strong propensities toward "typical" system states and patterns (i.e., states that we can learn to recognize through experience). These "typical" system states and patterns may be no "better" or "worse" than other alternatives in some intrinsic sense, but they have the advantage of being known and familiar. They allow us to learn and to form a vision of the future in spite of the great uncertainties in the system. This knowledge gives us the ability to adapt and provides the foundation for rational investment in the resource.

## COLLECTIVE LEARNING IN COMPLEX ADAPTIVE SYSTEMS

The complexity of these systems—their size, spatial distribution, multiple scales, large number of components, continuous change, and other factors—create circumstances in which no one individual or group could hope to adequately address the learning problem. The problem is a collective problem and, as such, is dependent on social organization and process. By collective learning, I mean simply the way we (collectively) accumulate observations of a phenomenon such as patterns in the ocean, the way we interpret and articulate those observations (convert them to knowledge), and the way we remember that knowledge. From a resource management perspective, the problem in a common-pool, complex system is learning enough to develop a convincing rationale for individual and collective restraint. This is as much a social problem as a scientific problem.[13] In fact, it is the difficulty of the collective learning in a complex environment that weaves the social and scientific problems into an inseparable matrix.

The social side of the problem has two closely related facets that are pertinent to the problem of collective learning. The first has to do with the institutions—especially the processes and the rights structures—that give rise to a rationale for stewardship and an incentive to learn. The second facet of the learning problem is the organization, or architecture, of those institutions. This second aspect is related most closely to an institutional attribute that Ostrom (1990) calls congruence. Overall this aspect of the problem has received much less attention than the others. Yet, given the complexity and associated uncertainty of these systems, it is critical to the social ability to efficiently acquire, analyze, and respond to changes in the system.

Nearly always, the literature on common-pool institutions assumes relatively complete (if stochastic) biological knowledge operating in a Newtonian world. This is most obvious in the economics literature, but it is also a pervasive assumption (even if unstated) in the other social science writings in this area. I don't think the fundamental outlook of either economics or the other social sciences is challenged by a complex systems approach, but the particular kinds of solutions—the institutions and so on—suggested vary dramatically. This is much more true for economics than for the other social sciences because economists tend to employ, and translate into policy, prescriptions derived from analytical models that emphasize optimizing or maximizing behavior on the basis of full, or nearly, full knowledge. Much of the literature on co-management tends to view at least the human environment as complex, and for that reason alone has tended away from the neat analytical conclusions of economists (e.g. Ostrom, 1990, 1997; Pinkerton, 1989; McCay, this volume:Chapter 11). I suggest that a complex systems approach provides a strong theoretical basis that is consistent with most of the important conclusions about the structure of rights and institutional organization contained in the co-management literature.

## Organization Question

To conceptualize this problem, I'll turn to the ideas about the organization of complex systems originally put forward by Simon (1962, 1996).[14] These ideas have been adopted by many others working in complex systems (e.g., O'Neill et al., 1986; Pattee, 1973) and are implicit in most of the aggregation schema used in economics and ecology.[15] They provide a fruitful conceptual foundation for addressing the collective learning problem or, what is nearly equivalent, the problem of organization of management institutions.

Simon proposed a compellingly simple generalization about the organization of complex systems—one that makes few assumptions about causal relationships. Namely, Simon proposed that these systems are organized hierarchically and partitioned into nearly decomposable (or independent) subsystems. The key element in Simon's scheme is the nearly decomposable subsystem. He defines the boundaries of such subsystems in terms of rates of interactions—within each subsystem, rates of interaction are high; between, rates of interaction are lower. In terms of the previous discussion, each subsystem might contain fast and slow variables (reflecting a hierarchy of process), but there is also a tendency for larger scale subsystems in the hierarchy to react more slowly than smaller scale subsystems.

In both natural and artificial systems, near decomposability creates a tendency for efficient use of information, robustness, and resilience (Simon, 1962). A complex computer program, for example, is intractable if organized as a seamless, tightly integrated whole. Even if one were able to construct a seamless program, any small change thereafter would be extraordinarily difficult to implement and any unanticipated bug would be nearly impossible to chase down if everything were connected to everything else. As a practical matter it is possible to conceive, construct, and debug a complex program only if that program is organized in a series of loosely connected, usually nested, nearly decomposable subroutines within which groups of highly interactive variables are brought together. This hierarchical structure with nearly independent components is not only a necessary conceptual tool for the program's creation, but also a functional aspect that affects its operating resiliency or stability.

All large natural and artificial systems face difficult problems of coordination nearly all of which are solved by finding ways to maintain the advantages of decomposability (Low et al., in press). Business organizations have to be broken up into divisions, each with considerable autonomy if they are to operate with even a modicum of efficiency. Here also the organizational problem is to group together activities with strong interactions and to tie them to the rest of the organization only when those activities impact or need to be coordinated with the activities of the rest of the firm. By doing this, the firm is able to assign particular decision-making responsibilities to that part of the firm with the most pertinent knowledge.[16] This allows the firm to better monitor for accountability, and reward on the basis of contribution to firm goals. Avoiding disharmonious incentives, such as might arise when responsibility is unclear and accountability difficult, is a major problem for firms because it has the potential to seriously attenuate intrafirm coordination and the achievement of firmwide goals (e.g. Hurwicz, 1972; Williamson, 1986; Demsetz, 1993; Rosen, 1993).

The federalist political system under which the United States is organized also creates large numbers of relatively independent local authorities such as towns, counties, states, and the national government, all neatly arranged in a spatially nested hierarchy. The U.S. Constitution, similar constitutions at the state level, and law govern interactions within this well-defined hierarchy. But they also govern many other specialized units and agreements whose purpose is to address interactions whose patterns of occurrence do not conform to the "normal" nested hierarchy. For example, states are part of the federal union but also members of various associations or agreements among states organized for particular purposes, such as the Atlantic States Marine Fisheries Commission. In all these systems the connections between units are generally loose but, on the whole, lead to coordinated activities (Ostrom, 1991).

An important benefit of this form of organization is that the scale of operation of each component of the organization can always be chosen (for efficiency or other reasons) so that it matches the scale of the activity in question, that is, the scale at which the impacts from an activity generate consequences (costs and benefits). So local activities are assigned to local authority, regional to regional, and so on. The other side of that same coin is that the governance of activities at a local scale that might generate costs for neighbors can always be shifted to a higher scale, where wider than local impacts can be handled. When activities do not interact along the neat lines of a spatially nested hierarchy, arrangements can be made for ad hoc components tailored to the structure of that particular problem. In economic terms this kind of polycentric organization is equivalent to the internalization of spatially related externalities, or if one prefers, to the minimization of the transaction costs necessary to resolve spatially relevant externalities.

These ideas apply to natural as well as social systems (O'Neill et al., 1986; Pattee, 1973; Walker, 1992, 1995). Simon and other authors describe ecosystems, the human body, and living organisms in general in terms of nearly decomposable subsystems. From this perspective a straightforward (i.e., simple hierarchical) view of an ocean ecosystem translates into a world of spatially discrete but not completely independent subsystems connected horizontally and aggregated into larger, nested subsystems.

This is consistent with the modern treatment of scale and space in ecology (e.g., MacArthur and Wilson, 1967; O'Neill et al., 1986; Levin, 1992, 1999; Hanski and Gilpin, 1997) and, as I'll describe, with our ability to organize a collective learning process. It is not consistent with the species-centered approach of conventional management. One of the principal reasons for suggesting this alternative conceptual approach is that the conventional approach does not lend itself to a practical way to manage ecosystems. In other words, when the complexity of the ocean is simplified by looking at individual species, we may blind ourselves to much of the feedback in the system. Just as important, an ecosystem conception based on a species-centered approach only makes sense if one could conceive of "modeling" all the biotic and abiotic interactions in the system. The massive impracticality of such an undertaking leaves one with, at best, ad hoc adjustments to the conventional approach (see, for example, National Research Council, 1999).

One might think of patches (or nearly decomposable subsystems) of biological activity at a fairly small scale, which are replicated in a similar, but generally novel way, at other locations at the same scale. Patches might be expected to arise because of heterogeneity in the environment. Bottom and coastal typography, currents, wind, and a host of other factors create areas of up-welling, windrows, eddies, and a variety of other features that tend to concentrate biological activity. Patches are separated in space by areas in which the density of organisms and the rate of interactions are relatively low. The flows (e.g., drift and migration) between these patches or subsystems define the phenomenon peculiar to the scale of the subsystem at the next, more aggregate layer in the hierarchy (Levin, 1999).

In other words, an aggregation, or clustering, of subsystems defines a larger scale subsystem. Changes in the composition of the organisms and other biological activity in subsystems (or patches) as well as the differences between subsystems at the same and different scales is the information that is read and interpreted as patterns. Furthermore, from a process-oriented perspective (i.e., observing the nonspecies specific energy flows), rates of interaction may vary considerably over the course of an annual cycle (e.g., photosynthesis), with certain functions such as herbivory and predation occurring at high rates only at those times of the year when migratory species find the local availability of nutrients or prey sufficient to be present (O'Neill et al., 1986).

The high rates of interaction within subsystems are important because they encompass a large part of the feedback about human and other perturbations (Levin, 1999; Levins, 1992). If we are to ever understand the patterns in the system and the kinds of restraint that are appropriate, we have to be able to capture this feedback. If each subsystem were completely independent of other subsystems, it would contain all possible feedback relevant to its own dynamics, even though that feedback might be a very ambiguous or difficult-to-decipher reflection of the patterns generated by the subsystem itself. But, because subsystems are connected to other subsystems, some feedback escapes from the local system (due to migration and drift). This "lost" feedback is potentially subject to capture at the next highest scale in the system, where it emerges as a separate aggregate phenomenon. For that capture to be meaningful, however, there must be some sort of cross-scale network that can acquire information about and make sense of the aggregate phenomenon. Consequently, to the extent that learning about the results of human interventions is possible, capturing the feedback at the various subsystem levels and between levels is necessary if we are to learn about the proximate results of interventions.

This strongly suggests the efficacy of a multiscale institution whose organization and activities parallel the organization and activities of the natural system. The fundamental rationale for this parallelism rests on the assumed nature of feedback within the natural system and the need within the social system to organize in a way that increases the likelihood of acquiring the information necessary for learning. The presumption is that when the "receptors" in the social system are aligned with feedback in the natural system, information costs are reduced, the possibilities for learning and adaptation are increased, and, of course, the ability to cope with uncertainty is strongly enhanced.

Resource mobility is usually one of the reasons cited for centralized approaches to ecosystem management. Nevertheless, mobility is one of the reasons the (simple) nested, or (multiple nested) polycentric, form of organization is important, especially from the collective learning perspective. Centralized approaches, as tend to be employed with single-species theory, obscure through aggregation and averaging a large part of the spatial and temporal behavior—the patterns—of the system. But the spatial and temporal incidence of events at a broad scale and their correlation with events in local subsystems are a large part of what we recognize as patterns. Even at the local scale aggregation probably obscures the source of many changes in the system (Holling, 1987; Levin, 1992). Current single-species attempts to manage over the range of the stock and to assess the status of the resource principally on the basis of aggregate measures of a very small part of the system—individual populations—essentially mask a large part of the local-aggregate patterns (of both populations and processes) one would expect to be relevant to an understanding of the system.

The problem of learning to recognize patterns is very much a problem of capturing system behavior and changes at a multitude of scales and locations. For scientific purposes it is often sufficient to isolate a particular scale of interest, holding everything higher in the hierarchy constant and treating the variations in lower level subsystems as noise around an average (Ahl and Allen, 1996; O'Neill et al., 1986; Simon, 1996). Resource management, however, does not have the luxury of attending to a single scale. To make the observations and conduct the analysis for management requires an information network spanning units at the same scale and reaching into units at higher and lower scales—a nested hierarchical structure or most probably a polycentric structure. Such a network is necessary to learn from local experience about local and nearby phenomena and is equally important for learning about the spatial and temporal attributes of phenomena that emerge at a larger than local scale. For example, the full extent of a migration pattern may only be observable at a particular large scale, but understanding its direction and timing (including especially exceptions to the general pattern) are often functions of more local phenomena, such as the availability of food. Similarly, understanding of local phenomena is clearly enhanced by knowledge of larger scale events. Thus, changes in aggregate phenomenon are better understood when combined with knowledge of the smaller scale factors from which they emerge, and smaller scale events are better understood in the context of the larger scale factors that contain them (Berkes, this volume:Chapter 9; O'Neill et al., 1986; Rosa 1998b; Young, this volume:Chapter 8).

This kind of organization has other important implications for the collective learning problem. Refer to the short list of learning difficulties presented earlier and consider a multiscale natural environment with many similar but not identical subsystems and a parallel human organization. First, we are likely to find that the slow rate at which we can gain experience with complex systems can be greatly accelerated if we can pool and compare observations of subsystems. Experiences in similar, proximate subsystems can be aggregated into a relevant collective experience applicable to the scale of those subsystems. This is probably as close as we can come to controlled experiments in these systems (Walters, 1986), but it is possible to learn a lot this way. Furthermore, we can usually accelerate the systemwide adoption of new rules or procedures by first adopting and tailoring them at the most applicable local level (i.e., one that encompassed all the costs and benefits of the change). In a heterogeneous environment, attempts to accomplish the same end might be completely stymied by the need to satisfy all local conditions simultaneously.[17]

An informative example is the way in which municipalities within a state, or states within the nation, actively compare and contrast one another's experiences in various realms. Experiments, new methods of doing something, the response to a natural or economic disturbance—whatever happens in one jurisdiction can be followed, modified, and applied in other jurisdictions. These information flows often do not occur within a simple nested hierarchy. Cities and states and, for that matter, all kinds of similar governing units tend to maintain collective organizations for the purpose of articulating their collective experience and developing new ways to operate. This information is then disseminated among members of the organization through publications, model legislation, personal discussions, conferences, and a variety of other networking activities (Levitt and March, 1995).

Importantly, the value of such information to a particular locality can only be assessed by someone (or a group) with a reasonably detailed knowledge, especially including a history, of changes in local factors. Model legislation, for example, is just that; it is usually constructed on the basis of collective (averaged) experience of a small group of early adopters, but with the expectation that it will be modified by localities so that it better fits local circumstances. Localities, in effect, introduce the lower scale "noise" necessary to tailor model legislation to the peculiarities of local circumstances. The same local peculiarities mean the collective value of numerous local experiments (i.e., their aggregate effect) can only be assessed with relatively particularized knowledge of local and aggregate circumstances.

The greater the number of relevant parties that can be brought into the deliberation, the greater the likelihood that common patterns can be identified with confidence even though the circumstances around each locality's experience may differ. Small numbers, of course, always leave open the strong possibility that unknown factors special to a locality may be responsible for a particular result and, thereby, the value of the collective knowledge that can be acquired from that experience might be diminished. Nevertheless, in circumstances where there are many similar, redundant local units, a locality can learn from the experiences of a small number of similar units to the extent that it understands those dissimilarities. It can compensate for perceived differences between itself and the other and can adapt its behavior (or an experiment) in ways that are thought to assure a better result (Dietz and Stern, 1998; Low et al., in press). Knowledge of local conditions can penetrate the ignorance embedded in averages.

Another way of looking at the learning advantages conferred by decentralization is in terms of the ability to avoid possibly persistent maladaptive policies. In the conventional view of the scientific process, little thought is given to this problem because it is assumed that the ability to validate theory or policies will select out ones that are maladaptive. But the ambiguity of evidence in complex systems seriously attenuates these selection pressures. As a result, as Gell-Mann (1994:296-305) points out, there is a tendency to substitute external criteria, ones that do not necessarily reflect the adaptive value of the policy. For example, in the absence of clear evidence one way or another, criteria appear that might select for policies that tend to reinforce the power of particular individuals or groups or an agency or a religious or scientific dogma. One could probably trace all sorts of organizational ills, from continuing ineffective policies to serious corruption, to this basic problem. Perhaps the only reasonable institutional response to this problem is to maintain independent (nearly decomposable) local governing units. Their ability to probe different policies and to remain skeptical without great cost is one of the few ways there might be to constraint persistent maladaptive policies, or viewed more positively, to assure the continuing evolution of the institution.

## Organization, Rights, and Incentives

Individual incentives are generally the most important factor in the transformation of organizational structure into outcomes (Williamson, 1986; Pfeffer, 1995). Incentives are important for rule compliance and stewardship, as is usually emphasized, but in the context of complex systems they have a particularly strong bearing on the collective learning problem and the feasibility of developing restraining rules. An organization capable of undertaking the kind of learning problem inherent in ecosystem management is, nevertheless, likely to be highly impaired if the individuals who comprise the organization do not have incentives consistent with the goals of the organization. The state, for example, can always (at great cost) use the threat of force to produce compliance with its rules; but there is little it can do, short of providing self-interested incentives, to produce forthcoming engagement in the processes of collective learning and rule development.

The formation of incentives depends on property or quasi-property rights (as is usually emphasized) and also on a set of circumstances that creates feedback and the ability to respond to that feedback (Hurwicz, 1972; Libecap, 1995). If the resource rights possessed by a decision maker provide no way to obtain feedback or means of control or influence over the resource, even if the interests of the rights holder are coincident with the long-term maintenance of the productivity of the resource, action producing that end is not likely to be forthcoming simply because the appropriate action cannot be identified. Because feedback in single-species theory is assumed to be straightforward and somewhat obvious (at least to social scientists), this problematic aspect of incentive formation, it seems, is usually assumed away without much thought.

From the perspective of a complex, multiscale system, however, the ability to detect, understand, and act on feedback in a way that reinforces a species-specific right is clearly a major problem. If, as argued here, we are likely to have only modest ability to control outcomes in these kinds of systems, then the best we may be able to do is assure the existence of the conditions necessary to produce familiar patterns in the system, that is, the maintenance and protection of the slow-growing structures in the system. In addition, the convenient analytical fiction of a single neurophysical system operating in a simple environment is also seriously misleading. Observations of multiple factors must be made at multiple scales and locations; the resulting information must be transferred to some sort of deliberative/analytical forum and then transformed into a decision to take action or impose restraint. This difficult process by itself is likely to impair the ability to adapt successfully.

From the perspective of individual incentives, this impaired and modest control argues strongly against species-specific rights. Such rights would provide no incentive to protect common resources, such as habitat, necessary for the sustainability of more than one species. Neither would any incentive exist to acquire or provide information that might contribute to the identification of system patterns, also a common resource. And most importantly, species-specific rights would create strong incentives against the creation of rules that might be eminently sensible from a system perspective but of negative or no value to someone holding species-specific rights, such as, restraints on harvesting that protect someone else's resource relevant habitat.

Consequently, broad rights are much more likely to generate an expectation of a beneficial return to restraint because they conform to the modest level of control that we can exercise. We may find it very difficult to predict species-specific outcomes in these systems, but it is fairly easy to be confident about broad outcomes. For example, temporarily or permanently closing an area to fishing is likely to lead to greater standing biomass in that area, but the quantitative composition of that biomass by particular species may be impossible to predict. A person with narrowly specified rights may or may not benefit from such a policy, depending on the particular outcome. A person with broad rights and the ability to adapt is almost certain to benefit regardless of the particularities and for that reason is much more likely to be predisposed to agree to restraining measures of this sort.

For example, in 1994 a large part of Georges Bank was closed for the purpose of restoring cod stocks. Surveys in 1999 revealed little recovery of cod but a bonanza of scallop growth (Murawski et al., 2000). Holders of rights to fish cod might be very skeptical of additional proposals to close areas because they realize they might never or only occasionally be on the winning end. Holders of rights to other fisheries, those that benefited in this instance, on reflection might come to the same conclusion because cod or dogfish or something entirely unexpected might bloom the next time. If, on the other hand, these fishermen had held rights that allowed them to exploit the systemwide benefits of closures (i.e., multiple-species rights), they would have had strong economic incentives to accept closures, and the experience of the closure on Georges Bank would have reinforced that incentive even more.

Fortunately, rights do not have to be predicated on particular cause-and-effect relationships. Agricultural land rights, for example, are not based upon any particular biological relationship; the rights are valuable because they allow the owner to employ any of a large number of known biological relationships. As climate, market, and known biological relationships change, the owner of agricultural land rights is free to adapt. What is important about the right is that a large class of phenomena about which we do and do not have knowledge of causal relationships and whose impacts are contained within the boundaries of the property are placed under the owner's potential control. There is, as a result, a strong incentive to learn and adapt in a way that is consistent with profitability (and presumably the social interest in conservation). Even lacking the kind of biological control that is possible with agricultural rights, broad rights in ocean ecosystems allow the owner to adapt to changes in the market and the environment. So long as the variation in the environment conforms with the kinds of patterns expected in the system, individuals can make the preparations (investments) necessary for successful adaptation.[18] This capability, combined with information flowing from the collective network, generates the individual's expectation that restraint is likely to be beneficial.

The problem, then, is that there is very little species-specific control, especially over the period in which abundance is dependent on recruitment. But this is the period that is relevant to sustainability. Broad rights that correspond with the dimension at which there is at least limited control—all species at the (sub)system level, or perhaps, the functional group—align individual incentives with the need to learn about and maintain ecosystem function. However, individual rights are not the only key to learning.

Individuals clearly learn from the experience of others and construct their view of the ecosystem through a complex interplay of their own and others' experiences at their "own" scale, and how that fits into the aggregate picture that is conveyed to them by individuals or by organizations operating at a larger scale (Michael, 1995; Parson and Clark, 1995). The organizational problem is to place individuals inside a network that is capable of generating appropriate feedback. For all the reasons discussed to this point, hierarchical and, by necessity, representative governance structures are most likely to be able to convey to individuals the collective experience at all scales in the system—that is, most likely to provide feedback about system patterns.

These same governance organizations also provide the mechanisms for attaching meaning to observations, for deliberation, and for taking ameliorative action. These capabilities are essential to the understanding of the ecosystem. They are capabilities that can be partitioned into their nearly decomposable tasks, but cannot be isolated from the system as a whole. In other words, given the mobility of resources in the system, the rights and the incentives of a person operating at a low level in the hierarchy are dependent on information generated at the same and at higher levels. Patterns at all scales and the efficacy of rules also at all scales are of interest to the individual.

In short, in a complex system, the creation of individual incentives that might lead to collective restraint involves the identification of system patterns, the formation of a broad, not narrowly specified, vision of the future, and the ability to adapt to that future. Given all the difficulties of learning discussed thus far a rights system that relies on only individual learning is likely to be untenable. The collective learning process has to be an enterprise whose organization parallels the structure of feedback in the system. The tight local coupling on the ecosystem side that Levin (1999) refers to has to be captured by tight local coupling on the social side. There have to be broad, relatively stable networks that link multiple localities. A collective deliberation facilitates and converts those deliberations into meaningful restraint or a process that can lead to meaningful restraint (Dietz, 1994; Dietz and Stern, 1998). Thus, a rights system that relies on only individual learning is likely to be untenable.

The individual's perception of the environment and the formation of his incentives are intimately dependent on this governance process. Inclusion within a stable network of discussion, being a part of the experience and analysis of a broad array of individuals, learning the likely response of others to changes in rules, and having a vote or substantial role in the decision process all contribute to the alignment of individual incentives. However, if this process is not organized so that it can capture feedback about the effect of human interventions, the incentives and the actual behavior of individuals and groups is not likely to lead to conservation. Externalities will persist.

On the other hand, to the extent that individual incentives can be aligned with the social goal of conservation (or sustainability), the state is relieved, by and large, of the need to rely on its police powers and threats of force in order to ensure individual behavior. Administrative and enforcement costs are reduced and the scope of feasible rules is expanded.[19] Most important, however, is the change in the kinds of information strategies individuals (and groups) find it in their interest to pursue. In the typical top-down administrative approach to management, individuals (or groups) rarely find it in their interests to be forthcoming with information. All sorts of exaggerations, games, lies, dissembling, and other behavior is encouraged because there is generally only a limited and costly ability for others to verify such (mis-)information and generally no penalty—and often a reward—for its introduction into the public process.

This kind of behavior always will be difficult to constrain in a complex environment; however, when management organization and resource rights are designed with the problem of learning in mind and actually lead to "tight local coupling" in the form of social networks, problems of information verification can be reduced and the costs of dissembling increased. Individual and collective learning can be encouraged. This increases the feasibility of conducting a constructive "analytical deliberation," arriving at a shared vision of the future and aligning individual incentives.

This kind of institutional arrangement, which I believe is principally consistent with decentralized, democratic governance, does not resolve scientific uncertainty but it does create a constructive environment in which the collective pursuit of useful knowledge can take place. This may appear to be a woefully complicated process, but it is nothing more than what we accomplish in our everyday governance. Society and the economy are extremely complex, multi-scale, rapidly changing systems in which we've learned to govern ourselves.

## SUMMARY

Finding ways to effectively restrain human activity in complex ecosystems has been very difficult. A large part of the problem arises from scientific uncertainty, which is often used as a pretext for not making hard political decisions for conservation. This chapter suggests we have wrongly characterized our knowledge of the natural environment and, consequently, have viewed the uncertainty and learning problem as if it were a typical engineering problem. As a result, we have created institutions and administrative procedures ill adapted to a solution of the conservation problem.

Usually we assume we are dealing with a classical Newtonian system in which cause-and-effect relationships are stable, or at least can be treated as if they were. In systems that truly conform to this assumption, the normal procedures of science can lead to understanding and reliable prediction. From the social point of view, repeated successful prediction generates trust even when there may be a lack of understanding among affected nonscientists. It also creates the circumstances for effective accountability and provides the rationale for reliance on expert-staffed institutions for the resolution of science-related problems.

Complex adaptive systems do not lend themselves to long-term prediction consistent with the needs of sustainability because of their changing, complex, and usually nonlinear causal relationships. We may be able to understand the structure and dynamics of these systems without being able to predict anything but broad patterns, or propensities, to use Ulanowicz' (1997) terminology. This is a fundamentally different and important characteristic when compared with Newtonian systems; it raises two closely related social problems: (1) How do we collectively learn what kinds of restraint will work when the time-honored reductionist process of "predict → test → learn → revise → and predict again" by which we hone our understanding cannot be followed; and (2) in this kind of environment, what kinds of institutions are necessary to best facilitate learning, accountability, and incentive alignment?

Holland (1998) suggests that learning in this kind of environment is based on the identification of recurring system patterns. The checker board game that he uses as an example of pattern learning is a relatively simple example of a complex adaptive system. It presents a limited and stable set of possible system states and patterns; the criteria for successful intervention in the system are fairly clear and the time and resource costs of learning are relatively low.

When this same learning problem is applied to ecosystems, especially those in which humans play an active or dominant role, such as fisheries, the complexity and extent of the environment transforms the learning problem. Patterns in this kind of system, I suggest, are best understood in terms of the differing time steps of variables in the system. The relative stability of slower changing variables, such as habitat, constrains and limits the range of patterns that appear in the more quickly changing aspects of the system, such as the size of populations. It may be possible to ameliorate, or minimize, the learning problem through policies meant to affect the range of patterns we encounter. However, we will always be faced with a multiscale system in which observation is costly, analysis is difficult, and prediction about specific results of our intervention in the environment is not possible. This is not the kind of environment in which it is easy to build an atmosphere of credibility and trust. For all these reasons, learning in this kind of environment is very much a collective enterprise that has to be mediated by institutions. The design of those institutions is important.

An institution's success in minimizing the cost and difficulty of observation and analysis depends principally on its ability to capture the feedback in the system it governs. To do this well, the organization of institutions must take on a hierarchical structure that reflects the patchy, multiscale hierarchical structure of the natural system. At each level in the hierarchy, institutions must be "positioned" so that their boundaries correspond as much as possible in terms of scale and location to the boundaries of strong interactions in the biological system. There must be connections (information flows) between locations at the same scale and between higher and lower scales as in the ecosystem.

The purpose of this parallelism is to align the "receptors" of the institution as much as possible with the spatial patterns of feedback in the system. In a situation with a crazy quilt of social boundaries that bear no resemblance to ecological boundaries, it might be possible to disaggregate and reaggregate observations in a way that made ecological sense, if analysis and observation were costless. However, noncongruent boundaries are much more likely to simply compound, or even confound, the learning process. A parallel structure, on the other hand, minimizes observational and analytical problems and, if across-scale and between-scale connections exist, provides for a flow of information that can be used to generate an understanding of processes at various scales and locations.

A very important—the dominant—aspect of the collective learning problem is the need to extend the process of learning down to the individual level. Individual incentives—and, importantly, the willingness to enter into restraining agreements—have to be based on a perception of a beneficial connection between restrained current actions and future states of the natural system. In a complex system, in which it is difficult to predict the future state of system components (e.g., species abundance), this would appear difficult to achieve. Nevertheless, so long as individuals are in a position to adapt to changes in system states, the connection between current and (expected) future states does not have to be mechanically precise. It is sufficient that the resulting (expected) future state(s) are positioned within the set of patterns that characterize the typical system and that individuals are in a position, technologically and legally, to adapt to those new states when they appear (i.e., not tied to the fate of particular species). Under these circumstances the probability of a positive economic outcome for the individual is very high and, as a result, so also is the rationality of entering into restraining agreements.

## NOTES

1 I have in mind here people like Schaefer, Gulland, Ricker, Cushing, Berverton, and Holt—scientists whose work during the 1950s and 1960s formulated the still-extant structure of fisheries population dynamics.

2 By path dependency I mean the tendency to become locked into a particular (in this case) theoretical approach (Waldrop, 1992). In this instance I would hypothesize that the inability to depart from a particular path stems from the great difficulty that attends any attempt to validate or invalidate theory in this area. Over time programs, data collection, equipment, careers, and legal authority all become more and more tailored to the approach; change becomes more difficult and the inability to validate obscures all but the most compelling reasons to change.

3 This is not too surprising when one realizes that the Canadian and U.S. scientists were the same people who had worked for ICNAF.

4 Just as this paper was being sent to the editor, I became aware of an article by Brodziak et al. (2001) that claims a stock relationship is discernable in 14 Georges Banks stocks.

5 The Fisheries Conservation and Management Act of 1977 established eight regional fisheries management councils that act as advisory bodies to NMFS. NMFS is located within the National Oceanic and Atmospheric Administration (NOAA) within the Department of Commerce. Council members are appointed by the Secretary of Commerce from a set of nominees supplied by governors of relevant states. Generally, there appears to be an attempt to appoint representatives of the major stakeholders. The regional councils appear to have more weight than the usual federal advisory committee. So long as their advice conforms reasonably with a set of national standards, NMFS/NOAA/ Commerce is more or less constrained to follow.

6 This interpretation is the result of my observations as a member, and sometimes chair, of the Scientific and Statistical Advisory Committee of the New England Fisheries Management Council.

7  See Rosa, 1998a for a thorough review of the study of risk.

8  However, I should note that two reviewers of this paper believe I am overly confident about these areas.

9  An earlier version of this perspective appears in Wilson et al. (1994) and Wilson et al. (1991). Also see Fogarty (1995) and Hilborn and Gunderson (1996) for disagreements with that perspective.

10  A problem I have with this terminology concerns the often asymmetric or episodic rates of change in many environmental variables. Biotic habitat may take a long time to build up and, once built, may persist for a long time (as might fit the definition of a slow variable), but it is also possible that that habitat could be destroyed by humans or a storm or internal dynamics in a very short time (Holling, 1987).

11  But Holling (1973, 1987) and Gunderson et al. (1995) argue that the accumulation of energy in older age structures (e.g., old-growth forests, woody scrub lands) can set the stage for dramatic system shifts through processes such as fire, suggesting the familiar is not easy to attain for long.

12  See also Samuel (1959).

13  The discussion that follows puts the emphasis on the social organization problem rather than the scientific problem. This does not mean the scientific problem is not important; it is simply not my first interest here.

14  It may be appropriate to attribute these ideas solely to Simon. See Pattee (1973) and O'Neill et al. (1986) and even the Federalist Papers (V. Ostrom, 1991). Interestingly, much of the work in corporate learning also traces back to Simon and Barnard, as does work on bounded rationality (see Williamson, 1995). Perhaps these questions are inevitable once one starts looking at the world as if it was a complex, adapting system rather than a stable clockwork mechanism.

15  They do not, however, conform to the aggregation from species to system implicit in species-centered population approaches such as used conventionally in fisheries management. I believe it is generally recognized that aggregation to the system from a species base presents intractable measurement and modeling problems.

16  Significantly, these arguments have little weight in circumstances where production is completely routine. The lack of change in local situations means it is possible for central authorities to acquire the knowledge necessary to direct and control such operations (Williamson, 1986).

17  For approximately 25 years, the legislature of the state of Maine tried unsuccessfully to adopt statewide trap (or effort) limits for the lobster fishery. Then the legislature created seven local jurisdictions, giving each limited local powers. Within a year of their creation, each jurisdiction adopted a trap (effort) limit.

18  This assumes a system in which there is considerable niche overlap, compensation among species, a relatively stable system energy input, and broad acceptance of species in the market (Wilson, 1990).

19  When willing compliance is low, only those rules that are enforceable with low cost to the state are feasible. These rules are not by any means likely to be the best rules for achieving conservation.

# REFERENCES

Acheson, J.M.
   1988  Patterns of gear changes in the Maine fishing industry. *Maritime Anthropological Studies* 1:49-65
Ahl, V., and T.F.H. Allen
   1996  Hierarchy Theory: A Vision, Vocabulary, and Epistemology. New York: Columbia University Press.
Allen, T.F.H., and T.B. Starr
   1982  *Hierarchy.* Chicago: University of Chicago Press.

Ames, E.
  1998    Cod and haddock spawning grounds in the gulf of Maine. In *The Implications of Localized Fishery Stocks*, I. Hunt von Herbing, I. Kornfield, M. Tupper, and J. Wilson, eds. New York: Natural Resource, Agriculture, and Engineering Service.

Appell, D.
  2001    The New Uncertainty Principle: For complex environmental issues, science learns to take a backseat to political precaution. *Scientific American* 284:18-19.

Boreman, J., B.S. Nakashima, J.A. Wilson, and R.L. Kendall, eds.
  1999    Northwest Atlantic Groundfish: Perspectives on a Fishery Collaspse. Bethesda, MD: American Fisheries Society.

Brodziak, J.K.T., W.J. Overholtz, and P.J. Rago
  2001    Does spawning stock affect recruitment of New England groundfish? *Canadian Journal of Fishery and Aquatic Sciences* 58: 306-318.

Costanza, R., and T. Maxwell
  1994    Resolution and predictability: an approach to the scaling problem. *Landscape Ecology* 9: 47-57.

Demsetz, H.
  1993    The theory of the firm revisited. In *The Nature of the Firm: Origins, Evolution, and Development*, O.E. Williamson and S.G. Winter, eds. New York: Oxford University Press.

Dickie, L.M., and J.E. Valdivia G.
  1981    Investigations cooperativa de la anchoveta y su ecosistema (ICANE) between Peru and Canada: A summary report. *Boletin Instituto del Mar del Peru* Vol. Extraordinario: XIII-XXIII.

Dietz, T.
  1994    What should we do? Human ecology and collective decision making. Human Ecology Review 1:301-309.

Dietz, T., and P.C. Stern
  1998    Science, values and biodiversity. *Bioscience* 48(6):441-444.

Finlayson, A.C.
  1994    *Fishing for Truth: a sociological analysis of northern cod stock assessments from 1997 to 1990.* St. Johns, Nfld.: Institute of Social and Economic Research, Memorial University of Newfoundland.

Fogarty, M.
  1995    Chaos, complexity, and community management of fisheries: an appraisal. *Marine Policy*. 19:437-444.

Gell-Mann, M.
  1994    *The Quark and the Jaguar.* New York: W.H. Freeman and Company.

Gunderson, L.H., C.S. Holling, and S.S. Light
  1995    Barriers and Bridges to the Renewal of Ecosystems and Institutions. New York: Columbia University Press.

Hall, C.A.S.
  1988    An assessment of several of the historically most influential theoretical models used in ecology and of the data provided in their support. *Ecological Modelling* 43:5-31.

Halliday, R.G., and A.T. Pinhorn
  1990    The delimitation of fishing areas in the northwest Atlantic. *Journal of Northwest Atlantic Fishery Science* 10:1-51.

Hanski, I.A., and M.E. Gilpin
  1997    Metapopulation Biology: Ecology, Genetics, and Evolution. San Diego: Academic Press.

Hilborn, R., and D. Gunderson
  1996    Chaos and paradigms for fisheries management. *Marine Policy* 20:87-89

Hilborn, R., and C.J. Walters
   1992   *Quantitative Fisheries Stock Assessment: Choice, Dynamics, and Uncertainty.* New York: Chapman and Hall.
Holland, J.
   1998   *Emergence.* Cambridge, Eng.: Perseus Books.
Holling, C.S.
   1973   Resilience and stability of ecological systems. *Annual Review of Ecology and Systematics* 4:1-23.
   1987   Simplifying the complex: The paradigms of ecological function and structure. *European Journal of Operational Research* 30:139-146.
Hurwicz, L.
   1972   On informationally decentralized systems. Pp. 297-336 in *Decision and Organization*, C.B. McGuire and R. Radner, eds. Amsterdam: North-Holland Publishing Company.
Hutchings, J.A.
   1996   Spatial and temporal variation in the density of northern cod and a review of hypotheses for the stock's collapse. *Canadian Journal of Fishery and Aquatic Sciences* 53:943-962.
Kuhn, T.S.
   1962   *The Structure of Scientific Revolutions.* Chicago: University of Chicago Press.
Levin, S.
   1992   The problem of pattern and scale in ecology. *Ecology* 73:1943-1967.
   1999   *Fragile Dominion.* Cambridge, Eng.: Perseus Books.
Levins, R.
   1992   Evolutionary Ecology Looks at Environmentalism. Unpublished paper delivered at the Symposium on Science, Reason and Modern Democracy, Michigan State University, East Lansing, MI, May 1.
Levitt, B. and J.G. March.
   1995   Chester I. Barnard and the Intelligence of Learning. Pp. 11-37 in *Organization Theory: From Chester Barnard to the Present and Beyond.* O. Williamson, ed. New York: Oxford Univ. Press.
Libecap, G.
   1995   The conditions for successful collective action. In *Local Commons and Global Interdependence: Heterogeneity and Cooperation in Two Domains*, R. Keohane and E. Ostrom, eds. London: Sage.
Low, B. E. Ostrom, C. Simon, and J. Wilson
   in      Redundancy and diversity: Do they influence optimal management? In Fikret Berkes,
   press   Johan Colding, and Carol Folke (Eds.), *Navigating Nature's Dynamics: Building Resilience for Adaptive Capacity in Social Ecological Systems.* Cambridge, UK: Cambridge University Press.
Ludwig, D., R. Hilborn, and C.J. Walters
   1993   Uncertainty, resource exploitation, and conservation: Lessons from history. *Science* 260:17, 36.
MacArthur, R.H., and E.O. Wilson
   1967   *The Theory of Island Biogeography.* Princeton: Princeton University Press.
Michael, D.N.
   1995   Barriers and bridges to learning in a turbulent human ecology. Pp. 461-485 in *Barriers and Bridges to the Renewal of Ecosystems and Institutions*, L.H. Gunderson, C.S. Holling, and S.S. Light, eds. New York: Columbia University Press.
Myers, R.A., N.J. Barrowman, J.A. Hutchings, and A.A. Rosenberg
   1995   Population dynamics of exploited fish stocks at low population levels. *Science* 269: 1106-1108.

Murawski, S.A., R. Brown, and L. Hendrickson
    2000   Large-scale closed areas as a fishery-management tool in temperate marine systems: The Georges Bank experience. *Bulletin of Marine Science* 66:775-798.
National Oceanic and Atmospheric Administration (NOAA)
    1986   *Fishery Management Study.* Washington, DC: U.S. Department of Commerce.
    1989   50 CFR Part 602, Guidelines for the Preparation of Fishery Management Plans Under the FCMA. Washington, DC: U.S. Department of Commerce.
National Research Council
    1999   *Sustaining Marine Fisheries.* Committee on Ecosystem Management for Sustainable Marine Fisheries. Ocean Studies Board. Commission on Geosciences, Environment, and Resources. Washington, DC: National Academy Press.
O'Neill, R.V., D.L. DeAngelis, J.B. Waide, and T.F.H. Allen
    1986   *A Hierarchical Concept of Ecosystems.* Princeton: Princeton University Press.
Ostrom, E.
    1990   *Governing the Commons: The Evolution of Institutions for Collective Action.* New York: Cambridge University Press.
    1997   A Behavioral Approach to the Rational Choice Theory of Collective Action. Presidential Address to the American Political Science Association annual meetings, August 28-31.
Ostrom, V.
    1991   Polycentricity: The structural basis of self-governing systems. In *The Meaning of American Federalism: Constituting a Self-Governing Society*, V. Ostrom, ed. San Francisco: ICS Press.
Pahl-Wostl, C.
    1995   *The Dynamic Nature of Ecosystems: Chaos and Order Entwined.* Chichester, Eng.: John Wiley & Sons.
Palsson, G.
    2000   "Finding one's sea legs": Learning, the process of enskilment, and integrating fishers and their knowledge into fisheries and science and management. In *Finding Our Sea Legs: Linking Fishery People and Their Knowledge with Science and Management*, B. Neis and L. Felt, eds. St. John's, Newfoundland: Institute for Social and Economic Research Press.
Parson, E.A., and W.C. Clark
    1995   Sustainable development as social learning: Theoretical perspectives and practical challenges for the design of a research program. In *Barriers and Bridges to the Renewal of Ecosystems and Institutions*, Gunderson, L.H., C.S. Holling, and S.S. Light, eds. New York: Columbia University Press.
Pattee, H.H.
    1973   *Hierarchy Theory: The Challenge of Complex Systems.* New York: George Braziller.
Pauly, D., V. Christensen, J. Dalsgaard, R. Froeese, and F. Torres, Jr.
    1998   Fishing down marine foodwebs. *Science* 279:860-863.
Pfeffer, J.
    1995   Incentives in organizations: The importance of social relations. Pp. 72-97 in *Organization Theory: From Chester Barnard to the Present and Beyond*, O.E. Williamson, ed. New York: Oxford University Press.
Pinkerton, E.
    1989   *Co-operative Management of Local Fisheries: New Directions for Improved Management and Community Development.* Vancouver: University of British Columbia Press.
Rosa, E.A.
    1998a  Metatheoretical foundations for post-normal risk. *Journal of Risk Research* 1:15-44.
    1998b  Comments on commentary by Ravetz and Funtowicz: 'Old-fashioned hypertext'. *Journal of Risk Research* 1:111-115.

Rose, G.A., B. DeYoung, D.W. Kulka, S.V. Goddard, and G.L. Fletcher
    2000   Distribution shifts and overfishing the northern cod (Gadus morhua): A review from the ocean. *Canadian Journal of Fishery and Aquatic Sciences* 57:644-664.
Rosen, S.
    1993   Transactions costs and internal labor markets. In *The Nature of the Firm: Origins, Evolution, and Development*, O.E. Williamson and S.G. Winter, eds. New York: Oxford University Press.
Rosenberg, A.A., M.J. Fogarty, M.P. Sissenwine, J.R. Beddington, and J.G. Shepherd
    1993   Achieving sustainable use of renewable resources. *Science* 262:828-829.
Samuel, A.L.
    1959   Some studies in machine learning using the game of checkers. In *Computers and Thought*, E.A. Feigenbaum and J. Feldman, eds. New York: McGraw-Hill.
Scott, A.
    1992   Obstacles to fishery self government. *Marine Resource Economics* 8(3):187-199.
Simon, H.
    1962   The architecture of complexity. *Proceedings of the American Philosophical Society* 106:467-482.
    1996   *The Sciences of the Artificial.* 3rd ed. Cambridge, MA: MIT Press.
Smith, M.E.
    1990   Chaos in fisheries management. *Marine Anthropological Studies* 3(2):1-13.
Stephenson, R.L.
    1998   Consideration of localized stocks in management. A case statement and a case study. In *The Implications of Localized Fishery Stocks*, I. Hunt von Herbing, I. Kornfield, M. Tupper, and J. Wilson, eds. New York: Natural Resource, Agriculture, and Engineering Service.
Ulanowicz, R.
    1997   *Ecology, the Ascendent Perspective.* New York: Columbia University Press.
Waldrop, M.M.
    1992   Complexity: The Emerging Science at the Edge of Order and Chaos. New York: Simon and Schuster.
Walker, B.H.
    1992   Biodiversity and ecological redundancy. *Conservation Biology* 6:18-23.
    1995   Conserving biological diversity through ecosystem resilience. *Conservation Biology.* 9:747-752.
Walters, C.J.
    1986   Adaptive Management of Renewable Resources. New York: McGraw-Hill.
    1998   Evaluation of quota management policies for developing fisheries. *Canadian Journal of Fishery and Aquatic Sciences* 55:2691-2705.
Watling, L., and E. Norse
    1996   Disturbance of the seabed by mobile fishing gear: A comparison to forest clear-cutting. *Conservation Biology* 12: 1180-1197.
Williamson, O.E.
    1986   *The Economic Institutions of Capitalism: Firms, Markets, Relational Contracting.* New York: Free Press.
    1995   *Organization Theory: From Chester Barnard to the Present and Beyond.* New York: Oxford University Press.
Wilson, J.A.
    1990   Fishing for knowledge. *Land Economics* 66:12-29.
Wilson, J. A., B. Low, R. Costanza, and E. Ostrom
    1999   Scale misperceptions and the spatial dynamics of a social-ecological system. *Ecological Economics* 31(2) (November): 243-257.

Wilson, J.A., J. French, P. Kleben, S.R. McKay, and R. Townsend
    1991    Chaotic dynamics in a multiple species fishery: A model of community predation. *Ecological Modelling* 58:303-322.
Wilson, J.A., J.M. Acheson, M. Metcalfe, and P. Kelban
    1994    Chaos, complexity and community management of fisheries. *Marine Policy* 8:291-305.
Wroblewski, J.S.
    1998    Substocks of northern cod and localized fisheries in Trinity Bay, East Newfoundland and in Gilbert Bay, Southern Labrador. In *The Implications of Localized Fishery Stocks*, I. Hunt von Herbing, I. Kornfield, M. Tupper, and J. Wilson, eds. New York: Natural Resource, Agriculture, and Engineering Service.

# 11

# Emergence of Institutions for the Commons: Contexts, Situations, and Events

*Bonnie J. McCay*

For my part in this common endeavor, I was asked to consider the emergence of institutions for common-pool resources and "the commons." My goal is to use the topic of emergence to present ideas and research that modify and supplement the neo-institutional effort, providing ideas about new directions for common property research. The notion of "situated choice" frames my discussion of emergence. Although closely linked to the neo-institutionalist endeavor through the focus on choice, it is tied even more closely to a critical perspective on commons research that emphasizes the embeddedness of the individual and rational choice in larger contexts and in particular situations that can only be known through investigations into history, political dynamics and social structure, culture, and ecology. Consequently, in addition to an effort to think through what might be involved in the emergence of institutions for the commons, I address larger methodological and theoretical issues.

My ideas about institutions and the commons owe a great deal to the large body of institutionalist and rational choice literature that informs the rest of this volume and the collective effort behind it. Underpinned by a "rational action model" (Dietz, 1994) of human behavior and the mechanics of "free rider" and "prisoners' dilemma" situations, much of the "neo-institutionalist" work on common-pool resources has focused on incentive structures and group dynamics that change the perceived costs and benefits to individuals to favor more cooperative action (e.g., Bromley, 1992; Agrawal, this volume:Chapter 2).

The cultural, historical, and ecological approach that I advocate calls for a somewhat different perspective on institutions than is currently dominant in common-pool resource studies. Institutions are more than "rules of the game in society" (North, 1990:3). Rules and rule making have proved a fruitful focus of inquiry in understanding commons institutions (e.g., Bromley, 1989; Ostrom, 1990). Rules, law, and governance are major institutions affecting human behavior. However, many social scientists see institutions as including not only rules but also norms and values (McCay and Jentoft, 1998; Scott, 1995), and at the very least as including both rules and the patterns of behavior that may or may not be shaped by rules and lead to changes in them (Leach et al., 1997). Accordingly, the emergence of institutions for the commons should include not only rules and governance systems but also new and changed patterns of behavior and norms and values. For example, changing perceptions of the environment or patterns of supply and demand can change human behavior on a fairly large scale without involving the social dynamics and political behavior involved in making and changing rules. Consequently, I assume a broader conception of institutions that includes patterned behavior as well as rules and that locates institutions as major features of the cultural, cognitive, and ecological realms within which acting and decision-making individuals and social groups are embedded.

In emphasizing the importance of "situation" and "context," I join those who believe that a fuller and more satisfactory account would include the possibility of irrational and arational action and of motivations beyond narrowly pecuniary ones. It also would rely less on methodological individualism than the classic neo-institutional approaches do. Methodological individualism starts with the individual as the heuristic in understanding the behavior of groups. It frames "commons" questions as ones that are about the bases of cooperation or about how individual motivations and actions affect the collective. So far so good, but these frames also marginalize huge sets of phenomena that concern interrelations among collectivities as well as how the choices and actions of individuals are embedded in, influenced by, and constitutive of larger social and cultural phenomena (Peters, 1987). A more cultural and historical approach in human ecology sees "commons" questions as ones about competition and collaboration among social entities; the embeddedness of individual and social action; and the historical, political, sociocultural, and ecological specificity of human-environment interactions and institutions. It suspends or at least calls into question the methodological individualism that is associated with rational action models. In theory all institutions and social actions could be reduced to the individual level. However, reducing complex local situations and local and larger institutions to individuals is not always necessary or appropriate for adequate explanation, the requirements for which are contingent on the question being asked and the particulars of the phenomenon being studied.

## STRUCTURE OF THE CHAPTER

In the first section of this chapter I focus on the assigned "emergence" question, using the notion of "situated choice" to underscore the importance of contexts and situations when attempting to explain the behavior of people faced with choices related to common-pool resources. The discussion builds on work done in health psychology and risk studies, but it should be clear that it calls for a far more social, political, and ecological perspective than usually found in those research traditions or in the neo-institutionalist tradition of common-pool resource studies. For example, I discuss matters such as the role of culture in appraisals of environmental problems, why incrementalism or "muddling through" helps in the provision of institutions, and the importance of physical and social spaces and open communication for deliberation about common problems. The second section of the chapter reviews alternatives to the neo-institutionalist paradigm for understanding commons problems. I begin by introducing two approaches in human ecology that may be helpful in understanding commons problems and the emergence of institutions. The older one is the "economics of flexibility" or "response process" approach developed and used mostly during the 1970s. The newer one is "political ecology." I then discuss the broader set of historical, social constructionist, and "embeddedness" perspectives that underpin many critiques of common-pool resource studies and the importance of being specific and critical about key concepts, in this case "community." The third section brings together social constructionism and "event ecology," emphasizing the methodological points shared by otherwise seemingly strange bedfellows. Among the shared perspectives is concern about adopting a priori any particular theory or hypothesis if the goal is to understand and explain human-environment interactions.

## SITUATED RATIONAL CHOICE AND THE EMERGENCE OF INSTITUTIONS FOR THE COMMONS

A start toward bringing together the rational choice approach and theoretical and methodological approaches in the social sciences that emphasize context and sociality may be found in the notion of situated or embedded rational choice. Rational choices are embedded in situations or contexts that structure the preferences people have, the knowledge available to them, its quality and levels of uncertainty, the risks they face, the resources to which they have access, the people with whom they interact, and more, including the institutions—norms, rules, values, organizations, and patterns of behavior—that frame and structure their lives.

Neo-institutional models of behavior play a major role in this discussion of the emergence of institutions. The analytic and rhetorical power of such models cannot be denied. The idea of "*situated* rational choice" is that rational choice is affected strongly by the situation of the individual or other decision-making entity, with situation defined in social, cultural, political, and ecological terms as relevant to contexts that are specified in historical, geographic, and other ways. It is an incremental move toward an analytic orientation that gives stronger methodological and theoretical weight to the complexity, history and dynamics, and interactive features of social and environmental phenomena (e.g., McCay and Acheson, 1987a; Leach et al., 1997).

The study of risk perception and human behavior (Chess et al., 1995; Gardner and Stern, 1996; National Research Council, 1996), has been helpful to me in thinking about the emergence of common-pool resource institutions. Problems affecting the sustainability of natural resources or the viability of livelihoods based on those resources may be viewed as situations of risk: the risk of losing access to and the use of something valuable and essential to the life of a person, a family, a community. Many common-pool resource issues are classic ones of risk, such as the contamination of water, air, and soil. But the concept of risk can be extended to changes in the condition of renewable natural resources as well. How people—individuals, organizations, communities, bodies of experts—are affected by and perceive those risks is critical to whether and how they respond, including responses that affect the emergence of institutions for reducing or preventing those risks.

Take, for example, the risk of illness and death from exposure to natural or anthropogenic sources of the gas radon. The kind of question typically asked in this research tradition concerns voluntary individual action: Why do some people voluntarily test their homes and make structural changes to reduce their exposure to radon, and others do not? What leads people to adopt precautions to protect themselves from threats of exposure to radon?

The answer is interesting and important to the notion of situated rational choice. According to Weinstein and Sandman (1992), it all depends: Some people do not know or understand the threat of radon in their homes; others know but do not see it as serious to themselves personally; others are at a different stage, seeking or perhaps assessing information about what can be done about the problem; yet others have concluded that they can or cannot afford to address the problem, given their resources and what they know about it. Only a few are actually receptive to educational campaigns. This is a good example of the experimental and observational research done by health psychologists that shows that the responses of people to perceived risks to their lives and health are contingent on the so-called "stage" they are in with respect to recognizing the problem and adopting some precaution or taking other action (Weinstein et al., 1998). Although the notion of stages implies an unfolding process of linked events, it also can be viewed as a decision-tree or a step-wise series of situations. The idea of "situation" and "situated rational choice" applies to the "stages" or decision-points, which may or may not be part of a predictable process (cf. Vayda et al., 1991). It is reasonable to conclude that answering questions about what leads people to change how they act—including actions that affect institutions with respect to common-pool resources—similarly requires analysis of their situations.

Translated into the domain of concern about common-pool resources and related environmental problems, the theory becomes the following: Depending on their situations, some people simply may be unaware of environmental problems; others may be aware but not convinced they can do anything about them; and others simply may not have the resources required to do something about them or may reckon that the effort is not worth it, given costs and other obligations. Some may be in a situation and have the interests and resources that affect their decisions about whether to participate in or support collective action, whether through existing political forums or through social movements (Stern et al., 1999). As will be shown, addressing the question of emergence of institutions this way can lead to important insights and expanded arenas of inquiry.

The following ideas about the emergence of institutions that relate to the use and management of common-pool resources thus are guided by a situated rational choice perspective—what is reasonable for an individual, or group of individuals, given the situation. However, the emphasis on situations also leads to a more social, political, and ecological perspective on the nature and explanatory importance of those situations and their contexts.

### The Step-Wise Model of Situated Rational Choice

The emergence of institutions for governance of common-pool resources will depend on several step-wise conditions. To begin, is a problem calling for institutional change actually recognized by the people involved, particularly the people with the resources and power required to make changes? How serious is this problem compared with other issues as well as with past experience? Will it merit being put on the agenda of individuals, households, firms, social movement organizations, government agencies, or other actors?

*Recognition of a Serious Problem*

Attributes of the resource or environmental system make a big difference to these situations: Can people really know what is happening? Do they perceive changes in the environment that may signal problems with common-pool resources? Can they distinguish transient and local from persistent and large-scale problems? For example, for the Koyukon people of northern Alaska, moose and caribou differ significantly with implications for how people think about and act toward them. Moose are less migratory, and more territorial, and tend to be found alone or in small groups; consequently, people know more about particular moose and their habits, and the moose are less likely than caribou to be hunted by different groups of people (see Nelson, 1983). For the Miskito Indians of the Atlantic coast of Nicaragua, the presence, absence, and abundance of sea turtles off their shores seemed to have little to do with their own behavior, even when they began intensive commercial harvesting (see Nietschmann, 1973). Decline in turtle catches was interpreted as being due to the fact that turtles simply had gone somewhere else. This interpretation has a certain rationality given the fact that the sea turtles do migrate over huge areas and are at their most vulnerable, when egg laying, far from the shores of Nicaragua.

In addition, some kinds of common-pool resource problems are inherently difficult to perceive and assess, particularly those that are very diffuse, mostly invisible and intangible, and not easily associated with particular consequences. The most obvious examples are exposures to radiation, air and water pollution, and toxic emissions and wastes, where uncertainty plays a major role in the construction of personal understandings of risk but also in the political deliberative processes (Freudenberg, 1988).

Attributes of experience and social organization and political system also make a difference, for example, to the ability of common-pool resource users to communicate and teach others about what they see as a problem and to deliberate the seriousness of the problem in comparison with the past and other issues. The challenge is to get people's attention, to put it on the agenda. Social structure and culture can play a major role in determining which phenomena will be defined as risky as well as levels of risk and general notions about risk and the environment as discussed later in the chapter. They also influence the distribution of knowledge and expertise, whether widely shared or the closely guarded treasure of a few, as well as how effectively experts can communicate with the larger community.

Many cases of the nonemergence of self-governance can be due to difficulties at the level of problem recognition and placement on an "agenda." Some groups may not be able to appreciate the magnitude of the problems confronting them (such as declining productivity of an estuary or increased soil erosion due to grazing practices) because of the subtlety, novelty, or stochasticity of the ecological systems or because of imperfections in their monitoring systems. They may be unaware of or disinterested in the public goods (such as biodiversity or watershed quality) associated with their private uses (e.g., Gibson and Becker, 2000). If some people in the group do recognize the problem, they may or may not be able to communicate it effectively to others and get it onto the larger agenda, depending on their position in the social hierarchy, the social legitimacy of their knowledge versus other sources of knowledge, and other social-situational factors. Cultural understandings of human-environment relations can affect the definition of problems and the potential for solutions. In complex socioeconomic systems, some people are affected more than others, and differing interests and access to political power and communicative resources greatly affect the agenda. Examples include situations involving environmentalist nongovernmental organizations (NGOs), resource-dependent local communities, and extraction companies, but actual situations are likely to be even more complex and nuanced (for example, elite members of a local community making special deals with either the NGOs or the companies) (Sawyer, 1996). The agenda can be shaped heavily by national ideology and politics as well, as is the case for interpretations of grasslands decline in China and Inner Mongolia (Williams, 2000). In addition, where there is a high degree of uncertainty about the environmental problems, as is often the case in fisheries as well as in many toxic exposure situations, there is even more scope for conflict and opportunistic behavior by special interests (Wilson and McCay, 1999).

Finally, in some situations, the predicted response of most people and social entities is "so what?" During civil war or a famine, protecting a forest or water supply is not likely to galvanize action. The critical and scarce resource is the ability to survive. "So what" can also be the response when new opportunities arise quickly, before existing institutions can respond to them (or are overwhelmed by them). For example, if most people in a community are making money from the destructive practice of dynamiting fish on a coral reef, protecting that reef is not likely to happen unless someone can provide alternative resources and motivations (see Alcala and Russ, 1990).

## Determining Cause and Effect

Once on the agenda, a whole new set of questions arises. Do people see and accept any cause-and-effect or action-and-consequence relationship between their behavior and the environment issue at hand? (This also affects whether the problem gets onto the agenda.) If they do, is the situation viewed as something that can be corrected or that is "too far gone"?

In many situations, because of culture, past experience, or the inherent disconnects between perceived action and perceived consequences, people do not accept that their actions or the actions of other people have any real effect on the resources in question, either as causes of problems or as potential sources of solutions. Carrier (1987) shows this for Ponam Islanders of Papua New Guinea, who believed that God, not people, caused change in fish, shellfish, and turtles, and thus were unwilling to accept the need to change their harvesting practices being promulgated by people concerned about major declines in some of these resources. Similarly, many New England fishermen have resisted changes in fishery management because they were convinced that chaotic-like processes in nature had long resulted in cycles of abundance and decline, and thus that restrictions on their catches would do little good (Smith, 1990; see Wilson, this volume:Chapter 10).

The role played by such dismissals or suspicions of human agency is likely to be greater with respect to resources that are difficult to monitor (i.e., fast-moving fish versus stationary shellfish; or fish versus trees). Other ecological factors are important as well, such as variability and uncertainty. As noted already, features of the natural world influence whether people are able to accurately see what is happening to a common-pool resource, much less appraise the effects of human activities on it and predict what happens next. However, one should not focus too much on features of the natural environment at the expense of recognizing ethno-ecological and cosmological differences in knowledge systems and philosophy. These differences are found among academic cultures as well. Nearly a generation of postmodern critical theory and analysis has shown how much our representations of the natural and social worlds are shaped by social facts and cultural preconceptions (e.g., Soulé and Lease, 1995). They are very imperfect mirrors of a reality we are hard pressed to know about, much less to care for.

Culture plays a major role in how people assign causation and link events to consequences. One fairly well-developed way of incorporating culture into this kind of analysis is the "cultural theory" of anthropologist Douglas and political scientist Wildavsky (1982). They posit deep-rooted "cultural biases" that affect how people see cause and effect and appropriate action, as well as whether people will be concerned about the natural environment and be likely to act on that concern. They identified egalitarianism, hierarchy, individualism, and fatalism as the generic biases, which are distributed differently within and among societies and cultures. The biases express broad differences in the understanding of causes and consequences of environmental change and of the proper way of dealing with them, including reliance on authorities (hierarchy), individual behavior (individualism), possible collective action (egalitarianism), and leaving it to fate. Subject to much criticism (Rosa, 1996; Johnson and Griffith, 1996; cf. Stern et al., 1999), this approach nonetheless highlights the importance of culture. It is also another reminder of the evolving multidisciplinary area of research on risk perception and behavior, which articulates with the common property research tradition at several points.

One of the dangers of "cultural theory" lies in esssentialism: In these accounts, people are or are not "individualists" or "fatalists," and so on. Situation specificity should apply here, too, to capture the differences and changes in culture apparent in a particular situation. I noted that many New England fishermen have resisted changes in fishery management because they were convinced that chaotic-like processes in nature had long resulted in cycles of abundance and decline, and thus that restrictions on their catches would do little good (Smith, 1990). It is quite possible that this perception of nature, and especially its rhetorical use in public forums, was socially constructed in the course of decades-long conflicts over fisheries management in New England (see Miller and van Maanen, 1979) as well as in encounters with nature. Certainly the expression of this perception in the contexts depicted by Smith (1990) was skeptical and oppositional. In recent years, the use of such skeptical ideologies in adversarial encounters has begun to decline, as the contours of conflict have shifted such that New England fishermen are more likely to accept their role in the decline of fish stocks and seek a greater role in research and management (see Wilson, this volume:Chapter 10). From another, related analytic perspective, the cultural dimension is less about an overall, holistic "culture" than about how particular problems are framed, or socially constructed. To many fishermen in the New England and Mid-Atlantic regions, such problems appear to be framed more in terms of the need to protect their livelihoods against intrusive outsiders, but for conservationists it was framed as the need to protect the fish populations in the context of what they viewed as a situation of industry "capture" of the regulatory institutions.

Moving on to the next step: If there is acceptance of a serious problem and the possibility that human behavior has contributed to it, another question that arises is whether the problem is "too far gone" by the time it is recognized and accepted (Ostrom, 2001). Members of the community may decide that they can do nothing about it. And doing something about it may prove very difficult. Hanna's (1995) analysis of user participation in fishery management in the Pacific coast of the United States showed the difficulties of sustaining cooperation where the natural resource had declined sharply.

In sum, institutions for common-pool resource management may or may not arise depending on whether people accept that human behavior is a cause of problems, agree on whether some kind of regulation or other institutional change is called for, and believe the situation is not too far gone to do something about it.

### What to Do and Whether It Is Worth Doing

In theory, even though people may be in a situation of recognizing and being concerned about a salient risk or environmental problem, nothing will happen unless they see possible solutions to the problem that they can take, individually or collectively, and then, whether they can weigh the costs and benefits of the alternatives and act on them. One or more of the alternatives must be seen as affordable and potentially effective to be considered worthwhile (Weinstein et al., 1998). Moreover, accepting human agency (one's own or someone else's) as a cause of common-pool resource or environmental problems does not necessarily mean acceptance of the need for institutional changes. Unless the institutional frameworks already exist, these changes can be very costly, and there may be considerable uncertainty about whether existing or new measures will actually work.

For many common-pool resources, particularly the wild ones we often call "natural resources," there is a high level of uncertainty about their behavior and dynamics. In addition, in situations dominated by bureaucratic structures, the issue of whether something will work may get lost because of lack of will and resources to plan for evaluation and adaptation of measures undertaken as well as because of conflict (Lee, 1993). Conflict is a major problem. A typical social response to perceptions of scarcity or other manifestations of trouble with a common resource is to exclude others from using that resource (Oakerson, 1992). This immediately raises the likelihood of conflict. As Bruce (1999) has shown in an overview of challenges to common property institutions for forest management (see also Pendzich et al., 1994), these and other conflicts, including internal ones, can defeat attempts to create or change appropriate management institutions. Because of competing claims and interests, managing the commons often is tinged with fear and violence as well as competing and discordant uses.

To sum the argument to this point, let me address the topic of this chapter in terms of how to explain the *non*emergence of institutions for managing the commons. Nonemergence may come about because some people simply are unaware of environmental or common-pool resource problems. That is their situation. Others may be aware but not convinced they can do anything about them, given their situations. In some situations the problem is inability to come up with acceptable and reasonable ways to deal with the problematic conditions. And in others, it may be a matter of people not having the resources required to do something about the problem or reckoning that the effort is not worth it, given costs and other competing obligations, not to mention fear of reprisal from those with other interests and inability to resolve conflicts.

## Building on Existing Institutions

From a rational choice perspective, the existence of institutions that can be adapted for new purposes may be extremely important to the emergence of self-governance of common-pool resources (Ostrom, 1990). They can lower transaction costs, providing the decision-making structures, enforcement powers, experiences, and cultural expectations that otherwise might have to be created anew and at great economic and political expense. Accordingly, the emergence of institutions is as likely as not to be a case of adapting or redirecting institutions that already exist and were created for other purposes. One example from the Shetland Islands is a community-based thrift institution that has become the vehicle for an innovative method of ensuring community benefits from privatized fishing rights (Goodlad, 1999).

It is tempting to suggest that institutions for managing the commons are more likely to be ones that had their genesis in situations of conflicting claims to common-pool resources than ones that came about in situations in which people became aware of depletion or degradation per se. Most of the "sea tenure" institutions in fisheries (Cordell, 1989) were constructed in response to user conflicts rather than resource sustainability concerns. Rules, norms, and other institutions mitigate conflict by coordinating the use of fishing grounds and techniques. They also are created to protect groups against other groups, through creating exclusive territories (Acheson, 1987) or restricting the use of particular techniques and outlawing waste disposal in fishing grounds (Stocks, 1987). This is not to deny the existence and value of conservation-oriented behaviors. In fisheries the value of many such institutions has been amply documented, but with the interesting finding that in hardly any cases is the amount of catch actually controlled, in contrast with controls over access, timing, spacing, and other factors (Acheson and Wilson, 1996; Schlager, 1994). Hence, "indigenous conservation" is actually "indigenous conflict management" in many cases. Polunin (1984) makes a similar argument for the many and various systems of complex sea tenure arrangements in Indonesia and New Guinea, in the context of concern about overreliance on these "indigenous" sea tenure institutions for a task toward which many were not designed: preventing resource decline.

Many cases of indigenous groups trying to create institutions for the commons are also good reminders of the danger of assuming that "conservation" or protecting the "sustainability" of local resources is always or properly the goal. As shown in many parts of Latin America, struggles to claim or gain recognition for common property by indigenous groups are often struggles for territory and for cultural identity vis-à-vis other claimants (Bruce, 1999:53). Prolonged political and other forms of conflict often are required before these groups gain the legal and political recognition they seek (Pendzich et al., 1994). In many cases, the goal of attempts to create institutions for the commons is less finding ways to address local resource scarcities or environmental problems than to protect against incursions from outsiders or to claim, or reassert, cultural identities and political power. Whether success in achieving those goals provides the wherewithal and motivation to develop appropriate internal rules for managing the commons for sustainability is another question. Arguably, it is a critical step, the basis for the boundary definition, local autonomy, and other "design principles" of managing local commons (Ostrom, 1990), but it may or may not lead to management beyond the exclusion of outsiders.

Conflicts do, of course, come about because of the very scarcities or threats to common resources that may prompt people to create or change institutions, making it difficult to separate conflict from conservation. A common response to resource scarcity is to try to exclude others (Oakerson, 1992). Those others may dispute the claims, leading to conflict. Indeed, the entire process of creating institutions for the commons can be highly conflictual, and finding ways to effectively resolve conflicts can be a critical task. As Bruce (1999:53) notes, "Disputing can harass and exhaust, and ultimately lead to the dissolution of common property institutions."

The development of institutions for conflict management and attempts to convert them to conservation purposes can be seen at national and international levels, too. In the course of the Law of the Sea proceedings of the 1970s and 1980s, nations eagerly grabbed 200 nautical miles off their coasts as exclusive territory or "extended economic zones" (EEZs) while paying little attention to the requirement that they manage their own fisheries as well as restrict outsiders' fisheries in the new EEZs (Hoel, 2000). Regional institutions and organizations also have developed from similar bases, and the challenge today is their reorientation for sustainable resource use and conservation (Hall, 1998; Noonan, 1998).

There is also "indigenous marketing management." McCay (1980) and Berkes and Pocock (1987) report on fisheries cooperatives that do limit catches, exceptions to the finding noted earlier. In those cases, the intended reason for controlling the amount of fish caught and landed by member fishers is to prevent oversupply to local markets rather than to prevent overfishing. The goal for common-pool resource management concerns prices and other market conditions, not levels of abundance of the resource per se. One question that arises from this observation is whether this is a legitimate case of common-pool resource management. Much of the literature seems to assume that the goal is sustainable levels of resource available for use (i.e., "conservation"), but "management" is a broader concept. The emergence of institutions for common-pool resource management that focus on specific marketing or other economic issues should not be marginalized simply because resource conservation is not a principal intention. Rather, more respect should be accorded to systems like these that manage to develop some degree of self-regulation, in the context of shared rights and competition, and often, if incidental to the main intended purpose, support biological conservation goals as well (for an example see McCay, 1980:37). Some of the institutions that emerged because of specific and immediate economic or conflict resolution needs provide experience and infrastructure that may be used to handle other common-pool resource problems, including protecting fish stocks from overfishing and their habitats from destruction.

## Horizontal and Vertical Linkages

Thus far my discussion might be read to assume isolation of a group from the outside world; it assumes that the sources of motivation for common-pool resource institutions must be internal, and it says nothing about situations where rules are imposed on common-pool resource users. External agency and resources may make all the difference. Even without accepting causal linkages between human agency and the environment, people may be persuaded to accept institutional changes because of other benefits. In many rural areas of the world, government agencies and NGOs seeking paths to sustainable development are able to convince people to cooperate with a project in "community-based" natural resource management as long as side benefits such as jobs or improved access to health care are available. External resources and actors can play an extremely important role, interacting with internal and local ones, in creating civic arenas or forums, social and political spaces for deliberation.

Broadening the analytic scope to include much larger vertical and horizontal linkages among social entities, one sees that forces external to local communities of common-pool resource users play an extremely important role in institutional change. With outside help, local communities can imitate and adapt what they see others doing and the models for change created and promulgated by government and nongovernment organizations. Thus, for example, in the tropical rain forest region of coastal Ecuador studied by Rudel (2000), one community, Playa de Oro, came up with an ecotourism-oriented sustainable forestry program using foreign technical assistance. Within a few years, many surrounding communities have begun to develop similar programs.

Although the situations and choices of the individual decision makers are important in determining whether a community will be receptive to such opportunities, helpful explanation of the adoption of such institutional changes can be done without reference to the individual, looking instead at factors such the availability of appropriate technological approaches, the use of demonstration projects, and the existence of a network of individuals and organizations committed to the process. Much institutional activity depends on the actions of parliaments and presidents, of kings and county officials, and of organized groups and governmental forest and fisheries agencies—as well as networks and coalitions of nongovernmental, domestic, and international environment and development organizations (Stonich, 1996). Consequently, the question of emergence also should be directed at these actors and levels.

Recently, Bates and Rudel (2000) did this in a cross-national study of how to explain the creation of parks in the forested humid tropics. They theorize a general process, beginning with perception of a threat to wild areas or biodiversity because of the activities of coalitions of companies, politicians, and landowners. Whether this provokes the creation of counter-coalitions on the part of environmentalists and others depends in part on how compelling the problem appears to be. Take the case of deforestation. If deforestation rates are very rapid, and if relatively little is left, then the chances of a major response are greater than where the country has many rich forests left. Thus the ultimate cause of park creation is the mobilization of a countercoalition. But that is not enough for an adequate explanation. What happens next depends on whether the governments sense broad popular support and whether they have resources to create and manage parks. It is equally important to focus on proximate causes such as the political conditions that support park creation, which may include various manifestations of "green imperialism," including political pressure of conservation groups on international lending organizations or regional development banks, or the activities of groups such as the Nature Conservancy, which use the "power of the purse" to shape the political agendas of environmental groups. One also would need to examine the social and ecological conditions affecting the behavior of those groups.

Clearly "the outside" is very important. In some poor rural regions of developing countries, there are fewer local organizations and other features of civic society than in wealthier areas (Esman and Uphoff, 1984). Consequently, the emergence of institutions often means involving insiders and outsiders, resource users and development workers or resource managers. "Co-management" is one way to talk about this, but it may be too narrow, if it implies as it has in fisheries contexts—a simple arrangement between a group of resource users and a government agency. Interests and issues are often more diverse.

Emergence of viable commons institutions may thus depend on the creation of large multistakeholder organizations, or "encompassing organizations," as

Rudel (2000) discusses in his study of Ecuadorian attempts at sustainable development. In that case, a "coordinating unit" was created (using a model tried out in Mexico) that represented local communities, timber companies, government agencies, environmental NGOs, and foreign assistance groups. It became a forum for discussion and debate on sustainable forestry issues, and a civic arena for bargaining and making compromises and tradeoffs, as well as communication. For example, the small local communities were able to improve the terms of trade with the timber companies because they could exchange information on deals offered and cooperate in demanding better prices. The timber companies also benefit by getting the communities to agree on a workable policy for sales of timber land. Watershed associations are excellent examples in the North, facilitated by the existence of a strong civic society.

## Muddling Through

When common-pool resource users are faced with the need to invest time, energy, money, and other resources in developing or changing self-governing institutions, the rational choice of free-rider strategies can overwhelm the effort. A "privileged group" may be able to counteract free riding by investing enough to provide benefits and eventually cajole others into contributing—or change the rules in ways that further marginalize or exclude most of the free riders. That is a side benefit of social stratification or unequal distribution of wealth and power that can make a great difference to the emergence of common-pool resource institutions. However, another way out of this collective action bind most likely available to groups with relatively equal power is to make institutional changes in small, incremental steps, starting small and cheap, the so-called "muddling through" method of public policy making (Lindblom, 1959). Ostrom (1990) showed this in her analysis of the efforts at collective action among private and public water rights holders in the Los Angeles metropolitan area. Small steps have low initial costs and the prospect of early successes, which can change the decision-making environment (Ostrom, 1990:137): "Each institutional change transformed the structure of incentives within which future strategic decisions would be made."

A second benefit of "still muddling, not yet through," as Lindblom called it in 1979, is that a go-slow, incremental approach to problem solving may be a very wise strategy vis-à-vis complicated and highly uncertain ecological systems. This was a major lesson we learned when engaged in a program intended to restore productivity to shellfish in New Jersey's bays (McCay, 1988). Given the high level of ignorance and uncertainty concerning clam biology and estuarine hydrodynamics in the area, we found that an incremental approach, where we acted without full prior examination of the situation and alternatives, was very helpful. Although we failed to increase the productivity of clams in the bay by the method we selected, we also reduced ignorance and uncertainty because our method was designed to allow us to learn more about causes of declining productivity and to refine both goals and means. When "muddling through" is combined with efforts to learn and the capacity to adapt, or "adaptive management" (Walters, 1986), it can contribute to the emergence of effective common-pool resource institutions.

## OLD AND NEW DIRECTIONS IN COMMON-POOL RESOURCE STUDIES

The previous discussion is influenced heavily by mainstream and also by less well-known and emerging traditions in studies of common property and, more generally, human ecology. My goal in the rest of the chapter is to highlight the less familiar and newer traditions that have influenced my own thoughts and are of potential interest to other scholars and practitioners.

### Actor-Focused Ecology and the Economics of Flexibility

The value of muddling through processes, through which initial changes are small, relatively cheap, and not necessarily informed by consideration of larger values and goals, is similar to an argument made by Bateson (1963, 1972) and Slobodkin (Slobodkin and Rapoport, 1974; Slobodkin, 1968) concerning the "economics of flexibility" in evolution and adaptation. From that perspective, developed by human ecologists in the 1970s, responding adaptively involves not only deploying resources to cope with the immediate problem, but also leaving reserves (the source of flexibility) for future contingencies (Vayda and McCay, 1975:294). Minimal, less costly, and more reversible responses are predicted to occur first. If an environmental problem worsens or is not adequately met by the initial responses, "deeper," most costly, and less reversible responses take over (McCay, 1978). In other words, there's no point mustering the troops if you can survive by ignoring the problem or, if necessary, scare away the intruder yourself. But you may not survive, much less deal with problems like paying the rent, unless the troops are mustered. As stated by McCay (1978:415-416):

> Within the "economics of flexibility" theory, minimal responses to perturbation may be valuable in providing a built-in time lag for evaluating the magnitude, duration, and other characteristics of problems, as well as the effectiveness of solutions. They therefore minimize the chance that costly and irreversible responses are activated for what might turn out to be trivial or transient problems. The implied cautiousness might also be adaptive for human actors who tend to define inherently complex problems in terms of narrow solutions on hand…and thus, as in the case of "technological fix" solutions to natural hazards…create new problems for themselves and others. However, if environmental problems persist, the costs of diversification strategies…may increase…for the actors. They are then expected to make decisions leading to increased commitment to

one or another course of action. If adaptive, the shift to "intensification" response strategies reduces some of those costs and helps restore "flexibility" to actors and their social units....

The approach was welcomed by ecological anthropologists for several reasons, including the fact that it fit nicely into a more actor-based ecology paralleling the neo-Darwinian shift in biology, in contrast with prior tendencies to relegate individuals and social groups to passive roles within cultures or systems (Vayda and McCay, 1975, 1977). The economics of flexibility can be translated into the topic at hand. For instance, if simple adjustments in technology compensate for decline in common resources, there is little reason to bother with the task of creating and changing regulatory institutions, particularly as that task can divert resources from other important issues such as finding food and shelter for one's family. On the other hand, if those technological changes do not work, or if the environmental problem worsens or expands its scope, "deeper" or more costly changes are more likely to take effect, such as those implied in personal decisions to create or join social movements or social agreements to create, implement, and enforce regulations. To the extent that they work, they then free up the "lower level" capacities to respond to other and new issues. Governing institutions, like all leadership, are successful when they allow people to return to doing what they do best.

This line of thinking corresponds in broad outline with the work of economic historians on conditions for changes in property rights and other institutions, with a focus on transaction costs in relation to changing technology, population pressure, and other facts affecting costs and benefits of creating and maintaining new institutions (e.g., Anderson and Hill, 1977; Libecap, 1986; North, 1981). But there are some differences. As developed by human ecologists, including geographers and others (Grossman, 1977), a focus on responses to natural hazards has led to generalizations about how individual and social responses may be expected to relate to environmental variables. Temporal pattern is one class of environmental variables: The magnitude, speed of onset, duration, and relative novelty of environmental changes might be expected to affect the levels and kinds of responses (Barton, 1969). Spatial patterns also affect responses. An excellent example is the geographer Waddell's (1975) analysis of how the Fringe Enga people of the New Guinea highlands coped with recurrent, and sometimes severe, plant-killing frosts. In addition, Vayda (1976) used the approach in his study of war in three Oceanian societies; Lees (1974) developed it in her analysis of the development of hydraulic control institutions and technology; Rudel (1980) examined automobile-related responses in the United States to the energy shortage of the early 1970s from this perspective; and Morren (1980) used it to analyze the pattern of responses to a drought in Great Britain. I used it in my analyses of responses to fisheries decline in Newfoundland (McCay, 1978, 1979) and New Jersey (McCay, 1981).

The "economics of flexibility" provides a general predictive framework for relating the systemic "depth" of responses undertaken to various features of the environment, such as the scale, scope, and duration of the environmental hazards or risks involved. Its use also can lead to a sharper focus on institutional and political issues. For example, in my study of responses to decline in fish abundance in Newfoundland (McCay, 1978), I noted that through the 1960s, responses were initially at the level of individuals and households, and that two general strategies could be discerned: diversification (expressed through occupational pluralism, deploying different fishing techniques, and so on) and intensification (investing more in one activity). The logic of the "economics of flexibility" suggested that diversification should be the primary strategy because intensification requires greater investments and can lock people and their organizations into particular, "deeper" modes of response, becoming nearly irreversible. However, as the problem continued, and worsened, individuals and households were more likely to make intensification types of responses, including going on welfare, moving away, and buying bigger boats. There were also important social and institutional responses, including a "rural development" movement that led to the formation of a fisheries cooperative and other groups, organized to address the problems facing the fishing-dependent households and communities.

There seemed to be a graduated series of responses that articulated with worsening environmental problems as predicted. But it was puzzling that people had invested so much in the bigger boats, which seemed only to add to the problem of declining fish catches. In other words, those investing in the larger vessels appeared to have been moving too rapidly and too inappropriately, given the situation and what one might predict from any theory that emphasizes cautiousness in the face of uncertain environmental change. Exploring why that happened led to appreciation of the role of various social entities and economic and political actors outside the local community in decisions made about the future of the community. For example, the social movement that led to the creation of the cooperative and decisions to invest in new fishing technology both were shaped heavily by the involvement of government agencies, university extension workers, and particular "outsiders," including film makers, in local matters. Such involvement, however, was also a source of the loss of local power to formulate problems and solutions, particularly the direction that the fishery would take.

The "economics of flexibility" or "response process" approach lends itself to a more open-ended methodology than implied by the task set out in relation to what we have come to know about common-pool resource research. The question is not what causes the emergence of or changes in governing rules, but rather how people with or without their institutionalized patterns of behavior respond to an environmental event (for example, a killing frost) or a series of events (for example, a series of poor fishing seasons). Responses may be small adjustments in individual behavior or in the deployment of household resources, or major investments in new technologies and resource procurement strategies. Responses may take the form of organized social action, whether raids on neighboring groups or concerted efforts to get assistance from a government agency, or perhaps attempts to come up with effective community-based institutions for dealing with an environmental problem or its consequences. But they may not. At one time I suggested using the term "people ecology" (McCay, 1978) as a way to signal the need to avoid a priori prescriptions and assumptions about the units of action or significant factors implied in other terms used in anthropology and geography, such as cultural ecology, population ecology, systems ecology, and political ecology. People ecology was intended to suggest the value of leaving open the possibility that the significant units may be individuals, households, or various other social entities, ranging from voluntary associations and transient networks to political units such as municipalities and nations. These change in relation to changes in environmental and social situations or contexts, including the local culture and the larger political economy.

The "economics of flexibility" approach is but one of several sources of middle-range theory for understanding relationships between environmental phenomena and human behavior. Evolutionary ecology is another, particularly as it has evolved to apply optimization models and the predictions of "optimal foraging theory" to human populations (Dyson-Hudson and Smith, 1978; Smith and Winterhalder, 1992). Others are microeconomics and decision-making theory, which are, incidently, closely related to optimization models in evolutionary ecology (Rapport and Turner, 1977). The "economics of flexibility" approach shares economizing assumptions but differs by viewing flexibility, rather than efficiency or optimization, as the proximate goal of adaptive processes, and by viewing survival, rather than inclusive fitness, as the ultimate goal (Slobodkin and Rapoport, 1974). Flexibility means "uncommitted potentiality for change" (Bateson, 1972:497). The approach also is more abstract and inclusive in its application than evolutionary ecology, having been used by students of tribal horticulturists, agrarian peasants, "postpeasant" fishermen, and industrial society, not just hunter-gatherers.

## Political Ecology

Reporting on my use of the economics of flexibility approach in a study of the fishing strategies and illegal behavior of some New Jersey fishers, I argued that it is useful in stimulating a transition from studying narrowly defined environmental interactions to studying the politics of environmental problems or political ecology. However, I also posed the question: "Why use it at all when a straightforward analysis of social and political processes might do as well? The same question can be asked of optimal foraging theory: Why use it when a straightforward analysis of decision making, entrepreneurship, and their constraints might do as well, or better?" (McCay, 1981:376). Adopting theoretical approaches developed for quite specific problems and topics in evolutionary biology and ecology bears the risk of overly "naturalizing" the complex cultural, social, and political phenomena of human ecology, or, put another way, defying the parsimony principles of Occam's Razor.

Responding to questions like mine, geographers, anthropologists, and other social scientists have crafted the term and enterprise of "political ecology," which is a general rubric for a wide set of approaches that make more explicit the role of human institutions and social, economic, and political forces in shaping both environmental problems and the ways people are affected by them and deal with them (Blaikie and Brookfield, 1987; Greenberg and Park, 1994). For example, so far in this chapter I've moved from the microscale of the rational individual to the local scale of a community of resource users in some particular place to the mesoscale of mediating and encompassing institutions. Only implicit so far are macroscale phenomena—the off-site structures and institutions of power and authority; demographic and ecological changes; and the workings of other social forces including political action and social movements, which Goldman (1998:45) and others insist should be foregrounded in any discussion of local-level institutions and commons problems. Approaches that foreground relationships of power and authority, domination and resistance, and so forth often are subsumed under the label "political ecology."

Stepping back to the step-wise model for situated rational choice, one can see limitations of the health psychology model because it is limited mainly to the study of voluntary individual action—which is also the focus of much work in common-pool resource studies, for example, on free-rider disincentives to cooperative action on the part of individuals or individual entities. To properly account for how people respond to common-pool resource challenges, we need to know more about institutions and the deliberative processes that lead to their emergence and change. What led municipalities, states, and federal agencies to provide educational information, establish standards, and give subsidies for radon-protective home construction? The now-widespread rules and norms in North America against smoking in public places, airplanes, and many private spaces may be considered institutions for common-pool resource management, protecting many people from exposure to the risk of secondhand tobacco smoke. What accounts for variation in the pace and process of development of these rules and norms? Closer to the image that many people have when they talk about common-pool resources, what accounts for the decision of a group of fishing vessel owners to create and maintain complex rules about how much fish can be landed, as I observed when studying a New Jersey commercial fishery in the late 1970s and early 1980s (McCay, 1980)? Or what accounts for the development of new principles and understandings about global environmental problems that influence both voluntary actions and intergovernmental agreements? Issues of power and politics loom large when questions are reframed this way.

Political ecology also may be understood as calling for greater emphasis on local politics concerning common-pool resources and the environment. In the course of their critical review of the use of a simplistic, generic notion of community in relation to conservation and development, Agrawal and Gibson (1999:629) argue for a more "political" approach, "focusing on the multiple interests and actors within communities, on how these actors influence decision-making, and on the internal and external institutions that shape the decision-making process." In common-pool resource studies, political ecology also is expressed through

- increased focus on the workings of power as well as differentiation by gender, age, class or caste, ethnicity, and other factors, within common-pool resource-using communities (Leach et al., 1997);
- greater attention to the power dynamics among communities and between them and the institutions and organizations within which they are embedded or to which they are linked, or taking meso- and macro-scale perspectives to understand what is happening at the local level (e.g., Goldman, 1998; Mosse, 1997); and
- more sensitivity to the exercise of power in the production of knowledge about common-pool resourcs (Taylor, 1998).

A political approach includes critical reflection and research on the practice of common-pool resource research and analysis itself. Goldman (1998) and Mosse (1997) are among those who suggest relationships between the approaches taken by common-pool resource researchers and the agendas and interests of various political actors. Brosius et al. (1998) emphasize the "external" and raise questions about how the paradigm of community, and community-based management, is "worked into politically varied plans and programs in disparate sites" (see also Zerner, 2000). The ideas of community, plus ideas of territory, customary law, and locality, are central to common property studies and the community-based resource management movement. A major difference is between conservationists and development organizations, the one emphasizing protecting biodiversity and habitat integrity, the other local participation. Spokespersons for indigenous peoples add values such as respect for local rights, knowledge, and culture. How these ideas materialize in actual cases depends on the workings of transnational as well as national and local actors and institutions, and they have different constructions or ways of "imagining" community-based resource management that play out in the lives and ecologies of local places (West, 2000). Peter Taylor (1998) emphasizes the rhetorical tactics used in common-pool resource discourse and their political implications, from a perspective that emphasizes the social construction of science itself. The "modernist" orientation of many common-pool resource approaches, emphasizing control and hierarchy, and sharp discontinuities between the natural world and the human experience, is also criticized in favor of more "postmodernist" (or poststructuralist [Escobar, 1996]) approaches that recognize the need to break down nature/culture dichotomies, the social construction of both nature and culture, the indeterminacies and contingencies of socionatural systems, and the need for more pragmatic approaches that neither rely on nor reinforce dichotomies between nature and culture (Descola and Pálsson, 1996; McCay, 2000a; Pálsson, in press).

## Embeddedness

Much of what goes under the label of "political ecology" is influenced strongly not only by Marxist and other political economy approaches in the social sciences but also by theoretical developments that emphasize the social embeddedness and cultural construction of seemingly individual, economic, and natural phenomena.

In social theory, "embeddedness" (Polanyi, 1944) is a way of resolving the discrepancy between agency and structure-based approaches, which remains a key issue in many of the social sciences. Speaking directly to the problem as it appears in common-pool resource studies, Peters (1987:178) defines it as follows: "To avoid these polemic extremes we argue for the social embeddedness of a commons. It is an error to suppose that an individual calculus can explain a commons system; rather, one has to understand the socially and politically embedded commons to explain the individual calculus." Agency-based theories, which dominate common-pool resource studies and have a long and strong history in the social sciences, see society as the aggregation of independent individual behaviors and often assume that these behaviors express the rational pursuit of utility on the part of those individuals. Structure-based theories include Marxist and other political economy approaches but also Durkheimian sociology. They emphasize the role of supra-individual social forces and groups in society, resisting reduction to individuals and utility functions. Hence, at one extreme we have the image of self-seeking individuals who, faced with a common-pool resource or public good, can only defect or free ride. At the other extreme is the romanticized society or local community imbued with the moral economy of "the commons" as belonging to and cared for by everyone but besieged by larger forces, such as commercialization and capitalism. Surely there is a more realistic middle ground, as suggested by experimental results (Ostrom, 1998) and as called for by many social theorists who wish to integrate both agency and structure. The social construction approach discussed in the last part of the chapter is a move in that direction. So is "embeddedness."

A "thin" version of embeddedness is found in the work of sociologists and others who emphasize social networks (Swedberg and Granovetter, 1992). Actors or decision makers are embedded in social networks or patterned interactions (which can be construed as institutions). Granovetter (1992, 1985) argues that the agency approach sees actors as "undersocialized," pursuing only their own interests, and the structural approach sees actors as "oversocialized" products of their class or group. If social structure is seen instead as patterned interactions among actors, or social networks, then we can see that structure influences individuals in patterned ways, but we can also see that the individuals have agency, that they are more than just representatives of social categories (Wilson and McCay, 1999).

A "thicker" and more ethnographic perspective adds the missing elements of meaning and communicative content (Emirbayer and Goodwin, 1994) as well as a stronger sense of group differentiation and identity and contests over power and meaning. It "gives...interdependence a more specific conceptualization, one that includes the structure of relations (of which the individual commoners are a part), the differentiation among groups, and the set of shared and/or competing meanings and values associated with a particular commons and its use" (Peters, 1987:178). Peters' case study of Botswana's rangelands is an impressive indicator of the analytic potential of such an approach. She shows that this perspective, when applied to a situation of multiple links and claims to use of the commons, can result in a radical redefinition of the commons problem: "The 'dilemmas' of a commons emerge not from an absence of social ties between the individual user and others [as postulated in most common-pool resource studies], but from competing rights and claims to legitimate use" (Peters, 1987:178). As noted in the introduction to the book in which Peters' paper appeared, "Commons dilemmas must be explained in terms of the dynamics of conflict and competition between different social groups located in history and social systems rather than between the rational economizing individual unspecified and the group also unspecified" (McCay and Acheson, 1987a:22).

Embeddedness has several other analytic functions. In work focused on fisheries and institutions such as co-management, the metaphor of "embeddedness" has been used to emphasize the potentials for coordinated and cooperative action on the part of resource users, who otherwise are thought of as inherent free riders or, worse, opportunistic "foxes in the henhouse," but who are linked with each other through webs of significance and histories of association within communities (Jentoft et al., 1998; Mccay, 1996; McCay and Jentoft, 1996, 1998). There it points to the role of the values and culture of an embedding community. In a somewhat different sense (Giddens, 1990), the notion of embeddedness is used to distinguish local communities in terms of the extent to which particular activities—for example, fish harvesting and processing—are embedded in or disembedded from the larger local community due to the globalization of production and marketing and other processes (Apostle et al., 1998).

Most important, the metaphor of embeddedness is a way to communicate the importance of specifying the historical, geographic, ecological, and social situations and contexts of individuals and groups. The notion of embeddedness thus emphasizes the need for fine-grained, long-term historical and ethnographic research on particular common-pool resource situations and their contexts. Specifying the embedding context allows for a focus on cultural and social phenomena as sources of institutional creation and change without having to reduce social action to individual choice alone. At the same time, it recognizes the agency of the individual embedded within such phenomena, and particularly the agency involved in the social process of interpreting and recreating the natural and social environments (Helgason and Pálsson, 1997).

## Deliberation, Discourse, and Embeddedness

Communication is central to social relations and culture; it is also central to the question of how people respond to environmental problems and risks. Psychological research has revealed interesting patterns in how individuals perceive risk, but these patterns are not necessarily what one would find if studying how people behave when confronted with practical problems of environmental management (Renn et al., 1996). People in communities talk to each other and to outsiders about particular risks, and that is what creates their meaningful, action-prompting perceptions of those risks. Cognition is linguistic, not calculative, and language and communication are at the core of perception and decision making (Dietz, 1994). Talking, discursive behavior, and the meanings construed from talk and action as shaped by identity and power—these are the stuff of social relations and culture. Discourse analysis and research on the social and cultural dimensions of communication are thus part of common-pool resource studies.

Emergence of institutions for the commons requires situations with the possibility of truly open and constructive deliberation as much as it calls for decision-making structures that are able to overcome free-rider and other perverse incentives that plague situations involving the provision of public goods. Often underappreciated is how hard it is sometimes to come up with good solutions to common-pool resource problems. Given the "bounded rationality" of the human mind (Simon, 1983), and the inclination toward "muddling through" when faced with difficult policy choices (Lindblom, 1959, 1979), the alternatives available for institutional change are likely to be quite limited, based in large part on the kinds of things people have already done for the same or other problems. Therefore, in some cases the critical factor may be the ability to share experiences and ideas among members of the group, as well as with other groups, in order to "get out of the box." Doing this requires some kind of deliberative forum where information can be shared and conflicts and ideas aired. The existence of a political, social, and physical space for learning from and arguing with one another is one important "design principle" that should not be taken for granted. In many nations and at many times, political repression makes it nearly impossible to find and use places for talking and arguing about the commons, and economic deprivation can make it difficult for most people to come to them if they are allowed. These problems can occur within local communities as well.

Another criterion is the nature of discourse within such a deliberative forum. The nature and functioning of the process of discussing and deciding on solutions to a perceived common-pool resource problem is obviously critical. To what extent does a particular forum or ongoing deliberation about a commons problem meet the requirements of "rational communication," or of open and honest exchange and deliberation (Habermas, 1984; Dryzek, 1987)? It is well known that cooperative solutions require communication, trust, and reciprocity, but what are the sources of trust and reciprocity and the conditions for effective communication, and what sustains and reproduces these conditions (Hajer, 1995)? They may be affected by many things, including local leadership, the distribution of wealth, the structure of power and authority, the existence of other institutions, and relationships with outside governmental and nongovernmental groups, all of which, in theory, alter the possibilities for communicative rationality.

The question becomes to what extent are the decisions due to open and honest exchange and deliberation, or instead the result of the exercise of the "governing mechanisms" of money and political power and authority, on the one hand, or of prestige and social influence, on the other?[1] What are the social and ecological consequences? Rational communication involves trust, information exchange, and joint problem solving. It works through convincing each other that something is true or right, in contrast with the roles of money, power, and authority, in forcing some to agree with others. Rational communication is heavily dependent on shared background assumptions, or embeddedness in a common world view or culture. As Wilson and I have argued (Wilson and McCay, 1999), if all participants are situated or embedded in similar cultures, social structures, and experiences, they are more likely to be able to engage in rational communication. If not, money, power, or influence "talk."

This sociological approach to communication and decision making has ecological meaning as well. We have argued that in situations where environmental variables have high uncertainty and variability, institutions based on rational communication (and prestige and influence, to some extent) work better than ones based on the governing mechanisms of money and authority (Wilson and McCay, 1999). On the other hand, where the scale of the common-pool resource problems is very large, they may be difficult to resolve without recourse to the constraints of bureaucratic rules, property rights, and other "anchoring institutions" that express the roles of money and authority in social deliberation.

The degree to which deliberation is embedded in local culture, social relations, and experiences, the scale of environmental and common-pool resource problems, and the extent to which conclusions are reached through "communicative rationality" or through the exercise of institutions anchored by the forces of power, authority, and money are thus important conditions for the emergence of self-governing institutions for common-pool resource problems.

## Being Specific and Critical About Community

The emergence of institutions for common-pool resource management is widely assumed to depend on either government or market or community—and much of the critical work done in this regard keeps returning to community. The notion that local communities can and should play a major role in conservation and environmental management has been adopted by many important players in the world, from the World Bank to international NGOs such as the World Wide Fund for Nature (WWF). It is the one unifying message of the network of scholars and practitioners centered on the International Association for the Study of Common Property (IASCP), as well as the larger and more diffuse network of people devoted to community-based sustainable development and natural resource management. The reason is simple: "even well-funded coercive conservation generally fails" (Agrawal and Gibson, 1999:632; Peluso, 1993).

The general idea is that where people who live and/or work together share a sense of identity and belonging (therefore some notion of boundaries and membership criteria), where they share some level of dependence on or caring for the resources in question (or streams of income coming from those resources), and where they also share many norms and goals, they are more likely to be able to develop institutions appropriate to deal with the challenges they face in using common-pool resources. In other words, they are more likely to overcome the self-interested obstacles to collective action, the free-ridership temptation, the appeal of cashing in and defecting. Sharing a sense of identity and belonging also likely means having some shared history, even ancestry, and some expectation of a shared future. These are critical to the development of the trust and reciprocity known to be essential to developing cooperative relationships (Ostrom, 1998). These elements are found in the definition of community provided by Singleton and Taylor (1992; see also Taylor, 1982), who add the notion of mutual vulnerability. Community is measured by the presence, absence, or strength of shared beliefs and preferences; some stability in membership; some expectation of future interactions; and direct and multiple kinds of relationships among members. Mutual vulnerability refers to the extent to which members of the group can be affected by the contributions or withholdings of others; that is, the extent to which they are subject to peer pressure because they value the good opinion, friendship, or cooperation of others. Both attributes are essential conditions for the mutual monitoring and sanctioning that are widely acknowledged to be critical endogenous factors for managing local common resources. The explanatory link to solving the collective action problem is transaction costs in the models of Singleton and Taylor (1992:316) and Ostrom (1990, 1992). The more community there is, the lower the costs of getting information, bargaining, monitoring, and enforcement. Building on the argument made in discussing deliberation, discourse, and embeddedness, the more "community," the more likely people are to be able to communicate with each other about whether there are problems that need to be addressed and if so, what to do (Wilson and McCay, 1999).

Singleton and Taylor (1992) argue that community can bridge inequality and heterogeneity to some extent, although it also can be undermined by economic and social differences (see Agrawal and Gibson, 1999). They also theorize that the types of solutions that result will depend on the degree of community: At one extreme are fully decentralized, endogenous solutions, which depend on high degrees of community; at the other, solutions heavily dependent on the state, because of low degrees of community, and hybrids such as co-management.

There are communities and there are communities. Take three examples of communities featured in *The Question of the Commons* (McCay and Acheson, 1987b): the Swiss mountain village of Torbel studied by Netting (1976, 1981) and reanalyzed by Ostrom (1987); the coastal community of Teelin, Ireland, studied by Taylor (1987); and the lobstering communities of coastal Maine studied by Acheson (1987). The Swiss example is one of a very long history of formalized rules for use of common lands and resources such as mountain forests and the alp. It has become an icon in common property studies for the idea that local communities can manage common resources by themselves (as well as Netting's specific argument that variation in ecology and land use accounts for variation in property rights). But it is also one where there is an unusually well-defined and fairly rigid hierarchy of authority within the village, presumably linked to the larger political system and culture as well as the very intensive and restricted system of farming and the expectation that most young people will go elsewhere.

The Irish example is one where local people have decided not to have direct control over the key common resource, the salmon fishery. They do manage access to it in a rotational system, but they prefer to leave creation and enforcement of conservation rules to outside authorities. Is it no accident that the structure of the village is very egalitarian with extremely weak leadership? No one is willing to risk offending that particular social order: "The river would run red with blood," said a priest when explaining why villagers refused to support his idea of buying the rights to the salmon, so that the villagers would no longer be fishing illegally (Taylor, 1987). Hierarchical and oppositional relations are more between villagers and outsiders than within the village, and the locals seem to like it that way, even though it means that they bear the risks of being fined for illegal fishing and have no direct say in the management of salmon.

The third example, the Maine lobster fisheries, also has become an icon in common property studies because of data showing that the lobster fishers have constructed a system of territoriality, on their own, that counters the notion that all modern commercial fisheries are open access unless restricted by government action (Acheson, 1981). However, communities in the geographic and political sense have very little role in this. With the exception of a few offshore islands, lobstermen embark from communities that have no organized say in lobster management and are increasingly organized around the interests of tourism and ex-urbanites. Moreover, the law of the state protects the rights of everyone to participate. At the time studied, self-regulation took place very subtly and informally through the formation of what Acheson calls "lobster gangs," which perform gatekeeping and regulatory functions. Only the island communities were able to really enforce territories, to the inch. Moreover, the "gangs"—which might be considered occupational or functional communities—have not been able to prevent major increases in lobstering in recent years. Attempts to create lobster management zones within the framework of the state's management authority are thus building on dispersed and complex communities.

One would like to know why these societies are structured as they are and how that relates to the beliefs and norms of the people within them, the interplay of social forms, beliefs, and the capacity to govern uses of the natural environment, a challenge posed by Douglas in *How Institutions Think* (1986; see also Douglas, 1985). To some extent the dimensions of community identified by Singleton and Taylor (1992) might help: There's more "community" in the Swiss Alps community studied than in Maine lobstering communities because of closer interaction, more different kinds of social interactions, and a longer history in the alpine communities than in Maine's coastal communities. But how do we account for the Irish? The specifics of hierarchical and oppositional relationships of power and authority seem to make a difference, as do the specific histories of domination and resistance and more subtle cultural differences in, for example, how people think they should relate to their neighbors as well as to the environment. "Community" and its relationship to common-pool resource management is meaningless without further specification, without clearly positioning particular places and peoples within their environments, their histories, their cultures, as well as regional, national, and global relations of wealth and power.

## QUESTION-DRIVEN RESEARCH:
## SOCIAL CONSTRUCTIONISM AND EVENT ECOLOGY

Theoretical perspectives emphasizing embeddedness are closely tied to those emphasizing the "social construction of reality." The constructivist tradition in sociology and anthropology, triggered by a treatise on the sociology of science (Berger and Luckmann, 1966), has come to focus on the interpretive processes by which individual and corporate actors perceive their surroundings and act on those perceptions to continuously construct and reconstruct themselves and their environments (McLaughlin, no date). One important idea is that of "frames," or interconnected sets of socially constructed categories, that "provide a basis for forging shared meanings and coordinating social action" (McLaughlin, no date:6; Snow and Benford, 1992; Snow et al., 1986). Obviously, "frames" are likely to make a great difference in the "steps" outlined previously, such as whether people are aware of or prepared to act on their awareness of environmental problems. The focus on "framing" also has contributed to greater appreciation of the dynamics of social movements and their contributions to political and cultural change. The question about the emergence of institutions becomes one about when the "master frames" can no longer cope with changing conditions, forcing actors to question routinized assumptions and embark on a project of change (Krauss, 1983; McLaughlin, 1996). This body of work also addresses questions about the sources of legitimacy for new institutions in the social construction of organizations for dealing with common-pool resource and environmental problems (Snow et al., 1986). Such an approach could be useful in analyzing the rise and deployment of concepts such as community-based management and self-governance of the commons (Brosius et al., 1998; Taylor, 1998).

Steins and Edwards (1999) employ a social constructivist perspective to criticize and improve "standard common-pool resources theory." They point to the frequent assumption that common-pool resources are single-use resources even though they often have multiple and conflicting uses; the tendency to focus on factors that are internal to a resource-using community at the expense of external factors that affect the decisions of shareholders; and, finally, the failure to appreciate the role of processes "through which collective action is constructed (and reconstructed) by the shareholders" (Steins and Edwards, 1999:540). Steins and Edwards make their arguments through a case study of a situation in Ireland where an oyster-growers cooperative had been formed but most people refused to meet their obligations. They attempt to explain this free-riding behavior, and they find that explanation in the particular details of this situation and its larger context and longer history. They show the importance of careful analysis of processes of social construction of everyday reality and the environment. They contribute a stronger focus on the ways that critical elements of the situation under study, such as whether the institution is a "failure" or a "success," and actors' understandings of the political environment, are socially constructed. They also point to the ways these constructions or understandings vary among social actors and change over time and with experience. The process of collective action itself will reshape the networks, meanings, perceptions, and social experience that affect stakeholders' choices (Steins and Edwards, 1999:544). In closely related work, Selsky and Memon applied social constructionism to examine the emergence of institutions for dealing with port development issues in New Zealand (Memon and Selsky, 1998; Selsky and Memon, 2000). They suggest that studies of complex common-pool resource domains[2] such as ports may advance theory in ways that studies of simpler, single-use common-pool resource domains cannot. They are sites for multiple and indeterminate interactions among stakeholders, involving dynamics of power, conflict, and competition as well as collaboration and institutional innovation (Selsky and Memon, 2000). The constructivist orientation they use highlights the roles of various actors and relations among them in determining how property rights and other institutions are constructed and change in such domains, as well as the emergence of de facto rights and local rules as against de juris rights and extra-local policy.

Steins and Edwards argue that using a common-pool resource "design principles" approach at the outset of the analysis would have made it harder to see and appreciate the role of contextual and external factors. Moreover, they argue that using a design-principles analytic model too easily results in generalizations that raise the question of "normativity." Design principles such as the high noticeability of cheating may be useful starting points for analysis, but it is important to ask whether a particular principle is indeed a condition in the case at hand and if so, whether or not it explains the phenomena observed, whether compliance or free riding.

One group of human ecologists has come to similar methodological positions. Vayda and Walters (1999) and McCay and Vayda (1996) emphasize the detriments to the task of explaining human-environment interactions created by the practice of starting with a priori assumptions about units of action, scales and levels of analysis, and appropriate explanatory tools (including the "tragedy of the commons" theory or meta-narrative). Like the social constructionists, they warn against the practice of bringing theories and models to particular issues in explanation (i.e., why deforestation proceeds at a high rate in certain areas, or why fish populations have declined in certain places), without careful bottom-up exploration of the events (what happened, where, who was involved, and why) and the situations and contexts of those events (which are important to their explanation) (see also McGuire, 1997; Walters and Vayda, 2001).

Both groups are calling for question-driven research rather than research organized around particular methods or hypotheses. In the Steins and Edwards case, the initial questions are similar to those of the neo-institutionalists working within the middle-range theories of Olson (1965), Ostrom (1990), and others: Why are some people free riding rather than contributing to a cooperative endeavor? However, they wish to answer that question for a specific instance of free riding, rather than for free riding in general. Their argument is that relying on those theories rather than examining the specifics of the case at hand may lead one to miss important causal factors. For Vayda and Walters, in the approach they now call "event ecology" (1999; Walters and Vayda, 2001),[3] the initial questions are more likely to concern environmental issues (such as the causes and consequences of tropical deforestation) than institutional or social questions, although the latter may become important to the analysis. Their causal historical approach to explanation requires moving from large questions, such as the causes and consequences of forest change, to more specific questions, such as the causes and consequences of particular instances of forest change, just as Steins and Edwards insist on explaining particular instances of free riding. The goal with highest priority is to address such specific questions rather than to evaluate the merits of theoretical claims, although it is possible that method and theory will benefit.

Like the social constructionism of Steins and Edwards (1999), the methodology of event ecology leads to a critique of the use of common-pool resource thinking as "question-begging," or overly "theory-driven," rather than explanatory and question driven. Questions about the conditions of human cooperation and defection (or free riding), and the workings of models with greater or lesser degrees of open access, communication, risk, and so forth, are intrinsically interesting, but they are not always adequate or relevant to the task of explaining particular social or environmental events. Like dependency theory or claims about globalization, they are meta-narratives that may or may not be appropriate to the explanatory task. This discussion assumes that explanation is not necessarily dependent on law-like generalizations; the causal historical approach to explanation, well-established in history but also acknowledged in the sciences, results in narratives linking particular events to causal antecedents and to consequences, using methods such as progressive contextualization (Vayda, 1983) and counterfactual reasoning (if x did not exist, would the results be the same?) (Walters and Vayda, 2001).

Just as Steins and Edwards (1999) argue that using a common-pool resource "design principles" approach at the outset of the analysis would have made it harder to see and appreciate the role of contextual and external factors, the human ecologists recommend avoiding a priori designation of appropriate models, theories, and methods. An important point is not to privilege any particular aspect of human-environment relations (political, social-structural, economic, cultural) nor any particular scale of those relations (local, regional, global) when beginning a study, letting importance emerge from empirical analysis.

From this line of reasoning, the very topic of this chapter is misleading. The appropriate question is not necessarily about the institutions appropriate for common-pool problems, but rather about the causes and consequences of particular human-environment situations, including institutions for managing the commons if and where they are relevant.

Take, for example, the general question of the role of pastoralists in the ecology of arid lands. Rather than arguing either that pastoralists tend to overexploit fragile grasslands, or that they have their own systems of regulating the use of such systems of production, the appropriate first step is to examine one or more particular cases of grassland ecology, coming up with concrete, clearly specified events to be explained. The initial question might be causes of decline in the quality of forage grasses in a particular arid region. A common-pool resource scholar might quickly jump to a study of the regulatory institutions of local tribal pastoralists, expecting from the middle-range theory that has developed (e.g., Ostrom, 1990) to find that relatively small, homogeneous groups with a long history in the region have developed rules and other institutions that help prevent overgrazing of common lands. However, this "jumps the gun." Perhaps changes in the quality and quantity of forage grasses have little to do with the pastoralists' herds. Perhaps they do. It may turn out that the patterns are heavily influenced by informal or formal rules and other institutions, in which cases those institutions are candidates for further study. But it may turn out that those changes in grazing activity that warrant the term "overgrazing" have little to do with local institutions, in comparison with changes in market demand, conflicts among pastoralist groups, expanded investment in livestock on the part of urban elites, or invasion of the grasslands by an exotic species. In that case, the investigator would put less effort into examining local institutions than in looking for causal connections with whichever of these other factors seems important.

Both social constructionism and event ecology emphasize the search for specific mechanisms of causation. Many evolutionary psychologists, ecologists, and anthropologists explain something by recourse to ultimate causes (i.e., inclusive fitness) or by recourse to consequences that are alleged to be the causes. Too often they fail to identify or study the mechanisms linking a behavior or institution and its purported causes and consequences (Vayda, 1995a, 1995b). To return to our hypothetical grazing case: A consequentialist (or "naive functionalist") account might say that the institution of tribal control over specific territories is explained by its function in preventing overgrazing in areas controlled by tribes. The relevant generalization might be that the lack of territorial control results in overexploitation of resources (the open-access hypothesis of "the tragedy of the commons"), or that all else being equal, a group of people will respond to signs of overexploitation by developing rules intended to cope with the problem (the community-based management hypothesis or "the comedy of the commons"). There might be more focused generalizations, such as the need for sufficient time for trial and error, for the capacity to monitor behavior and enforce rules, and for other "design principles." However, what is often absent or underdeveloped is an account of the mechanisms or causal linkages. How does the exercise of territorial control affect grazing patterns and intensities in particular cases? How, if at all, do changes in the condition of pasture lands trigger the use of institutional measures to control behavior? What were the conditions and events that led to the development of territorial control and that have influenced its maintenance or its breakdown or transformation?

The disagreement is not with the use of models but rather with the dominant role of models in research and explanation (Vayda 1995a; McCay and Vayda, 1996). A more judicious use would follow Schelling's (1978:89) admonition to ask "whether we need the model—whether the model gives us a head start in recognizing phenomena and the mechanisms that generate them and in knowing what to look for in the explanation of interesting phenomena." Optimal-foraging models of patch choice among Inuit hunter-gatherers may not be needed as much as investigations of foragers' daily decisions and actions and cognized bases for them, in relation to contextual changes (Beckerman, 1983:288-289). The economists' open-access model of problems in natural resource exploitation in fisheries (Gordon, 1954)—an important origin of modern thinking about the tragedy of the commons—may be very misleading or beside the point (McCay and Acheson, 1987a). In a particular situation, institutions affecting access or the distribution of rights to fish may or may not have played a role in causing or preventing resource decline. A good example is the recent tragic decline of Newfoundland's northern cod *(Gadus morhua)* populations and the fishing communities dependent on them. Causes for the decline and the failure of the populations to recover are still the focus of research and debate, but it is clear that fishing rights were quite restricted, and that much of the overfishing that took place was due to errors in scientific practice and the politics of interpreting scientific results (Finlayson and McCay, 1998; Hutchings et al., 1997). These are issues that would receive little attention if one relied solely on a bioeconomic model of fish population dynamics and incentives for entry into a fishery. Moreover, use of the "tragedy of the commons" model may blind investigators to other causes and justify the adoption of social policies that unfairly disadvantage the "commoners."

A much broader set of explanatory possibilities emerges if one is not constrained by a theory-driven agenda, including the "common property" model, or even a topic-driven agenda, including the admittedly fascinating study of rule-driven institutions for dealing with common-pool resources. Institutions affecting access and regulation of uses of common-pool resources may indeed be important causes of environmental change, and the study of them can be important in its own right, but if our ultimate goal is explaining the causes of environmental change, we should start elsewhere, keeping institutions in their situation-specific contexts.

## CONCLUSION: SPECIFYING THE COMMONS

It is widely appreciated that context is important to the choices and behavior of people, but the theoretical and empirical underpinning for that observation is woefully lacking. Ostrom reviewed the roles of trust, reputation, and reciprocity in enhancing levels of cooperation in structured and natural common-pool resource experiments and suggested some of the contextual factors that make a difference, including the size of the group and whether there is face-to-face communication, as well as information about past actions (Ostrom, 1998:15). The experimental and behaviorist approach she advocates and has developed addresses the question of collective action or cooperation. She calls for a concerted effort to develop "second-generation theory of boundedly rational and moral behavior" (Ostrom, 1998:16) that would focus on questions such as why levels of cooperation change and vary so greatly among individuals and situations and why specific configurations of situational conditions affect cooperation. Others in this volume contribute to the effort and questions she has outlined.

I have taken a somewhat different tack in order to draw attention to efforts by social theorists to address "commons" kinds of questions in ways that bring the social and contextual more directly into the analytic picture. One is the well-known approach that emphasizes the social construction of reality. Another is the "embeddedness" approach, which has been credited with finding an accommodation between the economic individualist and the social/structuralist ways of analyzing human behavior and institutions. Closely related is the set of approaches loosely called political ecology, which insert the macro-structural forces emphasized in political economy into studies of human-environmental relations, and which also emphasize the roles of discourse and power in the social construction of environmental and social realities—and in the construction and use of key terms, including "the commons" and "community." Context is much more than group size, the nature of communication, and group history, especially if the question is not what causes people to cooperate or not but rather how to explain particular institutional and environmental outcomes. I have also identified two approaches that provide a stronger orientation toward the "ecological" side of the human-ecological set, the "economics of flexibility" theory and the methodology that Vayda and Walters (Walters and Vayda, 2001) have come to call "event ecology."

My argument is simple, although its implications for research are not. Explaining how people relate and respond to common-pool resources requires knowing more about their "situations" and how property rights and other institutions have been specified within those historical, ecological, and cultural situations. It requires specification of those situations and their broader contexts. These are essential elements of the frameworks within which decisions and actions concerning particular "commons" are embedded. Accordingly, the task of explaining the interactions between people and the common-pool resources in their environments requires looking not only at the decision-making calculi of individuals, but also a more fully specified account of who they are, what they have done, and what they will do in relation to those common-pool resources and in relation to governance issues. It requires documenting the events that lead to and follow from particular human-environmental interactions and trying to explain causes and consequences. Depending on what appears significant to explaining such events and interactions, it may require investigating the social entities that represent them and that they help reproduce and alter (families, households, voluntary associations, ad hoc coalitions and action groups, professional societies, political parties, government agencies); their histories, values, resources, and social networks; the nature of the common-pool resource/environmental problems they face; the local, regional, and global economic and political forces that influence their behavior; their "webs of significance" or the cultural "filters" by which people perceive, construct, and understand common-pool resource/environmental problems; and the political, legal, cultural, and other institutions that mold and constrain their perceptions and interpretations and the options and incentives they face. These are some of the tasks that may be required to adequately explain "dramas of the commons," whether tragedy, comedy, romance, or just plain narratives of human ecology.

## NOTES

1 The concept of "governing mechanisms" is part of a respecification by Wilson of Habermas' communicative systems theory (see Wilson and McCay, 1999).

2 Selsky and Memon introduce the concept of domain to refer to the larger social field within which common-pool resources and the various groups involved in their use and regulation exist. They use "commons" to refer to "an enduring set of emergent local processes of resource mobilization and institution building with specific properties" (Selsky and Memon, 2000:6), and they distinguish between emergence in and emergence of a commons, arguing that a focus on domains allows study of the emergence of commons institutions.

3 "Event ecology" developed from the "response process" and "economics of flexibility" ap-

proach described earlier, as well as the method known as "progressive contextualization" (Vayda, 1983) and the epistemology of "people ecology" (McCay, 1978; see McCay, 2000b). Events are changes that are caused or that may cause and are part of causal chains, whereby one event may be involved in causing another (Walters and Vayda, 2001). This definition distinguishes events from facts, which are merely descriptive, and from factors, which imply causation but are frequently proposed through correlational analysis without adequate attention to the question of causal mechanisms.

# REFERENCES

Acheson, J.
    1981    The lobster fiefs: Economic and ecological effects of territoriality in the Maine lobster industry. *Human Ecology* 3(3):183-207.
    1987    The lobster fiefs revisited: Economic and ecologic effects of territoriality in Maine lobster fishing. Pp. 37-65 in *The Question of the Commons*, B. McCay and J. Acheson, eds. Tucson: University of Arizona Press.
Acheson, J.M., and J.A. Wilson
    1996    Order out of chaos: The case for parametric fisheries management. *American Anthropologist* 98(3):579-594.
Agrawal, A., and C.C. Gibson
    1999    Enchantment and disenchantment: the role of community in natural resource conservation. *World Development* 27(4):629-649.
Alcala, A.C., and G.R. Russ
    1990    A direct test of the effects of protective management on abundance and yield of tropical marine resources. *Journal du Conseil pour l'Exploration de la Mer* 46:40-47.
Anderson, T.L., and P.J. Hill
    1977    From free grass to fences: Transforming the commons of the American West. Pp. 200-216 in *Managing the Commons*, G. Hardin and J. Baden, eds. San Francisco: W.H. Freeman & Co.
Apostle, R., G. Barrett, P. Holm, S. Jentoft, L. Mazany, B. McCay, and K. Mikalsen
    1998    *Community, Market and State on the North Atlantic Rim: Challenges to Modernity in the Fisheries.* Toronto: University of Toronto Press.
Barton, A.H.
    1969    *Communities in Disaster: A Sociological Analysis of Collective Stress Situations.* New York: Doubleday.
Bates, D., and T.K. Rudel
    2000    The political economy of conserving tropical rain forests: A cross-national analysis. *Society and Natural Resources* 13:619-634.
Bateson, G.
    1963    The role of somatic change in evolution. *Evolution* 17:529-539.
    1972    *Steps to an Ecology of Mind.* New York: Ballantine.
Beckerman, S.
    1983    Carpe diem: An optimal foraging approach to Bari fishing and hunting. Pp. 269-299 in *Adaptive Responses of Native Amazonians*, R.B. Hames and W.T. Vickers, eds. New York: Academic Press.
Berger, P.L., and T. Luckmann
    1966    *The Social Construction of Reality: A Treatise in the Sociology of Knowledge.* New York: Anchor Books.
Berkes, F. and D. Pocock
    1987    Quota management and 'people problems': A case history of Canadian Lake Erie fisheries. *Transactions of the American Fisheries Society* 116:494-502.

Blaikie, P., and H. Brookfield
   1987   *Land Degradation and Society*. London: Methuen.
Bromley, D.W.
   1989   Institutional change and economic efficiency. *Journal of Economic Issues* 23(3):735-759.
Bromley, D.W., ed..
   1992   *Making the Commons Work: Theory, Practice, and Policy*. San Francisco: ICS Press.
Brosius, J.P., A. Lowenhaupt Tsing, and C. Zerner
   1998   Representing communities: Histories and politics of community-based natural resource management. *Society and Natural Resources* 11:157-168.
Bruce, J.W.
   1999   *Legal Bases for the Management of Forest Resources as Common Property. Community Forestry Note 14*. Rome: Food and Agriculture Organization of the United Nations.
Carrier, J.G.
   1987   Marine tenure and conservation in Papua New Guinea: Problems in interpretation. Pp. 142-167 in *The Question of the Commons*, B. McCay and J. Acheson, eds., Tucson: University of Arizona Press.
Chess, C., K.L. Salomone, and B.J. Hance
   1995   Improving risk communication in government: Research priorities. *Risk Analysis* 15(2): 127-135.
Cordell, J., ed.
   1989   *A Sea of Small Boats. Cultural Survival Report 26*. Cambridge, MA: Cultural Survival, Inc.
Descola, P., and G. Pálsson, eds.
   1996   *Nature and Society: Anthropological Perspectives*. London: Routledge.
Dietz, T.
   1994   "What should we do?" Human ecology and collective decision making. *Human Ecology Review 1*(Summer/Autumn):301-309.
Douglas, M.
   1985   *Risk Acceptability According to the Social Sciences*. Social Research Perspectives, 11. New York: Russell Sage Foundation.
   1986   *How Institutions Think*. Syracuse: Syracuse University Press.
Douglas, M., and A. Wildavsky
   1982   *Risk and Culture*. Berkeley: University of California Press.
Dryzek, J.S.
   1987   *Rational Ecology: Environment and Political Economy*. New York: Basil Blackwell.
Dyson-Hudson, R., and E.A. Smith
   1978   Human territoriality: An ecological reassessment. *American Anthropologist* 80:21-41.
Emirbayer, M., and J. Goodwin
   1994   Network analysis, culture, and the problem of agency. *American Journal of Sociology* 99:1411-1454.
Escobar, A.
   1996   Constructing nature: Elements for a poststructuralist political ecology. Pp. 46-68 in *Liberation Ecologies*, R. Peet and M. Watt, eds. London: Routledge.
Esman, M., and N. Uphoff
   1984   *Local Organizations: Intermediaries in Rural Development*. Ithaca, NY: Cornell University Press.
Finlayson, A.C., and B.J. McCay
   1998   Crossing the threshold of ecosystem resilience: The commercial extinction of northern cod. Pp. 311-337 in *Linking Social and Ecological Systems: Institutional Learning for Resilience*, C. Folke and F. Berkes, eds. Cambridge, Eng.: Cambridge University Press.

Freudenberg, W.R.
    1988    Perceived risk, real risk: Social science and the art of probabilistic risk assessment. *Science* 242:44-49.
Gardner, G.T. and P.C. Stern
    1996    *Environmental Problems and Human Behavior.* Boston: Allyn and Bacon.
Gibson, C.C. and C.D. Becker
    2000    A lack of institutional demand: Why a strong local community in western Ecuador fails to protect its forest. Pp. 135-161 in *People and Forests; Communities, Institutions, and Governance.* C.C. Gibson, M.A. McKean, and E. Ostrom, eds. Cambridge, MA: MIT Press.
Giddens, A.
    1990    *The Consequences of Modernity.* Stanford, CA: Stanford University Press.
Goldman, M.
    1998    Inventing the commons: Theories and practices of the commons professional. Pp. 20-53 in *Privatizing Nature: Political Struggles for the Global Commons*, M. Goldman, ed., New Brunswick, NJ: Rutgers University Press.
Goodlad, J.
    1999    Industry Perspectives on Rights-Based Management: The Shetland Experience. Unpublished paper presented to Fish Rights Conference, November 11-19, Fremantle, Western Australia, November. Available at http://www.fishrights/org.
Gordon, H.S.
    1954    The economic theory of a common property resource: The fishery. *Journal of Political Economy* 62:124-142.
Granovetter, M.
    1985    Economic action and social structure: The problem of embeddedness. *American Journal of Sociology* 91:481-510.
    1992    Economic action and social structure: The problem of embeddedness. Pp. 53-81 in *The Sociology of Economic Life*, M. Granovetter and R. Swedberg, eds. Boulder, CO: Westview Press.
Greenberg, J.B., and T.K. Park
    1994    Political ecology. *Journal of Political Ecology* 1(1):1-12.
Grossman, L.
    1977    Man-environment relationships in anthropology and geography. *Annals of the Association of American Geographers* 67:126-144.
Habermas, J.
    1984    *The Theory of Communicative Action.* Volume 1: reason and the Rationalization of Society. T. McCarthy, Trans. Boston: Beacon Press.
Hajer, M.A.
    1995    *The Politics of Environmental Discourse: Ecological Modernization and the Policy Process.* Oxford, Eng.: Clarendon Press.
Hall, C.
    1998    Institutional solutions for governing the global commons: Design, factors, and effectiveness. *Journal of Environment and Development* 7(2):86-114.
Hanna, S.
    1995    User participation and fishery management performance within the Pacific Fishery Management Council. *Ocean and Coastal Management* 28(1-3):23-44.
Helgason, A. and G. Pálsson
    1997    Contested commodities: The moral landscape of modernist regimes. *Journal of the Royal Anthropological Institute (incorporating Man)* 3(3):451-471.
Hoel, A.H.
    2000    Performance of Exclusive Economic Zones. Institutional Dimensions of Global Environmental Change Project, Dartmouth College, Hanover, NH. Available: http://www.dartmouth.edu/~idgec.

Hutchings, J.A., C.Walters, and R.L. Haedrich
    1997   Is scientific inquiry incompatible with government information control? *Canadian Journal of Fisheries and Aquatic Sciences* 54:1198-1210.
Jentoft, S., B.J. McCay, and D. Wilson
    1998   Social theory and fisheries co-management. *Marine Policy* 22(4/5):423-436.
Johnson, J.C., and D.C. Griffith
    1996   Pollution, food safety, and the distribution of knowledge. *Human Ecology* 24(1):87-108.
Krauss, C.
    1983   The elusive process of citizen activism. *Social Policy* 14 (Fall):50-55.
Leach, M., R. Mearns, and I. Scoones
    1997   *Environmental Entitlements: A Framework for Understanding the Institutional Dynamics of Environmental Change.* IDS Working Paper. Sussex, Eng.: Institute of Development Studies, University of Sussex.
Lee, K.N.
    1993   *Compass and Gyroscope: Integrating Science and Politics for the Environment.* Washington, DC: Island Press.
Lees, S.H.
    1974   Hydraulic development as a process of response. *Human Ecology* 2:159-175.
Libecap, G.
    1986   Property rights in economic history: Implications for research. *Explorations in Economic History* 23:227-252.
Lindblom, C.E.
    1959   The science of "muddling through." *Public Administration Review* 19:79-88.
    1979   Still muddling, not yet through. *Public Administration Review* 39:517-526.
McCay, B.J.
    1978   Systems ecology, people ecology, and the anthropology of fishing communities. *Human Ecology* 6(4):397-422.
    1979   'Fish is scarce': Fisheries modernization on Fogo Island, Newfoundland. Pp. 155-189 in *North Atlantic Maritime Cultures*, R. Andersen, ed. The Hague: Mouton.
    1980   A fishermen's cooperative, limited: Indigenous resource management in a complex society. *Anthropological Quarterly* 53:29-38.
    1981   Optimal foragers or political actors? Ecological changes in a New Jersey fishery. *American Ethnologist* 8:356-382.
    1988   Muddling through the clam Beds: Cooperative management of New Jersey's hard clam spawner sanctuaries. *Journal of Shellfish Research* 7(2):327-340.
    1996   Foxes and others in the henhouse? Environmentalists and the fishing industry in the U.S. Regional Council System. Pp. 380-390 in *Fisheries Resource Utilization and Policy; Proceedings of the World Fisheries Congress, Theme 2*, R.M. Meyer, C. Zhang, M.L. Windsor, B. McCay, L. Hacek, and R. Much, eds. New Delhi: Oxford & IB. Publishing Co. Pvt. Inc.
    2000a  Post-modernism and the management of natural and common resources, Presidential address to the International Association for the Study of Common Property, Part I. *Common Property Resource Digest* 54 (September):1-6.
    2000b  The Evolutionary Ecology of Ecologies in Anthropology. Unpublished paper presented at the Annual Meeting of the American Anthropological Association, San Francisco, November 15-20.
McCay, B.J., and J.A. Acheson
    1987a  Human ecology of the commons. Pp. 1-34 in *The Question of the Commons*, B. McCay and J. Acheson, eds. Tucson: University of Arizona Press.
    1987b  *The Question of the Commons.* Tucson: University of Arizona Press.
McCay, B.J., and S. Jentoft
    1996   From the bottom up: Participatory issues in fisheries management. *Society & Natural Resources* 9:237-250.

1998    Market or community failure? Critical perspectives on common property research. *Human Organization* 57(1):21-29.

McCay, B.J., and A.P. Vayda
1996    Question-Driven Research in Ecological Anthropology: Lessons from Applied Research in Fisheries, Forestry, and Conservation. Unpublished manuscript, Department of Human Ecology, Rutgers University.

McGuire, T.R.
1997    The last northern cod. *Journal of Political Ecology* 4: 41-54.

McLaughlin, P.
1996    Resource mobilization and density dependence in cooperative purchasing associations in Saskatchewan, Canada. *Rural Sociology* 61(2):326-348.
no      Toward an Ecology of Social Action: Merging the Ecological and Constructivist Tradi-
date    tions. Unpublished manuscript, Department of Anthropology and Sociology, Hobart and William Smith Colleges, Geneva, NY.

Memon, P.A., and J. Selsky
1998    Institutional design for the co-management of an urban harbour in New Zealand. *Society and Natural Resources* 11(4):587-602.

Miller, Marc L. and John van Maanen
1979    "Boats don't fish, people do": Some ethnographic notes on the Federal management of fisheries in Gloucester. Human Organization 38 (4): 377-385.

Morren, G.R.B., Jr.
1980    The rural ecology of the British drought of 1975-76. *Human Ecology* 8(1):33-63.

Mosse, D.
1997    The symbolic making of a common property resource: History, ecology and locality in a tank-irrigated landscape in South India. *Development and Change* 28:467-504.

National Research Council
1996    *Understanding Risk: Informing Decisions in a Democratic Society.* Committee on Risk Characterization. P.C. Stern and H.V. Fineberg, eds. Washington, DC: National Academy Press.

Nelson, R.K.
1983    *Make Prayers to the Raven: A Koyukon View of the Northern Forest.* Chicago: University of Chicago Press.

Netting, R.M.
1976    What alpine peasants have in common: Observations on communal tenure in a Swiss village. *Human Ecology* 4:135-146.
1981    *Balancing on an Alp: Ecological Change and Continuity in a Swiss Mountain Community.* Cambridge, Eng.: Cambridge University Press.

Nietschmann, B.
1973    *Between Land and Water: The Subsistence Ecology of the Miskito Indians, Eastern Nicaragua.* New York: Seminar Press.

Noonan, D.S.
1998    International fisheries management institutions: Europe and the South Pacific. Pp. 165-177 in *Managing the Commons.* 2d ed., J.A. Baden and D.S. Noonan, eds. Bloomington: Indiana University Press.

North, D.
1981    *Structure and Change in Economic History.* New York: W.W. Norton.
1990    *Institutions, Institutional Change and Economic Performance.* Cambridge, Eng.: Cambridge University Press.

Oakerson, R.J.
1992    Analyzing the commons: A framework. In *Making the Commons Work: Theory, Practice, and Policy*, D. Bromley, ed., San Francisco: ICS Press.

Olson, M.
   1965    *The Logic of Collective Action*. Cambridge, MA: Harvard University Press.
Ostrom, E.
   1987    Institutional arrangements for resolving the commons dilemma: Some contending approaches. Pp. 250-265 in *The Question of the Commons*, B.J. McCay and J.M. Acheson, eds. Tucson: University of Arizona Press.
   1990    *Governing the Commons: The Evolution of Institutions for Collective Action*. Cambridge, Eng.: Cambridge University Press.
   1992    Community and the endogenous solution of commons problems. *Journal of Theoretical Politics* 4(3):343-351.
   1998    A behavioral approach to the rational choice theory of collective action, presidential address, American Political Science Association, 1987. *American Political Science Review* 92(1):1-22.
   2001    Reformulating the commons. In *The Commons Revisited: An Americas Perspective*, J. Burger, R. Norgaard, E. Ostrom, D. Policansky, and B. Goldstein, ed. Washington, DC: Island Press.
Pálsson, G.
   in       Nature and society in the age of postmodernity. In *Culture/Power/History/Nature: Ecol-*
   press    *ogies for a New Millennium,* A. Biersack and J. Greenberg, eds. Oxford, Eng.: Oxford University Press.
Peluso, N.L.
   1993    Coercing conservation? The politics of state resource control. *Global Environmental Change* 3(2):199-217.
Pendzich, D., G. Thomas, and T. Wohigenant, eds.
   1994    *The Role of Alternative Conflict Management in Community Forestry*. Forests, Trees and People Working Paper No. 1. Rome: Food and Agriculture Organization of the United Nations.
Peters, P.E.
   1987    Embedded systems and rooted models: The grazing lands of Botswana and the commons debate. Pp. 171-194 in *The Question of the Commons*, B. McCay and J. Acheson, eds., Tucson: University of Arizona Press.
Polanyi, K.
   1944    *The Great Transformation*. Boston: Beacon Press.
Polunin, N.
   1984    Do traditional marine tenure systems conserve? Indonesian and New Guinean evidence. Pp. 267-283 in *Maritime Institutions in the Western Pacific*. Senri Ethnological Studies No. 17, K. Ruddle and T. Akimichi, eds. Osaka: National Museum of Ethnology.
Rapport, D.J., and J.E. Turner
   1977    Economic models in ecology. *Science* 195:367-373.
Renn, O., T. Webler, and H. Kastenholz
   1996    Perception of uncertainty: Lessons for risk management and communication. Pp. 205-226 in *Scientific Uncertainty and Its Influence on the Public Communications Process*, V.H. Sublet, V.T. Covello, and T.L. Tinker, eds. Dordrecht, Neth.: Kluwer Academic Publishers.
Rosa, E.A.
   1996    Metatheoretical foundations for post-normal risk. *Journal of Risk Research* 1(1):15-44.
Rudel, T.K.
   1980    Social responses to commodity shortages: The 1973-19i4 gasoline crisis. *Human Ecology* 8(3):193-212.
   2000    Organizing for sustainable development: Conservation organizations and the struggle to protect tropical rain forests in Esmeraldas, Ecuador. *Ambio* 29(2):78-82.

Sawyer, S.
   1996   Indigenous initiatives and petroleum politics in the Ecuadorian Amazon. *Cultural Survival Quarterly* 20(1):26-30.
Schelling, T.C.
   1978   *Micromotives and Macrobehavior.* New York: W.W. Norton.
Schlager, E.
   1994   Fishers' institutional responses to common-pool resource dilemmas. Pp. 247-265 in *Rules, Games, and Common-Pool Resources.* E. Ostrom, R. Gardner, and J. Walker, eds. Ann Arbor: University of Michigan Press.
Scott, W.R.
   1995   *Institutions and Organizations.* Thousand Oaks, CA: Sage.
Selsky, J.W., and P.A. Memon
   2000   Emergent Commons: Local Responses in Complex Common-Pool Resources Systems. Unpublished paper presented at 8th Biennial Conference of the International Association for the Study of Common Property (IASCP), Bloomington, IN, June 1-4.
Simon, H.A.
   1983   *Reason in Human Affairs.* Stanford, CA: Stanford University Press.
Singleton, S., and M. Taylor
   1992   Common property, collection action and community. *Journal of Theoretical Politics* 4(3):309-324.
Slobodkin, L.B.
   1968   Toward a predictive theory of evolution. Pp. 187-205 in *Population Biology and Evolution*, R.C. Lewontin, ed. Syracuse: Syracuse University Press.
Slobodkin, L.B., and A. Rapoport
   1974   An optimal strategy of evolution. *Quarterly Review of Biology* 49:181-200.
Smith, E.A., and B. Winterhalder
   1992   *Evolutionary Ecology and Human Behavior.* New York: Aldine de Gruyter.
Smith, M.E.
   1990   Chaos in fisheries management. *MAST/Maritime Anthropological Studies* 3(2):1-13.
Snow, D.A., and R.D. Benford
   1992   Master frames and cycles of protest. Pp. 133-155 in *Frontiers in Social Movement Theory*, A.D. Morris, and C.M. Mueller, eds. New Haven and London: Yale University Press.
Snow, D.A., E.B. Rockford, Jr., S.K. Worden, and R.D. Benford
   1986   Frame alignment processes, micromobilization and movement participation. *American Sociological Review* 51:464-481.
Soulé, M.E., and G. Lease, eds.
   1995   *Reinventing Nature? Responses to Postmodern Deconstruction.* Washington, DC: Island Press.
Steins, N.A., and V.M. Edwards
   1999   Collective action in common-pool resource management: The contribution of a social constructivist perspective to existing theory. *Society & Natural Resources* 12:539-557.
Stern, P.C., T. Dietz, T. Abel, G. A. Guagnono, and L. Kalof
   1999   A Value-Belief-Norm Theory of Support for Social Movements: The Case of Environmentalism. *Human Ecology Review* 6(2): 81-97.
Stocks, A.
   1987   Resource management in an Amazon Varzea Lake ecosystem: The Cocamilla case. Pp. 108-120 in *The Question of the Commons*, B. McCay and J. Acheson, eds. Tucson: University of Arizona Press.
Stonich, S.C.
   1996   Reclaiming the commons; Grassroots resistance and revitalization in Honduras. *Cultural Survival Quarterly* 20(1):31-35.

Swedberg, R., and M. Granovetter
   1992   Introduction. Pp. 1-26 in *The Sociology of Economic Life*, M. Granovetter and R. Swedberg, eds. Boulder, CO: Westview Press.
Taylor, L.
   1987   'The river would run red with blood': Community and common property in an Irish fishing settlement. Pp. 290-307 in *The Question of the Commons*, B. McCay and J. Acheson, eds. Tucson: University of Arizona Press.
Taylor, M.
   1982   *Community, Anarchy and Liberty.* Cambridge, Eng.: Cambridge University Press.
Taylor, P.
   1998   How does the commons become tragic? Simple models as complex socio-political constructions. *Science as Culture* 7(4):449-464.
Vayda, A.P.
   1976   *War in Ecological Perspective: Persistence, Change, and Adaptive Processes in Three Oceanian Societies.* New York: Plenum Press.
   1983   Progressive contextualization: Methods for research in human ecology. *Human Ecology* 11(3):265-281.
   1995a  Failures of explanation in Darwinian ecological anthropology: Part I. *Philosophy of the Social Sciences* 25(2):219-249.
   1995b  Failures of explanation in Darwinian ecological anthropology: Part II. *Philosophy of the Social Sciences* 25(3):360-375.
Vayda, A.P., and B.J. McCay
   1975   New directions in ecology and ecological anthropology. *Annual Review of Anthropology* 4:293-306.
   1977   Problems in the identification of environmental problems. Pp. 411-418 in *Subsistence and Surficial: Rural Ecology in the Pacific*, T.P. Baylor-Smith and. R.G.A. Feachem, eds. London: Academic Press.
Vayda, A.P., B.J. McCay, and C. Eghenter
   1991   Concepts of process in social science explanations. *Philosophy of the Social Sciences*, 21(3):318-331.
Vayda, A.P., and B.B. Walters
   1999   Against political ecology. *Human Ecology* 27(1):167-179.
Waddell, E.
   1975   How the Enga cope with frost: Responses to climatic perturbations in the Central Highlands of New Guinea. *Human Ecology* 3:249-273.
Walters, B.B. and A.P. Vayda
   2001   Event Ecology in the Philippines: Methods and Explanations in the Study of Human Actions and Their Environmental Effects. Unpublished manuscript; revised version of paper presented by Walters at the 97th Annual Meeting of the American Association of Geographers, New York, NY, February 27-March 3.
Walters, C.J.
   1986   *Adaptive Management of Renewable Resources.* New York: Macmillan.
Weinstein, N., and P. Sandman
   1992   A model of the precaution adoption process: Evidence from home radon testing. *Health Psychology* 11:170-180.
Weinstein, N.D., A. Rothman, and S. Sutton
   1998   Stage theories of health behavior. *Health Psychology* 17:290-299.
West, C.P.
   2000   The Practices, Ideologies, and Consequences of Conservation and Development in Papua New Guinea. Unpublished Ph.D. thesis, Anthropology, Rutgers-the State University, New Brunswick, NJ.

Williams, D.M.
  2000    Representation of nature on the Mongolian steppe: An investigation of scientific knowl-
          edge construction. *American Anthropologist* 102(3): 503-519
Wilson, D.C., and B.J. McCay
  1999    Embeddedness and Governance Mechanisms: An Approach to the Study of Institutions.
          Unpublished manuscript, August.
Zerner, C., ed.
  2000    *People, Plants, & Justice: The Politics of Nature Conservation.* New York: Columbia
          University Press.

# 12

# An Evolutionary Theory
# of Commons Management

*Peter J. Richerson, Robert Boyd, and Brian Paciotti*

ommon property and common-pool resources dilemmas are examples of
the broader problem of cooperation, a problem that has long interested
evolutionists. In both the *Origin* and *Descent of Man*, Darwin worried
about how his theory might handle cases such as the social insects in which individuals sacrificed their chances to reproduce by aiding others. Darwin could see
that such sacrifices ordinarily would not be favored by natural selection. He argued that honeybees and humans were similar: Among honeybees a sterile worker
who sacrificed her own reproduction for the good of the hive would enjoy a
vicarious reproductive success through her sibling reproductives. Humans, Darwin (1874:178-179) thought, competed tribe against tribe as well as individually,
and the "social and moral faculties" evolved under the influence of group competition:

> It must not be forgotten that although a high standard of morality gives but slight
> or no advantage to each individual man and his children over other men of the
> tribe, yet that an increase in the number of well-endowed men and an advance-
> ment in the standard of morality will certainly give an immense advantage to
> one tribe over another. A tribe including many members who, from possessing
> in a high degree the spirit of patriotism, fidelity, obedience, courage, and sympa-
> thy, were always ready to aid one another, and to sacrifice themselves for the
> common good, would be victorious over most other tribes; and this would be
> natural selection.

More than a century has passed since Darwin wrote, but the debate among
evolutionary social scientists and biologists is still framed in similar terms—the
conflict between individual and prosocial behavior guided by selection on individuals versus selection on groups. In the meantime, social scientists have devel-

oped parallel theories of cooperation—rational choice theory takes an individual-istic approach while functionalism analyzes the prosocial aspects of institutions.

In this chapter we review the evolutionary theory relevant to the question of human cooperation and compare the results to other theoretical perspectives. Then we review some of our own work distilling a compound explanation that we believe gives a plausible account of human cooperation and selfishness. This account leans heavily on group selection on *cultural* variation but also includes lower level forces driven by both micro-prosocial and purely selfish motives. Next, we review the empirical literature in commons management. Although much work remains to be done on the problem, we conclude that the existing evidence is consistent with our account. Then, we use our hypothesis to derive lessons for applied research in institution building for commons management. On the one hand, the theory of cultural group selection suggests that humans have cooperative sentiments usually assumed to be absent in rational choice theories. On the other hand, the slow rate at which cooperative institutions evolve suggests that considerable friction will afflict our ability to grow up commons manage-ment institutions if they do not already exist and to readapt existing institutions to rapid technological and economic change. A better understanding of the way cooperative institutions arise in the long run promises better tools to foster their more rapid evolution when needed and to regulate their performance as neces-sary.

## THEORIES OF COOPERATION

Our ideas about cooperation are drawn from many sources. Folk sources include diverse religious doctrines, norms and customs, and folk psychology. Anthropologists and historians document an immense diversity of human social organizations and most of these are accompanied by moral justifications, if often contested ones. Johnson and Earle (1987) provide a good introduction to the vast body of data collected by sociocultural anthropologists. The cross-cultural study of commons management is already a well-advanced field drawing on the disci-plines of anthropology, political science, and economics (Agrawal, this vol-ume:Chapter 2; Baland and Platteau, 1996; Bardhan and Dayton-Johnson, this volume:Chapter 3; Berkes, this volume:Chapter 9; McCay, this volume:Chapter 11; (Ostrom, 1998).

### Human Cooperation Is Extensive and Diverse

Human cooperation has a number of features begging explanation:

• *Humans are prone to cooperate, even with strangers.* Thus many people cooperate in anonymous one-shot prisoners' dilemma (PD) games (Marwell and Ames, 1981), and often vote altruistically (Sears and Funk, 1990). People begin contributing substantially to public goods sectors in economic experiments (Falk et al., this volume:Chapter 5; Kopelman et al., this volume:Chapter 4; Ostrom, 1998). The experimental results accord with common experience. Most of us have traveled in foreign cities, even poor foreign cities filled with strange people for whom our possessions and spending money are worth a small fortune, and found risk of robbery and commercial chicanery to be small.

- *Cooperation is contingent on many things.* Not everyone cooperates. Aid to distressed victims increases substantially if a potential altruist's empathy is engaged (Batson, 1991). Being able to discuss a game beforehand and to make promises to cooperate affect success (Dawes et al., 1990). The size of the resource, technology for exclusion and exploitation of the resource, and similar gritty details affect whether cooperation in commons management arises (Ostrom, 1990:202-204). Scientific findings again correspond well to personal experience. Sometimes we cooperate enthusiastically, sometimes reluctantly, and sometimes not at all. People vary considerably in their willingness to cooperate even under the same environmental conditions.

- *Institutions matter.* People from different societies behave differently because their habits have been inculcated by long participation in societies with different institutions. In repeated play common property experiments, initial defections induce further defections until the contribution to the public-goods sector approaches zero. However, if players are allowed to exercise strategies they might use in the real world, for example to punish those who defect, participation in the commons stabilizes (Fehr and Tyran, 1996). The strategies for successfully managing commons are generally institutionalized in sets of rules that have legitimacy in the eyes of the participants (Ostrom, 1990:Chapter 2). Families, local communities, employers, nations, and governments all tap our loyalties with rewards and punishments and greatly influence our behavior.

- *Institutions are the product of evolution.* The elegant studies by Nisbett's group show how people's affective and cognitive styles become intimately entwined with their social institutions (Cohen and Vandello, 2001; Nisbett and Cohen, 1996; Nisbett et al., in press). Because such complex traditions are so deeply ingrained, they are slow both to emerge and to decay. Many commons management institutions have considerable time depths (Ostrom, 1990:Chapter 3). Throughout most of human history, institutional change was so slow as to be nearly imperceptible by individuals. Today, change is rapid enough to be perceptible. Even universities, impeded as they are by conservative faculties deeply suspicious of change, change measurably on the time scale of a generation.

- *Variation in institutions is huge.* Already with its very short list of societies and games, the experimental ethnography approach of Henrich et al. (2001) and Nisbett et al. (in press) has uncovered striking differences. The cross-cultural commons work has uncovered much more, suggesting that a rich trove awaits the experimentalists. Agrawal (this volume:Chapter 2) describes the large number of conditions (38 and counting) that have been shown to affect whether local cooperation in commons management arises. Plausibly, design complexity, coordination equilibria, and other phenomena generate multiple evolutionary equilibria and much historical contingency in the evolution of particular institutions (Boyd and Richerson, 1992c). We all have at least some experience of how differently different communities, different universities, and different countries solve the same problems.

### Evolutionary Models Can Explain
### the Nature of Preferences and Institutions

These facts present a challenge to rational actor theories. High levels of cooperation are difficult to reconcile with the usual assumption of self-regarding preferences, and the diversity of institutional solutions is a challenge to any theory based on a universal human nature. The "second generation" bounded rational choice theory championed by Ostrom (1998), and the "situated" rational choice characterized by McCay (this volume:Chapter 11), address these challenges from within the rational choice tradition. These approaches add a psychological basis and institutional constraints to the standard rational choice theory. Although psychological and social structures are invoked to explain individual behavior and its variation, an explanation for psychology and social structure is not part of the theory.

Evolutionary theory permits us to address the origin of preferences. A number of economists have noted the neat fit between evolutionary theory and economic theory (Becker, 1976; Hirshleifer, 1977). Evolution, they observed, explains what organisms want, and economics explains how they should go about getting what they want. Without evolution, preferences are exogenous, to be estimated empirically, but not explained. To do a satisfactory job of explaining human social behavior, we need to expand the spare concept of preferences to include the conceptually richer properties of individuals and institutions of bounded and situated rationality. Then, to explain why humans have the unusual forms of social behavior depicted in our list of stylized facts, we need to appeal, we believe, to the special properties of *cultural* evolution.

Evolutionary models have both intellectual and practical payoffs. The intellectual payoff is that evolutionary models link answers to contemporary puzzles to crucial long time-scale processes. The most important economic phenomenon of the past 500 years is the rise of capitalist economies and their tremendous impact on every aspect of human life. Expanding the time scale a bit, the most important phenomena of the past 10 millennia are the evolution of ever more complex social systems and ever more sophisticated technology following the origins of agriculture. A real explanation of both current behavior and its variation must be linked to such long-run processes, where the times to reach evolutionary equilibria are measured in millennia. More practically, the dynamism of the contemporary world creates major stresses on the institutions that are used to manage commons. Evolutionary theory often will be useful because it will lead to an understanding of how to accelerate institutional evolution to better track rapid technological and economic change. (For an analogous argument in the context of medical practice, see Nesse and Williams, 1995.)

### Evolutionary Models Account for the Processes That Shape Heritable Genetic and Cultural Variation Through Time

Evolutionary explanations are *recursive*. Individual behavior results from an interaction of inherited attributes and environmental contingencies. In most species genes are the main inherited attributes, but in humans inherited cultural information is also important. Individuals with different inherited attributes may develop different behaviors in the same environment. Every generation, evolutionary processes—natural selection is the prototype—impose environmental effects on individuals as they live out their lives. Cumulated over the whole population, these effects change the pool of inherited information, so that the inherited attributes of individuals in the next generation differ, usually subtly, from the attributes in the previous generation. Over evolutionary time, a lineage cycles through the recursive pattern of causal processes once per generation, more or less gradually shaping the gene pool and thus the succession of individuals that draw samples of genes from it. Statistics that describe the pool of inherited attributes, such as gene frequencies, are basic state variables of evolutionary analysis. They are what change over time.

Note that in a recursive model, we explain individual behavior and population-level processes in the same model. Individual behavior depends, in any given generation, on the gene pool from which inherited attributes are sampled. The pool of inherited attributes depends in turn on what happens to a population of individuals as they express those attributes. Evolutionary biologists have a long list of processes that change the gene frequencies, including natural selection, mutation, and genetic drift. However, no organism experiences natural selection. They either live or die; reproduce or fail to reproduce. If, in a particular environment, some *types* of individuals do better than others and if this variation has a heritable basis, then *we* label as "natural selection" the resulting changes in gene frequencies. We use abstract categories like selection to describe such specific events because we wish to build up, concrete case by concrete case, some useful generalizations about evolutionary process. Few would argue that evolutionary biology is the poorer for investing effort in the generalizing project.

Although the processes that lead to cultural change are very different from those that lead to genetic change, their logic is the same. For example, the cultural generation time is short in the case of ideas that spread rapidly, but modeling rapidly evolving cultural phenomena like semiconductor technology presents no special problems (Boyd and Richerson, 1985:68-69). Similarly, human choices include ones that modify inherited attributes directly rather than indirectly by natural selection. These "Lamarckian" effects are added easily to models, and the models remain evolutionary so long as rationality remains bounded. The degenerate case, of course, needs no recursion because everything happens in the first generation (instantly in a typical rational choice model). Evolutionary models are a natural extension of the concept of bounded rational choice. They help explain how the innate and cultural constraints on choice and on rationality arise (Boyd and Richerson, 1993).

## Evolution is Multilevel

Evolutionary theory is always *multilevel;* at a minimum it keeps track of properties of individuals, like their genotypes, and of the population, such as the frequency of a particular gene. Other levels may also be important. Phenotypes are derived from many genes interacting with each other and the environment. Populations may be structured, perhaps divided into social groups with limited exchanges of members. Thus, evolutionary theories are systemic, integrating every part of biology. In principle, everything that goes into causing change through time plays its proper part in the theory.

This in-principle completeness led Mayr (1982) to speak of "proximate" and "ultimate" causes in biology. Proximate causes are those that physiologists and biochemists generally treat by asking *how* an organism functions. These are the causes produced by individuals with attributes interacting with environments and producing effects on them. Do humans use innate cooperative propensities to solve commons problems or do they have only self-interested innate motives? Or are the causes more complex than either proposal? Ultimate causes are evolutionary. The ultimate cause of an organism's behavior is the history of evolution that shaped the gene pool from which our samples of innate attributes are drawn. Evolutionary analyses answer *why* questions. Why do human communities typically solve at least some of the commons dilemmas and other cooperation problems on a scale unknown in other apes and monkeys? Human-reared chimpanzees are capable of many human behaviors, but they nevertheless retain many chimp behaviors and cannot act as full members of a human community (Temerlin, 1975). Thus we know that humans have different innate influences on their behavior than chimpanzees, and these must have arisen in the course of the two species' divergence from our common ancestor.

In Darwinian evolutionary theories, the ultimate sources of cooperative behavior are classically categorized into three evolutionary processes operating at different levels of organization.

• *Individual-level selection.* Individuals and the variants they carry are obviously a locus of selection. Selection at this level favors selfish individuals who are evolved to maximize their own survival and reproductive success. Pairs of self-interested actors can cooperate when they interact repeatedly (Axelrod and Hamilton, 1981; Trivers, 1971). Alexander (1987) argued that such reciprocal cooperation also can explain complex human social systems, but most formal modeling studies make this proposal doubtful (Boyd and Richerson, 1988, 1989; Leimar and Hammerstein, 2001; Nowak and Sigmund, 1998).

• *Kin selection.* Hamilton's (1964) papers showing that kin should cooperate to the extent that they share genes identical by common descent offer one of the theoretical foundations of sociobiology. Kin selection can lead to cooperative social systems of a remarkable scale, as illustrated the colonies of termites, ants, and some bees and wasps. However, most animal societies are small because individuals have few close relatives. It is the fecundity of insects, and in one case rodents, that permits a single queen to produce huge numbers of sterile workers and hence large, complex societies composed of close relatives (Campbell, 1983).

• *Group selection.* Selection can act on any pattern of heritable variation that exists (Price, 1970). Darwin's model of the evolution of cooperation by inter-tribal competition is perfectly plausible, as far as it goes. The problem is that genetic variation between groups other than kin groups is hard to maintain unless the migration between groups is very small or unless some very powerful force generates between-group variation (Aoki, 1982; Boorman and Levitt, 1980; Eshel, 1972; Levin and Kilmer, 1974; Rogers, 1990; Slatkin and Wade, 1978; Wilson, 1983). In the case of altruistic traits, selection will tend to favor selfish individuals in all groups, tending to aid migration in reducing variation between groups. The success of kin selection in accounting for the most conspicuous and highly organized animal societies (except humans) has convinced most, but by no means all, evolutionary biologists that group selection is of modest importance in nature (see Sober and Wilson, 1998, for a group selectionist's eye view of the controversy).

We could make this picture much more complex by adding higher and lower levels and cross-cutting forms of structure. Many examples from human societies will occur to the reader, such as gender. Indeed, Rice (1996) has demonstrated elegantly that selection on genes expressed in the different sexes sets up a profound conflict of interest between these genes. If female *Drosophila* are prevented from evolving defenses, male genes will evolve that seriously degrade female fitness. The genome is full of such conflicts, usually muted by the fact that an individual's genes are forced by the evolved biology of complex organisms to all have an equal shot at being represented in one's offspring. Our own bodies are a group-selected community of genes organized by elaborate "institutions" to ensure fairness in genetic transmission, such as the lottery of meiosis that gives each chromosome of a pair a fair chance at entering the functional gamete (Maynard Smith and Szathmáry, 1995).

## Culture Evolves

In theorizing about human evolution, we must include processes affecting *culture* in our list of evolutionary processes alongside those that affect genes. Culture is a system of inheritance. We acquire behavior by imitating other individuals much as we get our genes from our parents. A fancy capacity for high-fidelity imitation is one of the most important derived characters distinguishing us from our primate relatives (Tomasello, 1999). We are also an unusually docile animal (Simon, 1990) and unusually sensitive to expressions of approval and disapproval by parents and others (Baum, 1994:218-219). Thus parents, teachers, and peers can rapidly, easily, and accurately shape our behavior compared to training other animals using more expensive material rewards and punishments. Finally, once children acquire language, parents and others can communicate new ideas quite economically. Our own contribution to the study of human behavior is a series of mathematical models in the Darwinian style of what we take to be the fundamental processes of cultural evolution (e.g., Boyd and Richerson, 1985). The application of Darwinian methods to the study of cultural evolution was advocated forcefully by Campbell (1965, 1975). Cavalli-Sforza and Feldman (1973) constructed the first mathematical models to analyze cultural recursions (see also Durham, 1991).

The list of processes that shape cultural change includes:

• *Biases.* Humans do not passively imitate whatever they observe. Rather, cultural transmission is biased by decision rules that individuals apply to the variants they observe or try out. The rules behind such selective imitation may be innate or the result of earlier imitation or a mixture of both. Many types of rules might be used to bias imitation. Individuals may try out a behavior and let reinforcement guide acceptance or rejection. Or they may use various rules of thumb to reduce the need for costly trials and punishing errors. The use of a conformist rule of the form "when in Rome do as the Romans do" is an example that is important in our hypothesis about the origins of cooperative tendencies in human behavior.

• *Nonrandom variation.* Genetic innovations (mutations, recombinations) are random with respect to what is adaptive. Human individual innovation is guided by many of the same rules that are applied to biasing ready-made cultural alternatives. Bias and learning rules have the effect of increasing the rate of evolution relative to what can be accomplished by random mutation, recombination, and natural selection. We believe that culture originated in the human lineage as an adaptation to the Plio-Pleistocene ice-age climate deterioration, which included much rapid, high-amplitude variation of just the sort that would favor adaptation by biased innovation and imitation (Richerson and Boyd, 2000).

• *Natural selection.* Because selection operates on any form of heritable variation and imitation and teaching are forms of inheritance, selection will influence cultural as well as genetic evolution. However, selection on culture is liable to favor behaviors different from those favored by selection on genes. Because we often imitate peers, culture is liable to selection at the subindividual level, potentially favoring pathogenic cultural variants—selfish memes (Blackmore, 1999). On the other hand, rules like conformist imitation have the opposite effect. By tending to suppress cultural variation within groups such rules protect variation between them, potentially exposing our cultural variation to much stronger group selection effects than our genetic variation (Henrich and Boyd, 1998; Soltis et al., 1995). Human patterns of cooperation may owe much to cultural group selection.

## Evolutionary Models Are Consistent with a Wide Variety of Theories

Evolutionary theory prescribes a method, not an answer, and a wide range of particular hypotheses can be cast in an evolutionary framework. If population-level processes are important, we can set up a system for keeping track of heritable variation, and the processes that change it through time. Darwinism as a method is not at all committed to any particular picture of how evolution works or what it produces.

The view that many social scientists have of Darwinism is influenced too heavily by the work of human sociobiologists. Many things can be said in defense of this enterprise (Borgerhoff-Mulder et al., 1997) and much useful work goes on under its major research programs, human behavioral ecology (Cronk et al., 2000) and evolutionary psychology (Barkow et al., 1992). However, these research programs have two major weaknesses: neglect of culture and a taboo against group selection.

Sociobiologists typically assume that culture is a strictly proximate phenomenon, akin to individual learning (e.g., Alexander, 1979), or constrained so strongly by genes as to be virtually proximate (Wilson, 1998). As Alexander (1979:80) puts it, "Cultural novelties do not replicate or spread themselves, even indirectly. They are replicated as a consequence of the behavior of vehicles of gene replication." Commons institutions are deeply rooted in cultural traditions. Theoretical models show that the processes of cultural evolution can behave differently in critical respects from those only including genes. If such effects are important in the real world, neglecting them is a bad bet to get the approximately correct answers we hope to win using evolutionary theory.

Most evolutionary biologists believe that group beneficial behavior is always a side effect of individual payoffs. We have already noted the problems with maintaining variation between groups in theory and the seeming success of alternative explanations. Persuaded by the biologist's arguments, most social science scholars from the Darwinian tradition have followed the argument forcefully articulated by Williams (1966) and have anathematized group selection.[1] However, *cultural* variation is more plausibly susceptible to group selection than is genetic variation. For example, if people use a somewhat conformist bias in acquiring important social behaviors, the variation between groups needed for group selection to operate is protected from the variance-reducing force of migration between groups (Boyd and Richerson, 1985:Chapter 7). We believe considerable evidence supports the hypothesis that cultural group selection has played an important role in human social evolution (Richerson and Boyd, 2001).

## Evolutionary Models Are Widely Used in the Social Sciences

Although evolutionary tools are not yet commonplace in the study of human behavior, the general approach we advocate has a long history (Campbell, 1965, 1975) and several vigorous currently active branches. We mentioned evolutionary psychology and human behavioral ecology already. Others include evolutionary economics (Alchian, 1950; Day and Chen, 1993; Gintis, 2000; Hodgson, 1993; Witt, 1992), evolutionary sociology (Dietz and Burns, 1992; Luhmann, 1982; Maryanski and Turner, 1992; McLaughlin, 1988), evolutionary organization science (Baum and McKelvey, 1999; Hannan and Freeman, 1989), evolutionary epistemology (Callebaut and Pinxten, 1987; Derksen, 1998; Hull, 1988), evolutionary behavior analysis (Baum, 1994), and applied mathematics (Vose, 1999). The concepts of the meme (Blackmore, 1999), of complex adaptive systems (Holland, 1995), and of universal Darwinism (Dennett, 1996) have attracted much attention. Some of the most interesting evidence for the importance of evolutionary theory in the study of culture comes from the not infrequent reinvention of basic Darwinism when scholars in the social sciences find themselves in need of it. Empirical research traditions with strongly Darwinian overtones include historical linguistics (Mallory, 1989), sociolinguistics (Labov, 1973), studies of the diffusion of innovations (Rogers, 1995), human social learning theory (Bandura, 1986), experimental cultural evolution (Insko et al., 1983), and religious demography (Roof and McKinney, 1987). Weingart and colleagues (1997) attempt a comprehensive survey of the issues involved in integrating the historically abiological and non-Darwinian theories of the social sciences with Darwinian theory from biology.

## EVOLUTION OF COOPERATIVE INSTITUTIONS

Here we summarize a theory of institutional evolution that we have developed elsewhere in more detail (Richerson and Boyd, 1998, 1999, 2001). The theory is rooted in a mathematical analysis of the processes of cultural evolution and is, we argue in these papers, consistent with much empirical data. We make limited claims for our particular hypotheses, although we think that the thrust of the empirical data as summarized by the stylized facts already noted is much harder on current alternatives. We make a much stronger claim that a dual gene-culture theory of some kind will be necessary to account for the evolution of human cooperative institutions.

Understanding the evolution of contemporary human cooperation requires attention to two different time scales. First, a long period of evolution in the

Pleistocene shaped the innate "social instincts" that underpin modern human behavior. During this period, much genetic change occurred as a result of humans
living in groups with social institutions *heavily influenced by culture*, including
group-selected culture (Richerson and Boyd, 2000). On this time scale genes and
culture *coevolve*, and cultural evolution is plausibly a leading rather than lagging
partner in this process. Then, only about 10,000 years ago, the origins of agricultural subsistence systems laid the basis for revolutionary changes in the scale of
social systems. The evidence suggests that genetic changes in the social instincts
over the past 10,000 years are insignificant. Rather, the evolution of complex
societies has involved the relatively slow cultural accumulation of institutional
"work-arounds." These take advantage of a psychology evolved to cooperate with
distantly related and unrelated individuals belonging to the same symbolically
marked tribe while coping more or less successfully with the fact that these social
systems are larger, more anonymous, and more hierarchical than the tribal scale
ones of the late Pleistocene (Richerson and Boyd, 1998, 1999).

## Tribal Social Instincts Hypothesis

Our hypothesis is premised on the idea that group selection plays a more
important role in shaping culturally transmitted variation than it does in shaping
genetic variation. As a result, humans have lived in social environments characterized by high levels of cooperation for as long as culture has played an important role in human development. To judge from the other living apes, our remote
ancestors had only rudimentary culture (Tomasello, 1999) and lacked cooperation on a scale larger than groups of close kin (Boehm, 1999). The difficulty of
constructing theoretical models of group selection on genes favoring cooperation
matches neatly with the empirical evidence that cooperation in most social animals is limited to kin groups. In contrast, rapid cultural adaptation can lead to
ample variation among groups whenever multiple stable social equilibria exist,
due to conformist social learning, symbolically marked boundaries, or moralistic
enforcement of norms (Boyd and Richerson, 1992a). Such models of group selection are relatively powerful because they only require the social, not physical,
extinction of groups. Formal theoretical models suggest that conformism is an
adaptive heuristic for biasing imitation under a wide variety of conditions (Boyd
and Richerson, 1985:Chapter 7; Henrich and Boyd, 1998; Simon, 1990). Similarly, symbolic group marking arises for adaptive reasons in cultural evolution
models in which either ecological differences or different solutions to games of
coordination make the imitation of behaviors common in neighboring groups
maladaptive in one's own group (Boyd and Richerson, 1987; McElreath et al., no
date). Models of moralistic punishment (Boyd and Richerson, 1992c) lead to
multiple stable social equilibria and to reductions in noncooperative strategies if
punishment is prosocial. A consequence, we believe, is that a growing reliance on
cultural evolution led to larger, more cooperative societies among humans over
the past 250,000 years or so.

Consistent with this argument, late Pleistocene human societies were organized on a tribal scale (Bettinger, 1991:203-205; Richerson and Boyd, 1998). To judge from the ethnographic study of living hunter-gatherers, tribes were composed of several non-co-resident bands speaking the same dialect and numbering in the aggregate a few hundred to a few thousand people. Tribal-level institutions typically maintained peace between bands, made provision for emergency aid to fellow tribe members, celebrated communal rituals, defended the tribe against predatory raids by neighbor tribes (and often a specific territory from encroachment by other tribes), and legitimated the punishment of tribal miscreants. Institutions for making collective consensus decisions about war, peace, resource exploitation, institutional changes, and the like existed. Egalitarian social relations between males were maintained by the collaboration of potential subordinates to curb the impulse of the ambitious and skilled to dominate or exploit others (Boehm, 1999). Some ethnographically known hunter-gatherer societies, such as those of California and the Northwest Coast, had stronger leadership institutions and considerable inequality, and some late Pleistocene societies could have resembled them (Price and Brown, 1985). Our argument only requires that the central tendency of Pleistocene and post-Pleistocene societies differs sharply on these dimensions. Some sense of belonging to a delimited group was typical. Political, economic, and cultural alliance with culturally similar, or even not-so-similar, tribes was common. On the other hand, tribes often had hereditary enemies. The rule of law extended to a rather limited number of people by modern standards and self-help violence was commonly needed to secure justice even within societies when custom, public opinion, and weak leadership failed to find solutions to problems (Horowitz, 1990). The strength of such institutions and details of their implementation were likely highly variable (Kelly, 1995) if ethnographic hunter-gatherers are any indication. Unlike complex societies, division of labor (except between men, women, and different age groups) was modest.

We believe that the human capacity to live in tribes evolved by the coevolution of genes and culture. Rudimentary cooperative institutions created by cultural group selection would have favored genotypes that were better able to live in more cooperative groups. At first, such populations would have been only slightly more cooperative than typical nonhuman primates. However, genetic changes, such as a more docile temperament, would allow the cultural evolution of more sophisticated institutions that in turn enlarged the scale of cooperation. These rounds of coevolutionary change continued until eventually people were equipped with capacities for cooperation with distantly related people, emotional attachments to symbolically marked groups, and willingness to punish others for transgression of group rules. Mechanisms by which cultural institutions might exert forces tugging in this direction are not far to seek. Cultural norms affect mate choice and people seeking mates are likely to discriminate against genotypes that are incapable of conforming to cultural norms (Richerson and Boyd, 1989). People unable to control their self-serving aggression ended up exiled or executed in small-scale societies and in prison in contemporary ones. People whose social skills embarrass their families have a hard time attracting mates. Of course, selfish and nepotistic impulses never were suppressed entirely; our genetically transmitted evolved psychology shapes human cultures, and as a result cultural adaptations often still serve the ancient imperatives of inclusive genetic fitness. However, cultural evolution also creates new selective environments that cause *cultural imperatives to be built into our genes.*

Paleoanthropologists believe that human cultures were essentially modern by the Upper Paleolithic, 50,000 years ago (Klein, 1999). So even if the cultural group selection process began as late as the Upper Paleolithic, such social section easily could have had extensive effects on the evolution of human genes by this process. More likely, Upper Paleolithic societies were the culmination of a long period of coevolutionary increases in a tendency toward tribal social life.

We suppose that the resulting "tribal instincts" are something like principles in the Chomskian linguists' "principles and parameters" view of language (Pinker, 1994). The innate principles furnish people with basic predispositions, emotional capacities, and social dispositions that are implemented in practice through highly variable cultural institutions, the parameters. People are innately prepared to act as members of tribes; but culture tells us how to recognize who belongs to our tribes; what schedules of aid, praise, and punishment are due to tribal fellows; and how the tribe is to deal with other tribes—allies, enemies, and clients. The division of labor between innate and culturally acquired elements is poorly understood and theory gives little guidance about the nature of the synergies and tradeoffs that must regulate the evolution of our psychology (Richerson and Boyd, 2000). The fact that even human-reared apes cannot be socialized to behave like humans guarantees that some elements are innate. Contrariwise, the diversity and sometimes rapid change of social institutions guarantees that much of our social life is governed by culturally transmitted rules, skills, and even emotions. We beg the reader's indulgence for the necessarily brief and assertive nature of our argument here. The rationale and the ethnographic support for the tribal instincts hypothesis are laid out in more detail in Richerson and Boyd (1998, 1999). The same authors, (Richerson and Boyd, 2001) review a broad spectrum of empirical evidence supporting the hypothesis.

## Work-Around Hypothesis

Contemporary human societies differ drastically from tribal societies in which our social instincts evolved. Pleistocene hunter-gatherer societies were small and egalitarian and lacked powerful leaders. Modern societies are large and inegalitarian and have coercive leadership institutions (Boehm, 1993). If the social instincts hypothesis is correct, social instincts are part building blocks and part constraints on the evolution of complex social systems (Salter, 1995). To evolve large-scale, complex social systems, cultural strategies take advantage of whatever support the instincts offer. For example, families willingly take on the essential roles of biological reproduction and primary socialization. At the same time, cultural evolution must cope with a psychology evolved for life in quite different sorts of societies. Appropriate larger scale institutions must regulate small-group subversion of large-group favoring rules. To do this, cultural evolution often makes use of "work-arounds"—mobilizing tribal instincts for new purposes. For example, large national and international (e.g., great religions) institutions develop ideologies of symbolically marked inclusion that often fairly successfully engage the tribal instincts on a much larger than tribal scale. Such work-arounds are often awkward compromises, as is illustrated by the existence of contemporary societies handicapped by few loyalties outside the family (Banfield, 1958) or by destructive loyalties to relatively small tribes (West, 1941).

The most important cultural innovations required to support complex societies are command and control institutions that can systematically organize cooperation, coordination, and a division of labor in societies consisting of hundreds of thousands to hundreds of millions of people. Command and control institutions lead to more productive economies, more internal security, and better resistance to external aggression. Note that command and control are separable concepts. Command may aim at quite limited control. For example, a predatory conquest state may use command almost exclusively for the extraction of portable wealth, not for prosocial projects. Institutions often exert control without commands. Markets, most famously, control behavior by price signals from a diffuse world of anonymous buyers and sellers. Market enthusiasts do sometimes forget that command systems generally are needed to make markets function, ranging from mandatory use of calibrated weights and measures to central banks (Dahrendorf, 1968:Chapter 8). The main types of work-arounds seem to be the ones described in the following subsections.

*Coercive Dominance*

The cynics' favorite mechanism for creating complex societies is command backed up by force. The conflict model of state formation has this character (Carneiro, 1970), as does Hardin's (1968) recipe for commons management.

Elements of coercive dominance are no doubt necessary to make complex societies work. Tribally legitimated self-help violence is a limited and expensive means of prosocial coercion. Complex human societies have to supplement the moralistic solidarity of tribal societies with formal police institutions. Otherwise, the large-scale benefits of cooperation, coordination, and division of labor would cease to exist in the face of selfish temptations to expropriate them by individuals, nepotists, cabals of reciprocators, organized predatory bands, and classes or castes with special access to means of coercion. At the same time, the need for organized coercion as an ultimate sanction creates roles, classes, and subcultures with the power to turn coercion to narrow advantage. Social institutions of some sort must police the police so that they will act in the larger interest to a measurable degree. Such policing is never perfect and, in the worst cases, can be very poor. The fact that leadership in complex systems always has at least some economic inequality suggests that narrow interests, rooted in individual selfishness, kinship, and, often, the tribal solidarity of the elite, always exert an influence. The use of coercion in complex societies offers excellent examples of the imperfections in social arrangements traceable to the ultimately irresolvable tension of selfish and prosocial instincts.

Although coercive, exploitative elites are common enough, there are two reasons to suspect that no complex society can be based purely on coercion. The first problem is that coercion of any great mass of subordinates requires that the elite class or caste be itself a complex, cooperative venture. The second problem with pure coercion is that defeated and exploited peoples seldom accept subjugation as a permanent state of affairs without costly protest. Deep feelings of injustice generated by manifestly inequitable social arrangements move people to desperate acts, driving the cost of dominance to levels that cripple societies in the short run and often cannot be sustained in the long run (Insko et al., 1983; Kennedy, 1987). Durable conquests, such as those leading to the modern European national states, Han China, or the Roman Empire, leaven raw coercion with more prosocial institutions. The Confucian system in China and the Roman legal system in the West were far more sophisticated and durable institutions than the highly coercive systems sometimes set up by predatory conquerors and even domestic elites.

The modern commons literature has taken up this theme from its inception in Hardin's (1968) article, but even more so in his later work (e.g., Hardin, 1978; see also Low, 1996). The underlying model is one of selfish rationality that requires a leviathan to motivate self-interested actors to conserve commons. We think this analysis is flatly self-contradictory. Leviathans can't be drummed up simply because they would be useful; they must evolve. If evolution produces self-interested actors that need leviathans, then any leviathans will be selfish too, and so they may conserve commons in their own interest, but not in the interest of anyone else. In the modern world, there are many kleptocratic leviathans— Mobutu, Suharto, Marcos—men who take advantage of weak national institutions to exploit commons for their own narrow ends, and preside over corrupt bureaucracies that cannot even manage efficiently in the kleptocrat's self-interest—everyone cheats as much as they can. No one sensible person desires this kind of leviathan. Coercive elites can manage commons efficiently only if they are embedded in fundamentally prosocial institutions. A process like cultural group selection acting in the past and in the present puts the possibility of prosocial attitudes and institutions to work. In fact, costly prosocial behavior is common. Resistance to kleptocrats is often newsworthy, as their abuses of human rights are generally conspicuous and heavy handed. Not inconsiderable numbers of people resist such governments at the very real risk of brutal and often deadly repression.

## Segmentary Hierarchy

Late Pleistocene societies were undoubtedly segmentary in the sense that supra-band ethnolinguistic units served social functions, although presumably they lacked much formal political organization. The segmentary principle can serve the need for more command and control by hardening up lines of authority without disrupting the face-to-face nature of proximal leadership present in egalitarian societies. The Polynesian ranked lineage system illustrates how making political offices formally hereditary according to a kinship formula can help deepen and strengthen a command and control hierarchy (Kirch, 1984; Sahlins, 1963). A common method of deepening and strengthening the hierarchy of command and control in complex societies is to construct a nested hierarchy of offices, using various mixtures of ascription and achievement principles to staff the offices. Each level of the hierarchy replicates the structure of a hunting and gathering band. A leader at any level interacts mainly with a few near-equals at the next level down in the system. New leaders usually are recruited from the ranks of subleaders, often tapping informal leaders at that level. As Eibl-Eibesfeldt (1989) remarks, even high-ranking leaders in modern hierarchies adopt much of the humble headman's deferential approach to leadership.

Commons management institutions sometimes make use of segmentation. Hundley (1992) describes the importation of Spanish water management customs into the Northern Mexican borderlands, including California. According to Hundley, the Royal decrees sought to establish a Spanish economy in the New World to support other Spanish institutions. These decrees included an elaborate section on water management, codified as the *Plan of Pitic*, a model water ordinance. Water management was to be the responsibility of town councils. The details of management were left to the town under a few basic principles. First, no individuals were to have independent rights; water was to be managed as common property of the duly constituted town. Second, in times of scarcity, water was to be divided equitably among all users. Royal authorities were to resolve any disputes that escaped local management, such as disputes between upstream and downstream users according to the same two principles. Thus, the division of authority between town and royal officials was carefully crafted. The plan was consciously modeled on the successful Iberian tradition of local management of water, the modern manifestations of which Ostrom (1990:69-82) discusses.

The hierarchical nesting of social units in complex societies gives rise to appreciable inefficiencies (Miller, 1992). In practice, brutal sheriffs, incompetent lords, venal priests, and their ilk degrade the effectiveness of social organizations in complex societies. Squires (1986), elaborating on Tullock (1965), dissects the problems and potentials of modern hierarchical bureaucracies to perform consistently with leaders' intentions. Leaders in complex societies must convey orders downward, not just seek consensus among their comrades. Devolving substantial leadership responsibility to subleaders far down the chain of command is necessary to create small-scale leaders with face-to-face legitimacy. However, it potentially generates great friction if lower level leaders either come to have different objectives than the upper leadership or are seen by followers as equally helpless pawns of remote leaders. Stratification often creates rigid boundaries so that natural leaders are denied promotion above a certain level, resulting in inefficient use of human resources and a fertile source of resentment to fuel social discontent.

Young (this volume:Chapter 8), Berkes (this volume:Chapter 9), and Baland and Platteau (1996:Chapter 13) devote considerable attention to the problem of vertical linkages between small-scale commons management institutions and the larger ones in which they are necessarily embedded in a complex society. Kleptocratic behavior frequently infects the whole political and bureaucratic system. In states with inefficient national-level institutions, corruption often exists up and down the chain of command (Baland and Platteau, 1996:235 ff). Commons management bureaucracies, even in relatively successful democracies such as India, often legislate away tribal-scale commons management systems and replace them with bureaucracies that do a much worse job. Tightly organized, large command and control bureaucracies only function properly when the institutions that regulate their behavior favor efficiency and honesty. Otherwise, the ever-present selfish, nepotistic, and tribal-scale motives will support the emergence of corruption at every level of the hierarchy.

These authors identify two sets of issues. Looked at from the bottom up, higher level interference in the affairs of local communities can be catastrophic, but, from the top down, is at the same time often important for proper function. Catastrophes occur when, through ignorance or malevolence, larger scale institutions damage or destroy small-scale ones. Success is achieved, as in the *Plan of Pitic*, when the roles of higher and lower levels are complementary and when their interests largely coincide. We would only stress more than these authors that the most important feature of small-scale institutions is that they can tap most directly, free of problematical work-arounds, the tribal social instincts. High degrees of cooperation, buttressed by nuanced systems of monitoring and punishment, make for high-morale, highly effective systems. Self-interest not only does not explain such cooperation, but also may be dangerous if used in an effort to strengthen or change institutions. We believe that hierarchical systems cannot dispense with tribal solidarity at any level without losing important elements of function. This is a claim worth testing, as it is a linchpin of our hypothesis but inessential to those based on rational choice, in which hierarchical organization serves merely communication and monitoring function. On our view there is much more to segmentary hierarchies than a telephone tree down and surveillance information up.

On the other hand, failure to properly articulate tribal-scale units is often highly pathological. Tribal societies often must live with chronic insecurity because of intertribal conflicts. One of us once attended the *Palio*, a horse race in Siena in which each ward, or *contrada*, in this small Tuscan city sponsors a horse. The voluntary contributions necessary to pay the rider, finance the necessary bribes, and host the victory party amount to half a million dollars. The contrada clearly evoke the tribal social instincts: They each have a totem—the dragon, the giraffe, special colors, rituals, and so on. The race excites a tremendous, passionate rivalry. One can easily imagine a medieval Siena in which swords clanged and wardmen died, just as they do or did in warfare between New Guinea tribes (Rumsey, 1999), Greek city-states (Runciman, 1998), inner city street gangs (Jankowski, 1991), and ethnic militias. Natural resources are frequently sources of conflict that can lead to violence in the absence of superordinate institutions to resolve disputes. "Wars" between fishermen from different ports occur occasionally despite modern justice services. When fishermen from different nations are involved, fish wars cause major diplomatic tangles even between otherwise friendly nations. The three fish wars that occurred between Britain and Iceland over cod fishing rights after the Second World War (Kurlansky, 1998), and the ethnic-controlled fisheries in 19th-century California, included vigorous defense of each group's territory (Baland and Platteau, 1996:328). Territory defense is an ancient function of tribes, to judge from its high frequency in ethnographically known hunter-gatherers (Cashdan, 1992) and territory incursion is a frequent cause of violent conflict.

## Exploitation of Symbolic Systems

The high population density, division of labor, and improved communication made possible by the innovations of complex societies increased the scope for elaborating symbolic systems. The development of monumental architecture to serve mass ritual performances is one of the oldest archeological markers of emerging complexity. Usually an established church or less formal ideological umbrella supports a complex society's institutions. At the same time, complex societies extensively exploit the symbolic ingroup instinct to delimit a quite diverse array of culturally defined subgroups, within which a good deal of cooperation is routinely achieved. Ethnic group-like sentiments in military organizations often are reinforced most strongly at the level of 1,000 to 10,000 or so men (British and German regiments, U.S. divisions) (Kellett, 1982). Typical civilian symbolically marked units include nations, regions (e.g., Swiss cantons), organized tribal elements (Garthwaite, 1993), ethnic diasporas (Curtin, 1984), castes (Gadgil and Malhotra, 1983; Srinivas, 1962), large economic enterprises (Fukuyama, 1995), civic organizations (Putnam, 1993), and many others (Stern, 1995).

How units as large as modern nations can tap the tribal social instincts is an interesting problem. Anderson (1991) argues that literate communities, and the social organizations revolving around them (e.g., Latin-literates and the Catholic Church), lend themselves to creating "imagined communities" that in turn elicit significant commitment from members of the community. Because tribal societies were often large enough that some members were not known personally to any given person, common membership sometimes would have to be established by the mutual discovery of shared cultural understandings. The advent of mass literacy and print media—Anderson stresses newspapers—made it possible for all speakers of a given vernacular to have confidence that every reader of the same or related newspapers shared many cultural understandings, especially when organizational structures such as colonial government or business activities really did give speakers some institutions in common. Nationalist ideologists quickly discovered the utility of newspapers for building several variants of imagined communities, making nations the dominant quasi-tribal institution in most of the modern world. If Wolfe (1965) is right, mass media also can be the basis of a rich diversity of imagined subcommunities using vehicles such as specialized magazines, newsletters, and, nowadays, web sites. Subcommunities of the imagined type are often important for commons management, ranging from environmental pressure groups to professional communities with a role in environmental management.

Many problems and conflicts revolve around symbolically marked groups in complex societies. Official dogmas often stultify desirable innovations and lead to bitter conflicts with heretics. Marked subgroups often have enough tribal cohesion to organize at the expense of the larger social system. The frequent seizure of power by the military in states with weak institutions of civil governance is probably a byproduct of the fact that military training and segmentation, often based on some form of patriotic ideology, are conducive to the formation of *relatively* effective large-scale institutions. Wherever groups of people interact routinely, they are liable to develop a tribal ethos. In stratified societies, powerful groups readily evolve self-justifying ideologies that buttress treatment of subordinate groups that ranges from neglectful to atrocious. White Southerners had elaborate theories to justify slavery and Jim Crow and Westerners found brutal treatment of Indians legitimate and necessary. The parties and interest groups that vie to sway public policy in democracies have well-developed rationalizations for their selfish behavior. A major difficulty with loyalties induced by appeals to shared symbolic culture is the very language-like productivity possible with this system. Dialect markers of social subgroups emerge rapidly along social fault lines (Labov, 1973). Charismatic innovators regularly launch new belief and prestige systems, which sometimes make radical claims on the allegiance of new members, sometimes make large claims at the expense of existing institutions, and sometimes grow explosively. Or, contrariwise, larger loyalties can arise, as in the case of modern nationalisms overriding smaller scale loyalties, sometimes for better, sometimes for worse. The ongoing evolution of social systems can evolve in unpredictable, maladaptive directions by such processes (Putnam, 2000). The worldwide growth of fundamentalist sects that challenge the institutions of modern states is a contemporary example (Marty and Appleby, 1991; Roof and McKinney, 1987). Ongoing cultural evolution is impossible to control, at least completely.

The literature on commons management is rich in cases where tribal-scale institutions effectively govern commons. Gadgil and Guha (1992) describe the village-level management of forests and other commons by villages in traditional India and contrasts the successes of the traditional regime with failures under the bureaucratic institutions brought by the British and retained by independent India. Ruttan (1998) describes the successful management of a pearl-shell fishery by a village community. Acheson (1988) describes the management of a fishery by local fishermen. Ostrom's (1990:Chapter 3) cases all describe village-scale institutions. She mentions the existence of clear boundaries and sophisticated institutions for monitoring commons and assessing punishments to transgressors. She also notes that higher authorities have to leave local communities sufficient autonomy to exercise such institutions. The review by Baland and Platteau (1996:Part II) of many cases of local-level management of commons underscores these points. Bardhan and Dayton-Johnson (this volume:Chapter 3) note that egalitarian village-scale systems often have more successful commons management institutions than ones with an inegalitarian distribution of income.

So far as we can tell, the literature on commons management institutions has not yet tackled the precise role of *symbolically marked* groups in commons management. The fact that commons frequently are managed effectively by tribal-scale groups might be only because the scale of resources being managed is small and/or because efficient policing of commons requires clearly signifying who is and who is not entitled to participate in the commons, resulting in clearly defined boundaries (Ostrom, 1990:91). We believe that emotional bonds of the individual to the group frequently buttress these rational choice effects. One of us has observed that the Altiplano villagers around Lake Titicaca have distinctive costumes, especially women's but also sometimes men's. These villagers also effectively manage lake commons despite opposition from Peruvian authorities (LeVieil, 1987). We suspect that around the world, tribal-scale communities often have a sense of pride in their local corporate community, exemplified by wearing its "colors," which helps generate levels of cooperation and trust that are efficacious in providing many kinds of public goods. Experimentalists do not seem to have used symbolic marking of groups to test for whether such effects stimulate cooperation in public goods contexts (but see Kramer and Brewer, (1984). In the classic minimal group experiments of Tajfel (1981; see also Turner, 1995), very simple grouping and symbolic labeling of subjects caused substantial discrimination in favor of ingroup members. This experimental evidence dovetails nicely with the field data, very superficially reviewed in the two previous paragraphs. We predict that if experimental subjects are led to believe they are playing a commons game with any even thinly plausible ingroup, rates of participation in common property economy will rise significantly above base rates. If the game has even a minimal element of competition between symbolically marked groups, such as a nominal or symbolic prize for most money earned, participation should be especially high.

## Legitimate Institutions

In small-scale egalitarian societies, individuals have considerable autonomy, considerable voice in community affairs, and can enforce fair, responsive—even self-effacing—behavior by leaders (Boehm, 1999). At their most functional, symbolic institutions, a regime of tolerably fair laws and customs, effective leadership, and smooth articulation of social segments can roughly simulate these conditions in complex societies. Rationally administered bureaucracies, lively markets, the protection of socially beneficial property rights, widespread participation in public affairs, and the like provide public and private goods efficiently, along with a considerable amount of individual autonomy. Many individuals in modern societies feel themselves part of culturally labeled tribal-scale groups, such as local political party organizations, that have influence on the remotest leaders. In older complex societies, village councils, local notables, tribal chieftains, or religious leaders often hold courts open to humble petitioners. These local leaders in turn represent their communities to higher authorities. To obtain low-cost compliance with management decisions, ruling elites have to convince citizens that these decisions are in the interests of the larger community. As long as most individuals trust that existing institutions are reasonably legitimate and that any perceived needs for reform are achievable by means of ordinary political activities, there is considerable scope for large-scale collective social action.

However, legitimate institutions, and trust of them, are the result of an evolutionary history and are neither easy to manage or engineer. The social distance between different classes, castes, occupational groups, and regions is objectively great. Narrowly interested tribal-scale institutions abound in such societies, as we have seen. Some of these groups have access to sources of power that they are tempted to use for parochial ends. Such groups include, but are not restricted to, elites. The police may abuse their power. Petty administrators may victimize ordinary citizens and cheat their bosses too. Ethnic political machines may evict historic elites from office but use chicanery to avoid enlarging their coalition.

Without trust in institutions, conflict replaces cooperation along fault lines where trust breaks down. Empirically, the limits of the trusting community define the universe of easy cooperation (Fukuyama, 1995). At worst, trust does not extend outside family (Banfield, 1958) and potential for cooperation on a larger scale is almost entirely foregone. Such communities are unhappy as well as poor. Trust varies considerably in complex societies, and variation in trust seems to be the main cause of differences in happiness across societies (Inglehart and Rabier, 1986). Even the most efficient legitimate institutions are prey to manipulation by small-scale organizations and cabals, the so-called special interests of modern democracies. Putnam's (1993) contrast between civic institutions in Northern and Southern Italy illustrates the difference that a tradition of functional institutions can make. The democratic form of the state, pioneered by Western Europeans in the past couple of centuries, is a powerful means of creating generally legitimate institutions. Its success attracts imitation all around the world. The halting growth of the democratic state in countries ranging from Germany to those in Sub-Saharan Africa is testimony that legitimate institutions cannot be drummed up out of the ground just by adopting a constitution. Where democracy has struck root outside of the European cultural orbit, it is distinctively fitted to the new cultural milieu, as in India and Japan.

Legitimate institutions have a huge role to play in commons management. One of us has had considerable positive experience with the burgeoning system of Cooperative Resource Management Committees (CRMCs) that bring local, state, and federal agencies together regularly with interested citizens and citizen groups to deal with their joint commons (Richerson, with Lake County, California's Clear Lake Watershed CRMC). Although the resolutions of such committees have no weight of law at all, in the Clear Lake case they usually represent a strong consensus of the participants and thus often generate appropriate action. The most conspicuous absentee from the process at Clear Lake has been the U.S. Environmental Protection Agency (EPA), whose Superfund Program has charge of cleaning up a large abandoned mercury mine on the shore of the lake. Levels of trust even between technical professionals at EPA and other agencies are very low. From this one case, it is impossible to decide whether EPA's poor reputation is simply a result of nonparticipation or if nonparticipation itself is part of a wider malaise in the agency. Some evidence suggests that the culture of EPA derives more from the norms and habits of the legal community than from the engineering and science community, mainly because of choices made by its first administrator, William Ruckelshaus (Richerson, 1988). As a result, the agency has trouble attracting and retaining the highest caliber technical staff and hence has trouble dealing professionally with technical issues when they arise.

Hundley (1992) describes the many institutions created to manage the California water commons. On the small scale, towns created water companies, entrepreneurs created mutual water companies and platted the accompanying town, and farmers organized irrigation districts. On the medium scale, growing cities, especially Los Angeles and San Francisco, organized municipal water companies that seized water rights on distant drainages and built long aqueducts to the city. On the largest scale, the Federal Central Valley Project and the State California Water Project routed southward most of the flow of the state's largest river, the Sacramento. All of the large projects and many smaller ones were intensely controversial, and had to survive votes in legislatures, city councils, and boards of supervisors. Most faced general elections to approve bonds for construction financing. Many had to survive legal challenges. Chicanery was common, although often by public servants acting in what they believed was the general interest. Self-interested malfeasance was also common. Large landowners zealously exploited economies of scale in manipulating government decisions in their own favor. Despite bitter reversals, such as the then-new Sierra Club's failure to save the Hetch Hetchy Valley from San Francisco's dam, few losers stepped outside of the realm of legal forms of resistance. The citizens of the Owens Valley became so embittered at Los Angeles' massive diversion of water into its aqueduct that they dynamited the main pipeline on several occasions. The publicity resulting from these acts portrayed Los Angeles in such a bad light that the city ultimately bought out not water right holders but all of the private landholders in the Valley.

Thus, successful commons management on any scale requires a system of legitimate institutions. Where these do not exist, appropriate organizations may arise spontaneously at the tribal level, especially if the state does not actively interfere. In cases where the scale of the problem is larger, the whole panoply of work-arounds must act with enough efficiency to create large-scale management systems, such as ministries of the environment. When such bureaucracies work well, they are likely to adopt some tribal attributes. Individuals will have high loyalty to the organization and a deep commitment to making it function. In many societies, these institutions remain distressingly lacking in such attributes. Indeed, the contemporary enthusiasm for conservation-and-development projects to protect biodiversity in poor countries is an effort to cope with weakness in national institutions, which are the backbone of biodiversity conservation in the wealthy nations. The institutional basis for managing the global commons is still, of course, quite problematic.

## REPRISE: TESTING THE HYPOTHESES

How much confidence should we have in the tribal social instincts and work-around hypotheses? We argue elsewhere that much evidence from a number of domains is more consistent with the tribal social instincts hypothesis than with its best articulated competitors (Richerson and Boyd, 1998, 2001; Boyd and Richerson, no date). Soltis et al. (1995) used data on group extinctions in Highland New Guinea to estimate potential rates of group selection. The details of New Guinea extinctions are consistent with assumptions made in our conformity-based model of cultural group selection. Kelly (1985) and Knauft (1985; 1993) provide particularly good case studies describing the operation of cultural expansions at the expense of one group by another and pinpointing the institutional reasons for the group fitness differentials. We have tested the work-around hypothesis by drawing on the analytical history of the performance of World War II armies (Richerson and Boyd, 1999).

We think the empirical data on commons management institutions also conform to the patterns predicted by these hypotheses. In particular, both field and experimental evidence show that people cooperate in ways that are hard to reconcile with the behavior of selfish actors. We believe that cultural group selection is the best existing explanation for why humans but not other species can organize cooperation among nonrelatives on a considerable scale. Evidence from the commons literature suggests that people are neither individualist nor prosocial rational actors by nature. Given sufficient rationality and prosocial impulses, humans might leap immediately to solutions to commons dilemmas. The evidence suggests instead that we are dependent on culturally evolved institutions to make cooperation work. Institutions encode rules for operating commons that are neither innate nor learned on the spot but are cultural traditions. Successes and failures seem always to involve an institutional dimension. Some societies have evolved work-arounds that permit reasonably functional environment ministries, while others struggle.

In another sense our hypotheses are very poorly tested. The systematic application of modern evolutionary theory to human behavior is scarcely a quarter century old. The variety of evolutionary theories we can imagine is rather large, especially if cultural evolution and gene-culture coevolution play important roles. Our particular choices in formulating the tribal instincts and work-around hypotheses seem sensible to us in light of the evidence, but only a small part of the space of all possible theories is yet explored. For example, Campbell (1983) argued that simpler societies were built on the basis of kinship and reciprocity and that cultural group selection became important only with the rise of complex societies in the past few thousand years. We think the evidence supports the idea that hunting and gathering societies commonly cooperated on scales too large to be explained by reciprocity and kinship alone, but of course we have no direct data on the social organization of Pleistocene societies.

## OUTSTANDING QUESTIONS

The most important payoff to better theory is that better theory poses new, interesting, and practically important questions for further research. We think the dual inheritance evolutionary theory does these things.

We believe evolutionary theory might provide helpful directions for future research in four general areas.

### The Problem of Complexity and Diversity

Commons institutions are functional, complex, and unique. They appear to be deeply embedded parts of cultures and hence to have an evolutionary history of some depth. There are a myriad of ways to organize commons management (Agrawal, this volume:Chapter 2). The dominant hypothesis to explain such diversity has been the more and less advanced hypothesis. Modernist reformers

portray formal state control over natural resources as the superior modern successor to less formal, traditional, *ancien regime* commons institutions. Their local
diversity and cultural embeddedness are testimony to suboptimality on this view.
Overenthusiastic modernists unduly neglect alternative hypotheses. Complex design problems in artificial systems are known to have many optima, some of
which are more or less equally functional. We argue that biological and cultural
systems are similar (Boyd and Richerson, 1992b). As myopic evolutionary processes locally improve the function of complex systems, they explore a complex
adaptive landscape, some coming to equilibrium on less functional local peaks
than others. Large, simple jumps may unravel quite functional institutions without putting into place all the parts of a complex alternative, as students of commons institutions repeatedly have observed. The failures of outside reformers
who advocate major change to "more advanced" institutions are common.

A major task before us is to map out the proximal details of how institutions
fostering cooperation work and how evolutionary processes have shaped these
details. Traditional ethnographic investigations were a fine start on this project,
but more critical and quantitative methods are needed to describe function and
process in more detail (e.g., Edgerton, 1971). Ostrom's (1990) analysis of commons management, based on ethnographic and historical sources, asked many of
the right questions. We believe the evolution-inspired experimental comparative
ethnography pioneered by Henrich et al. (2001) and Nisbett et al. (in press) provide important insights. In even the most atomistic human societies, people have
some propensity to fairness in economic interchanges that can aid their transition
to the modern world. The indications that social organization is deeply entangled
with styles of thinking suggest that complex, historically contingent evolution
does indeed create considerable evolutionary inertia in institutions. We recommend our list of work-arounds as a practical tool in assessing the strengths and
weaknesses of commons institutions. For example, Young (this volume:Chapter
8) and Berkes (this volume:Chapter 9) argue that cross-scale linkages are important sources of both friction and necessary interplay using much the same terms
as our discussion of the segmentary hierarchy work-around.

## How Flexible Are Cooperative Institutions?

Putnam's (1993) contrast between Northern and Southern Italy suggests that
some institutional systems respond more quickly to changing opportunities than
others. Plausibly, an open political system that operates by either rough consensus or more formal voting is better adapted to solve a wide variety of public
goods problems by using legitimate institutions to formulate plans of action adaptable to new circumstances than is a regime lacking a measure of, or interest in,
popular needs and wants. Boehm (1993, 1996) argues that hunter-gatherers commonly make adaptive collective decisions by open discussion and consensus formation. Recall Inglehart and Rabier's (1986) finding that the strongest correlate

of reported happiness and satisfaction with life in the developed world (mostly Europe) is expressed levels of trust in one's fellow citizens. The happiest countries are relatively small, highly democratic societies like Sweden, Holland, and Switzerland that, we conjecture, retain strong participatory institutions at the tribal scale, however sophisticated they are in other ways (it would be hard to find a society more sophisticated than, say, Holland).

Open political systems seem to be among the most flexible of institutions for so many purposes because they maintain such a high level of local esprit and trust. Innovative ways of tapping these systems, such as Cooperative Resource Management Committees, seem to provide healthy cross-level linkages between the higher level bureaucracy and the local community. They are likely to fail either when consensus cannot be achieved at the local level or when local consensus is not acceptable to powerful actors beyond the local level; this seems to have been the case with the Quincy Library Group's consensus on logging/biodiversity conflicts in its local area. The visible precedent-setting nature of the Quincy Library exercise is perhaps not a fair test of the concept because it attracted very close scrutiny by national-level interest groups in a regionally highly polarized arena. Cooperative Resource Management Committees of our personal acquaintance operate much closer to the ground and can make local consensus work.

Other institutions have some of the same properties. Many economists claim that the market is one of the most general tools of all in managing human behavior. Tietenberg (this volume:Chapter 6) and Rose (this volume:Chapter 7) discuss the strengths and weaknesses of tradable permits as means for managing environmental resources. Tradable permits are resisted by those generically suspicious of market solutions, but to our way of thinking the most severe problem is the large amounts of wealth such rights create. Well administered by competent, honest bureaucracies, such systems have much promise. They seem, however, to be of little use in places where administering institutions are inefficient or corrupt. Crony capitalism systems will not administer such systems honestly any more than they honestly administer current commons by regulation. One again we stress Dahrendorf's (1968) point that efficient markets are the result of efficient, honest institutions, not somehow direct products of human nature set free, as some market ideologues would have us believe. Against this argument, Baland and Platteau (1996:134) review ideas suggesting that market economies cause erosion in moral norms. Henrich et al.'s (2001) data suggest the opposite. People from groups with experience with market institutions usually make fair offers in the ultimatum game, perhaps because experience in markets teaches participants that strangers are generally fair dealers. The rapid change that often accompanies market penetration to formerly isolated village societies is more likely, we suggest, the culprit in destabilizing traditional commons institutions than markets per se.

## How Rapidly Can New Institutions Emerge and Spread?

The spread of complex social institutions by diffusion is arguably more difficult than the diffusion of technological innovations. The pace of innovation of institutions is likely to be relatively slow for several reasons. We have already mentioned the problem of complex design inhibiting the easy optimization of institutions. Similarly, many coordination payoff structures will cause societies to reach a variety of equilibria, some of which are relatively inefficient but also difficult to improve (Sugden, 1986). Some models of cultural group selection are quite hostile to the exchange of innovations between groups because the between-group migration necessary to carry them from one group to the other also causes mixing and lowering of the between-group variance that group selection needs to operate (Soltis et al., 1995). The data and models reviewed in Soltis et al. suggest that it would take on the order of a millennium for an institutional innovation to spread from the innovators to the bulk of the societies in a region. Other models of cultural group selection make the necessary cross-cultural borrowing more plausible (Boyd and Richerson, no date). This model shows that the existence of multiple stable states due to the existence of games of coordination does not necessarily inhibit the rapid spread of the most successful solution from group to group.

Other problems may make the diffusion of successful institutions hard. Social institutions violate four of the conditions that tend to facilitate the diffusion of useful innovations (Rogers, 1995). Foreign social institutions are often (1) not compatible with existing institutions, (2) complex, (3) difficult to observe, and (4) difficult to try out on a small scale. For such reasons, some commentators view the evolution of social institutions as a much more likely rate-limiting step than technology in the evolution of more intensive economies. For example, North and Thomas (1973) argue that new and better systems of property rights set off the modern industrial revolution rather than the easier task of technical invention itself. A difficult revolution in property rights likely also is necessary for intensive hunting and gathering and agriculture to occur (Bettinger, 1999). Slow diffusion also means that historical differences in social organization can be quite persistent, even though one form of organization is inferior. As a result, the comparative history of the social institutions of intensifying societies exhibits many examples of societies getting a persistent competitive advantage over others in one dimension or another because they possess an institutional innovation that their competitors do not acquire. For example, the Chinese merit-based bureaucratic system of government was established at the expense of the landed aristocracy, beginning in the Han dynasty (2,200 B.P.) and completed in the Tang (1,400 B.P.) (Fairbank, 1992). This system has become widespread elsewhere only in the modern era and is still operated quite imperfectly in many societies.

Consistent with such ideas, the evolution of institutions in fact has been relatively slow. More than 10 millennia separate us from our Pleistocene tribal ancestors. We argue elsewhere (Richerson et al., 2001) that the transition from the harsh, highly variable climate regime of the last ice age to the much more benign regime of the Holocene set off a competitive footrace that consistently has favored more efficient subsistence and better organization of social systems. The fact that the human race has not yet reached equilibrium with the economic and social-organizational potential made possible by the benign climate of the Holocene (Richerson and Boyd, in press) is testimony to the relatively stately pace of cultural evolution. Even if equilibrium is at hand (Fukuyama, 1992), 10 millennia is a long time to get here! The pace of institutional evolution seems to have accelerated toward the present, no doubt because of the spread of literacy, mass communications, and science and social science. Foreign customs are much more transparent than they once were, and scholars often make more or less sophisticated comparative appraisals of the diversity of social experiments that come to their attention. Even so, institutional revolutions are apt to be frustratingly slow. For example, the conversion of Russia from a socialist one-party state to a market economy and elective democracy is far from a success after more than a decade of work.

The study of the rates of cultural evolution prevailing in the modern world and a sophisticated dissection of the processes that regulate those rates is a project in its infancy. In evolutionary biology, the coin-of-the-realm study of evolution is a quantitative estimate of the rate of evolution of a character and an attribution of the causes of change to particular processes such as natural selection and migration (e.g., Endler, 1986). Although such experiments are not commonly done by social scientists, plenty of examples exist to indicate that the project is perfectly feasible (Weingart et al., 1997:292-297). One of the most sophisticated literatures of this sort is the "policy learning/advocacy coalition" approach to studying policy change (Sabatier and Jenkins-Smith, 1993). Several of the studies applying this approach have been studies of commons policy issues. Obviously, applied institutional development agencies would benefit enormously from a sound knowledge of the comparative natural history of institutional evolution. The practical problem is to help a society with weak institutions acquire more functional ones of a specific orientation. The record indicates that inept interventions can do more harm than good, but good interventions also occur (Baland and Platteau, 1996: 243-245, 279-283).

## Is Small-Scale Cultural Evolution a Problem or a Resource?

Societies have political institutions of varying degrees of complexity for aggregating individual-level beliefs and desires to produce collectively desired outcomes (Boehm, 1996; Turner, 1995). In the limit, collective decision-making systems cause us to endow institutions such as the state with many of the attributes of an individual rational actor, although both theory (Arrow, 1963) and practical experience suggest that reaching sensible collective decisions is fraught with problems. Collective decisions, whether representative and rational or not, often have such durable effects as to constitute a form of cultural evolution. For example, the U.S. Constitution has shaped the political culture of the country for two centuries. The linkage of individual and small group-level culture with larger scale collective institutions is a complex problem with causal arrows running up and down the organizational hierarchy. The possibility of making collective decisions at all depends on some sufficient number of individual actors having norms and beliefs that support the institutions involved. If authors like Putnam (1993) are correct, the evolution of grassroots political culture is necessary to make higher levels of decision making work well. The ongoing evolution of beliefs and norms may act in concert with collective policy decisions, but some degree of friction is routine. The overextension of the state regulation of commons can wreck successful village-level systems, and the ideological and behavioral conformity demanded of all citizens by state authorities in authoritarian systems like Hapsburg, Spain, and Austria can damage the social capital on which sound policy making ultimately rests (Gambetta, 1993).

Many groups in developed nations are organized to advocate relatively narrow interests, or at least interests that seem narrow to those with other convictions. For example, wilderness advocates are accused of locking up vast tracts of land for their own pleasure, at the cost of excluding less hardy recreators and harming the interests of extractive resource users (usually claimed to be sustainable or otherwise harmless). The nature of passionate ingroups being what it is, such mud often sticks. Some of the opposition to dealing sensibly with global climate deterioration issues in the United States comes from Christians with apocalyptical beliefs. If the Second Coming is near, global climate change is either irrelevant or perhaps part of God's plan for the End Days. By some accounts, a growing appeal of ideologies with little patience with science (and likely, scientific management of natural resources) is a world-wide problem (Marty and Appleby, 1991). Developing wise large-scale policy to manage, but not overmanage or mismanage, cultural change is perhaps the most difficult and sensitive problem of statecraft. We are not convinced that much science can yet be brought to bear on the question of what cultural trends are threats and what are not by any criterion of judgment.

A few systems for collectively managed cultural evolution do stand out as possible examples of the application of sensible collective decision making to cultural change. In contemporary open societies, the harnessing of science to the public policy-making process via government-sponsored science at research institutions and research universities works splendidly when the science is tractable and social consensus as to directions to take are strong. Some other models are worth exploring. For example, Dupuy (1977) analyzed the history and operations of the Prussian and then German General Staff from the early 19th century to mid-20th century and argues that this institution typically outperformed its competitors in learning lessons from past successes and failures and applying them to reforms. One of the main reasons the German General Staff worked so well was that the prestigious and rather scholarly staff officers routinely served in line roles and earned the respect of line officers. In a few disciplines—engineering, economics—the flow of personnel from academic to practical line and staff roles is perhaps routine enough to resemble an informal general staff. In most disciplines academic and practitioner roles are mutually exclusive, practically speaking. The various agricultural extension services and other applied science organizations could be prospected for models. A practical scheme to "grow" innovative commons management institutions is perhaps only an inspirational innovation or two away from practicality. The two senior authors, who have had considerable, interesting, and rewarding experience as staffers in applied science and policy contexts, must admit that they found no way in the end to combine such work with an academic career.

## CONCLUSIONS

In this chapter, we have tried to tie together the literature on the evolution of cooperation with the literature on commons management institutions. We believe an interesting parallel exists between the sophisticated bounded rationality models necessary to account for the behavior of people toward commons and dual inheritance or gene-culture coevolutionary theory. People behave in experiments and in the field as if they have strong—perhaps innate—dispositions to cooperate, although dispositions vary considerably from person to person, society to society, and time to time. The variation is best explained by the existence of complex cultural traditions of social behavior, the collective results of which we call social institutions. Our ability to organize cooperation on a scale considerably larger than predicted by theory based on unconstrained selfish rationality, or by most evolutionary mechanisms, is one of the most striking features of our species. Another striking feature is our extraordinary facility for imitation and teaching. Our main hypothesis is that the co-occurrence of culture and cooperation in our species is not a coincidence. Group selection on cultural variation provides a plausible mechanism by which large-scale cooperation might arise. Cultural group selection is a slow process, at least in some models we have studied, so supplementary processes are likely to be more important in the shorter run evolution of cooperative institutions.

The cooperative dispositions, cultural or innate, favored originally by cultural group selection or some similar process will inevitably act as biases of cultural innovation and transmission. All else equal, people will tend to favor innovations that seem fair, that are efficient producers of public goods, and that contribute to their ingroup's position relative to competing outgroups. As team sports show, people play games of cooperation for fun. We can even organize institutions to promote desirable institutional evolution, ranging from research universities and political parties to village assemblies. Of course, people are hardly perfect paragons of cooperation. Our mixture of altruistic and selfish propensities varies across cultures but neither element is ever suppressed entirely. Gene-culture coevolution theory has a natural account of our conditional and incomplete altruism. At root, reproductive competition between the cooperators in human societies means that selection on genes still acts strongly to favor behavior enhancing inclusive fitness. Group selection on culture can only partially mitigate selfish and nepotistic impulses, not eliminate them.

Aside from providing an ultimate explanation for the patterns of cooperation we observe in humans, we hope the application of evolutionary theory to the understanding of commons institutions will lead to means to improve commons management. If our particular evolutionary theory is correct, we have good news and bad news for the practitioner. The good news is that we have much better raw material to work with improving commons management than the selfish rationality theorists think we have. The bad news is that institutions to capitalize on our prosocial instincts and traditions evolve relatively slowly and uncertainly. Regress is possible as well as progress. Cooperation within groups is all too often devoted to unhelpful if not destructive conflicts with other groups, as in the conflict between rivalrous national goals and the regulation of the global commons.

The new theory of the commons already understands all these things. Evolutionary theory offers a program for investigating just how institutions do evolve. We have outlined a little of the complexity possible when several different evolutionary processes can be at work, some stronger and some weaker, and all depending, at least to some extent, on the case at hand. The products of evolution are not only complex but also diverse. Exploring the tempo and mode of cultural evolution is a long-term project. After all, biologists are still at work on organic evolution a century and a half after Darwin, and they're still having plenty of fun. Of course, they have so many species to work on and we are only one, albeit a more than ordinarily diverse and complex one. In some ways cultural evolution is easier to study than organic evolution. Cultures change faster than gene pools. Historians and anthropologists have compiled vast amounts of qualitative information about our evolution and diversity and some innovative scholars have produced quantitative data. We believe that all the empirical methods needed to study cultural evolution have been used effectively in some specialized application or another, even if they are not yet in every social scientist's toolkit. We believe there is nothing to lose—and everything to gain—by developing and verifying a rigorous evolutionary theory of human behavior.

## NOTE

1 Several prominent modern Darwinians—Hamilton (1975), E.O. Wilson (1975:561-562), Alexander (1987:169), and Eibl-Eibesfeldt (1982)—have given serious consideration to group selection as a force *in the special case* of human ultrasociality. They are impressed, as we are, by the organization of human populations into units that engage in sustained, lethal combat with other groups, not to mention other forms of cooperation. The trouble with a straightforward group selection hypoth-

esis is our mating system. We do not build up concentrations of intrademic relatedness like social insects, and few demic boundaries are without considerable intermarriage. Moreover, the details of human combat are more lethal to the hypothesis of genetic group selection than to the human participants. For some of the most violent groups among simple societies, wife capture is one of the main motives for raids on neighbors, a process that hardly could be better designed to erase genetic variation between groups.

# REFERENCES

Acheson, J.M.
   1988    *The Lobster Gangs of Maine.* Hanover, NH: University Press of New England.
Alchian, A.A.
   1950    Uncertainty, evolution and economic theory. *Journal of Political Economy* 58:211-222.
Alexander, R.D.
   1979    *Darwinism and Human Affairs.* Seattle: University of Washington Press.
   1987    *The Biology of Moral Systems.* Hawthorne, NY: Aldine de Gruyter.
Anderson, B.R.
   1991    *Imagined Communities: Reflections on the Origin and Spread of Nationalism.* Rev. and
           extended ed. London: Verso.
Aoki, K.
   1982    A condition for group selection to prevail over counteracting individual selection. *Evolution* 36:832-842.
Arrow, K.J.
   1963    *Social Choice and Individual Values.* 2d ed. New Haven, CT: Yale University Press.
Axelrod, R., and W.D. Hamilton
   1981    The evolution of cooperation. *Science* 211:1390-1396.
Baland, J.M., and J.P. Platteau
   1996    *Halting Degradation of Natural Resources: Is There a Role for Rural Communities?* Oxford, Eng.: Oxford University Press.
Bandura, A.
   1986    *Social Foundations of Thought and Action: A Social Cognitive Theory.* Englewood Cliffs,
           NJ: Prentice-Hall.
Banfield, E.C.
   1958    *The Moral Basis of a Backward Society.* Glencoe, IL: Free Press.
Barkow, J.H., L. Cosmides, and J. Tooby
   1992    *The Adapted Mind: Evolutionary Psychology and the Generation of Culture.* New York:
           Oxford University Press.
Batson, C.D.
   1991    *The Altruism Question: Toward a Social Psychological Answer.* Hillsdale, NJ: Lawrence
           Erlbaum Associates.
Baum, J.A.C., and B. McKelvey, eds.
   1999    *Variations in Organization Science: In Honor of Donald T. Campbell.* Thousand Oaks,
           CA: Sage Publications.
Baum, W.B.
   1994    *Understanding Behaviorism: Science, Behavior, and Culture.* New York: HarperCollins.
Becker, G.S.
   1976    Altruism, egoism, and genetic fitness: Economics and sociobiology. *Journal of Economic
           Literature* 14:817-826.
Bettinger, R.L.
   1991    *Hunter-Gatherers: Archaeological and Evolutionary Theory.* New York: Plenum Press.

1999    From traveler to processor: Regional trajectories of hunter-gatherer sedentism in the Inyo-Mono Region, California. Pp. 39-55 in *Settlement Pattern Studies in the Americas: Fifty Years since Virú*, B.R. Billman and G.M. Feinman, eds. Washington, DC: Smithsonian Institution Press.

Blackmore, S.
1999    *The Meme Machine.* Oxford, Eng.: Oxford University Press.

Boehm, C.
1993    Egalitarian behavior and reverse dominance hierarchy. *Current Anthropology* 34:227-254.
1996    Emergency decisions, cultural-selection mechanics, and group selection. *Current Anthropology* 37:763-766.
1999    *Hierarchy in the Forest: The Evolution of Egalitarian Behavior.* Cambridge, MA: Harvard University Press.

Boorman, S.A., and P.R. Levitt
1980    *The Genetics of Altruism.* New York: Academic Press.

Borgerhoff-Mulder, M., P.J. Richerson, N.W. Thornhill, and E. Voland
1997    The place of behavioral ecological anthropology in evolutionary science. Pp. 253-282 in *Human by Nature: Between Biology and the Social Sciences*, P. Weingart, ed. Mahwah, NJ: Lawrence Erlbaum Associates.

Boyd, R., and P.J. Richerson
1985    *Culture and the Evolutionary Process.* Chicago: University of Chicago Press.
1987    The evolution of ethnic markers. *Cultural Anthropology* 2:65-79.
1988    The evolution of reciprocity in sizable groups. *Journal of Theoretical Biology* 132:337-356.
1989    The evolution of indirect reciprocity. *Social Networks* 11:213-236.
1992a  Group selection among alternative evolutionarily stable strategies. *Journal of Theoretical Biology* 145:331-342.
1992b  How microevolutionary processes give rise to history. Pp. 179-209 in *History and Evolution*, M.H. Nitecki and D.V. Nitecki, eds. Albany: State University of New York Press.
1992c  Punishment allows the evolution of cooperation (or anything else) in sizable groups. *Ethology and Sociobiology* 13:171-195.
1993    Rationality, imitation, and tradition. Pp. 131-149 in *Nonlinear Dynamics and Evolutionary Economics*, R.H. Day and P. Chen, eds. New York: Oxford University Press.
no       Group beneficial norms can spread rapidly in a structured population. Submitted to *Journal of Theoretical Biology*.
date

Callebaut, W., and R. Pinxten
1987    *Evolutionary Epistemology: A Multiparadigm Program.* Dordrecht, Neth.: Reidel.

Campbell, D.T.
1965    Variation and selective retention in socio-cultural evolution. Pp. 19-49 in *Social Change in Developing Areas: A Reinterpretation of Evolutionary Theory*, H.R. Barringer, G.I. Blanksten, and R.W. Mack, eds. Cambridge, MA: Schenkman Publishing Company.
1975    On the conflicts between biological and social evolution and between psychology and moral tradition. *American Psychologist* 30:1103-1126.
1983    The two distinct routes beyond kin selection to ultrasociality: Implications for the humanities and social sciences. Pp. 11-39 in *The Nature of Prosocial Development: Theories and Strategies*, D.L. Bridgeman, ed. New York: Academic Press.

Carneiro, R.L.
1970    A theory for the origin of the state. *Science* 169:733-738.

Cashdan, E.
1992    Spatial organization and habitat use. Pp. in *Evolutionary Ecology and Human Behavior*, B. Winterhalder and E.A. Smith, eds. New York: Aldine de Gruyter.

Cavalli-Sforza, L.L., and M.W. Feldman
    1973   Models for cultural inheritance: I. Group mean and within group variation. *Theoretical Population Biology* 4:42-55.

Cohen, D., and J. Vandello
    2001   Honor and "faking" honorability. Pp. 163-185 in *The Evolution and the Capacity for Commitment*, R.M. Nesse, ed., New York: Russell Sage.

Cronk, L., N.A. Chagnon, and W. Irons
    2000   *Adaptation and Human Behavior: An Anthropological Perspective.* New York: Aldine de Gruyter.

Curtin, P.D.
    1984   *Cross-Cultural Trade in World History.* Cambridge, Eng.: Cambridge University Press.

Dahrendorf, R.
    1968   *Essays in the Theory of Society.* Stanford, CA: Stanford University Press.

Darwin, C.
    1874   *The Descent of Man and Selection in Relation to Sex.* (2d ed.) New York: American Home Library.

Dawes, R.M., A.J.C. van de Kragt, and J.M. Orbell
    1990   Cooperation for the benefit of us—not me or my conscience. In *Beyond Self-Interest*, J.J. Mansbridge, ed. Chicago: University of Chicago Press.

Day, R.H., and P. Chen
    1993   *Nonlinear Dynamics and Evolutionary Economics.* New York: Oxford University Press.

Dennett, D.C.
    1996   *Kinds of Minds: Toward an Understanding of Consciousness.* New York: Basic Books.

Derksen, A.A.
    1998   *The Promise of Evolutionary Epistemology.* Tilburg, Neth.: Tilburg University Press.

Dietz, T., and T.R. Burns
    1992   Human agency and the evolutionary dynamics of culture. *Acta Sociologica* 35:187-200.

Dupuy, T.N.
    1977   *A Genius for War: The German Army and General Staff, 1807-1945.* Englewood Cliffs, NJ: Prentice-Hall.

Durham, W.H.
    1991   *Coevolution: Genes, Culture, and Human Diversity.* Stanford, CA: Stanford University Press.

Edgerton, R.B.
    1971   *The Individual in Cultural Adaptation: A Study of Four East African Peoples.* Berkeley: University of California Press.

Eibl-Eibesfeldt, I.
    1982   Warfare, Man's indoctrinability, and group selection. *Zeitschrift für Tierpsychologie* 67:177-198.
    1989   *Human Ethology.* New York: Aldine de Gruyter.

Endler, J.A.
    1986   *Natural Selection in the Wild.* Princeton, NJ: Princeton University Press.

Eshel, I.
    1972   On the neighborhood effect and the evolution of altruistic traits. *Theoretical Population Biology* 3:258-277.

Fairbank, J.K.
    1992   *China: A New History.* Cambridge, MA: Harvard University Press.

Fehr, E., and J.-R. Tyran
    1996   Institutions and reciprocal fairness. *Nordic Journal of Political Economy* 23:133144.

Fukuyama, F.
    1992   *The End of History and the Last Man.* New York: Free Press.

*437*

1995 *Trust: Social Virtues and the Creation of Prosperity.* New York: Free Press.
Gadgil, M., and R. Guha
1992 *This Fissured Land: An Ecological History of India.* Delhi: Oxford University Press.
Gadgil, M., and C. Malhotra
1983 Adaptive significance of the Indian caste system: An ecological perspective. *Annals of Human Biology* 10:465-478.
Gambetta, D.
1993 *The Sicilian Mafia.* Cambridge, MA: Harvard University Press.
Garthwaite, G.R.
1993 Reimagined internal frontiers: Tribes and nationalism—Bakhtiyari and Kurds. Pp. 130-148 in *Russia's Muslim Frontiers: New Directions in Cross-Cultural Analysis,* D.F. Eickelman, ed. Bloomington: Indiana University Press.
Gintis, H.
2000 *Game Theory Evolving: A Problem-Centered Introduction to Modeling Strategic Behavior.* Princeton, NJ: Princeton University Press.
Hamilton, W.D.
1964 Genetic evolution of social behavior I, II. *Journal of Theoretical Biology* 7:1-52.
Hannan, M.T., and J. Freeman
1989 *Organizational Ecology.* Cambridge, MA: Harvard University Press.
Hardin, G.
1968 The tragedy of the commons. *Science* 162:1243-1248.
1978 Political requirements for preserving our common heritage. In *Wildlife and America: Contributions to an Understanding of American Wildlife and Its Conservation,* H.P. Brokaw, ed. Washington, DC: Council on Environmental Quality.
Henrich, J., and R. Boyd
1998 the evolution of conformist transmission and the emergence of between-group differences. *Evolution and Human Behavior* 19:215-241.
Henrich, J., R. Boyd, S. Bowles, E. Camerer, E. Fehr, H. Gintis, and R. McElreath
2001 Reciprocity and punishment in fifteen small-scale societies. *American Economic Review* 91:73-78.
Hirshleifer, J.
1977 Economics from a biological viewpoint. *Journal of Law and Economics* 20:1-52.
Hodgson, G.M.
1993 *Economics and Evolution: Bringing Life Back into Economics.* Ann Arbor: University of Michigan Press.
Holland, J.H.
1995 *Hidden Order: How Adaptation Builds Complexity.* Reading, MA: Addison-Wesley.
Horowitz, A.V.
1990 *The Logic of Social Control.* New York: Plenum Press.
Hull, D.L.
1988 *Science as a Process: An Evolutionary Account of the Social and Conceptual Development of Science.* Chicago: University of Chicago Press.
Hundley, N.
1992 *The Great Thirst: Californians and Water, 1770s-1990s.* Berkeley: University of California Press.
Inglehart, R., and J.-R. Rabier
1986 Aspirations adapt to situations—but why are the Belgians so much happier than the French? A cross-cultural analysis of the subjective quality of life. Pp. 1-56 in *Research on the Quality of Life,* F.M. Andrews, ed. Ann Arbor: Survey Research Center, Institute for Social Research, University of Michigan.

Insko, C.A., R. Gilmore, S. Drenen, A. Lipsitz, D. Moehl, and J. Thibaut
    1983    Trade versus expropriation in open groups: A comparison of two types of social power. *Journal of Personality and Social Psychology* 44:977-999.
Jankowski, M.S.
    1991    *Islands in the Street: Gangs and American Urban Society.* Berkeley: University of California Press.
Johnson, A.W., and T.K. Earle
    1987    *The Evolution of Human Societies: From Foraging Group to Agrarian State.* Stanford, CA: Stanford University Press.
Kellett, A.
    1982    *Combat Motivation: The Behavior of Soldiers in Battle.* Boston: Kluwer.
Kelly, R.C.
    1985    *The Nuer Conquest: The Structure & Development of an Expansionist System.* Ann Arbor: University of Michigan Press.
Kelly, R.L.
    1995    *The Foraging Spectrum: Diversity in Hunter-Gatherer Lifeways.* Washington, DC: Smithsonian Institution Press.
Kennedy, P.M.
    1987    *The Rise and Fall of the Great Powers: Economic Change and Military Conflict from 1500 to 2000.* 1st ed. New York: Random House.
Kirch, P.V.
    1984    *The Evolution of the Polynesian Chiefdoms.* Cambridge, Eng.: Cambridge University Press.
Klein, R.G.
    1999    *The Human Career: Human Biological and Cultural Origins.* 2d ed. Chicago: University of Chicago Press.
Knauft, B.M.
    1985    *Good Company and Violence: Sorcery and Social Action in a Lowland New Guinea Society.* Berkeley: University of California Press.
    1993    *South Coast New Guinea Cultures: History, Comparison, Dialectic.* Cambridge, Eng.: Cambridge University Press.
Kramer, R.M., and M.B. Brewer
    1984    Effects of group identity on resource use in a simulated commons dilemma. *Journal of Personality and Social Psychology* 46:1044-1057.
Kurlansky, M.
    1998    *Cod: A Biography of the Fish That Changed the World.* London: J. Cape.
Labov, W.
    1973    *Sociolinguistic Patterns.* Philadelphia: University of Pennsylvania Press.
Leimar, O., and P. Hammerstein
    2001    Evolution of cooperation through indirect reciprocity. *Proceedings of the Royal Society of London* B268:745-753.
LeVieil, D.P.
    1987    *Territorial Use-Rights in Fishing (Turfs) and the Management of Small-Scale Fisheries: The Case of Lake Titicaca (Peru).* Unpublished dissertation, Ph.D., University of British Columbia.
Levin, B.R., and W.L. Kilmer
    1974    Interdemic selection and the evolution of altruism: A computer simulation study. *Evolution* 28:527-545.
Low, B.S.
    1996    Behavioral ecology of conservation in traditional societies. *Human Nature* 7:353-379.
Luhmann, N.
    1982    *The Differentiation of Society.* New York: Columbia University Press.

Mallory, J.P.
 1989   *In Search of the Indo-Europeans: Language, Archaeology, and Myth.* New York: Thames and Hudson.
Marty, M.E., and R.S. Appleby
 1991   *Fundamentalisms Observed.* Chicago: University of Chicago Press.
Marwell, G., and R.E. Ames
 1981   Economist's free ride: Does anyone else? *Journal of Public Economics* 15:295-310.
Maryanski, A., and J.H. Turner
 1992   *The Social Cage: Human Nature and the Evolution of Society.* Stanford, CA: Stanford University Press.
Maynard Smith, J., and E. Szathmáry
 1995   *The Major Transitions in Evolution.* Oxford, Eng.: W.H. Freeman Spektrum.
Mayr, E.
 1982   *The Growth of Biological Thought: Diversity, Evolution, and Inheritance.* Cambridge, MA: Harvard University Press.
McElreath, R., R. Boyd, and P.J. Richerson
 no     Shared norms can lead to the evolution of ethnic markers. Submitted to *Current Anthro-*
 date   *pology.*
McLaughlin, P.
 1988   Essentialism, population thinking and the environment. *Environment, Technology and Society* 52:4-8.
Miller, G.J.
 1992   *Managerial Dilemmas: The Political Economy of Hierarchy.* Cambridge, Eng.: Cambridge University Press.
Nesse, R.M., and G.C. Williams
 1995   *Why We Get Sick: The New Science of Darwinian Medicine.* 1st ed. New York: Times Books.
Nisbett, R.E., and D. Cohen
 1996   *Culture of Honor: The Psychology of Violence in the South.* Boulder, CO: Westview Press.
Nisbett, R.E., K. Peng, I. Choi, and A. Norenzayan
 in     Culture and systems of thought: Holistic and analytic cognition. *Psychological Review.*
 press
North, D., and R.P. Thomas
 1973   *The Rise of the Western World: A New Economic History.* Cambridge, Eng.: Cambridge University Press.
Nowak, M.A., and K. Sigmund
 1998   Evolution of indirect reciprocity by image scoring. *Nature* 393:573-577.
Ostrom, E.
 1990   *Governing the Commons: The Evolution of Institutions for Collective Action.* Cambridge, Eng.: Cambridge University Press.
 1998   A behavioral approach to the rational choice theory of collective action. *American Political Science Review* 92:1-22.
Pinker, S.
 1994   *The Language Instinct: How the Mind Creates Language.* New York: William Morrow.
Price, G.R.
 1970   Selection and covariance. *Nature* 277:520-521.
Price, T.D., and J.A. Brown
 1985   *Prehistoric Hunter-Gatherers: The Emergence of Cultural Complexity.* Orlando: Academic Press.
Putnam, R.D.
 1993   *Making Democracy Work: Civic Traditions in Modern Italy.* Princeton, NJ: Princeton University Press.

2000    *Bowling Alone: The Collapse and Revival of American Community.* New York: Simon & Schuster.

Rice, W.R.
1996    Sexually antagonistic male adaptation triggered by experimental arrest of female evolution. *Nature* 381:232-234.

Richerson, P.J.
1988    Improving cooperation between EPA and the universities: Some hypotheses. Pp. 296-318 in *Science, Universities and the Environment.* Chicago: Institute of Government and Public Affairs, University of Illinois.

Richerson, P.J., and R. Boyd
1989    The role of evolved predispositions in cultural evolution: Or sociobiology meets Pascal's wager. *Ethnology and Sociobiology* 10:195-219.
1998    The evolution of human ultrasociality. Pp. 71-95 in *Indoctrinability, Ideology, and Warfare: Evolutionary Perspectives*, I. Eibl-Eibesfeldt and F.K. Salter, eds. New York: Berghahn Books.
1999    Complex societies - The evolutionary origins of a crude superorganism. *Human Nature* 10:253-289.
2000    Built for speed. Pleistocene climate variation and the origin of human culture. *Perspectives in Ethology* 13:1-45.
2001    The evolution of subjective commitment to groups: A tribal instincts hypothesis. Pp. 186-220 In *Evolution and the Capacity for Commitment*, R.M. Nesse, eds. New York: Russell Sage Foundation.
in      Institutional evolution in the Holocene: The rise of complex societies. *Proceedings of the*
press   *British Academy.*

Richerson, P.J., R. Boyd, and R.L. Bettinger
2001    Was agriculture impossible during the Pleistocene but mandatory during the Holocene? A climate change hypothesis. *American Antiquity* 66:387-411.

Rogers, A.R.
1990    Group selection by selective emigration: The effects of migration and kin structure. *American Naturalist* 135:398-413.

Rogers, E.M.
1995    *Diffusion of Innovations.* 4th ed. New York: Free Press.

Roof, W.C., and W. McKinney
1987    *American Mainline Religion: Its Changing Shape and Future.* New Brunswick, NJ: Rutgers University Press.

Rumsey, A.
1999    Social segmentation, voting, and violence in Papua New Guinea. *Contemporary Pacific* 11:305-333.

Runciman, W.G.
1998    Greek hoplites, warrior culture, and indirect bias. *Journal of the Royal Anthropological Institute* 4:731-751.

Ruttan, L.M.
1998    Closing the commons: Cooperation for gain or restraint? *Human Ecology* 26:43-66.

Sabatier, P.A., and H.C. Jenkins-Smith
1993    *Policy Change and Learning: An Advocacy Coalition Approach.* Boulder, CO: Westview Press.

Sahlins, M.
1963    Poor man, rich man, big-man, chief: Political types in Melanesia and Polynesia. *Comparative Studies in Sociology and History* 5:285-303.

Salter, F.K.
1995    *Emotions in Command: A Naturalistic Study of Institutional Dominance.* Oxford, Eng.: Oxford University Press.

Sears, D.O., and C.L. Funk
  1990   Self interest in Americans' political opinions. Pp. 142-170 in *Beyond Self-Interest*, J.J. Mansbridge, ed. Chicago: University of Chicago Press.
Simon, H.A.
  1990   A mechanism for social selection and successful altruism. *Science* 250:1665-1668.
Slatkin, M., and M.J. Wade
  1978   Group selection on a quantitative character. *Proceedings of the National Academy of Sciences USA* 75:3531-3534.
Sober, E., and D.S. Wilson
  1998   *Unto Others: The Evolution and Psychology of Unselfish Behavior.* Cambridge, MA: Harvard University Press.
Soltis, J., R. Boyd, and P.J. Richerson
  1995   Can group-functional behaviors evolve by cultural group selection: An empirical test. *Current Anthropology* 36:473-494.
Squires, A.M.
  1986   *The Tender Ship: Governmental Management of Technological Change.* Boston: Birkhäuser.
Srinivas, M.N.
  1962   *Caste in Modern India, and Other Essays.* Bombay: Asia Publishing House.
Stern, P.C.
  1995   Why do people sacrifice for their nations? *Political Psychology* 16:217-235.
Sugden, R.
  1986   *The Economics of Rights, Co-Operation, and Welfare.* Oxford, Eng.: B. Blackwell.
Tajfel, H.
  1981   *Human Groups and Social Categories: Studies in Social Psychology.* Cambridge, Eng.: Cambridge University Press.
Temerlin, M.K.
  1975   *Lucy: Growing up Human, a Chimpanzee Daughter in a Psychotherapist's Family.* Palo Alto, CA: Science and Behavior Books.
Tomasello, M.
  1999   *The Cultural Origins of Human Cognition.* Cambridge, MA: Harvard University Press.
Trivers, R.L.
  1971   The evolution of reciprocal altruism. *Quarterly Review of Biology* 46:35-57.
Tullock, G.
  1965   *The Politics of Bureaucracy.* Washington, DC: Public Affairs Press.
Turner, J.H.
  1995   *Macrodynamics: Toward a Theory on the Organization of Human Populations.* New Brunswick, NJ: Rutgers University Press.
Vose, M.D.
  1999   *The Simple Genetic Algorithm: Foundations and Theory.* Cambridge, MA: MIT Press.
Weingart, P., S.D. Mitchell, P.J. Richerson, and S. Maasen
  1997   *Human by Nature: Between Biology and the Social Sciences.* Mahwah, NJ: Lawrence Erlbaum Associates.
West, R.
  1941   *Black Lamb and Grey Falcon.* New York: Penguin.
Williams, G.C.
  1966   *Adaptation and Natural Selection: A Critique of Some Current Evolutionary Thought.* Princeton, NJ: Princeton University Press.
Wilson, D.S.
  1983   The group selection controversy: history and current status. *Annual Review of Ecology and Systematics* 14:159-188.

Wilson, E.O.
    1975   *Sociobiology: A New Synthesis.* Cambridge, MA: Harvard University Press.
    1998   *Consilience: The Unity of Knowledge.* New York: Knopf.
Witt, U.
    1992   *Explaining Process and Change: Approaches to Evolutionary Economics.* Ann Arbor: University of Michigan Press.
Wolfe, T.
    1965   *The Kandy-Kolored Tangerine-Flake Streamline Baby.* New York: Farrar Straus and Giroux.

# CONCLUSION

# 13

# Knowledge and Questions
# After 15 Years of Research

*Paul C. Stern, Thomas Dietz, Nives Dolšak, Elinor Ostrom,
and Susan Stonich*

T he study of institutions for managing common-pool resources has ma-
tured considerably since 1985. This chapter assesses the progress of the
field as a scientific enterprise, characterizes what has been learned over
the past decade and a half, and identifies a set of key research directions for the
next decade of research. We find that the field is making marked progress along a
trajectory of development that is common to many maturing areas in the social
sciences. Some of the advances have practical value for natural resource managers, though knowledge has not progressed to a point at which managers can be
offered detailed guidance. And of course, practical guidance must be based on an
understanding of both the scientific knowledge base and the local situation. In
this chapter we summarize some key lessons from recent research, discuss seven
major challenges of institutional design, identify important directions for future
research, including key understudied issues, and note ways that the field can benefit from linkages to several related fields of social science research.

## PROGRESS OF THE FIELD

Research on institutional designs for common-pool resource management
has followed a development path that is similar to many other fields of social
science that investigate complex real-world phenomena and develop knowledge
intended to be useful for managing those phenomena. These fields seek to understand phenomena that are multivariate, path-dependent (i.e., historically contin-

We are indebted to James Acheson, Kai Lee, Ronald Mitchell, and the chapter authors of this
volume for insightful discussions and written comments on drafts of this chapter.

gent), and reflexive (i.e., alterable in important ways by the process of studying them). Many of the processes are hard to study with field experiments and care must be taken in generalizing from laboratory experiments. Thus, establishing causation is always a challenge. The complexity of the phenomena also means that models based on nonexperimental data have many parameters to be estimated relative to the number of observations available. Other fields facing this problem include international conflict resolution (Stern and Druckman, 2000) and comparative politics and sociology (King et al., 1994; Ragin, 1987, 2000; Ragin and Becker, 1992). Program evaluation has a long history of dealing with these issues (see, e.g., Cook and Campbell, 1979; Chen, 1990; Chen and Rossi, 1992; Weiss, 1998). Progress in such fields depends on reducing bewildering arrays of phenomena, each with multiple attributes that may be important, into manageable sets of measurable variables. It also depends on developing theory that specifies relationships among the variables, including identification of causal relationships among variables that can be manipulated intentionally. The development path in such fields typically involves at least four elements, all of which are evident in common-pool resource management research.

## Development and Differentiation of Typologies

Typologies are needed to classify the central phenomena under study, the outcomes worthy of investigation, and the factors both internal and external to the central phenomena that shape those phenomena and their effects on larger systems. Without a shared language that differentiates key concepts, theoretical progress is impossible. An example is the increasingly familiar classification of property rights institutions into four major types: individual property, government property, group property, and open access (the absence of rights to exclude) (e.g., Feeny et al., 1990). These types have been further differentiated into subtypes (e.g., Tietenberg, Chapter 6, on subtypes of private property). Another example is the classification of factors affecting institutional functioning into attributes of resources, attributes of appropriators, and attributes of institutions, and of each of these classes into subtypes (see Agrawal, Chapter 2). It can be useful to subdivide these even further. For example, Bardhan and Dayton-Johnson (Chapter 3) identify several kinds of heterogeneity among resource appropriators and conclude that economic heterogeneity and social heterogeneity have independent effects and operate through different causal mechanisms (incentives versus norms). Typologies allow researchers to focus attention on a tractable number of variables and then to state and systematically examine research hypotheses about them.[1]

## Contingent Generalizations

A second element of development is a shift from bivariate research hypotheses to contingent or conditional ones. For example, it has become clear that no

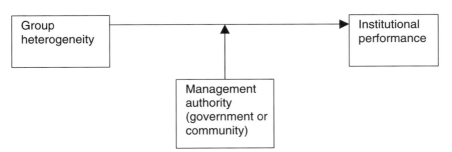

FIGURE 13-1 Schematic representation of an empirically supported contingent relationship between group heterogeneity and institutional performance.

single institutional form is best at maintaining resources across a wide range of environmental and social conditions. Researchers have begun to propose hypotheses about conditions under which particular institutional forms are likely to be successful. Similarly, research has shown that simple bivariate relationships of the sustainability of resource management with the size, heterogeneity, and poverty of the user group may be positive, negative, or curvilinear, depending on contextual factors (Agrawal, Chapter 2). Researchers have responded by developing and testing hypotheses that take these contingencies into account.

Studies with large numbers of cases (large-*n* studies) are particularly useful for generating such hypotheses because they allow regularities to be observed in subsamples that differ in factors that change the effect of other variables—the factors that make conclusions contingent. For example, Tang (1992, cited by Bardhan and Dayton-Johnson, Chapter 3) reports that heterogeneity among resource appropriators is associated with poorer performance in irrigation systems that are managed by government agencies, but not in community-managed systems (see Figure 13-1). Apparently, some community-managed systems are able to develop rules of allocation and cost sharing that meet the challenges of heterogeneity, while agency-managed systems are not. Similarly, Varughese and Ostrom (2001) show that various forms of heterogeneity within forest user groups depend for their effects on collective action on the specific form of organization established by the group. Identifying this difference in the effects of heterogeneity requires cases that differ in their degree of heterogeneity among both community-managed and government-managed systems.[2]

### Causal Analysis

A third element of development is a shift from correlational to causal analysis. Researchers hypothesize and search for causal paths or mechanisms that can account for and explain observed associations. These causal models include interactions such as the one just noted where the effect of one independent variable on the dependent variable changes with the value of a third variable. For example, Bardhan and Dayton-Johnson (Chapter 3) theorize that the effects of heterogeneity may follow several causal paths. "Olson effects" (Olson, 1965; see causal path (a) in Figure 13-2) operate when certain resource users have enough at stake, and enough wealth, to maintain the resource on their own even though there are free riders. Two alternative causal paths, (b) and (c), have negative effects on resources and, according to Bardhan and Dayton-Johnson, are more often consistent with the evidence on the functioning of irrigation systems.

Another example of causal models comes from experimental research attempting to understand why communication within a resource user group fosters cooperative outcomes. This research suggests three possible causal mechanisms. Communication may increase group identity or solidarity, create the perception of a consensus to cooperate, or result in actual commitments to cooperate, which function as shared norms to which members adhere (see Kopelman et al., Chapter

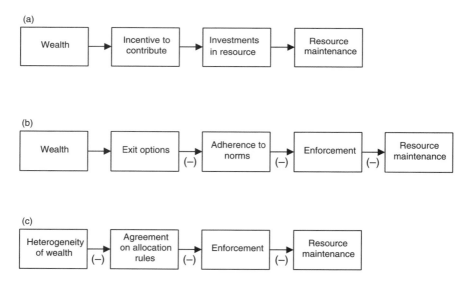

FIGURE 13-2 Three causal paths describing hypothesized effects of wealth or weath inequality on maintenance of common-pool resources.
NOTE: Minus signs (–) signify hypothesized negative effects of wealth or wealth inequality on the variable to the right of the arrow.

4). Experimental researchers have been working to understand whether all or only some of these mechanisms are important for understanding the observed communication effect. One advantage of the experimental approach compared to the analysis of an even larger number of case studies from the field is the ability of experimental researchers to structure the values of hypothesized causal variables so as to obtain clear estimates of their effects (Gintis, 2000). The experiment allows critical simplifications not possible with field data. Recent experiments conducted in field settings with Colombian villagers who are responsible for managing local common-pool resources provide complementary evidence to that generated in experimental laboratories located in universities (Cárdenas et al., 2000).

Research on communication and group norms is part of a larger effort to build causal models that explain how characteristics of resources and resource-using groups link through social institutions to produce outcomes for resource systems. Figure 13-3 presents a schematic model that specifies some such links in detail, focusing particularly on the roles of monitoring and enforcement of existing rules as mediating factors. The model is generally consistent with available evidence. It is also partly speculative and incomplete (e.g., it does not represent a full range of effects of communication nor does it address how some of these independent variables affect each other and may affect the likelihood of self-organization in the first place).

Models such as that in Figure 13-3 do much to advance theory and practice. They move understanding forward from correlation to causation. In doing so they greatly reduce the number of variables and hypotheses to be examined. Such models, when empirically verified, create an importance ranking among the variables: Some emerge as important because of strong direct effects on the sustainability of the resource and other outcomes of concern. Others are important only for their indirect effects. For example, properties of resources and resource users affect resource outcomes only indirectly, mainly by influencing the costs of monitoring and enforcement. Of course, in making policy or designing institutions, the ease with which a variable can be changed, and the consequences of those changes on issues other than commons management, often will be as important as the size of the direct or indirect effect of a variable on commons sustainability. In addition, because institutional design choices are normally the result of negotiation among political actors, the technical characteristics of the options are weighed in the context of their political acceptability.

Models like that in Figure 13-3 can also advance understanding by grouping variables and making connections to related fields of study. In the model shown in Figure 13-3, communication, dense social networks, and practices of reciprocity all affect outcomes through exactly the same causal paths. This suggests that these variables may be considered as multiple indicators of a single underlying construct—perhaps what has been called strength of community (Singleton and Taylor, 1992; Gardner and Stern, 1996), social ties (Petrzelka and Bell, 2000), or

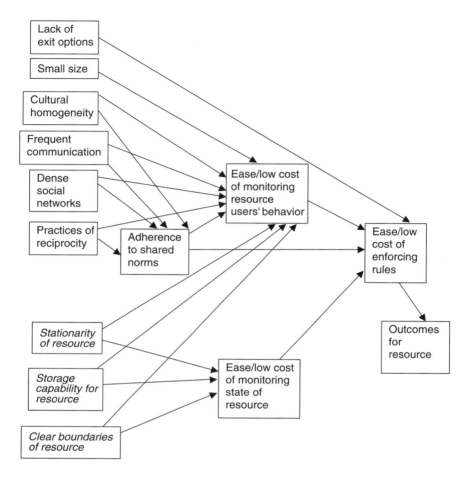

FIGURE 13-3 A schematic causal model postulating ways that costs of monitoring and enforcement mediate between characteristics of resources and resource users and outcomes of resource management institutions.
NOTE: Boxes in top part of left column refer to characteristics of resource users and their groups; boxes in bottom part with italic text refer to characteristics of resources. All arrows indicate posited positive relationships.

social capital (Putnam et al., 1993; Ostrom and Ahn, 2001; but see Abel and Stephen, 2000). It also suggests that what has been learned in research on these constructs may be relevant to problems of designing resource management institutions.

    Causal models can be useful to practitioners by helping them to identify potential ways to intervene to produce desired effects. This model redirects attention from variables on the left of the figure, none of which can be changed directly by institutional design, toward features that are more amenable to institutional solutions, in the middle of the figure.[3]

*Integration of Research Results*

The fourth element in the development of research is the integration of results from various research methods, each of which has its own contribution to offer—and its own limitations. Causal models of the sort described in the previous section are one form of integration, but here we are also referring to formal methods of integration and making sense of cross-study comparisons. Controlled experimental research (see Kopelman et al., Chapter 4) provides the strongest evidence for establishing relationships of cause and effect. But it is hard to apply to understanding complex phenomena like resource management institutions because they are hard to simulate realistically in the laboratory and because opportunities for field experiments are limited. Experiments seem to be most useful for understanding influences on the behavior of individuals and small groups that can be simulated in the laboratory. Because experiments must almost always be carried out in simulated resource-use situations, however, their external validity—that is, their relevance beyond the simulation setting—is always open to question.

Case studies have been the most frequently used method in studies of resource institutions. Careful case studies can provide deep understanding of realistic settings. It is difficult, however, to generalize from any single case, with all its contextual and historical uniqueness, to other situations. Careful comparisons across cases, such as was done in the studies reviewed by Agrawal (Chapter 2), can better distinguish phenomena unique to a single case from those with some generality. But as long as case study authors use a wide diversity of theoretical approaches and thus collect data that are not comparable across studies, rarely are there enough cases available with similar variables to support strong generalizations—the data usually leave room for alternative interpretations.

Researchers sometimes turn to multivariate data sets of moderate size to provide stronger evidence. Bardhan and Dayton-Johnson (Chapter 3) report on the results of some such multivariate studies, and databases are being developed that will support future studies of this type (see Gibson et al., 2000a; Poteete and Ostrom, 2001). This research strategy adds breadth not available from individual or small-*n* case studies, but it is limited by the range and quality of measures available for all cases in the data sets. Sometimes, variables of theoretical importance are not measured at all in a data set or can be measured only by using rough proxies. For example, Dayton-Johnson (2000) uses the number of villages where irrigators live as a measure of social heterogeneity. This measures spatial heterogeneity, but may not measure social or economic heterogeneity.

Another important research method involves the use of formal deductive theory, typically the theory of games (e.g., Falk et al., Chapter 5), of rational action (e.g., Bardhan and Dayton-Johnson, Chapter 3), or of optimal allocation (e.g., Tietenberg, Chapter 6). Formal theory has the virtue of precision, although the relevance of any particular formulation to practical situations can be determined only by empirical evidence. Put another way, deductions from theory generate hypotheses to explore with empirical methods. If observations do not square with theory, the theory can sometimes be elaborated to account for the data, thus generating new insights. An example is Tietenberg's explanation (Chapter 6) of why tradable permits seem to work better for controlling emission of air pollutants than for controlling the use of fisheries and water resources. Using simple economic models, advocates have promoted tradable permits for all three resource types, but experience calls attention to differences, particularly in the importance of negative externalities. Fishers of nonregulated species and downstream water users often are harmed by tradable-permit institutions. In contrast, the permitting systems for air pollutants do not seem to have produced externalities that have disrupted these institutions. In this instance, case studies reveal the need for theorists and institutional designers to give more attention to negative externalities produced by permit holders.

Because no research method is definitive, knowledge is best advanced by combining research methods in a strategy that is often referred to as "triangulation" (e.g., Campbell and Fiske, 1959) or "critical multiplism" (Cook, 1985, 1993). Results from using one method may offer hypotheses to explore with other methods, answer questions another method cannot answer, or call into question consistent conclusions from another method. A growing body of literature is intended to facilitate integration across methods and even hybridization of them (Bennett and George, 2001; King et al., 1994; McGinnis, 2000; Ragin, 1987, 2000; Ragin and Becker, 1992). Also relevant to research integration are methods of meta-analysis (e.g., Glass et al., 1981; Rosenthal, 1984; Petitti, 2000). All these kinds of exchange contribute to knowledge. Since the mid-1980s, increased communication and integration across methods of common-pool resource management research has benefitted the field.

*Toward a Conceptual Framework*

As this volume makes evident, researchers continue to identify variables that may be important for understanding and controlling the effects of resource management institutions. Agrawal (Chapter 2) identifies more than 30 such variables taken from a broad examination of the literature—and this list will surely get longer as research continues. As Agrawal notes, such a long list of variables creates significant challenges for research because of the large number of possible associations and causal relationships that must be examined. As he also notes, the development of theory presents one way through the thicket of possible hypotheses. Theory can potentially limit the number of theoretically meaningful propositions that are worth examining.

Theoretical propositions are beginning to emerge from recent research, as illustrated by the propositions presented by the relationships in Figures 13-1, 13-2, and 13-3. Other propositions, drawn from Chapters 2 through 12, are listed in the Appendix. We believe it is useful at this time to suggest a general conceptual framework within which such propositions can be placed. The classes of variables identified by Agrawal and others can be arranged into four broad functional categories defined by their possible theoretical relationships:

- Possible *interventions*, or independent variables. These are influential factors, including attributes of institutions, that can be altered by policy intervention over the short run.
- *Outcomes*, or dependent variables. These are things of importance to resource users that may be affected by resource conditions, resource use, and interventions.
- *Contingencies*, or moderator variables. These are factors that are out of the practical control of short-run policy interventions but that may determine how an intervention affects an outcome.
- *Mediators*, or intervening variables. These are factors that may affect outcomes but that may in turn be affected by interventions, subject to contingencies.

The typical relationships among these types of variables are represented schematically in the causal model shown in Figure 13-4. A grouping of variables from the commons literature into the four categories appears in Box 13-1.[4] This framework and causal model highlights some points that may be worth special attention in future research and practical analysis. One is that interventions often affect outcomes only indirectly through their effects on key intervening variables. The immediate policy challenge is often to influence variables such as the ease of monitoring the resource or adherence to group norms. The model highlights three tasks for theory: (1) to clarify how key intervening variables affect outcomes; (2) to identify the contingencies under which those mediators become critical; and (3) to identify the conditions under which particular interventions can successfully influence them. The framework also suggests that outcomes of interest depend on a variety of policy variables, not only on the design of resource management institutions. Thus, there may be more ways to achieve desired objectives than are immediately apparent.

Figure 13-3 can be seen as an elaboration and specification of the general model. It shows a variety of contingencies along the left edge and postulates their effects on a set of mediators (the center of the figure), all of which in turn affect outcomes. Figure 13-3 adds theoretical specificity by identifying key contingencies and mediators and by postulating causal links among the mediators. It does not, however, postulate effects of interventions on the variables in the figure.

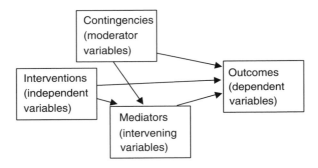

FIGURE 13-4 Schematic causal model showing typical relationships among variable types.

We believe the framework described can advance theory by helping to focus on kinds of propositions likely to have theoretical significance. We also believe it has practical potential because it clearly distinguishes types of variables that are possible policy levers (the independent variables) from two other types of variables that may also affect outcomes: mediators, which are appropriate targets of policy intervention, and contingencies, which must be taken into account in making policy choices, even though policy cannot quickly change them, because the outcome of an intervention may depend on the state of these variables when the intervention is tried.

The framework is incomplete and is not useful for all purposes. For example, it may not prove very useful for understanding the challenges of designing linkages among institutions. Nevertheless, we hope it will prove useful for advancing understanding.

## LESSONS LEARNED

As the previous section suggests, research since 1985 has changed the shape of the field and increased the sophistication of understanding, but it has not always produced definitive answers to practical questions. This section summarizes a few key substantive lessons that have gained solid support. We then discuss the practical relevance of knowledge so far developed, drawing extensively on the previous chapters.

This summary is of necessity quite selective and cannot touch all the key points raised in the dozen chapters that precede this one. To offer the reader further help in mining the rich lode of ideas developed in those chapters, the Appendix to this chapter contains a longer list of key findings or propositions and notes the chapters in which they can be found.

**BOX 13-1**
**Functional Classification of Variables from the Commons Literature with Examples within Each Type**
### Interventions (Independent Variables)

*Institutional arrangements regarding resource base* (e.g., property rights regime for resource, simplicity of rules, graduated sanctions, accountability of monitors, coordination with institutions at other scales or in other regions)
*Other institutional arrangements* (e.g., development, tax, investment policy; political representation rules)
*Technology choices* (e.g., decision to adopt new monitoring technology)

### Contingencies (Moderator Variables)

*Resource system characteristics* (e.g., size, boundaries, mobility of resource, storage, predictability)
*User characteristics* (e.g., population, boundaries, social capital, leadership, heterogeneities, prevalence of honesty, interdependence, poverty)
*Relationships between characteristics of resources and users* (see Box 2-5)
*Institutional forms at other scales or in other regions* (e.g., state support for local rules, nesting of institutions, international regimes)
*Available technology* (e.g., cost of technology for exclusion, monitoring)
Integration of resource base into global markets

### Mediators (Intervening Variables)

*Adherence of users to shared norms*
*Ease/cost of monitoring users' behavior*
*Ease/cost of monitoring state of resource*
*Ease/cost of enforcing rules*
*Users' understanding of rules and sanctions*

### Outcomes (Dependent Variables)

*Sustenance of the resource system (sustainability)*
*Durability of resource management institutions*
*Economic output of the resource system* (e.g., productivity, efficiency)
*Distribution of the economic output (equity)*
*Democratic control*

## Some Substantive Lessons

*The "Tragedy of the Commons" Model has Major Limitations*

The most basic lesson learned from studying actual common-pool resource management is that the metaphor of a "tragedy of the commons" is only apt under very special conditions. When resource users cannot communicate and have no way of developing trust in each other or in the management regime, they will tend to overuse or destroy their resource as the model predicts. Under more typical circumstances of resource use, however, users can communicate and have ways of developing trust. Under these conditions it is possible, though by no means certain, that they will agree on a set of rules (i.e., an institutional form) to govern their use patterns so as to sustain the resource and their own economic returns from it. Much of the research since 1985 can be understood as an effort to identify the factors affecting the likelihood that the resource users, by themselves or in conjunction with external authorities, will develop such rules, with accompanying incentives, and conform to the rules (Jensen, 2000).

*Three Conditions Are Necessary, but Not Sufficient, for Emergence of Self-Organized Institutions*

Research reviewed in the previous chapters identifies three basic conditions as necessary for resource appropriators to create and sustain effective resource management institutions. First, the resource must be salient enough to the users that they are willing to invest time and energy to create new institutions (Gibson, 2001). Second, users must have the autonomy to devise and change rules (that is, the external institutional environment must give or allow them this autonomy). Third, at least a subset of users must be able to engage in direct communication with each other, including the opportunity to bargain. Given these conditions, whether appropriators will organize, which institutional design they will choose, and the performance and survival of that design depend on specific characteristics of the resource, the resource users, and the repertoire of institutional rules considered.

*One Form Does Not Fit All*

The research clearly demonstrates that no particular institutional design can ensure successful management of all common-pool resources. Given ecological and social complexity, this finding should not be surprising. There are successes and failures with private property, government property, and community property institutions. What works best depends on specific characteristics of resources, resource users, external factors, details of institutional design, and the interactions among these factors. Thus, practitioners need to find an institutional form that fits the requirements of the biophysical system being used and the social context of the resource users. Fortunately for practical purposes, research has identified a great variety of institutional forms, thus expanding the practitioner's kit of tools from which to choose. Unfortunately, research has not yet matched the detailed characteristics of institutional forms to the characteristics of resources, resource users, and the context for which they are most suitable.

### "Success" Means Different Things to Different People

The common-pool resource research tradition began with questions about the sustainability of resources. This remains an important question for research, but it is not the only question for resource users. For them, the livelihoods and well-being of humans are often more important than any particular resource. Researchers who want to produce knowledge of practical value need to identify and examine the full range of outcome conditions that matter to the people who use, manage, and/or depend on the resource being studied. It is these people whose decisions the research will inform and who face important tradeoffs. Sometimes, one desirable outcome (e.g., sustainability or equity) can be achieved only by sacrificing another (e.g., efficiency). Institutions may be judged by how well they provide jobs and wealth, maintain good social relations in a community, provide access to resources from outside, and many other criteria in addition to resource sustainability. Research that ignores the multiplicity of valued outcomes is unlikely to produce realistic models for real decisions, which must take account of those varied outcomes.

### Indirect and Mediated Effects Are Important

Many of the characteristics of resources and resource users that have been hypothesized to affect the success of institutions (however defined) do so only contingently and indirectly. For example, as already noted, many effects are mediated by the costs of monitoring and rule enforcement. Understanding the indirect effects is important for making sense of the inconsistent bivariate associations that are reported in the literature. It is also important for institutional designers because, although in most cases they can do little to change characteristics of resources and resource users, they can often influence monitoring, enforcement, and other mediating factors in ways appropriate to the context they face.

## How Research Can Have Practical Value

When social science aspires to practical relevance, it must face the disjunction between its usual aim, which is to arrive at generalized propositions about the world, and the need of practitioners to act in highly specific but ever-changing circumstances. How can generalized knowledge be useful to these practitioners?

One field in which much thinking has been done about making social science knowledge useful is international relations. The following conclusions, drawn from that field (George, 1993; National Research Council, 2000), are also applicable to the field of common-pool resource management research.[5]

Practitioners always need many kinds of knowledge to achieve their objectives. Some types of essential knowledge are highly situation-specific and can come only from examining current features of particular situations—the forces in a particular location that are affecting a resource and resource users, and so forth. This can be called "time and place information" (Hayek, 1945). Other kinds of essential knowledge apply across situations. These forms of scientific knowledge tell scholars and practititoners what to expect with certain kinds of groups, in certain kinds of countries, or with certain resources. These kinds of knowledge are generic, that is, cross-situational, and therefore subject to improvement by systematic empirical studies.

The specific, contextually grounded problems practitioners must deal with are usually instantiations of generic problems of resource management. Although occurring in different contexts, these situations are encountered repeatedly. Examples include monitoring resources, enforcing rules, mediating disputes, and achieving cooperation. Practitioners typically consider several specific policy instruments and strategies for dealing with each of these generic problems. In this process they can benefit from the multiple types of knowledge about them.

First, *general conceptual models* identify the critical variables for dealing effectively with the phenomenon in question and the general logic associated with successful use of strategies or techniques to address a type of problem. For example, the theory of optimal allocation provides a general conceptual model for managing common-pool resources by creating institutions that help individuals clearly know their rights and duties and how these relate to sustainable management. A conceptual model provides a starting point for constructing a strategy for dealing with a particular situation. It assures that the practitioner is attentive to key dimensions of an issue and to the full range of institutional structures that might be brought to bear.

Second, practitioners need *conditional generalizations* about what favors the success of specific strategies under consideration. This kind of knowledge, as already noted, normally takes the form of statements of conditionality or contingency—that a strategy is effective under certain conditions but not others. Although conditional generalizations are not sufficient to determine which action to take, they are useful for diagnostic purposes. A practitioner can examine a situation to see whether favorable conditions exist or can be created for using a particular strategy or management approach. Good conditional generalizations enable a practitioner to increase the chances of making a good choice about whether and when to use a particular strategic intervention.

Third, practitioners need *knowledge about causal processes and mechanisms* that link the use of each strategy to outcomes. The effectiveness of pricing mechanisms, for example, is highly dependent on attributes of the resource. Rarely can one find a successful pricing mechanism for irrigation water in a system without reliable storage. If the amount of water available cannot be calculated, few farmers are willing to pay a price for an unknown quantity of water. On the other hand, where dams have been constructed and reliable measures of water quantity exist, farmers have been willing to engage in weekly water markets for centuries (see, for example, Maass and Anderson, 1986). Knowledge about such causal linkages is essential for monitoring the functioning of resource management institutions and for deciding whether they need additional support.

Fourth, in order to craft an appropriate strategy for a situation, practitioners need a *correct general understanding of the actors* whose behavior the strategy is designed to influence. To act effectively, it is necessary to see events from the perspective of those acting in the situation. Only by doing so can a practitioner diagnose a changing situation accurately and select appropriate ways of communicating with and influencing others. Faulty understanding of others is a major source of miscalculations leading to major errors in policy, avoidable catastrophes, and missed opportunities.

All these types of knowledge are generic in that they apply across situations that have the same characteristics. It is important to emphasize, however, that although such knowledge is useful, even indispensable, a practitioner also needs accurate time and place knowledge to act effectively. Skilled practitioners use their judgment to combine generic and specific knowledge in order to act in what are always unique decision situations, each with its own historical trajectory and current resource and institutional characteristics. The contributors to this volume have attempted to develop the first three kinds of knowledge described: (1) general conceptual models of resource management situations, (2) knowledge about the conditions favoring the success of particular institutional forms, and (3) knowledge about the causal processes that lead them to succeed or fail. In doing this they have had to grapple with other important but difficult issues: defining success, setting reasonable expectations and timelines for evaluating success, identifying indicators of success, and deciding how to make general inferences when historical evidence is imperfect and when one can never know what the outcome would have been if practitioners had acted differently or if events beyond their control had played out differently.

Some writers (e.g., Ostrom, 1990) have translated generic knowledge into sets of institutional design principles: generic advice about properties that should be designed into institutions to increase their chances of long-term success. These include principles such as clearly defining the boundaries of a resource, matching provision and appropriation rules to local conditions, participation of users in making future policies, devising ways of monitoring, using graduated sanctions, providing conflict resolution mechanisms, and recognizing the right to organize. This is a useful translation of the research literature into policy guidance, if practitioners understand that the design principles are provisional and likely to need refinement on the basis of improved knowledge (see Morrow and Hull, 1996; Asquith, 1999). Of course, the application of design principles is also filtered through the political processes through which institutional design decisions are made, so that these choices involve more than a straightforward application of generic knowledge in a specific situation. The applicability of design principles also changes over time and across contexts and thus proves to be contingent (Weinstein, 2000). For example, based on the discussion of the empirical cases presented at the 1985 Annapolis meeting discussed in Chapter 1, Ostrom (1986:611) proposed that institutions that had developed simple rules were more likely to survive. Specifically, the factor discussed was "The development of a clear-cut and unambiguous set of rules that all participants can know and agree upon." The logic on which this was based is that the "fewer rules used to organize activities, relative to the complexity of the activities, the more likely that individuals can understand them, remember them, and follow them, and the more likely that infractions will be interpreted by all as infractions" (Ostrom, 1986:611). This is obviously a highly contingent principle that has to be tailored to the complexity of the resource system itself, the cultural heterogeneity of the users, and their communication patterns. Slavish adoption of any stylized version of a design principle is unlikely to be a successful strategy (see Steins et al., 2000). Evidence exists that the "simple-rules" principle applies most strongly to institutions that engage large, diverse groups with weak community ties; small, tightly linked groups sometimes can function quite well with complex rules, provided that the users understand them well (Berkes, 1992).

How can generic knowledge be of practical value?[6] We do not expect that it will be prescriptive in the sense of providing a standard set of procedures that tells practitioners exactly what to do in particular situations. However, generic knowledge is useful to practitioners when they combine it with detailed knowledge about the situation at hand. Generic knowledge also has diagnostic value for practitioners. It describes the characteristics that determine the actions that will be effective. After a practitioner has accurately diagnosed a situation, knowledge about what works in which situations comes into play more strongly.

Even with a perfect diagnosis of a situation, however, there are several reasons why generic knowledge cannot be expected to provide detailed prescriptions for action. First, generic social science knowledge will never be as solidly established as, for example, a law of physics. For one thing, human actors can defy the laws said to govern their own behavior; for another, conditions continually change in ways that may invalidate conclusions from past experience. The principle of uniform laws across time and space, so central to the intellectual program of the physical sciences, is not realistic in developing theories about human behavior. Second, the many tradeoffs involved in any decision make general knowledge an imperfect guide to action. All the desired aspects of success cannot be achieved all at once, and choices must be based on tradeoff or compromise. Often, resource sustainability is not the only outcome relevant to practitioners. They must then weigh sustainability against other desired outcomes such as community governance and economic development.

Despite such limitations, we believe the kinds of knowledge developed in this volume, even with our current limited state of knowledge, will prove useful to resource management practitioners. They can help practitioners identify options for action they might not otherwise have considered, think through the implications of each course of action, and identify ways of monitoring to see if actions, once taken, remain on track.[7] However, one must recognize that practitioners may resist accepting conclusions developed by systematic analysis. Many practitioners mistrust such conclusions and prefer their own experiential knowledge and that developed by other practitioners. Although there is some wisdom in relying on what has worked and avoiding what has not, we have noted the difficulty of generalizing across contexts and time periods. This is as much a problem for the practitioner, who would rely on experience, as it is for the researcher. We believe that continued interaction between researchers and practitioners will, over time, improve mutual respect for and understanding of the kinds of knowledge that direct experience and systematic analysis taken together can provide. Bridging the gap between scholarship and practice remains an overriding challenge.

## Challenges of Institutional Design

Research has shown that the situation facing institutional designers is multifaceted, more than was previously appreciated. A greater variety of major institutional forms exist than were recognized in 1968, and each form has distinct subtypes. A good example in this volume is Tietenberg's (Chapter 6) effort to refine the simple notion of tradable permits—one private-property institutional form—to include a range of types of tradable rights and of institutions for addressing the negative externalities of regulating one or a few resources within a complex resource system. Such research clarifies the various institutional possibilities and their implications and helps increase the size of the practitioner's tool kit.

With such a large tool kit, it is difficult for practitioners to rely on their own experience alone in proposing institutional designs. This is probably a good thing because of the many pitfalls in reasoning from a limited basis of experience (Neustadt and May, 1984). Over time, research can help by developing systematic databases of experience and constructing a better map linking institutional forms to the conditions favoring their successful operation. This effort, however, will take a long time.

Meanwhile, it is useful to interpret available research results in terms of challenges facing institutional design—potential problems that must be addressed for resource management institutions to succeed. This section enumerates seven key challenges, discusses the conditions under which each one is especially critical, and notes some fairly robust strategies—the sort of design principles proposed by Ostrom (1990)—that have been proposed for meeting them.

*Low-Cost Enforcement of Rules*

Successful institutions are widely recognized to depend on the ability of users to devise rules for access to and maintenance of a common-pool resource and to sanction rule-breaching behavior. Much depends on whether these design characteristics can be achieved at reasonable cost and whether it is possible to get resource users to help provide for the costs (see Trawick, 1999; Ostrom, 2000). As Figure 13-3 indicates, the costs of enforcement, including adjudication of conflicts, are strongly influenced by several characteristics of resources and communities, as well as by the costs of monitoring. Enforcement looms large as a challenge when the resource users do not have the characteristics of strong communities that predispose them to adhere to shared norms. The availability of exit options for major resource users also heightens the enforcement challenge.

Allocation rules can affect the willingness of users to comply voluntarily, thus lowering enforcement costs. An example is the allocation of rights in the United States to use the atmosphere as a sink for sulfur dioxide. The major resource users (electric utilities) agreed to the new institutional design only because the rights to use the resource were allocated to them free of charge, while new users had to purchase their rights to use the resource (see Tietenberg, Chapter 6).

Researchers have identified several design principles that address the enforcement challenge. These include the support by higher authorities of the right of resource users to apply sanctions against those who break their rules, the establishment of clear definitions of who has rights of access, the drawing of clear boundaries around the resource, the importance of participation by resource users in devising the rules, the establishment of graduated sanctions for offenses, and the need for low-cost mechanisms of conflict resolution (Ostrom, 1990).

*Monitoring the Resource and Users' Compliance with Rules*

Both types of monitoring are essential for operating or enforcing any rules regulating common-pool resources (see McCay, Chapter 11; Tietenberg, Chapter 6; and Rose, Chapter 7). When resources and resource users' actions can be monitored reliably with relatively simple and inexpensive methods, it is relatively easy to protect the resources from overuse. When monitoring is difficult, unreliable, or requires sophisticated measurement technologies, it becomes a critical challenge. Figure 13-3 suggests several characteristics of resources and resource users that affect the ease or cost of monitoring; in principle, each of these may require a different kind of response. Factors not shown in the figure may also influence the cost of monitoring.

For example, Rose (Chapter 7) suggests that it is easier to monitor withdrawals of a "good" from a common-pool resource (extractive use) than deposits of a "bad" into a resource (putting pollutants into a sink). Therefore, in regard to monitoring, nonpoint forms of pollution may present a more serious challenge to institutional design than some forms of extraction or concentrated forms of pollution. Among pollutants, those that have a nonuniform pattern of effects (i.e., the effect depends on where the change in pollution levels occurs) present especially difficult challenges. Nonuniformity has been a major problem for devising rules for maintaining air quality (Tietenberg, 1974, 1980, 1990), and various institutional design features have been adopted to address it (Dolšak, 2000). Uniformly distributed pollutants with uniform effects also may present difficult challenges of a different form, as the example of greenhouse gas emissions to the atmosphere attests.

Critical tasks for institutional design therefore include understanding the requirements of monitoring, devising institutional and technological means of monitoring, and acquiring the necessary resources to carry these out. Successful institutions perform these tasks in ways appropriate to the situation. One robust strategy for success in monitoring is to make the monitors at least partly accountable to resource users (one of the design principles proposed by Ostrom, 1990). When monitors are hired exclusively by central governments, paid low wages, and sent to distant locations of little long-term interest to them, the temptation to extract illegal side payments may outweigh the benefits of undertaking efforts that are personally costly to them regarding their official responsibilities.

## Addressing Negative Externalities for Other Resources

Resource systems that do not affect the conditions of other resources are easier to manage than resources that are part of a complex, interactive system of resources. For example, the stocks of one species of fish may be affected by the quantity of other species harvested (reducing the number of predator species may increase the stock of a given species; reducing the amount of food species may reduce the stocks of a target species). Furthermore, the level of fish stocks may be affected by the quality of water, which is a function of the use of water as a pollution sink (Olsen and Shortle, 1996). When a resource is part of such a complex system, management requires information on the rest of the system, which is often available only after use of the resource for some time. More complex institutions are typically needed to manage the interconnected resource systems and the interrelated groups of users. Successful management may depend on regulating multiple species or even ecosystems, as well as an increased and more heterogeneous population of resource users; it also may require linkages among preexisting institutions with responsibilities for managing parts of the system. As Tietenberg and Rose point out in Chapters 6 and 7, it is very difficult to devise simple individual transferable property rights regimes (such as transferable environmental permits) that can manage such resources effectively.

*Reconciling Conflicting Values and Interests*

Experimental and survey research, as well as theoretical analyses, indicate that effective cooperation for resource management depends on the presence of a sufficient proportion of individuals in a group of appropriators who place a value on the group's well-being or who are willing to trust other group members to keep their promises most of the time (see Campbell, 1975; Axelrod, 1984; Sober and Wilson, 1998; Kopelman et al., Chapter 4; Richerson and Boyd, in press). The prevalence of supportive values and attitudes is associated with cultural traditions and strength of community among the appropriators, two factors that institutional designers cannot readily change. They should, however, be alert to a lack of supportive values and attitudes as a major challenge. One of the key challenges of institutional design is to cope effectively with heterogeneity among resource users with respect to predisposition to cooperate in the absence of clear sanctions.

A related challenge is the presence of conflicting values and interests among the appropriators. This challenge, ubiquitous when policy decisions are being made, is most severe when groups are economically and culturally heterogeneous, when members are heterogeneous in their relationships to the resource (e.g., upstream and downstream water users) (Lam, 1998; Tang, 1992), and when members differ in their degree of dependence on the resource (Berkes, 1992). It is also more disruptive when the dynamics of the resource are poorly understood (to be discussed). Resource designs based on market principles, such as tradable permits, are intended to address conflicting interests by facilitating tradeoffs; they may, however, be rejected by some resource users on equity grounds. Recommendations for developing low-cost conflict resolution mechanisms are intended to address the full range of conflicts of values and interests.

*Managing Resources with Imperfect Knowledge*

Empirical research suggests that it is easier to create and maintain institutions to manage resources whose dynamics are well understood (Gibson et al., 2000a; McCay, Chapter 11; Wilson, Chapter 10). Similar findings are reported in theoretical analyses (Olsen and Shortle, 1996; Pindyck, 1984, 1991). Unfortunately, many important common-pool resources, including ocean fisheries, tropical ecosystems, and the global climate, have poorly understood dynamics that create major management challenges.

Our understanding of the dynamics of a resource may be imperfect in two ways. In one instance, the variables affecting resource stocks are understood, but the relationship is probabilistic and/or highly nonlinear rather than deterministic and linear. Imperfect understanding presents its most serious challenge in a second instance, when the users do not know which variables affect the stocks of a resource or the nature (functional form) or strength of the relationships (see Wilson, Chapter 10; Rosa, 1998; Dietz et al., 2000). This situation makes monitoring difficult because it is not certain what or how frequently to monitor. The regulation of air pollution provides a good example. Airsheds are significantly affected by variables beyond the control of the institutional design (for example, wind velocity and direction, air temperature). In addition, variations on short time scales are consequential. As a result, monitoring is needed not only of long-term (e.g., annual) averages, but also on short time scales (e.g., hourly maximum limits).

Imperfect understanding also raises significant management problems because of different interpretations of data from monitoring. Scientific experts and resource users are likely to disagree about interpretation, especially when there is a significant divergence of values and interests among the appropriators (Dietz and Stern, 1998; National Research Council, 1996, 1999). The allocation of rights to resource flows and the transferability of these rights also can be problematic when knowledge is lacking about what the appropriate limits should be for resource use (see Tietenberg, Chapter 6; Rose, Chapter 7) and when diverse appropriators disagree about how cautious to be under uncertainty.

Wilson (Chapter 10) suggests that the more uncertain and variable the resource stocks are, the more effort needs to be put into frequent monitoring of the stocks, notification of users and managers when stocks are at levels that should be a cause for concern, and procedures for redefining the limits of resource withdrawals. Many researchers have concluded that uncertainty based in ignorance requires flexible institutions that adjust to improved understanding, allow users to quickly redefine the limits of resource use when the resource stocks require it, and incorporate low-cost conflict resolution methods. It is not yet clear, however, how best to design decision processes that can create the needed flexibility and responsiveness to conflicting demands (National Research Council, 1996; Wilson, Chapter 10).

## Establishing Appropriate Linkages among Institutions

Environmental systems do not neatly match the boundaries of the social systems within which they are managed. It is thus unlikely that the rules of any one social system will be adequate for resource management. It is necessary to link institutions both horizontally (across space) and vertically (across levels of organization). The need for vertical linkage is especially critical for resources of large size or high complexity or whose use results in extensive negative externalities for other common-pool resources (Karlsson, 2000). Higher level institutions may support the authority for local enforcement and provide resources for local monitoring or enforcement. When favorable conditions do not exist for local monitoring and enforcement, external authorities can help by providing information, long-term contracts, and enforcement mechanisms, taking into consideration the views of local resource users (Morrow and Hull, 1996). The most extreme challenges of linkage probably arise for global resource management (e.g., the atmosphere, the oceans, global biodiversity). Here, a global interest exists in managing resources that are directly affected by the local actions of individuals and organizations. Local actions are shaped by local, regional, national, and global institutions. The challenge is to design institutional forms capable of accommodating the demands of governance at all the relevant levels while sustaining resources.

The challenge of linkage, as Young (Chapter 8) and Berkes (Chapter 9) both point out, is not to identify an appropriate institutional level for resource management—institutions at different levels all may have essential contributions to make—but to determine how institutions at various levels can be vertically linked. Linking institutions vertically is a challenge because of the different objectives of governance at different levels. For example, spatially heterogeneous resources create divergent interests in different localities within higher levels of governance, and higher level institutions respond to different economic and political interests than local institutions. Ostrom (1990) proposed a nesting of institutions as a design principle for making needed linkages, but other approaches are also possible. Young and Berkes (Chapters 8 and 9) begin to conceptualize the issues and to propose arrangements that can maintain the benefits of each level of organization.

### Adapting to Change in Social and Environmental Conditions

The case-based research makes it clear that effective resource management institutions adapt to variation and change in the resources they manage and to changes in the resource user groups. However, despite much interest in adaptive management approaches (e.g., Holling, 1978; Lee, 1993; Berkes, Chapter 9), how institutions adapt has not received much systematic research attention. Institutional adaptation and flexibility are likely to become increasingly important for common-pool resource management because of increasing rates of change in the stocks of some resources and in the institutional environment, particularly at the international level. These issues are discussed further in the section on understudied issues at the end of the chapter.

## RESEARCH DIRECTIONS

Research on common-pool resource management institutions has made great strides since the 1980s. In the next decade, research can progress further by continuing in established directions and by addressing some key understudied issues. We believe the state of the field is such that an investment in these areas will produce results that are both sound science and useful to practitioners.

### Continuing the Systematic Development of Knowledge

We have noted that the field is following a path of development typical for this stage of the science. Progress is likely to accelerate if the research community pursues this path self-consciously. This implies some changes in styles of research and an increased coordination of the research community around theory development and testing.

## The Roles of Case Study Research

Case studies have contributed greatly to knowledge by documenting the limitations of the Tragedy of the Commons model, identifying key variables, and generating hypotheses. Case studies will continue to break new ground, particularly in investigations of "new commons" and interinstitutional linkages, as well as in participatory research (to be discussed). In such frontier research areas, in-depth observation is needed to uncover phenomena or variables that might be missed if researchers looked only at variables known to be important in well-studied areas of the field. However, it is now possible to use case study methods within theoretically driven research programs, for example, using methods of focused and structured case comparison (Bennett and George, in press) and theory-driven evaluation (Birckmayer and Weiss, 2000; Chen, 1990; Chen and Rossi, 1992). It is also possible to mine existing case studies by developing structured coding forms to extract common information about theoretically relevant variables and thus to test propositions (see Ragin, 1987, 2000; Ragin and Becker, 1992; Tang, 1992; Schlager, 1994). The results of such systematic assessments of previous case studies can also point toward critical questions for new case studies. All these strategies should become much more prominent in future uses of case study methods. However, it is important to remember that there are inherent limitations to case approaches, such as those created by the need to compare any case history with a counterfactual scenario based on what might have happened instead (see Tetlock and Belkin, 1996; Roese, 1997).

## Expanded Use of Multicase Comparative Methods for Investigating Contingent Hypotheses

Theory has developed to a point that it provides contingent generalizations that can be illuminated by multicase, multivariate research methods. This development implies an increased role in the next decade for relatively large-$n$, multivariate research as well as for the new case study-based methods already described. Efforts to develop large-$n$ multivariate databases (Agrawal and Yadama, 1997; Ostrom, 1998) provide essential infrastructure for the quantitative multivariate style of multicase research. Syntheses of existing research, such as those offered in several chapters of this volume, are also essential because they generate hypotheses that involve variables that are missing from the large databases but that can be investigated by focused case comparison methods.

## Development and Testing of Causal Models

Empirically supported causal hypotheses are emerging at an increasing rate. Experimental methods have long been useful for establishing cause-effect relationships, and they will continue to be useful, especially for studying variables that operate at the individual and small-group levels. Formal models based on game theory and related approaches continue to be fruitful, but their results will require empirical validation. In the next decade, the field will need to begin to use the developing multicase, multivariate data sets intensively for causal modeling, an approach that was rare in the early years of the field. In moving toward this approach, serious attention will have to be paid to the quality and independence of data on theoretically relevant variables, as well as to the development of time series for individual cases and, ideally, time series on many cases to allow the use of panel analysis methods.

## Increased Emphasis on Triangulation

The field will continue to benefit from communication among researchers from different methodological and disciplinary traditions, leading to findings that are robust across research methods. Triangulation of methods is most likely to occur in problem-oriented settings, such as the meetings of the International Association for the Study of Common Property and research projects focused on particular institutional design problems. Affirmative efforts may have to be made to bring together representatives of different research traditions that do not normally communicate. It is also important to encourage communication among research subcommunities focused on different resource types as a way to clarify the breadth of applicability of particular elements of theory. Doing this will be important for applying institutional design theory in new and unfamiliar settings.

## Improving Conceptual Categories

As theory develops and is tested against a breadth of data, it becomes possible both to refine concepts, adding more resolution, and to combine concepts. Tietenberg's investigation (Chapter 6) of tradable permits regimes provides good examples of how careful examination of these regimes led to refinement and differentiation in theory—for instance, to account for the fact that regimes that work well for air pollution do not work so well for fisheries. The discussion of Figure 13-3 earlier in this chapter suggests that cultural heterogeneity, communication, dense social networks, and practices of reciprocity may be part of a cluster of variables that reflect a single underlying construct, such as strength of community or social capital. Although these particular developments in theory may or may not prove fruitful, they do reflect a desirable direction in theory development, namely, that of refining conceptual categories to better reflect the regularities of experience.

## Refining Understanding of Institutional Design

The research community is converging on the idea that all institutional regimes must accomplish certain key tasks (e.g., creating common understanding and agreement on rules, monitoring, enforcement) in order to succeed. Resource managers need to understand these necessary tasks, the conditions under which each one becomes particularly difficult, and the ways that institutions can meet the challenges those conditions pose. Continued efforts by the research community to aid the understanding of resource managers are likely to advance theory as well as to provide practical value.

## Key Understudied Issues

In addition to the continued development of knowledge along the lines that are already central to the field and that are discussed in the previous section, we see four major substantive issues that call for intensified attention from researchers. These are (1) understanding the dynamics of resource management institutions, (2) extending insights to more kinds of common-pool resources, (3) understanding the effects of context on resource management institutions, and (4) understanding the role of linkages across institutions.

## Dynamics of Resource Management Institutions

An increasing need exists to understand the dynamics of resource management institutions—their evolution and adaptation, the ways they respond to problems of decision making and internal conflict, and the mechanisms that govern change in the institutions and in how they relate to resources (for studies of change in resource institutions, see Becker, 1999; Futemma et al., in press). Such analyses are essential for developing causal models that can explain why certain conditions favor or impede the effective operation of institutions in particular contexts. Because resource institutions must deal with changing environments, however, understanding process and change in institutions is also critical to the practical tasks of institutional design and operation. As the following discussion suggests, studying the dynamics of institutions also can forge links to other active fields of social science research.

*Deliberative processes in decision making.* It is common to think that scientific analysis of the state of a resource should be insulated from the conflicts involved in making decisions about its use—that the science should influence the decision but the conflicts and tradeoffs involved in making policy should not influence the science. The literature on common-pool resources often has noted the problem of political pressures on scientific analysis, as illustrated by Wilson in Chapter 10. Although public involvement in institutional design is a given in a democracy, when institutional success depends on the use of scientific and technical information about the environment, many perceive a tension between the imperatives of science and democracy.

The environmental policy literature has been addressing this issue for at least two decades. Although some have argued for a fairly strict separation between "risk assessment" and "risk management" (National Research Council, 1983, is usually cited in this regard), others have argued that public deliberative processes that include input from nonscientists must be integrated with scientific analysis to properly inform public policy (e.g., Cramer et al., 1980; Dietz, 1987). The latter view has entered the mainstream of the environmental policy literature (e.g., National Research Council, 1996, 1999; Commission on Risk Assessment and Risk Management, 1997; Environmental Protection Agency, 2000).

Although this perspective on the role of "interested and affected parties" (National Research Council, 1996) in scientific analysis is perhaps new to the literature on resource institutions (see, e.g., Berkes, Chapter 9), the latter literature has long emphasized the importance of participatory processes in institutional design (e.g., Ostrom, 1990, identifies participation as a principle of institutional design). Berkes notes several forms that such institutions may take when linked to national government, including co-management between local and national bodies and multistakeholder groups. McCay (Chapter 11) suggests that deliberative processes involving scientific experts and resource users may be an important tool for producing common-pool resource management regimes. Wilson (Chapter 10) provides an interesting example of such an arrangement. And as Tietenberg (Chapter 6) suggests, good design of tradable allowance systems requires both analyses of the structure of the market and deliberation about how to allocate permits initially, taking into account both market structure and social values about the allocation of wealth.

Many common-pool resource settings have all the characteristics of situations that benefit from broadly based analytic-deliberative decision processes: multidimensionality of outcomes, scientific uncertainty about the resource, value conflict and uncertainty among those involved, mistrust of some actors by others, and the need to act before scientific uncertainties can be resolved (Dietz and Stern, 1998). It therefore seems likely that the theory and practice of common-pool resource management can benefit from interchange with a growing body of work on public participation processes in environmental and technological decision making (e.g., Renn et al., 1995; Sclove, 1995; Chess et al., 1998; Chess and Purcell, 1999). This work can help illuminate the process variables through which features of institutional design come to influence outcomes and can suggest promising approaches to the process aspects of institutional design. Effective participation mechanisms are especially necessary when acting in the face of scientific

uncertainty and controversy and for accommodating diverse perspectives on resource management issues.

*Institutional learning.* Ostrom (1990) advised institutional designers to build in
procedures for changing rules, on the grounds that success in resource management often depends on the ability of institutions to learn (also see Wilson, Chapter 10). Learning depends on responsiveness to many kinds of information: information from monitoring resource bases and users' behavior, changes in basic
scientific understanding of the resource, and the information and cognitive frameworks of the resource users. Wilson's chapter illustrates the dangers of failure to
learn and of failure to take relevant sources of insight into account.

Although learning is essential, limited empirical research examines how resource management institutions learn. Therefore, little empirical basis exists for
advice on how to design institutions for learning. Several lines of research may
offer useful starting points, however. Wilson notes the relevance of research on
adaptive management (Holling, 1994). Theory and research on deliberative and
participatory processes, already mentioned, also offer insights. Also relevant are
growing bodies of theory and research on organizational adaptation to environments (e.g., Aldrich and Marsden, 1988) and on learning in policy systems
(Sabatier, 1999). Researchers and practitioners interested in making institutions
more adaptable may be able to take useful concepts off the shelf rather than starting from scratch. Finally, there is the matter of monitoring the learning process.
One of the lessons of decades of program evaluation research has been that policies (i.e., systems of rules) are best instituted as experiments (e.g., Campbell,
1969). They are unlikely to work perfectly when first tried, but they can be refined and improved if their effects are monitored and they are revised accordingly. The program evaluation literature is full of suggested methods that managers and participants in policies and programs can use for evaluating and
readjusting them (e.g., Cook and Campbell, 1979; Chen and Rossi, 1992; Weiss,
1998).

*Conflict management.* The need for low-cost methods of conflict management
has long been recognized in the resource management context (e.g., Ostrom,
1990), but little research attention has been given to this aspect of institutional
design (but see Blomquist, 1992). Challenges of conflict management are probably most severe when institutions govern people with heterogeneous values,
interests, and objectives and when knowledge is contested about how the resource
and the resource users will be affected by management decisions. How can such
challenges be met? Researchers and practitioners of resource management probably can gain useful insights from a voluminous literature on conflict management, particularly literature that deals with intergroup conflict and with conflict
management involving institutions at different levels of organization (e.g.,
Deutsch and Coleman, 2000; Fisher, 1997; National Research Council, 2000).

Some of this literature specifically addresses environmental policy and other policy conflicts (e.g., Susskind and Cruikshank, 1987; Wondolleck, 1988; Wondolleck and Yaffee, 1994).

*Emergence, adaptation, and evolution of institutions.* Researchers have only limited understanding of why self-organized resource institutions emerge where and when they do. We also have limited understanding of the processes that govern adaptation to changes in the institutions' social and biophysical environments. McCay (Chapter 11) draws on theory from psychology and human ecology to address the question of emergence. Richerson et al. (Chapter 12) suggest ways in which evolutionary approaches can shed light on both emergence and adaptation, as well as other related questions. Their evolutionary approach offers a theory-based explanation of the fact that human groups create and maintain self-governing resource management institutions that is an alternative to explanations based entirely on individual self-interest. A parallel approach in the organizational ecology literature uses an evolutionary logic to understand the population dynamics of organizational forms and may also provide useful concepts and tools (Hannan and Freeman, 1989; McLaughlin, 1998). An evolutionary analysis also generates new research hypotheses, as Chapter 12 shows. It may also be useful for opening up questions such as these: What factors shape the rates of evolution of systems of socially created rules (and therefore, the ability of human groups to adapt their institutions to rapidly changing environments)? What can be done to aid human groups with slowly evolving systems of rules living in rapidly changing environments? Does diversity among institutional forms in a population of human groups offer adaptive advantages for the population compared to a uniformity of institutions, as some theorists have argued? Which features of biophysical or social environments are conducive to speciation (creation of new forms) or extinction among institutional forms?

### Extending Insights to a Broader Array of Common-Pool Resources

Although the concept of common-pool resources is abstractly defined, much of the empirical base for theory consists of studies of local resources suitable for subsistence of local resource users. Over the past 15 years, researchers and practitioners have begun extending the insights from this research to other settings that fit the definition of common-pool resources but that are nevertheless quite different from those that have received the most research attention (Barkin and Shambaugh, 1999; Burger et al., 2001; Dolšak and Ostrom, in press).

An early extension was from resource extraction settings to pollution settings. Rose (Chapter 7), Tietenberg (Chapter 6), and Young (Chapter 8) all address the extent to which pollution settings may require different institutional forms from those that work well for extraction settings. Another extension was from local to global commons, such as the atmosphere and nitrogen cycling through the biosphere, that are showing signs of being threatened by human activities. Technologies now enable monitoring of changes in local environments that affect global commons, thus making it possible to design management institutions at various levels. The key research question is how and to what extent can the lessons from traditional commons be applied to the new commons. One aspect of this question, that of institutional linkages across scales, is discussed in more detail in a later section.

Other settings are being suggested as test beds for extending the insights from research on resource institutions. Some of these involve outputs of technological progress (the Internet, gene pools, human organ banks, the spectrum of frequencies used in telecommunications, public roads) and of new institutional arrangements (for example, budgets of corporations, countries, and international organizations). All of these fit the definition of common-pool resources.

Efforts to extend theory in such directions are likely to be fruitful in several ways. They may offer valuable insights for managing the resources in question. They test the generality of empirical findings of past research. And they are likely to lead to a questioning and refinement of existing knowledge about the conditions and processes affecting institutional success. For example, some research on international and global commons suggests, contrary to much past research on local commons (see Bardhan and Dayton-Johnson, Chapter 3), that heterogeneity of interests can provide a motive for trading across issues and can thereby increase the likelihood of cooperation (Martin, 1995). Attention to global commons also highlights the importance of types of heterogeneity that do not receive much attention in research on traditional commons. The relative shares of an international market obtained by multinational corporations, for example, may affect the ease of negotiating international treaties as well as the formulae used within them (e.g., Benedick, 1991).

Efforts to extend theory in new directions can also bring additional variables into focus. For example, an interesting feature of many global and technological commons that distinguishes them from local subsistence systems is that there can be near-total separation between those who gain the benefits and those who bear the risks. With many forms of regional or global pollution, the benefits of using the environment as a sink are to reduce costs of production for firms, which directly affects their profitability and may indirectly affect the price of the goods or services provided. Thus the benefits are concentrated in the owners of the firm, to a lesser extent the firm's workers, and to an even lesser extent, those who purchase the goods or services produced. But the associated risks, such as climate change and acid precipitation, are distributed to a very large population that may have only slight overlap with those who are receiving the benefits. The beneficiaries and those at risk may not live in the same nation, let alone in the same community. Research on such commons brings into focus a distributional issue that has received relatively little attention in research on small-scale commons but that may become increasingly important, even for small commons, as resource systems become more globally integrated. We return to this theme in the next section.

Another variable highlighted by attention to different kinds of resources is the rate of replenishment of the resource and its relationship to rates of use. Most existing research has focused on resource bases that can be significantly degraded on a time scale of years to decades and that can replenish or regenerate themselves on similar time scales. But, as noted in Chapter 1, resources differ greatly in their rates of replenishment. Some, like fossil fuel deposits, replenish at a rate so slow as to be effectively zero from the standpoint of social institutions—no institution has lasted long enough to wait for replenishment. Others, such as broadcast bandwidth and Internet traffic, replenish instantaneously when usage declines. The problem for these is crowding rather than degeneration of the resource base. Still other common-pool resources, such as ocean circulation in the North Atlantic, which maintains the mild climate of Western Europe, probably cannot be described accurately in terms of rates of degradation and replenishment. Degradation is thought to be a nonlinear function of physical processes affected by human activity, with no change until a threshold is crossed and then dramatic change; there may or may not be any possibility of restoring the resource after the threshold has been crossed.

Resources that replenish on different schedules provide different signals. For example, with slowly depleting nonrenewables, the price mechanism induces searches for substitutes, greater efficiency, and exploration to find new supplies. Instantly renewable resources are by definition forgiving in that they instantly reward changes in user behavior. Most resources that are renewable at moderate time scales of years to decades—the ones that have received the most research attention—involve ecological and/or hydrologic systems that are nonlinear and thus difficult to forecast accurately.[8] They are subject to rapid shifts from "business as usual" to "crisis" modes, which contributes to the management problem. So far, there is limited knowledge about how different replenishment schedules affect the willingness of resource users to organize and maintain management institutions. Empirical research suggests, however, that users of renewable resources pay close attention to withdrawal and replacement rates. There is some evidence that users are less likely to devise institutions to manage the resources if they estimate that the replacement rate grossly exceeds the withdrawal rate or the withdrawal rate exceeds the replacement rate (Berge and Stenseth, 1999). Also, small groups have been found to have much greater difficulty maintaining resources whose stocks depend on previous stocks and withdrawals than in maintaining time-independent resources (Herr et al., 1997).

## Effects of Social and Historical Context

Agrawal (Chapter 2) points out that researchers on institutions have paid much more attention to the characteristics and functioning of institutions than to the contexts within which institutions function. Recent interest in linkages to other institutions (see the next section) is an exception to this otherwise apt generalization. This section discusses a few important contextual influences on institutional functioning that deserve systematic attention in the coming years.

*The globalization syndrome.* "Globalization" is not a scientific concept. Nevertheless, certain aspects of the ongoing phenomena usually described by the term almost certainly influence the possibilities for designing effective resource institutions, even at the local level. These include:

- Enhanced integration and interdependence of ideas, cultures, people, and places that previously had been isolated from or independent of each other;
- Enhanced integration of people and communities into national and global markets;
- Integration of what had been local commons managed by informal, traditional systems into international and global economic and governance systems;
- Tensions between motives for economic integration and motives for political decentralization and devolution, especially in developing countries;
- Efforts by international institutions to impose standards and obligations on national governments; and
- A blurring of distinctions between local and global (e.g., the claim that tropical moist forests are a global management issue).

Some of these changes directly affect variables that in turn shape the effectiveness of governance institutions. For example, integration of resource users into world economies tends to make them less dependent on particular local resources, thus increasing their exit options with respect to local resources and management rules (see Figures 13-2 and 13-3 for some implications of increased exit options). Effective institutions that slowly evolved when a local economy was nonmonetized may be substantially challenged if a change to monetized transactions occurs rapidly. Developing rules and norms to offset the temptation to use monetized tax revenue for personal gain is always a challenge. If "taxes" have been collected in the form of labor and materials for many generations, no counteracting rules or norms will have developed for coping with the problem that parts of the public treasury secretly can be allocated for private purposes. Global demand tends to make it more difficult for local groups to control access. Broader commerce in ideas may allow local groups to learn more easily from each other and to transfer knowledge and skills of institution building. The net effects of all these changes is unknown and has barely been theorized or investigated.

Some aspects of globalization are creating new phenomena that are likely to become increasingly important for common-pool resource management. One is resistance to globalization at local, regional, national, and international/global levels. Local and national movements against the spread of genetically engineered crops, for protection of local rights to intellectual property (e.g., medicinal uses of local plants), and against global trade liberalization have spawned new social movement organizations, many of them concerned with maintaining local control over local resources or protecting local rights to use and manage commons (e.g., Burger et al., 2001). These organizations have asserted the right to participate in institutional design; their assent may be necessary for institutions to function. Also, they are linked across scale and place in ways that may help to spread design innovations.

Globalization phenomena are raising a range of new questions about the proper locus of governance. For example, it is becoming common for crop genetic resources to be developed by multinational corporations that control the genetic material and market it worldwide. The use of this material has major consequences, positive and negative, that concern national and local institutions. Another example is national government decisions in some countries to annul the rights of local resource users to govern wetlands and coastal zones in order to advance national economic development objectives tied to global markets (e.g., tourism, aquaculture) (see, e.g., Ganjanapan, 1998; Agrawal, 1999). Resources that had been managed locally have become contested terrain among local users, national governments, multinational corporations, international development banks, and social movements at various levels. It is noteworthy that some of the important actors on this list have so far received little attention in common-pool resource management research.

Another emerging phenomenon is a blurring of the distinctions between local and global commons and between traditional and new commons. For example, many new global networks of peasants, indigenous peoples, fishers, and others— whose primary objectives relate to access and control of local commons by face-to-face communities—operate as virtual communities linked together by new commons like the Internet. As another example, the destruction of a mangrove ecosystem in Thailand for the construction of a tourist resort or shrimp farm may be a loss of a traditional, local commons to the people who live there. However, from the perspective of international groups like the Mangrove Action Project and Conservation International, this destruction, along with similar acts elsewhere in Asia and Latin America, represents the degradation of a vital global commons.

*Other global social changes.* Other major social changes that may or may not be related to the globalization syndrome occur on a global level and form part of the context to which resource management institutions must adapt. The list of global social trends may change over time, but a current list should probably include political democratization within nation-states, privatization of government-held assets, the emergence of regional and global economic institutions, and the simultaneous devolution of political control to levels below the nation-state. These trends almost certainly affect the prospects for local resource management and for effectively linking resource institutions at different levels of social organization. They can be expected to play out differently in different countries and at different levels of social organization. The research community has hardly begun to address these important influences on resource institutions.

*Major demographic changes.* Now and over the next decades, we can expect to see continued, though slowing, growth of global population, rapid urbanization in developing countries (with the potential for reduced size or stability of rural communities), decreasing household sizes, increased participation of females in education and the labor force, and increased dependence of local resource users on remittances from relatives who have migrated off the land. These demographic changes seem likely to affect the resource management capacities of local groups and levels of concern about rural resources in national governments, perhaps both negatively. However, these hypotheses have not received much research attention.

*Technological change.* As Agrawal (Chapter 2) has noted, technological change is an important part of the context of resource management institutions. New technology may hasten degradation by enabling more effective harvesting of resources (e.g., better fishing equipment) or providing consumers with attractive products (e.g., all-terrain vehicles) that increase resource demands. It may also help prevent degradation by reducing pollution emissions and facilitating monitoring and enforcement. Of course, technological change is not exogenous to social institutions, though it may be exogenous to small local communities. Institutional designs may induce technological changes that either facilitate or impede achievement of an institution's objectives. Insights about induced innovation (e.g., Binswanger and Ruttan, 1978) have yet to be applied seriously in research on the design of resource management institutions.

*Historical context.* The theory of institutions for common-pool resource management has been remarkably ahistorical, considering the important contributions of case study research in the field. Yet it is clear that the options available for institutional design are historically contingent (see, for example, Tietenberg's discussion in Chapter 6 of the problem of initial allocation of tradable permits for air pollutants). The nature of such historical contingencies is an important topic for future research. This research can be aided immeasurably by the development of time-series data sets on resource management institutions.

### Institutional Linkages

We have already noted that a central challenge of institutional design is establishing appropriate links among institutions (Young et al., 1999). Although there are large literatures on resource institutions at small scales, as evidenced in this volume, and also substantial literatures focused on international to global scales (e.g., Krasner, 1983; Rittberger, 1993; Levy et al., 1995; Hasenclever et al., 1997), knowledge about how to meet the challenge of vertical linkage across scales is still rudimentary (but for a good recent study, see Grafton, 2000). The management of global commons especially highlights the need for appropriate vertical linkages, that is, links among institutions at different levels of social organization. This need is likely to become more acute as increased attention is given to devising institutions to implement international agreements to manage global commons such as the climate and biodiversity. Knowledge about how to meet the challenge of vertical linkage is still rudimentary, however. Practical needs are raising research questions faster than the research community can address them. For example, a problem in particular need of investigation is how to establish linkages at levels below the nation-state to meet management objectives set internationally.

Another problem related to vertical linkage is often referred to in terms of scaling up and down (Gibson et al., 2000b; International Human Dimensions Program, 1999; Young, 1997). It concerns whether lessons learned about institutions at one level of social organization transfer to other levels. The actors at different levels are not completely analogous, so transferability should be expected to be imperfect. For example, individuals using a local commons may create rules that they can enforce on their own and for which they may get support from government at higher levels. Nation-states also can enforce rules on their own, but they cannot turn to a world government for support. Also, although it may make some sense to think of individuals as having a single set of objectives and motives, it is a serious oversimplification to think of states in this way. The actions of each state are a result of various interests and the commitments of a state are much less easily converted into action than those of an individual. Scaling up and down is thus not a straightforward matter: the applicability of learning across scales is an important topic for empirical study.

The study of horizontal linkages, which Young (Chapter 8) notes but does not discuss in detail, is perhaps even less well understood than that of vertical linkages. Yet efforts to establish such linkages appear to be proliferating. For example, as noted earlier, part of the response to global commons problems and to the syndrome of globalization has been the creation of links between local resource user groups and supportive outside groups, including regional, national, and international social movement organizations that in turn link to resource user groups elsewhere. These networks, which often link institutions in the global North and South (roughly, temperate, high-income countries and tropical, low-income countries, respectively), create problems for coordinated action, as well as the obvious opportunities. Problems can arise when organizations based in the North and South have different objectives and goals and different ways of operating. For example, Northern groups are often concerned with global commons and global environmental governance, while the Southern groups they support often care much more about local livelihood issues.

## CONCLUSION

The study of the human uses of common-pool resources has made considerable progress since Hardin's 1968 article and the recognition in the mid-1980s that it represents a nascent scientific field. It has amassed a considerable body of data on actual common-pool resource use and has used those data to inform significant advances in theory and conceptualization. In particular, simple theoretical formulations such as that of Hardin have been replaced by more complex ones that more accurately reflect empirical reality. The field is identifying the critical variables determining the success of institutions in sustaining resources and meeting other objectives and it is beginning to develop useful explanatory models. It is integrating results from different disciplines, research methods, and resource types in support of improved theory and models, and it is beginning to offer useful input to the decisions of resource managers.

Of course, the field still has far to go. Theoretical development is at an early stage and a number of key questions are still unresolved. In addition, as we have noted, several key issues so far have received little examination. However, as we have also shown, it is now possible to identify clearly a set of research directions that will lead to major advances in theoretical and practical understanding. Pursuing these directions holds promise for advancing understanding of some of the central questions of social science and for providing the kinds of generic knowledge about institutional design that resource managers need to make wise choices.

We are optimistic about the future of commons research. As we have noted, this is a time of theoretical synthesis, methodological advance, and interdisciplinary integration. As our understanding of the commons becomes richer, we believe the commons perspective might be used fruitfully to illuminate a diversity of public policy problems. The commons perspective raises two key questions: Will the dynamics of commons apply in this situation? If so, will either the resource or the institutions that manage it fail, and if so, how? Already commons researchers have begun to ask these kinds of questions in new areas, such as electromagnetic bandwidth and traffic congestion. By pursuing these directions, the commons perspective may prove useful for designing adaptive management strategies for a range of human problems by providing a way of thinking about the dynamic interplay between institutions and resources.

## NOTES

1 It should also be noted that typologies can reify and lead to procrustean classifications if interpreted too rigidly.

2 Analyses that either ignored the effects of management form or treated it as an additive effect

would yield results about heterogeneity that are an artifact of the proportion of cases that were government and community managed.

3  Of course, the malleability of any of the factors depends on the context, including the technologies available. For example, governments can build reservoirs that change the storage capacity of an irrigation system. Malleability is time dependent—what is impossible to change in the short run may be relatively easy to change in the long run.

4  It is important to keep in mind that the classification offered is heuristic. Reasonable people might disagree about which variables fit into which categories, and the appropriate categorization will certainly change across contexts. Our hope is that these categories can sharpen thinking, not that they provided a definitive categorization of all variables important in the dynamics of the commons.

5  The cross-field applicability of these conclusions is best illustrated by the fact that many of the key sentences in the next seven paragraphs are taken directly or paraphrased closely from a previous National Research Council (2000:12-13) work on international conflict resolution.

6  Much of the language in this and the next two paragraphs is taken from National Research Council (2000:15); the ideas draw heavily on the work of George (1993).

7  Of course, monitoring is less than perfectly accurate and may be so inaccurate as to convey no information or even perverse information about the resource and the management institution. Management is of course much less likely to be successful in the absence of accurate monitoring, but how accurate is accurate enough depends on the context.

8  When the dynamics of a resource are nonlinear and unpredictable, the standard assumptions about the ability of prices and markets to allocate resources may not apply—price signals may not adequately indicate impending shortages.

## REFERENCES

Abel, T.D., and M. Stephen
   2000   The limits of civic environmentalism. *American Behavioral Scientist* 44:614-628.
Agrawal, A., with C. Britt and K. Kanel
   1999   *Decentralization in Nepal: A Comparative Analysis. A Report on the Participatory District Development Program.* San Francisco, CA: ICS Press.
Agrawal, A., and G.N. Yadama
   1997   How do local institutions mediate market and population pressures on resources?: Forest *panchayats* in Kumaon, India. *Development and Change* 28(3):435-465.
Aldrich, H.E., and P.V. Marsden
   1988   Environments and organizations. In *Handbook of Sociology*, N.J. Smellser, ed. Newbury Park, CA: Sage.
Asquith, N.M.
   1999   *How Should the World Bank Encourage Private Sector Investment in Biodiversity Conservation?* Durham, NC: Sanford Institute of Public Policy, Duke University.
Axelrod, R.
   1984   *The Evolution of Cooperation.* New York: Basic Books.
Barkin, J.S., and G.E. Shambaugh
   1999   *Anarchy and the Environment. The International Relations of Common Pool Resources.* Albany: State University of New York Press.
Becker, C.D.
   1999   Protecting a *garua* forest in Ecuador: The role of institutions and ecosystem valuation. *Ambio* 28(2):156-161.
Benedick, R.
   1991   *Ozone Diplomacy: New Directions in Safeguarding the Planet.* Cambridge, MA: Harvard University Press.

Bennett, A., and A.L. George
  in    *Case Study and Theory Development*. Cambridge, MA: MIT Press.
  press
Berge, E., and N.C. Stenseth, eds.
  1999   *Law and the Governance of Renewable Resources*. San Francisco, CA: ICS Press.
Berkes, F.
  1992   Success and failure in marine coastal fisheries of Turkey. Pp. 161-182 in *Making the Commons Work: Theory, Practice, and Policy*, D. Bromley et al., eds. San Francisco: ICS Press.
Binswanger, H.P., and V.W. Ruttan
  1978   *Induced Innovations: Technology Institutions, and Development*. Baltimore: Johns Hopkins University Press.
Birckmayer, J.D., and C.H. Weiss
  2000   Theory-based evaluation in practice: What do we learn? *Evaluation Review* 24:407-431.
Blomquist, W.
  1992   *Dividing the Waters: Governing Groundwater in Southern California*. San Francisco, CA: ICS Press.
Burger, J., E. Ostrom, R.B. Norgaard, D. Policansky, and B.D. Goldstein, eds.
  2001   *Protecting the Commons: A Framework for Resource Management in the Americas*. Washington, DC: Island Press.
Campbell, D.T.
  1969   Reforms as experiments. *American Psychologist* 24(4):409-429.
  1975   On the conflicts between biological and social evolution and between psychology and moral tradition. *American Psychologist* 30(11):1103-1126.
Campbell, D.T., and D. Fiske
  1959   Convergent and discriminant validation by the multitrait-multimethod matrix. *Psychological Bulletin* 56:81-105.
Cárdenas, J.-C., J.K. Stranlund, and C.E. Willis
  2000   Local environmental control and institutional crowding-out. *World Development* 29(10): 1719-1733.
Chen, H.
  1990   *Theory-Driven Evaluation*. Newbury Park, CA: Sage.
Chen, H., and P.H. Rossi, eds.
  1992   *Using Theory to Improve Program and Policy Evaluations*. New York: Greenwood.
Chess, C., T. Dietz, and M. Shannon
  1998   Who should deliberate when? *Human Ecology Review* 5:45-48.
Chess, C., and K. Purcell
  1999   Public participation and the environment: Do we know what works? *Environmental Science and Technology* 33:2685-2692.
Commission on Risk Assessment and Risk Management
  1997   *Framework for Environmental Health Risk Management*. Washington, DC: Presidential/Congressional Commission on Risk Assessment and Public Management.
Cook, T.D.
  1985   Post-positivist critical multiplism. Pp. 21-62 in *Social Science and Social Policy*, R.L. Shotland and M.M. Mark, eds. Beverly Hills, CA: Sage.
  1993   A quasi-sampling theory of the generalization of causal relationships. In *New Directions for Program Evaluation: Understanding Causes and Generalizing About Them*, L. Sechrest and A.G. Scott, eds. San Francisco: Jossey-Bass.
Cook, T.D., and D.T. Campbell
  1979   *Quasi-Experimentation: Designs and Analysis Issues for Social Research in Field Settings*. Boston: Houghton Mifflin.

Cramer, J.C., T. Dietz, and R. Johnston
    1980    Social impact assessment of regional plans: A review of methods and a recommended
            process. *Policy Sciences* 12:61-82.
Dayton-Johnson, J.
    2000    The determinants of collective action on the local commons: A model with evidence from
            Mexico. *Journal of Development Economics* 62:181-208.
Deutsch, M., and P. Coleman, eds.
    2000    *Handbook of Conflict Resolution: Theory and Practice.* San Francisco: Jossey-Bass.
Dietz, T.
    1987    Theory and method in social impact assessment. *Sociological Inquiry* 57:54-69.
Dietz, T., R.S. Frey, and E.A. Rosa
    2000    Risk, technology and society. In *The Environment and Society Reader*, R.S. Frey, ed. New
            York: Allyn and Bacon.
Dietz, T., and P.C. Stern
    1998    Science, values, and biodiversity. *BioScience* 48:441-444.
Dolšak, N.
    2000    Marketable Permits: Managing Local, Regional, and Global Commons. Doctoral Disserta-
            tion, Dissertation Series, No. 5. Center for the Study of Institutions, Population, and Envi-
            ronmental Change, Indiana University, Bloomington.
Dolšak, N., and E. Ostrom
    in      **The Commons in the Millennium: Challenges and Adaptations**. Cambridge, MA:
    press   MIT Press.
Environmental Protection Agency
    2000    Toward Integrated Environmental Decision-Making. Washington, DC: EPA Science Ad-
            visory Board.
Feeny, D., F. Berkes, B.J. McCay, and J.M. Acheson
    1990    The tragedy of the commons: Twenty-two years later. *Human Ecology* 18:1-19.
Fisher, R.J.
    1997    *Interactive Conflict Resolution.* Syracuse, NY: Syracuse University Press.
Futemma, C., F. de Castro, M.C. Silva-Forsberg, and E. Ostrom
    2001    The Emergence and Outcomes of Collective Actions: An Institutional and Ecosystem
            Approach. Working paper, Center for the Study of Institutions, Population, and Environ-
            mental Change, Indiana University, Bloomington.
Ganjanapan, A.
    1998    The politics of conservation and the complexity of local control of forests in the Northern
            Thai highlands. *Mountain Research and Development* 18(1):71-82.
Gardner, G.T., and P.C. Stern
    1996    *Environmental Problems and Human Behavior.* Needham Heights, MA: Allyn and Bacon.
George, A.L.
    1993    *Bridging the Gap: Theory and Practice in Foreign Policy.* Washington, DC: United States
            Institute of Peace Press.
Gibson, C.
    2001    Forest resources: Institutions for local governance in Guatemala. Pp. 71-89 in *Protecting
            the Commons: A Framework for Resource Management in the Americas*, J. Burger, E.
            Ostrom, R.B. Norgaard, D. Policansky, and B.D. Goldstein, eds. Washington, DC: Island
            Press.
Gibson, C., M. McKean, and E. Ostrom, eds.
    2000a   *People and Forests: Communities, Institutions, and Governance.* Cambridge, MA: MIT
            Press.
Gibson, C., E. Ostrom, and T. Ahn
    2000b   The concept of scale and the human dimensions of global change: A survey. *Ecological
            Economics* 32(2):217-239.

Gintis, H.
   2000   Beyond homo economicus: Evidence from experimental economics. *Ecological Economics* 35(3):311-322.
Glass, G.V., B. McGaw, and M.L. Smith
   1981   *Meta-Analysis in Social Research.* Beverly Hills, CA: Sage.
Grafton, R.Q.
   2000   Governance of the commons: A role for the state. *Land Economics* 76(4):504-517.
Hannan, M.T., and J. Freeman
   1989   *Organizational Ecology.* Cambridge, MA: Harvard University Press.
Hardin, G.
   1968   The tragedy of the commons. *Science* 162:1243-1248.
Hasenclever, A., P. Mayer, and V. Rittberger
   1997   *Theories of International Regimes.* Cambridge: Cambridge University Press.
Hayek, F.A.
   1945   The use of knowledge in society. *The American Economic Review* 35(4) (September):519-530.
Herr, A., R. Gardner, and J. Walker
   1997   An experimental study of time-independent and time-dependent externalities in the commons. *Games and Economic Behavior* 19:77-96.
Holling, C.S., ed.
   1978   *Adaptive Environmental Assessment and Management.* New York: Wiley.
Holling, C.S.
   1994   An ecologist's view of Malthusian conflict. Pp. 79-103 in *Population, Economic Development and the Environment*, K. Lindahl-Kiessling and H. Landberg, eds. New York: Oxford University Press.
International Human Dimensions Program
   1999   *Institutional Dimensions of Global Environmental Change.* Bonn, Ger.: International Human Dimensions Program.
Jensen, N.N.
   2000   Common sense and common-pool resources. *Bioscience* 50:638-644.
Karlsson, S.
   2000   *Multilayered Governments. Pesticides in the South-Environmental Concerns in a Globalized World.* Linköping, Sweden: Linköping Studies in Arts and Sciences, Linköping University.
King, G., R.O. Keohane, and S. Verba
   1994   *Designing Social Inquiry. Scientific Inference in Qualitative Research.* Princeton, NJ: Princeton University Press.
Krasner, S.D., ed.
   1983   *International Regimes.* Ithaca, NY: Cornell University Press.
Lam, W.F.
   1998   *Governing Irrigation Systems in Nepal: Institutions, Infrastructure, and Collective Action.* San Francisco, CA: ICS Press.
Lee, K.N.
   1993   *Compass and Gyroscope: Integrating Science and Politics for the Environment.* Washington, DC: Island Press.
Levy, M.A., O.R. Young, and M. Zürn
   1995   The study of international regimes. *European Journal of International Relations* 1:267-330.
Maass, A., and R.L. Anderson
   1986   *...and the Desert Shall Rejoice: Conflict, Growth and Justice in Arid Environments.* Malabar, FL: R.E. Krieger.

Martin, L.
  1995  Heterogeneity, linkage, and commons problems. Pp. 71-91 in *Local Commons and Global Interdependence*, R. Keohane and E. Ostrom, eds. London: Sage.
McGinnis, M., ed.
  2000  *Polycentric Games and Institutions: Readings from the Workshop in Political Theory and Policy Analysis.* Ann Arbor: University of Michigan Press.
McLaughlin, P.
  1998  Rethinking the agrarian question: The limits of essentialism and the promise of evolutionism. *Human Ecology Review* 5:25-39.
Morrow, C.E., and R.W. Hull
  1996  Donor-initiated common pool resource institutions: The case of the Yanesha Forestry Cooperative. *World Development* 24(10):1641-1657.
National Research Council
  1983  *Risk Assessment in the Federal Government: Managing the Process.* Washington, DC: National Academy Press.
  1996  *Understanding Risk: Informing Decisions in a Democratic Society.* Committee on Risk Characterization. P.C. Stern and H.V. Fineberg, eds. Washington, DC: National Academy Press.
  1999  *Perspectives on Biodiversity: Valuing Its Role in an Everchanging World.* Washington, DC: National Academy Press.
  2000  Conflict resolution in a changing world. Pp. 1-37 in National Research Council, *International Conflict Resolution after the Cold War*, Committee on International Conflict Resolution. P.C. Stern and D. Druckman, eds. Washington, DC: National Academy Press.
Neustadt, R.E., and E.R. May
  1984  *Thinking in Time: The Uses of History for Decision Makers.* New York: Free Press.
Olsen, J.R., and J.S. Shortle
  1996  The optimal control of emissions and renewable resource harvesting under uncertainty. *Environmental & Resource Economics* 7(2):97-115.
Olson, M.
  1965  *The Logic of Collective Action: Public Goods and the Theory of Groups.* Cambridge, MA: Harvard University Press.
Ostrom, E.
  1986  Issues of definition and theory: Some conclusions and hypotheses. Pp. 599-614 in *National Research Council Proceedings of the Conference on Common Property Resource Management.* Washington, DC: National Academy Press.
  1990  *Governing the Commons: The Evolution of Institutions for Collective Action.* New York: Cambridge University Press.
  1998  The international forestry resources and institutions research program: A methodology for relating human incentives and actions on forest cover and biodiversity. Pp. 1-28 in *Forest Biodiversity in North Central and South America, and the Caribbean: Research and Monitoring*, F. Dallmeier and J.A. Comiskey, eds. Man and the Biosphere Series, Vol. 21. Paris: United Nations Educational, Scientific, & Cultural Organization.
  2000  Collective action and the evolution of social norms. *Journal of Economic Perspectives* 14(3):137-158.
Ostrom, E., and T.K. Ahn
  2001  *A Social Science Perspective on Social Capital: Social Capital and Collective Action.* Report prepared for the Bundestag-Enquete Commission. Bloomington: Indiana University, Workshop in Political Theory and Policy Analysis.
Petitti, D.B.
  2000  *Meta-Analysis, Decision Analysis, and Cost-Effectiveness Analysis: Methods for Quantitative Synthesis in Medicine.* New York: Oxford University Press.

Petrzelka, P., and M.M. Bell
    2000    Rationality and solidarities: The social organization of common property resources in the
            Imdrhas Valley of Morocco. *Human Organization* 59(3):343-352.
Pindyck, R.S.
    1984    Uncertainty in the theory of renewable resource markets. *Review of Economic Studies*
            51:289-303.
    1991    Irreversibility, uncertainty, and investment. *Journal of Economic Literature* 31:1110-1149.
Poteete, A., and E. Ostrom
    in      An institutional approach to the study of forest resources. In *Influence of Human Impacts*
    press   *on Tropical Forest Biodiversity and Genetic Resources*, J. Poulson, ed. New York: CABI
            Publishing.
Putnam, R., R. Leonardi, and R.Y. Nanetti
    1993    *Making Democracy Work: Civic Traditions in Modern Italy*. Princeton, NJ: Princeton
            University Press.
Ragin, C.
    1987    *The Comparative Method: Moving Beyond Qualitative and Quantitative Strategies*. Ber-
            keley: University of California Press.
    2000    *Fuzzy-Set Social Science*. Chicago: University of Chicago Press.
Ragin, C., and H.S. Becker
    1992    *What is a Case? Exploring the Foundations of Social Inquiry*. Cambridge, Eng.: Cam-
            bridge University Press.
Renn, O., T. Webler, and P. Wiedemann
    1995    *Fairness and Competence in Citizen Participation: Evaluating Models for Environmental
            Discourse*. Dordrecht, Neth.: Kluwer.
Richerson, P.J., and R. Boyd
    in      The biology of commitment to groups: A tribal social instincts hypothesis. In *The Biology*
    press   *of Commitment*, R.M. Neese, ed. New York: Russell Sage Foundation.
Rittberger, V., ed.
    1993    *Regime Theory and International Relations*. Oxford: Clarendon Press.
Roese, N.J.
    1997    Counterfactual thinking. *Psychological Bulletin* 121:133-148.
Rosa, E.
    1998    Metatheoretical foundations for post-normal risk. *Journal of Risk Research* 1:15-44.
Rosenthal, R.
    1984    *Meta-Analytic Procedures for Social Research*. Beverly Hills, CA: Sage.
Sabatier, Paul, ed.
    1999    *Theories of the Policy Process*. Boulder, CO: Westview Press.
Schlager, E.
    1994    Fishers' institutional responses to common-pool resource dilemmas. Pp. 247-266 in *Rules,
            Games, and Common-Pool Resources*, E. Ostrom, R. Gardner, and J. Walker, eds. Ann
            Arbor: University of Michigan Press.
Sclove, R.
    1995    *Democracy and Technology*. New York: Guilford Press.
Singleton, S., and M. Taylor
    1992    Common property, collective action, and community. *Journal of Theoretical Politics*
            4:309-324.
Sober, E., and D.S. Wilson
    1998    *Unto Others: The Evolution and Psychology of Unselfish Behavior*. Cambridge, MA:
            Harvard University Press.
Steins, N.A., V.M. Edwards, and N. Röling
    2000    Redesigned principles for CPR theory. *The Common Property Resource Digest* 53(June):
            1-3.

Stern, P.C., and D. Druckman

    2000    Evaluating interventions in history: The case of international conflict resolution. Pp. 38-89 in National Research Council, *International Conflict Resolution after the Cold War*, P.C. Stern and D. Druckman, eds. Washington, DC: National Academy Press.

Susskind, L., and J. Cruikshank.

    1987    *Breaking the Impasse: Consensual Approaches to Resolving Public Disputes*. New York: Basic Books.

Tang, S.Y.

    1992    *Institutions, and Collective Action: Self-governance in Irrigation*. San Francisco, CA: ICS Press.

Tetlock, P.E., and A. Belkin, eds.

    1996    *Counterfactual Thought Experiments in World Politics*. Princeton, NJ: Princeton University Press.

Tietenberg, T.H.

    1974    The design of property rights for air pollution control. *Public Policy* 27(3):275-292.

    1980    Transferable discharge permits and the control of stationary source air pollution: A survey and synthesis. *Land Economics* 26(4):392-416.

    1990    Economic instruments for environmental regulation. *Oxford Review of Economic Policy* 6(1):17-33.

Trawick, P.

    1999    The moral economy of water: 'Comedy' and tragedy in the Andean commons. Working paper, Department of Anthropology, University of Kentucky, Lexington.

Varughese, G., and E. Ostrom

    2001    The contested role of heterogeneity in collective action: Some evidence from community forestry in Nepal. *World Development* 29(5):747-765.

Weinstein, M.S.

    2000    Pieces of the puzzle: Solutions for community-based fisheries management from native Canadians, Japanese cooperatives, and common property researchers. *Georgetown International Environmental Law Review* 12(2):375-412.

Weiss, C.H.

    1998    *Evaluation: Methods for Studying Programs and Policies*. Upper Saddle River, NJ: Prentice Hall.

Wondolleck, J. M.

    1988    *Public Lands Conflict and Resolution: Managing National Forest Disputes*. New York: Plenum Press.

Wondolleck, J.M., and S. Yaffee

    1994    *Building bridges across agency boundaries*. Seattle, WA: U.S. Department of Agriculture Forest Service Pacific Northwest Research Station.

Young, O.

    1997    *Global Governance: Drawing Insights from the Environmental Experience*. Cambridge, MA: MIT Press.

Young, O., A. Agrawal, L.A. King, P.H. Sand, A. Underdal, and M. Wesson

    1999    *Institutional Dimensions of Global Environmental Change Science Plan*. Report no. 9. Bonn: International Human Dimensions Program.

## APPENDIX TO CHAPTER 13

Table 13-A presents a collection of findings or propositions put forward by the authors of Chapters 2 to 12 of this volume. It is organized under five headings: Institutional Arrangements, Resource System Characteristics, Group and Individual Characteristics, External Environment, and Interaction among Factors. In terms of the schematic causal model of Figure 13-4, the first heading includes propositions that focus on interventions and the next three headings focus on contingencies. The last category, with one proposition in it, reflects the likelihood that interventions are shaped by contingencies.

Of necessity, the box is telegraphic, covering only the highlights and attempting to summarize careful arguments in single phrases. It is not a full summary of what we know. Rather, we view it as a guide back to the theoretical and substantive content of the 11 chapters that form the heart of the book.

TABLE 13-A  Hypotheses about Resource Management Institutions Proposed in Chapters 2 to 12

---

**Institutional Arrangements**

---

◆ Effective commons management is a cross-scale co-management process (local, governmental, national, supranational) that allocates specific tasks to the proper level of social organization and ensures that cross-scale interactions produce complementary actions rather than actions that interfere with or undermine one another (Ch. 6, Ch. 8, Ch. 9, Ch. 12).

◆ Higher level institutions lack sensitivity to the knowledge, rights, and interests of local stakeholders (Ch. 8).

◆ Rather than identifying the "appropriate institutional level," we need to examine how various institutional levels could be vertically linked (Ch. 8 and Ch. 9).

◆ Linking institutions vertically can result in tensions between benefits and costs of institutional arrangements at various levels. These tensions depend on the characteristics of the resource and of the resource users (Ch. 8).

◆ Successful commons management requires a system of resilient institutions that evolve over time and reflect dynamics of the ecosystem (and the goods and services they provide) (Ch. 12 and Ch. 9).

◆ Tradable permits are more successful in air pollution programs than in fisheries and water resources. The initial allocation problems of tradable permits are least intense for air pollution and most intense for fisheries (Ch. 6).

◆ Tradable permits are a flexible approach to resource management. Successful applications of tradable permits can simultaneously protect the resources and provide sustainable incomes for users (Ch. 6).

◆ Common property regimes easily evolve within close-knit relations and promote adaptation, long-term stability, and risk sharing (Ch. 7).

◆ Tradable environmental allowances apply to loose and stranger relations, and encourage investment, innovation, and commerce (Ch. 7).

## TABLE 13-A  Continued

◆ Tradable environmental allowances are less adaptive to natural environment and more adaptive to human demand; common property regimes are the reverse (Ch. 7).

◆ Nongovernmental or governmental organizations should undertake institutional development when individuals or small groups are unwilling and/or able to bear the required costs (Ch. 11).

◆ Institutions affect which type of individuals (selfish or reciprocal) are pivotal in social outcomes (Ch. 5).

◆ Sanctioning enhances cooperation when there are some reciprocators and the cost of sanctioning is not too high (Ch. 5).

◆ Communication enhances cooperation by facilitating coordination, by providing chances to express approval and disapproval, and by creating group identity (Ch. 4 and Ch. 5).

◆ Reciprocity, the essential element stimulating cooperation in multiple-time games, results from an individual's perception of relative payoffs and kindness of other individuals in the game (Ch. 5).

◆ Successful institutions find the right balance between incentives, social influence manipulation, and sanctions (Ch. 4).

### Resource System Characteristics

◆ Larger, simple, and single-focus resources with additive resource use are more easily managed by tradable permits than small, complex, and interactive resources with subtractive use (Ch. 7).

◆ Complex resource systems severely limit predictive ability but do not preclude understanding (Ch. 10).

◆ Local variations in biogeophysical conditions challenge unified institutional designs made at higher levels (Ch. 8).

◆ Resource characteristics are associated with both more or less co-management. Less co-management tends to occur in air pollution and other large-scale resources and more co-management tends to occur in groundwater basins and fisheries (Ch. 6).

### Group and Individual Characteristics

◆ Smaller groups more effectively evoke prosocial instincts than larger groups (Ch. 12) and are more likely to achieve cooperation (Ch. 4).

◆ Economic and social heterogeneity have independent effects that operate through different causal mechanisms (incentives versus norms). Either may hamper collective action when large start-up costs are involved (Ch. 3).

◆ Heterogeneity in power leads to defection and overharvesting (Ch. 4).

◆ People generally have a disposition to cooperate with each other, although dispositions vary considerably from person to person, society to society, and time to time. The variation is best explained by the existence of complex cultural traditions of social institutions (Ch. 12).

◆ Evolved prosocial tendencies among human beings combined with culturally evolved institutions make cooperation more likely and more effective (Ch. 12).

◆ Users are more likely to devise institutions governing resources if they have good information about the variables that affect the structure of a resource and its dynamics and the seriousness of resource depletion. Identifying factors that affect the dynamics of a resource and that can be manipulated by institutional design is important to the adaptation of institutions (Ch. 11).

◆ Individual differences (motives, trust and fear, gender, and culture) and nonindividual differences (institutional design, social structure, perception of the cause of resource depletion, and framing of the problem) affect individuals' decisions about the extent of resource use (Ch. 4).

## TABLE 13-A  Continued

◆ Interaction of economic incentives and social mechanisms affect the performance of institutions governing resources (Ch. 3).

**External Environment**

◆ Development in computer and information technology decreases monitoring costs and thereby improves institutional performance (Ch. 6).
◆ Not the market but the rapid change that accompanies market penetration is the reason for destabilizing traditional commons institutions (Ch. 12).

**Interaction among Factors**

◆ The emergence of common-pool resource institutions depends on collective-choice/rule-in-use and features of both the resources in question and their users (Ch. 11 and Ch. 2).

Compiled with the help of: T.K. Ahn, Jianxun Wang, Oyebade Kunle Oyerinde, Paul Aligica, Workshop in Political Theory and Policy Analysis, Indiana University.

# ABOUT THE CONTRIBUTORS

**ARUN AGRAWAL** is an associate professor of political science at Yale University. His research focuses on questions related to development, environmental conservation, indigenous knowledge, and the politics of decentralization. He is the author of *Greener Pastures: Politics, Markets, and Community among a Migrant Pastoral People* (1999, Oxford University Press) and the co-editor of *Agrarian Environments: Resources, Representations and Rule in India* (forthcoming, Duke University Press [with K. Sivaramakrishnan]) and *Communities and the Environment*. His published papers have appeared in journals such as *Comparative Political Studies, Development and Change, Journal of Asian Studies, Journal of Theoretical Politics, Politics and Society*, and *World Development*. He received a bachelor's degree in history from Delhi University, India, and a Ph.D. in political science from Duke University, Durham, North Carolina.

**PRANAB BARDHAN** is a professor of economics at the University of California, Berkeley, and the chief editor of the *Journal of Development Economics*. He has done both theoretical and field research in the areas of agrarian institutions, political economy, and international trade. His recent publications include *Development Microeconomics* (with C. Udry, Oxford University Press, 1999) and "Social Justice in a Global Economy" (The International Labor Organization's Nobel Peace Prize Lectures, given at Capetown, South Africa; ILO, Geneva, 2000). He is currently working on a book, *Scarcity, Conflicts and Cooperation: Essays in Institutional and Political Economy of Development*, to be published by MIT Press. He received a bachelor's degree from Presidency College, Calcutta, India, and a Ph.D. from Cambridge University, Cambridge, England.

**FIKRET BERKES** is a professor of natural resources at the University of Manitoba, Winnipeg, Canada, and a founding member and past president of the International Association for the Study of Common Property. With an academic background in both environmental and social sciences, Berkes' long-term research program has been the investigation of interrelations between societies and their resources. His main area of expertise is common property resources and community-based resource management. His publications include three recent books, *Managing Small-Scale Fisheries* (with F. Mahon, R. McConney, P. Pollnac, and R. Pomeroy, International Development Research Centre, Ottawa, 2001), *Sacred Ecology* (Taylor and Francis, 1999), and *Linking Social and Ecological Systems* (with C. Folke, Cambridge University Press, 1998).

**ROB BOYD** is a professor of anthropology who has taught at Duke and Emory Universities and has been at UCLA since 1986. He received a bachelor's degree in physics from the University of California, San Diego and a Ph.D. in ecology from the University of California, Davis. His research focuses on population models of culture as summarized in his book, co-authored with P.J. Richerson, *Culture and the Evolutionary Process*. He has also co-authored an introductory textbook in biological anthropology, *How Humans Evolved*, with his wife, Joan Silk. He and Joan have two children and live in Los Angeles. His hobbies are rock climbing and jogging.

**JEFF DAYTON-JOHNSON** is an assistant professor of economics and international development studies at Dalhousie University, Halifax, Canada. He specializes in the microeconomic analysis of economic development. His research has focused on cooperation in the use of local natural resource systems, the economic consequences of social cohesion, the efficient allocation of foreign aid, and the economics of culture. He is the author of *Social Cohesion & Economic Prosperity* (Lorimer, Toronto, 2001). He holds a B.A. (honors) in Latin American studies and a Ph.D. in economics from the University of California, Berkeley.

**THOMAS DIETZ** is College of Arts and Sciences Distinguished Professor and professor of environmental science and policy and sociology at George Mason University. He holds a bachelor of general studies from Kent State University and a Ph.D. in ecology from the University of California, Davis. He is a Fellow of the American Association for the Advancement of Science, a Danforth Fellow, and a past president of the Society for Human Ecology, and he has received the Distinguished Contribution Award from the Section on Environment, Technology and Society of the American Sociological Association. His research interests are in human ecology and cultural evolution. He has a longstanding program of scholarship on the relationship between science and democracy in environmental policy. Recent publications include *Environmentally Significant Consumption: Research*

*Directions* (National Academy Press, 1997), *New Tools for Environmental Policy: Education, Information and Voluntary Measures* (National Academy Press, forthcoming), and *Human Dimensions of Global Environmental Change* (MIT Press, forthcoming).

**NIVES DOLŠAK** is a postdoctoral research associate at the Workshop in Political Theory and Policy Analysis, Indiana University, Bloomington. Her research examines institutional challenges in governing common-pool resources at multiple levels of aggregation. Her Ph.D. dissertation focused on designing markets for common-pool resources, including global carbon dioxide emission and sequestration markets. Her published work includes an analysis of domestic sources of international cooperation in curbing global climate change. She is currently co-editing a book, *The Commons in the New Millennium: Challenges and Adaptation*, with E. Ostrom. She holds a B.A. in economics from the University of Ljubljana, Slovenia, and a joint Ph.D. from the School of Public and Environmental Affairs and the Department of Political Science, Indiana University, Bloomington.

**ARMIN FALK** is an assistant professor at the University of Zürich and CESifo research fellow and research affiliate in the Labour Economics Programme of the Centre for Economic Policy Research in London, U.K. He teaches classes in organizational theory, microeconomics, and game theory. His research addresses labor economics, behavioral economics, and experimental economics. His recent work studies the determinants of informal sanctions and the theoretical modeling of reciprocity. Another line of research concerns contractual incompleteness and the nature of market interactions. Falk is also interested in economic policy and issues of policy evaluation. He received his Ph.D. in economics from the University of Zürich in 1998.

**ERNST FEHR** is a professor in labor economics and social policy and director of the Institute for Empirical Research in Economics at the University of Zürich. He also directs the Ludwig Boltzmann Institute for the Analysis of Economic Growth in Vienna. He has published many articles in journals, including *American Economic Review*, *Econometrica*, *Journal of Political Economy*, and *Quarterly Journal of Economics*. He is on the editorial board of *Quarterly Journal of Economics*, *European Economic Review*, *Experimental Economics*, and *Journal of Socio-Economics*. In 1999 he won the Gossen Prize of the German Economic Association and in 2000 the Hicks-Tinbergen Medal of the European Economic Association. His research focuses on the interplay among social preferences, social norms, and strategic interactions and on the psychology and economics of incentives. Fehr graduated from the University of Vienna in 1980, where he earned his doctorate in 1986.

**URS FISCHBACHER** has a position at the Institute for Empirical Research in Economics at the University of Zürich. His main interests are social preferences and the economics of social interactions. He has published experimental studies that investigate the structure of fair behavior and works on game theoretical models of reciprocity (*A Theory of Reciprocity*, with A. Falk). He received his Ph.D. in mathematics from the University of Zürich.

**SHIRLI KOPELMAN** is a doctoral candidate in the Department of Management and Organizations at Kellogg Graduate School of Management, Northwestern University. Her research focuses on the influence of culture and social motives in conflict resolution, business negotiations, and social dilemmas. She is also interested in the influence of emotion on decision making. She recently contributed a chapter on emotion in negotiation to the *Blackwell Handbook in Social Psychology: Group Processes* (M.A. Hogg and R.S. Tindale, eds., Blackwell Publishers, Malden, MA, 2000) and an article on cross-cultural negotiations to *Negotiations Journal*. Her dissertation research focuses on cross-cultural resource dilemmas. She received her B.A. in psychology from The Hebrew University of Jerusalem and her M.S. in organizational behavior from Northwestern University.

**BONNIE J. MCCAY** is a professor of anthropology and ecology at Rutgers State University, New Brunswick, New Jersey, where she has taught since 1974. She has conducted field research in Newfoundland, Puerto Rico, and New Jersey. Among her books are *The Question of the Commons* (University of Arizona Press, 1987), *Community, State, and Market on the North Atlantic Rim* (University of Toronto Press, 1998), and *Oyster Wars and the Public Trust* (University of Arizona, 1998). She received a B.A. from Portland State University and a Ph.D. from Columbia University.

**DAVID M. MESSICK** is the Morris and Alice Kaplan Professor of Ethics and Decision in Management at the Kellogg Graduate School of Management of Northwestern University. He is also the director of the Ford Motor Company Center for Global Citizenship within the Kellogg School. A social psychologist by education (B.A., University of Delaware; M.A. and Ph.D., University of North Carolina, Chapel Hill), his research focuses on decision making in social contexts with an emphasis on the ethical aspects of decision making and information processing. He has published widely on topics ranging from social influence and self-serving biases to leadership and decision framing.

**ELINOR OSTROM** is the Arthur F. Bentley Professor of Political Science and co-director of the Workshop in Political Theory and Policy Analysis and of the Center for the Study of Institutions, Population and Environmental Change at Indiana University, where she has taught since 1965. Among her publications are

*Governing the Commons* (Cambridge University Press, 1990), *Rules, Games, and Common-Pool Resources* (with R. Gardner and J. Walker, University of Michigan Press, 1994), and *Institutions, Ecosystems, and Sustainability* (with R. Costanza, B.S. Low, and J. Wilson, Lewis Publishers, 2001). She has conducted field work in Nepal, Nigeria, Uganda, Bolivia, and the United States. She received the Frank E. Seidman Distinguished Award in Political Economy in 1997 and the Johan Skytte Prize in Political Science in 1999, and was elected to membership in the National Academy of Sciences in 2001. She received her B.A., M.A., and Ph.D. from the University of California, Los Angeles.

**BRIAN PACIOTTI** is a graduate student in the Ecology Graduate Group at the University of California, Davis. He is conducting thesis work on cultural differences in justice systems and criminal behavior. His work in Africa is aimed at understanding how an indigenous policing system in Tanzania functions and his work in the United States is aimed at understanding regional differences in homicide rates. He received his B.S. from Colorado State University.

**PETER J. RICHERSON** is a professor of human ecology in the Department of Environmental Science and Policy and Graduate Group in Ecology at the University of California, Davis. His main work, mainly in collaboration with Robert Boyd, is in the theoretical analysis of cultural evolution. Their book, *Culture and the Evolutionary Process* (Chicago, 1985), is their single largest contribution to the field. It received the Staley Prize of the School of American Research in 1989. References to their recent papers and book chapters can be found on their departmental Web pages. Richerson received his B.S. and Ph.D. from the University of California, Davis.

**CAROL M. ROSE** is the Gordon Bradford Tweedy Professor of Law and Organization at Yale University. She has also taught at several other universities, including Harvard, Stanford, Berkeley, Northwestern, and University of Chicago. Professor Rose's research focuses on history and theory of property, and on the relationships between property and environmental law. Her writings include two books, *Property and Persuasion* (1994, Westview Press) and *Perspectives on Property Law* (1995, with R.C. Ellickson and B.A. Ackerman, Aspen Law and Business), as well as numerous articles on traditional and modern property regimes, environmental and natural resource law, environmental ethics, and expropriation. She has degrees from Antioch College (B.A., philosophy), University of Chicago (M.A., political science; J.D., law), and Cornell University (Ph.D., history). She is on the Board of Editors of the Foundation Press and is a member of the American Academy of Arts and Sciences.

**PAUL C. STERN** is study director of two National Research Council committees: the Committee on the Human Dimensions of Global Change and the Committee to Review the Scientific Evidence on the Polygraph. His research interests include the determinants of environmentally significant behavior, particularly at the individual level, and participatory processes for informing environmental decision making. His recent books include *Environmental Problems and Human Behavior* (with G.T. Gardner, Allyn and Bacon, 1996), *Understanding Risk: Informing Decisions in a Democratic Society* (edited with H.V. Fineberg, National Academy Press, 1996), and *International Conflict Resolution after the Cold War* (edited with D. Druckman, National Academy Press, 2000). Stern received his B.A. from Amherst College and his M.A. and Ph.D. from Clark University.

**SUSAN C. STONICH** is a professor of anthropology and environmental studies and a member of the faculty of the Interdepartmental Graduate Program in Marine Science at the University of California, Santa Barbara, who also serves as Chair of the Environmental Studies Program. She received a B.S. in mathematics from Marquette University and a M.A. and Ph.D. in anthropology from the University of Kentucky. Her research has focused on the human and environmental consequences of economic development. Her earlier work examined the effects of agricultural development in highland areas of Latin America; her recent research has expanded to emphasize the impact of aquaculture and tourism development in coastal zones of Latin America and Asia. Her work now centers on the use of information and communications technologies in facilitating and hindering participatory development and environmental governance.

**TOM TIETENBERG** is the Mitchell Family Professor of Economics and director of the Environmental Studies Program at Colby College in Waterville, Maine. He is the author or editor of 11 books and more than 100 articles and essays on environmental and natural resource economics. Elected president of the Association of Environmental and Natural Resource Economists for 1987-88, he was the principal investigator for a United Nations study to facilitate the implementation of Article 17 of the Kyoto Protocol, an article that uses economic incentives to control climate change, *Greenhouse Gas Trading: Defining The Principles, Modalities, Rules and Guidelines for Verification, Reporting & Accountability* (Geneva: Switzerland, United Nations Conference on Trade and Development, 1998). He received a bachelor's degree in international affairs from the U.S. Air Force Academy and a Ph.D. in economics from the University of Wisconsin.

**ELKE U. WEBER** is a professor of management and psychology and co-director of the Center for the Decision Sciences at Columbia University. After receiving her Ph.D. in behavior and decision analysis from Harvard University in 1984, she taught in both the United States (University of Chicago, University of Illinois, Ohio State University) and Europe (Koblenz Graduate School of Corporate Management), and spent a year at the Center for Advanced Studies in the Behavioral Sciences at Stanford. She is an expert on behavioral models of judgment and decision making under risk and uncertainty. Recently she has been investigating psychologically appropriate ways to measure and model individual and cultural differences in risk taking, specifically in risky financial situations and environmental decision making and policy. Weber is president of the Society for Mathematical Psychology and past president of the Society for Judgment and Decision Making.

**J. MARK WEBER** is a doctoral student in Management and Organizations at the Kellogg Graduate School of Management, Northwestern University. His research interests include social dilemmas, trust, negotiations, the role of values in decision making, and social and organizational identity. His professional experiences include senior managerial and leadership roles in local government, the financial services industry, and not-for-profit organizations. He holds a B.A. in psychology from the University of Waterloo, an M.A. in social psychology from McGill University, and an M.B.A. from Wilfrid Laurier University.

**JAMES A. WILSON** is a professor of marine sciences and resource economics at the University of Maine. His research is directed toward an understanding of social and biological interactions, especially in commercial fisheries. He has worked in the Maine lobster fishery for more than 25 years, has chaired the Scientific and Statistical Committee of the New England Fisheries Council, and is vice president of the Maine Fishermen's Forum. Wilson received a B.A. in English from Lake Forest College and a Ph.D. in economics from the University of Wisconsin.

**ORAN R. YOUNG** is a professor of environmental studies and director of both the Institute of Arctic Studies and the Institute on International Environmental Governance at Dartmouth College. He is also Professor II of political science at the University of Tromsø in Norway. He chairs the Scientific Steering Committee of the international project on the Institutional Dimensions of Global Environmental Change and is chairman of the Board of Governors of the University of the Arctic. Among his many books are *Governance in World Affairs* (1999), *Creating Regimes: Arctic Accords and International Governance* (1998), *International Governance: Protecting the Environment in a Stateless Society* (1994), and *International Cooperation: Building Regimes for Natural Resources and the Environment* (1989).

# Index

theoretical models of reciprocity and fairness, 159–163
theoretical predictions, 163–176
Arctic Council, 308
Arctic Environmental Protection Strategy (AEPS), 280
Artisanal practices, 266
Assurance game, 12, 28n
Asymmetric equilibria, with inequity-averse subjects, 168, 188–190
Atlantic States Marine Fisheries Commission, 343
Atmosphere, global, 19
greenhouse gases released into, 3, 24
Authority rules, 98

**B**

Baland, facilitating conditions for institutional sustainability identified by, 54–55
Bandwidth, broadcasting, 22–23
Bangladesh Department of Fisheries, 305–306
Barbed wire, example of exclusion, 57
Barents Sea, 281–282
Bargaining power, of users, 15
Baseline issue, in design of tradable permits, 204
Behavior
explanation of, 38
prediction of, 37–38
Beneficiaries, impossibility of excluding, 19–20
Benefits, unfair distribution of, 66
Bering Sea, 275–276, 281–283
doughnut hole, 287n
Fishery Conservation Zone in, 275
Bering Sea Community Development Program, 210
Beverly-Qamanirjuaq Caribou Co-Management Board (Canada), 301, 303
Biases, in cultural evolution, 410
Biodiversity
loss of, 24
threats to, 23
Biological species, as resources, 22
Biophysical environment, as a sink, 29n
Biosphere, concern with, 28n
Boundary rules, 98
British North America Act, 267
Broadcasting bandwidth, 22–23

Bush, President George W., 73n
Bycatch, 213
Bycatch discard, 213

**C**

CAFF. *See* Conservation of Arctic Flora and Fauna
California water shortage, 131, 138, 203
CAMPFIRE program (Zimbabwe), 251
Canadian North, fur trade in, 297–298
Canal networks, 99–100
Capacity, in interplay between international and national environmental regimes, 276–279
Case study research, roles of, 467
Cattle herds, 56
Causal analysis, 67–68, 447–451
knowledge about processes and mechanisms, 458–459
links for institutional sustainability, illustrative sets of, 69–70
paths describing hypothesized effects of wealth or wealth inequality on the maintenance of common-pool resources, 448
Causal models
development and testing of, 468
specifying carefully, 66
Causation
contingent and multiple, 67
of perceptual factors influencing cooperation in commons dilemmas, 137–139
proximate, 408
in situated rational choice, determining, 367–369
ultimate, 408
CBMRs. *See* Community-based management regimes
CFCs. *See* Chlorofluorocarbons
Change
demographic, 477
global social, 476–477
in social and environmental conditions, adapting to, 466
technological, 477
Cheating, 143, 174, 388
Checkers, and learning in complex adaptive systems, 337–340
Chicken, game of, 12, 28n